UIQ 3:

The Complete Guide

UIQ 3:

The Complete Guide

John Holloway, Mark Wright

With
Matthew Hunt, Simon Judge

Reviewed by
Afsha Aziz, Andre Hopper, Andreas Johansson, Andrew Langstaff, Andy Leadbetter, Asker Brodersen, Ben Morris, Colin Ward, Dan Daly, Dan Ding, David Mery, Henrik Jersling, Jimmy Clutter, Jo Stichbury, Johan Ringdahl, John Gustafsson, Jonas Bengtsson, Lars Olsson, Louis Mehr, Mark Cawston, Mark Shackman, Markus Eliasson, Martin Ade-Hall, Matthew Hunt, Praveen Karadiguddi, Ray Cheung, Roy Vernon, Sébastien Peirone, Steven Rawlings, Tim Ocock, Toby Gray, Vladimir Marko

Supported by
Symbian Press

John Wiley & Sons, Ltd

Published by John Wiley & Sons Ltd, The Atrium, Southern Gate, Chichester, West Sussex PO19 8SQ, England

Telephone (+44) 1243 779777

Email (for orders and customer service enquiries): cs-books@wiley.co.uk
Visit our Home Page on www.wileyeurope.com or www.wiley.com

Other Wiley Editorial Offices

John Wiley & Sons Inc., 111 River Street, Hoboken, NJ 07030, USA

Jossey-Bass, 989 Market Street, San Francisco, CA 94103-1741, USA

Wiley-VCH Verlag GmbH, Boschstr. 12, D-69469 Weinheim, Germany

John Wiley & Sons Australia Ltd, 42 McDougall Street, Milton, Queensland 4064, Australia

John Wiley & Sons (Asia) Pte Ltd, 2 Clementi Loop #02-01, Jin Xing Distripark, Singapore 129809

John Wiley & Sons Canada Ltd, 6045 Freemont Blvd, Mississauga, Ontario, L5R 4J3, Canada

Wiley also publishes its books in a variety of electronic formats. Some content that appears in print may not be available in electronic books.

British Library Cataloguing in Publication Data

A catalogue record for this book is available from the British Library

ISBN: 978-0-470-69436-7

Typeset in 10/12pt Optima by Laserwords Private Limited, Chennai, India
Printed and bound in Great Britain by Bell & Bain, Glasgow
This book is printed on acid-free paper responsibly manufactured from sustainable forestry in which at least two trees are planted for each one used for paper production.

Contents

Foreword

Peter Molin, Chief Technology Officer, UIQ Technology

'I learn it from a book!' is my favorite quote by Manuel in *Fawlty Towers.*[1] As a matter of fact, I really believe that reading books is an excellent way of learning. Today, when 'everything' is available on the Internet, it is important to appreciate the power and magic of a book; besides providing factual information organized in a structured way, a book also gives you the possibility to browse around and pick the areas you want to learn more about. With a book you can sit in your favorite chair and let the author explain the concepts using a clear step-by-step approach. I am convinced that this book will make it easier and more enjoyable for you to learn about UIQ programming.

I am especially glad to see Mark Wright as the editor of this book. Mark and I worked together in the early days of defining the Quartz UI – a predecessor to UIQ – and making it all come together. Mark has had good help from many talented people within UIQ Technology, Sony Ericsson, Symbian and other companies to put this project together; not least John Holloway, who has written most of the content of this book, and is someone whom I hold in high esteem. John has been part of the Symbian world since its creation and I have struggled a lot with his chess game, which beats me more often than I am willing to confess to anybody!

Mobile phones are getting more advanced all the time and allow you to create amazing applications for them. Some very exciting UIQ phones have already reached the market and more are to come. A key aspect of UIQ programming is the framework to help you write flexible

---

[1] For more information on *Fawlty Towers* see, for instance, this tribute site: ***www.fawltysite.net***. As I write this, a video of Manuel saying 'I learn it from a book!' is available at ***www.youtube.com/watch?v=xX85Y5Zb7sw***.

applications that can run on various phones from Motorola and Sony Ericsson. This book will show you how to take advantage of that feature and many others, and provide you with a set of illustrative examples. The more complex concepts are explained in a way that will help you achieve your programming goals. I am also convinced that you will use some of the chapters in this book to support you in taking your first steps on UIQ and others to find out how everything works in detail when you need to optimize your application.

Finally, I wish you a happy reading, and look forward to the time when, after learning UIQ from this book, you will be one of the developers that will amaze us with great applications.

Peter Molin
Chief Technology Officer
UIQ Technology

Foreword

Mats Blomberg, Manager, Software Strategy, Sony Ericsson CTO Office

It's a great pleasure to see the first book written specifically for UIQ 3 developers. Having worked with Symbian OS since the very beginning in 1998, I've seen many different efforts to establish a credible platform for Open OS phones. It has been a long and winding road but we are very pleased to see the new UIQ 3 establishing its place in the mobile phone market and being adopted by top-tier mobile phone manufacturers.

UIQ was defined in the past as the *Quartz* UI meaning a fixed screen resolution of a quarter VGA assisted by a touchscreen for advanced user navigation. The first Symbian OS phone, the Ericsson R380, was more of a PDA with GSM phone call capability. Learning from that, Sony Ericsson created the first phone-oriented device, the P800, using the second release of Quartz, now named UIQ 2.1. Now we have the third generation of UIQ defined and several phones are already coming out, such as the P990, M600, W950, W960 and P1 from Sony Ericsson and the MOTO Z8 from Motorola. By the time you read this, I hope many more phones will be available. UIQ as a configurable UI and platform is more flexible than ever, enabling many different form factors while still maintaining a platform approach for application developers.

Developing for UIQ 3 has been somewhat challenging in the past. Learning new development tools, new C++ methods and new user interface paradigms has been tough, even for experienced application developers. The lack of a UIQ development guide has been obvious and has limited a wider uptake of the platform amongst developers.

Now, with this book, it is possible to get a head start on developing for UIQ, either from scratch or by porting an existing application from another platform to UIQ. There is a lot of commonality between UIQ 3 and S60 3rd Edition since they are both based upon the same Symbian

OS v9. Developers targeting different phone UIs but using Symbian OS benefit a lot from that commonality. Porting from other platforms, such as Palm and Microsoft, to UIQ 3 may benefit from the similar pen navigation UI on those platforms.

This book attempts to fill the gap between the Symbian Press books, covering generic Symbian OS topics, and the UIQ SDK, which contains the API reference guides. However we recommend that the reader be familiar with Symbian OS, for which we suggest the *Symbian OS Explained* book by Jo Stichbury. See ***developer.symbian.com/books*** for further Symbian Press titles.

Readers will find this book a comprehensive guide to developing applications. Starting with Symbian OS essentials and UIQ 3 basics, readers are guided through the development tools and SDK to create a first UIQ 3 application and to understand and use the many features of UIQ through detailed examples. Building an application and taking it through the Symbian Signed process is explained step by step. Advanced platform features in multimedia and telecommunications areas are covered as well.

Readers interested in porting existing applications to UIQ 3 will find expert guidance in Chapter 16. Porting from S60 and other platforms is covered.

Further books in the Symbian Press series provide complementary information in areas such as Platform Security and Multimedia.

I'm very pleased to see this book finally being written and would like to thank all authors, reviewers, sponsors and publishers putting their efforts together. It's a big milestone in the evolution of UIQ.

Happy reading!

About the Authors

John Holloway, Lead Author

With over two decades' experience of designing compact mobile software applications including a deep understanding of all the leading mobile operating systems, John Holloway is one of the most experienced mobile software architects in the world.

Having obtained a first class honors degree in computer engineering from City University, London, John commenced a ten-year career with Psion Computers, where he contributed significantly to the body of software code that now lies at the heart of Symbian OS.

John was a founder of Purple Software, one of the world's first mobile games companies, in 1995. Purple Software grew rapidly to become one of the industry's leading independent developer of software applications for mobile devices. His role as Chief Technical Officer culminated in the company being awarded the first ever BAFTA for a mobile computer game in 2002.

John is currently the Chief Technical Officer at ZingMagic Limited (***www.zingmagic.com***), a mobile games development company, specializing in games that require substantial artificial intelligence in order to play well.

John is also the Director of Client Software Applications at Mobrio Limited (***www.mobrio.com***), the online social networking and UGC services company. Mobrio designs, builds, manages and moderates online communities built around user-generated content for third-party brand owners.

Mark Wright

Mark joined Ericsson in 1997 and took a key role in defining the functional and user interface specifications for 'Communicator' class products. When Symbian was formed, Mark was seconded to the 'Quartz' team as Project Manager and managed the functional specification and definition of the GUI. The Ericsson Communicator Platform, shown at CeBIT 2000, was the first working UIQ device. It served to build interest in UIQ products and also enabled early developers to test applications on a real machine.

Mark then worked with the R380 and P800 products with specific responsibility for improving ease of use and the out-of-the-box customer experience. He established the 'over the air' setup of WAP and email for Ericsson phones.

Mark has extensive experience in communicating between different teams, notably in marketing and development. He has provided third-party support and written detailed white papers for the Sony Ericsson P800 smartphone and Sony Ericsson's range of PC Cards and broadband modems.

Mark is a graduate of Bath University with a BSc in electrical and electronic engineering.

Matthew Hunt

Matthew joined Sony Ericsson in 2002, working on the company's smartphone projects which include the P800, P900, P910, P990, M600 and W950. At Sony Ericsson, Matthew has worked to establish and build the Enterprise and Partner Support Team which is responsible for providing technical and development support for Sony Ericsson, Symbian and UIQ to software suppliers who are delivering software or services for Sony Ericsson smartphones.

Matthew studied software engineering at Manchester Metropolitan University, graduating in 1999. Following university, he went straight into a software development role for the GB Group in Chester writing data-processing applications in C++ for major utilities, finance and government organizations. This involved writing software for limited-resource devices, which sparked an interest in the world of the mobile phone.

Outside work, Matthew is a keen sailor, windsurfer, mountain biker and snowboarder. He has a small boat which is kept in the Solent and he enjoys competing in Cowes Week and the Round the Island Race. Matthew lives in Cheshire with his partner Chloe and dog Charlie.

Simon Judge

Simon is a freelance mobile developer and has worked in the mobile phone industry for over 11 years and IT for over 20 years.

Having obtained a BSc and MEng at UMIST, Simon joined Logica where he worked on several space and defense projects. Simon contracted at Vodafone and Oracle, working in object-oriented design and C++ before working full time in mobile development.

Since then, Simon has been responsible for many Symbian projects for companies including Pixology, Boots, Jessops and Philips Research. More recently, Sony Ericsson commissioned Simon to write UIQ tutorials for Sony Ericsson Developer World and he has worked at Symbian in the JAVA group that implements JSRs under Symbian. He is an Accredited Symbian Developer.

You can find out more at ***www.simonjudge.com***.

About this Book

This book is all about UIQ! It's aimed at C++ developers who wish to write applications for UIQ 3 or to port from other environments to UIQ 3.

We start with a brief history of UIQ and an overview of the current UIQ 3 platform, the UIQ 3 application suite and the development environment. Next, we explain how UIQ 3 supports a wide variety of different mobile phone styles in a single platform. We cover configuration and layout options for the user interface, which includes support for one-handed operation using softkeys and two-handed operation using a touchscreen. We explain how building blocks make it easy to lay out your application and introduce the UIQ Command Processing Framework (CPF). The CPF allows you to describe your commands in an abstract manner so that UIQ 3 can present them in an optimal way on different hardware configurations. For those of you familiar with UIQ 2, we summarize the main differences.

Now it's time to code! We describe the UIQ 3 SDK and Carbide.c++ and walk you through the steps to build the `QuickStart` example application.

UIQ 3 is based on Symbian OS v9, which introduces a new security architecture. We explain Platform Security and also provide an introduction to the key Symbian OS concepts that you will encounter as you work through the book. This serves as a useful refresher and reference, but is not intended to teach Symbian OS to newcomers.

Using a set of example applications, the core of the book, Chapters 6 to 11, discusses each aspect of UIQ 3 in detail. We recommend that all commercial grade applications should be Symbian Signed. Therefore, we build and Symbian Sign an application, `SignedApp`, in three phases. In Chapter 10, we create `SignedAppPhase1` to establish a basic application outline. In `SignedAppPhase2` we add simple file manager functions and assess the resulting code for defects against the Symbian Signed Test Criteria. In Chapter 11, we fix the defects in `SignedAppPhase3` and add some multimedia functions.

We cover the implications of Symbian Signed at relevant points throughout the book and provide a description of the scheme itself in Chapter 14.

Multimedia and communications programming in UIQ 3 follow standard Symbian OS practice. We provide a set of examples to demonstrate the typical functionality that you may need in your applications.

In Chapter 13, we show you how to localize your application and improve its performance and reliability. Chapter 15 guides you through testing, debugging and deploying your application.

Finally, we cover porting, with particular reference to Series 60 3rd Edition, Windows Mobile and Palm OS.

Who Is this Book for?

This book is primarily for the following types of reader:

- Experienced C++ programmers who are new to UIQ. If you are new to Symbian OS, we recommend that you also read *Symbian OS Explained*.
- Programmers writing applications for other Symbian OS platforms such as Series 60, who wish to port them to UIQ 3 or write new applications for UIQ 3.
- UIQ 2 programmers wishing to port applications to UIQ 3 or to write new applications in UIQ 3.
- UIQ 3 programmers who wish to extend their knowledge.

For those involved in the wider process of software development, but who are not directly coding applications, Chapters 1 and 2 explain the principles behind UIQ 3 and how it supports a diverse range of mobile phone hardware.

How to Use this Book

Besides reading the chapters in sequential order, the book can be used in other ways.

The applications provided in Chapters 6 to 13 provide example code using as much of the GUI as possible. If you want to learn more, for example, about the Command Processing Framework, you can go directly to that section, read about CPF and work with the `Commands` example applications.

The example applications frequently contain more than is described in the text. It would, for example, be very boring for the book to describe every building block in detail since, once a concept is understood, it is easily applied to the building block set provided in UIQ 3. However, the example applications cover most of the building blocks, giving you an opportunity to see them in an application and play with them in code.

Readers interested in porting may start with Chapter 16 and then refer to the rest of the book to see examples of UIQ 3 code explained in detail.

Conventions

To help you get the most from the text and keep track of what's happening, we have used a number of conventions throughout the book.

When we refer to a filename or to words you use in your code, such as variables, classes or functions, we use the style:

```
CQikAppUi::ConstructL()
ListView1.h
```

Blocks of code are presented like this:

```
void CAppSpecificUi::ConstructL()
  {
  CQikAppUi::ConstructL();
  // Create and set up an engine.
  iEngine=new(ELeave)CAppEngine(EQikCmdZoomLevel2);
  TBuf<KMaxListItemText>bb;
  const TInt KListView1Items=7;
  for (TInt i=0;i<KListView1Items;i++)
    {
    iEikonEnv->ReadResourceL(bb,R_STR_LIST_CONTENT_1+i);
    iEngine->SetListItem(bb);
    }
  // Create and set up the single view.
  CAppSpecificListView* q = new(ELeave)
  CAppSpecificListView(*this,iEngine);
  CleanupStack::PushL(q);
```

```
q->ConstructL();
AddViewL(*q);
CleanupStack::Pop(q);
}
```

Commands typed at a command prompt are shown like this:

```
abld build winscw udeb
```

From time to time, our experts have added tips. These are identified in the text as:

> **Pro tip:**
> Read all the tips for useful hints.

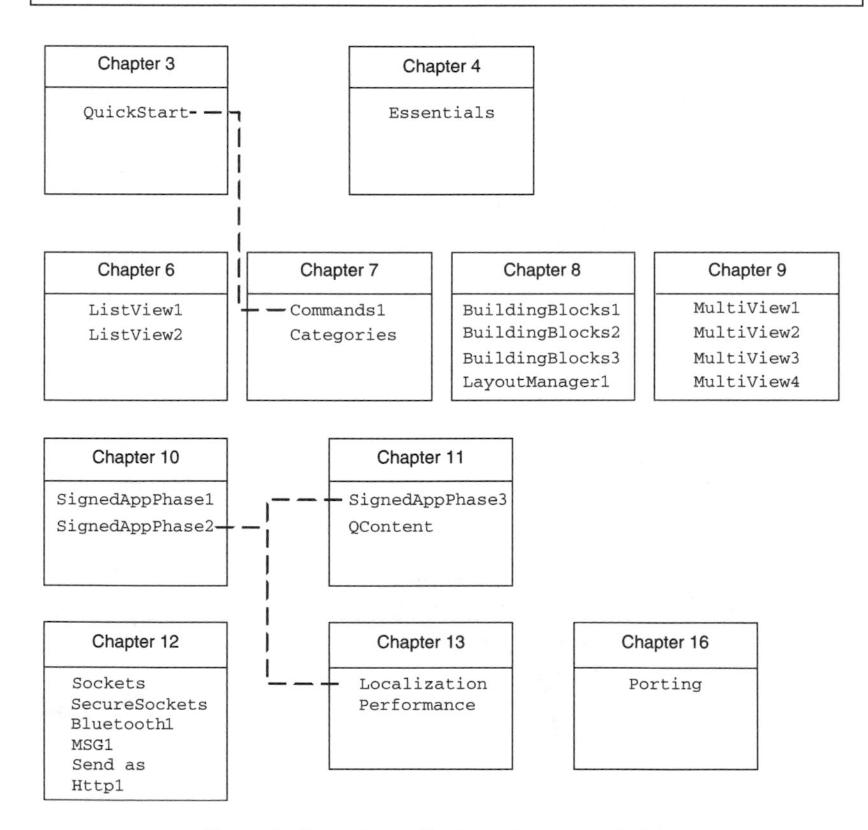

Figure 1 Example applications mapped to chapters

Example Application Code

The code for the example applications can be downloaded from ***books.uiq.com***, where you will also find a wiki based on this book.

There are extra comments in the code itself. You may find it useful to read both the code sections in the book and the actual source files.

Figure 1 helps you to link the example applications to the chapters in this book.

Acknowledgements

Authors' Acknowledgements

This book would not exist without the determination of Mikael Nerde, Thomas Bailey, Jens Greve and Ulf Wretling at Sony Ericsson to assemble the necessary people and resources. We are also indebted to the Symbian Press team, especially Freddie Gjertsen, Jo Stichbury and Satu McNabb, for their support in the project.

We are very grateful to our reviewers for diligently reading through the manuscripts, checking code and answering questions.

John would particularly like to thank his children Alex and Jenna, his wife Alison and all her book club members for their endless support and encouragement, along with all those he has worked with, past and present, especially those that have shown him the error of his ways; please don't stop!

Mark would like to thank his parents for suggesting a career in telecoms. Thanks also to the Communicators team in Ericsson for his first break in the mobile world and to the many talented and fun people at Sony Ericsson and UIQ with whom it is a pleasure to work.

Symbian Press Acknowledgements

Experience has taught us that it's hard to create a new book. Even harder when it must be created from nothing. Harder still when the topic to document is vast.

UIQ 3: The Complete Guide overcame all three difficulties and thus its existence is a testament to the hard work, commitment and sheer determination of those who worked to make it happen. We'd like to thank them all.

However, extra recognition is due, for their foresight and support of the project, to Mats Blomberg and Mikael Nerde of Sony Ericsson and, for their great endeavor, to Mark Wright and John Holloway.

1

Background

1.1 A Little History

UIQ started life in late 1998 when Ericsson, Psion and Symbian decided to work together to create the Quarter-VGA (portrait, 240 × 320 pixel) user interface for Symbian OS. A small team based in London, England and Ronneby, Sweden defined the functional content of the platform and set about designing the user interface. Ericsson provided design inputs from its research and development work and the R380 phone project while Psion brought along its considerable experience in the design of PDAs; third-party software development companies were also involved. The team set about designing the core interaction model and prototyping the important applications.

Having established a basic model that included an application launcher, a status bar, a menu structure and initial application outlines for Contacts, Phone and Agenda, it was tested using touchscreen tablet PCs. User feedback and further design work was incorporated so that gradually Quartz, as UIQ was then known, took shape (see Figure 1.1).

Symbian first announced Quartz at the Symbian Developer Conference in Santa Clara, California in February 2000. The following month, Ericsson unveiled the Ericsson Communicator Platform (Figure 1.2) at the CeBIT exhibition in Hannover, Germany.

In 2002, Quartz became UIQ and the first UIQ phone, the Sony Ericsson P800, was launched, based on Symbian OS v7.0 and UIQ 2.0. Motorola, BenQ and Arima have also developed UIQ-based mobile phones (see Figure 1.3).

1.2 About UIQ Technology

Ericsson originally set up a Mobile Applications Lab in Ronneby, southern Sweden in 1998, the same year that Symbian was established. In 1999, the

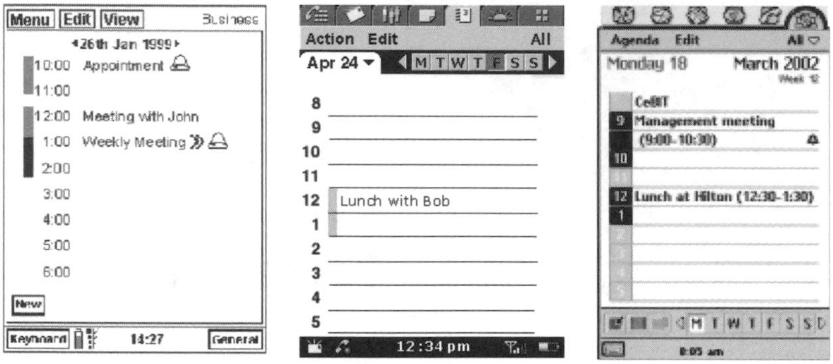

Menu Edit View Business
26th Jan 1999
10:00 Appointment
11:00
12:00 Meeting with John
1:00 Weekly Meeting
2:00
3:00
4:00
5:00
6:00
New
Keyboard 14:27 General

Action Edit All
Apr 24 M T W T F S S
8
9
10
11
12 Lunch with Bob
1
2
3
4
5
12:34 pm

Agenda Edit All
Monday 18 March 2002
Week 12
CeBIT
9 Management meeting
(9:00–10:30)
10
12 Lunch at Hilton (12:30–1:30)
1
M T W T F S S
8:05 am

Figure 1.1 Evolution of user interface

Figure 1.2 Ericsson Communicator Platform

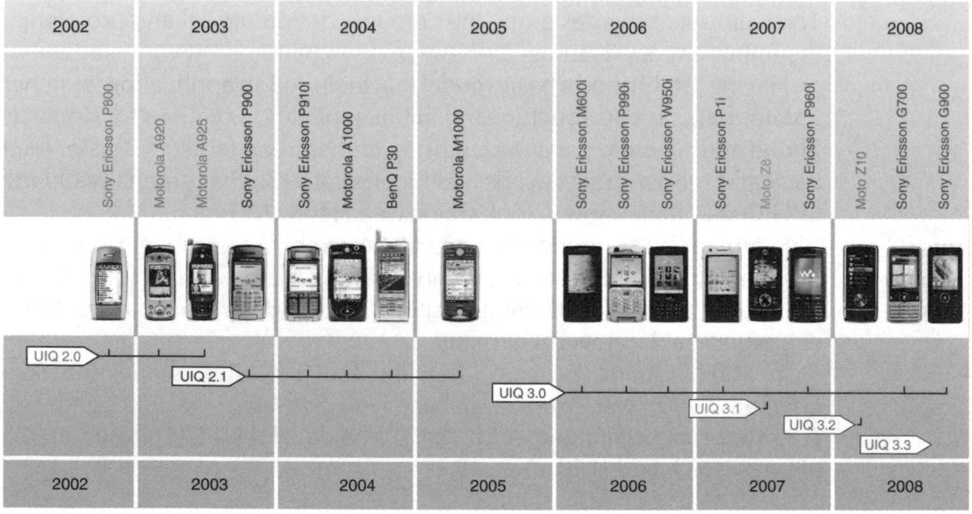

2002 2003 2004 2005 2006 2007 2008

Sony Ericsson P800
Motorola A920
Motorola A925
Sony Ericsson P900
Sony Ericsson P910i
Motorola A1000
BenQ P30
Motorola M1000
Sony Ericsson M600i
Sony Ericsson P990i
Sony Ericsson W950i
Sony Ericsson P1i
Moto Z8
Sony Ericsson P960i
Moto Z10
Sony Ericsson G700
Sony Ericsson G900

UIQ 2.0
UIQ 2.1
UIQ 3.0
UIQ 3.1
UIQ 3.2
UIQ 3.3

2002 2003 2004 2005 2006 2007 2008

Figure 1.3 UIQ timeline

lab became part of Symbian and work was focused on developing UIQ. In 2002, UIQ Technology AB was formed. UIQ Technology produces and licenses UIQ to mobile phone vendors around the world. In 2006, it was announced that Sony Ericsson was to acquire the company.

At the time of writing, November 2007, UIQ Technology is owned 50 % by Sony Ericsson and, subject to customary regulatory approval, 50 % by Motorola.

Both companies have agreed that UIQ will be independent of both vendor and chipset. In addition, UIQ will be licensed on equal terms to all mobile phone vendors in the industry. Sony Ericsson and Motorola are committed to expanding the shareholder base of UIQ to include other handset vendors.

1.3 Overview of UIQ 3

UIQ 3 is the latest major revision of the UIQ software platform, introducing major new features compared to the previous version, UIQ 2.1. The UIQ 3 software platform is pre-integrated and tested with Symbian OS v9, the global, industry-standard operating system designed for mobile phones. The UIQ 3 platform (see Figure 1.4) enhances the mobile phone user experience by enabling a large range of add-on software and content.

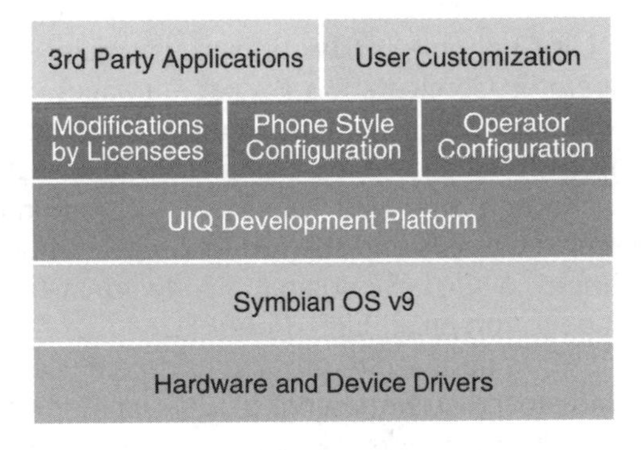

Figure 1.4 UIQ architecture

UIQ 3 makes it possible for a manufacturer to build a portfolio of mobile phones with diverse forms (see Figure 1.5) and characteristics, yet all based on a single codeline, reducing development and support costs. UIQ 3 can be customized to fit in with a manufacturer's style and user interface (UI) preferences. UIQ 3 was announced at the 3GSM event in February 2004. Models such as the MOTO Z8 and Sony Ericsson P990i, P1i and W960i (see Figure 1.6) demonstrate some of the different types of phone that are possible with UIQ 3.

Figure 1.5 Example UIQ phone form factors

Figure 1.6 UIQ 3 phones: MOTO Z8 and Sony Ericsson P990i, P1i and W960i

UIQ 3 phones can be further customized for mobile operators using the Operator Configuration Package. Images, settings, multimedia content, extra applications and much more can be preconfigured. In many cases, the customization can also be applied over the air.

Through the UIQ Developer Community, application developers receive the SDK and support to create compelling applications for UIQ 3 phones. A single Symbian OS Software Install Script (SIS) package can be made to run on all UIQ 3 phones.

UIQ is designed to provide excellent user interaction with easy access to advanced network services. User-centric design and testing has ensured that key user tasks can be completed quickly and smoothly, including those that involve multiple applications. Rich graphics and effects strengthen the user experience. Consistency in overall structure, layout and key assignments enables users to feel at home across different phone styles.

Symbian OS is a multi-tasking operating system, which means that a user can, for example, send an SMS while listening to music.

UIQ includes a suite of applications for messaging, web browsing, managing personal data and more. Data can be synchronized with PC applications or remotely using OMA Data Sync. The UIQ Messaging application includes SMS/EMS, MMS and email.

1.4 UIQ 3 Application Suite

The UIQ application suite includes all key functionality for a mobile phone. Mobile phone manufacturers have the freedom to add to or modify the delivered software.

The applications also make themselves available to third-party applications. For example, an application can access the Contacts database to find a phone number and can create a new meeting in Agenda.

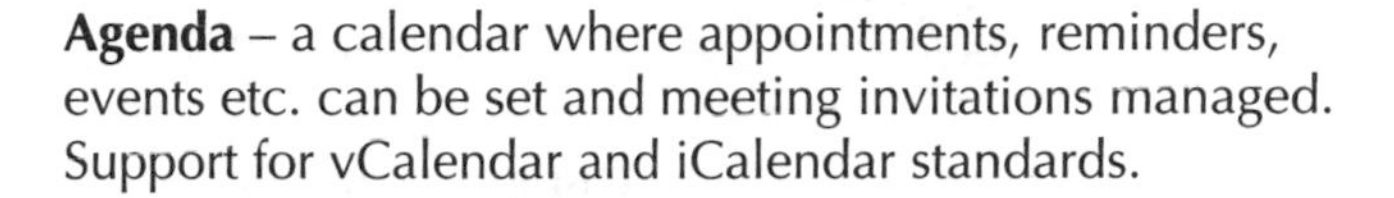

Agenda – a calendar where appointments, reminders, events etc. can be set and meeting invitations managed. Support for vCalendar and iCalendar standards.

Calculator – a basic desktop calculator with useful arithmetic functions and a memory.

Contacts – lets you store all your contacts in one place and categorize them into folders and groups. The application supports USIM/SIM card management, vCard standard 2.1 and integrated instant messaging as well as direct navigation links to instant messaging applications.

Messaging – a suite including email, SMS/EMS, MMS and beamed entries. Each email account (POP3, IMAP4, SMTP) has a separate inbox and all other message types are collected in a unified inbox. UIQ has a messaging type module (MTM) framework, which enables third-party developers to create support for additional messaging types, such as push email.

Jotter – for creating notes and drawing sketches. In common with other applications, it includes a 'Send as' function so that content can be easily shared.

Remote Synchronization – the application is compliant with OMA Data Sync 1.2 and is used to synchronize the user's data in Agenda/To do and Contacts with a remote server. There is also a plug-in framework for third-party developers to add synchronization support for other applications.

Telephony – UIQ supplies a dummy Phone application in the emulator; mobile phone manufacturers supply their own Phone application.

Time – an application used to display a clock and time, as well as set alarms and time zones, etc.

To do – for keeping track of tasks and creating to-do lists with priorities, dates, categories, etc. Supports the vCalendar standard.

Utilities – UIQ includes tools for managing and personalizing the phone: application installer, application launcher, control panel, file manager, task list and themes.

Viewers – UIQ has a set of viewers to view and store received files, including images, audio and themes. The appropriate viewer starts automatically when a file is selected.

Voice recorder – an application for sound recording and playback that can be used as a dictaphone.

Web – UIQ includes a full HTML web browser where multiple pages can be opened at the same time in separate windows. The browser supports a wide range of web standards.

1.5 Technologies and Features in the UIQ Platform

UIQ is continually updated and enhanced with the latest mobile technologies. *Product Description* and *Datasheet* documents on ***www.uiq.com*** provide full detail for each version. We outline some of the key technologies and features of UIQ in this section.

Bluetooth Technology

The Symbian OS Bluetooth stack is compliant with the Bluetooth specifications. UIQ Bluetooth extends the Symbian OS implementation with several components that provide additional profiles on top of Symbian OS: FTP Server, Object Push (OPP), and Personal Area Network (PAN), a user interface for Bluetooth settings and support for audio streaming (A2DP). UIQ also supplies utilities and a user interface for common Bluetooth tasks, such as pairing, authentication, and authorization.

Digital Rights Management (DRM)

UIQ enables downloading, rendering and installation of DRM-protected content, such as music, files, and themes. This includes support for: forward lock, separate delivery, combined delivery and ROAP triggers. The UIQ OMA download agent supports downloading content protected by OMA DRM v2.0.

Graphical Effects

- Animations: supported formats are MNG, GIF89a and SVG
- Color support: 8-bit to 24-bit color and 8-bit transparency
- Text styles: support for normal, outlined, and shadowed text; the color of each text style can be selected
- Semi-transparent windows: support for UI elements, such as menus and dialogs, to have rounded corners, for example, or let the screen area behind the window shine through
- SVG Tiny: the SVG Player instantly displays high-quality static images at any resolution, and runs animations smoothly. The SVG Tiny ICL Plug-in makes it possible for applications to get a still frame from an SVG file or animation and In-scene SVG allows applications to render SVG files in their UI.

Infrared

Legacy infrared is fully supported with Slow Infrared (SIR), allowing signaling rates of 9.6 kbps to 115.2 kbps, IrOBEX, IRCOMM, and IrTRANP.

Instant Messaging Integration API

UIQ offers a framework that allows third-party, instant messaging (IM) clients to be integrated into the UIQ Contacts application.

International Language Support

The UIQ platform is supplied in English but mobile phone manufacturers can produce phones in many languages. For example, UIQ supports ideographic languages such as Japanese, Chinese, and also right-to-left scripts and mirroring of controls and layouts for languages such as Arabic and Hebrew. All localizable data is separated from the code to facilitate localization.

Multi-Homing

UIQ supports multi-homing, which enables the user to have multiple connections open to different network services, such as Internet and MMS.

Operator Configuration Package

Mobile phone manufacturers and network operators need to brand their products and services, to provide a specific look and feel to their end users. Features and services must be easy to use. UIQ provides a solution called Operator Configuration Package that allows manufacturers to configure a UIQ phone with their own specific customization and to fulfill network operators' requirements.

Over-the-Air Provisioning

UIQ supports provisioning of the phone using OMA Device Management and OMA Client Provisioning, which allows a network operator to manage the device remotely, over-the-air (OTA).

Platform Security

UIQ 3 includes the platform security features of Symbian OS v9, which are designed to protect the interests of consumers, network operators and developers.

Virtual Private Networks

Virtual private networking (VPN) is supported using the industry-standard protocol IPSec.

1.6 UIQ 3 Development Platform

The UIQ Application Development Platform contains a set of components with corresponding APIs that are used to create applications for UIQ phones.

Application Framework

The application framework provides application structures and layout support for creating application UIs, for example, layout managers and building blocks. It also defines the standard UI behavior of applications and dialogs.

System Services

System Services supply UIQ-specific services to the applications, such as the Send as feature which can be seen as a shortcut to the Messaging application and to beaming functionality.

Rich GUI Toolkit

UI controls such as the status bar, input fields, menus and scrollbars are part of typical UIQ applications.

UIQ 3 SDK

The UIQ 3 Software Development Kit (SDK) enables you to write applications in C++ for devices based on UIQ 3. The UIQ 3 SDK comprises development tools, including the UIQ Emulator, public APIs, programming examples and documentation. With the UIQ emulator (hosted on a Windows-based PC), programs can be run and tested. The SDK also includes deployment tools that allow you to package your application for convenient delivery to end users.

Phone manufacturers can extend the UIQ 3 SDK with add-on packages. Such extensions typically contain phone-specific emulator skins, APIs and documentation.

Integrated Development Environments (IDE)

The standard IDE for UIQ 3 development is Carbide.c++ which is described in more detail in Chapter 3. Other development environments, such as Microsoft Visual Studio 2003, are also supported.

Java ME

UIQ provides a Java ME solution for Java applications. However, this book covers C++ programming only.

1.7 UIQ Ecosystem

UIQ Technology works with key players in the telecommunications world to promote and develop UIQ, a platform which is manufacturer-

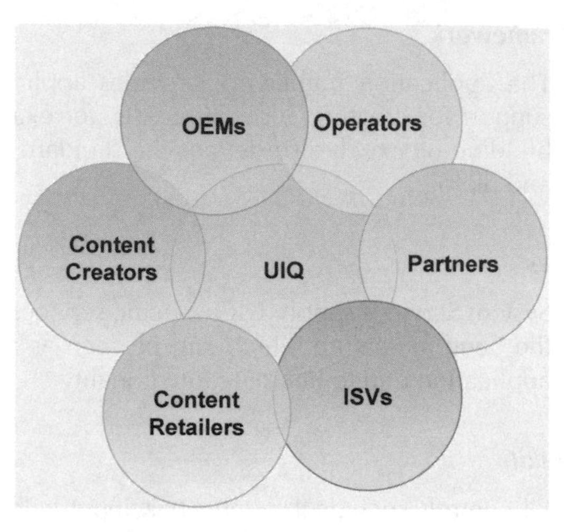

Figure 1.7 The UIQ ecosystem

independent and open to the industry. The UIQ ecosystem includes mobile phone manufacturers, mobile network operators, application developers and end users (see Figure 1.7).

UIQ works with manufacturers to develop an ever wider range of mobile phones meeting the needs of key target markets. UIQ participates in the third-party developer programs of mobile operators and works with them to understand requirements for applications, services and customization.

By actively engaging with all these parties, UIQ can help application developers understand the enterprise and consumer marketplace and develop valuable applications and services.

UIQ provides information and support to developers via the UIQ Developer Community, ***developer.uiq.com***, and also works closely with manufacturer support services such as Sony Ericsson Developer World and MOTODEV.

UIQ attends the key conferences and industry events, as well as organizing its own events such as *UIQ Developer Fast Track*. The *UIQ Alliance* partner program is a collaboration of key companies in the mobile industry who provide services and solutions including applications and training.

UIQ encourages developers to use the Symbian Signed process, which certifies the origin of an application and provides a benchmark level of quality assurance. The Symbian Signed catalog promotes signed applications to channel partners (operators, aggregators, distributors, mobile phone manufacturers, and so on). In Chapters 10 and 11 of this book, we build a simple file manager application, `SignedApp`, which passes the Symbian Signed process.

A rich API set, attractive UIQ phones, industry-standard development tools, and active developer and partner communities ensure UIQ is the mobile phone developer platform of choice.

2

UIQ 3 Basics

UIQ 3 enables mobile phone manufacturers to create a range of products that have different hardware characteristics but use the same codeline. You can write a UIQ 3 application and deploy it to all UIQ 3 phones via a single package (SIS file). This chapter explains the fundamental concepts so that you can understand how UIQ 3 delivers this flexibility.

2.1 UI Configuration

UIQ 3 supports a variety of screen and hardware configurations. These are defined using parameters. Many configurations are theoretically possible and more options may be added as new types of hardware and user interaction are developed. In practice, you will work with just a few UI configurations, as we explain later on.

2.1.1 Configuration Parameters

Screen Mode

Screen Mode defines the screen resolution (in pixels) and its position, portrait or landscape (see Table 2.1 and Figure 2.1).

UIQ 3 is optimized for these display resolutions, but a phone manufacturer may build a phone with a different screen resolution. Your application can find out the exact resolution from the screen device.

If the UI configuration changes from portrait to landscape, it also rotates the four navigation keys (up, down, left, right) so that they stay lined up correctly with the screen. For example, up becomes left. Typically, the phone operates in portrait mode and applications that want to use landscape switch the display to that mode, generally via a command.

Table 2.1 Screen resolution

Screen Mode	Display Size (width x height)	Position
Portrait	240 × 320 (QVGA)	Portrait
Landscape	320 × 240 (QVGA)	Landscape
Small	240 × 256	Portrait
Small Landscape	240 × 256	Landscape

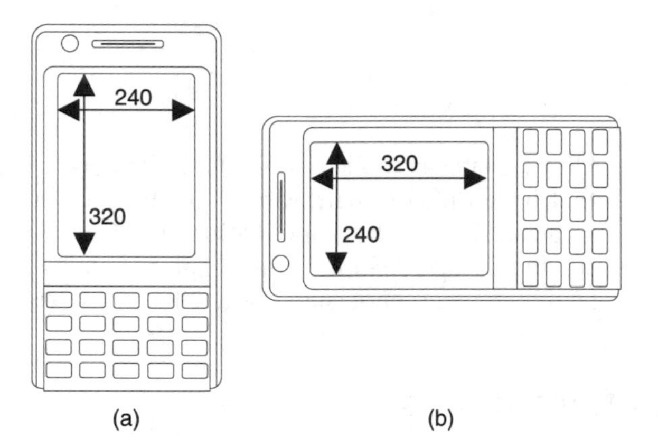

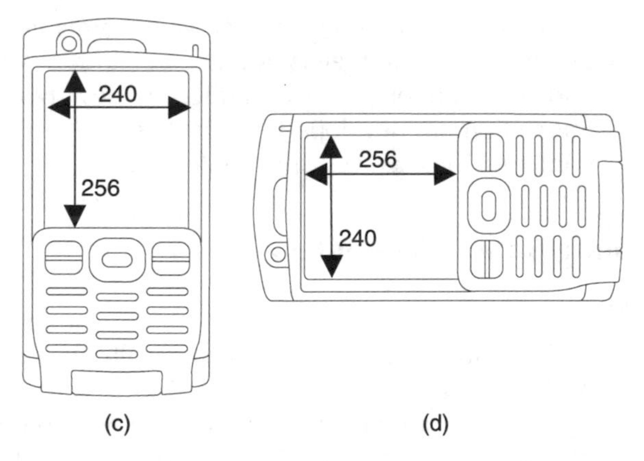

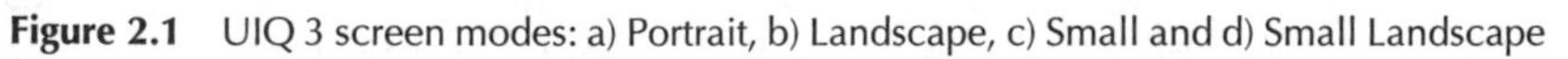
Figure 2.1 UIQ 3 screen modes: a) Portrait, b) Landscape, c) Small and d) Small Landscape

Screen Orientation

UIQ 3 supports inversion of the screen by 180 degrees (see Figure 2.2). What this means in practice is that, where supported, the user may choose to see Landscape mode by turning the phone to the left or to the right.

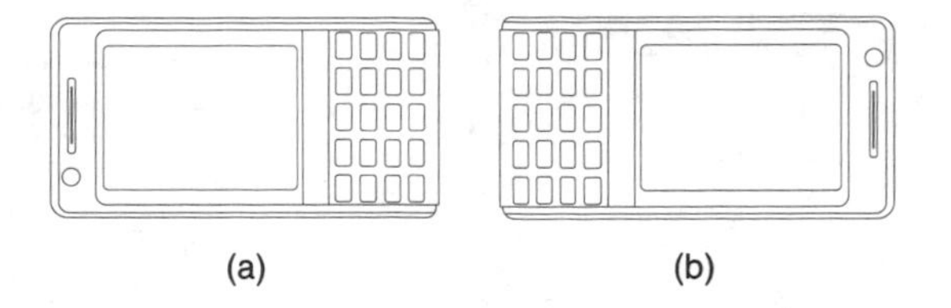

Figure 2.2 Screen orientation: a) normal and b) inverted by 180°

Your application does not need to do anything special to support inverted mode, it is handled by UIQ 3. The four-way navigation keys are transposed so that they operate correctly in the new mode.

Touchscreen

The touchscreen parameter indicates the presence of a digitizer, typically a touchscreen layer applied to the front of the screen.

Interaction Style

Interaction style defines the way the user works with the phone:

- **Softkey Style** is optimized for one-handed use without a touchscreen.
- **Softkey Style Touch** is based on Softkey Style, but allows both one-handed use via softkeys and two-handed use via a touchscreen.
- **Pen Style** is optimized for two-handed use. The phone is held in one hand and applications are used by means of a stylus held in the other hand. A touchscreen is required for this mode. Hardware buttons can also be used to interact with applications.

2.1.2 UIQ 3 Screen Layout

The UIQ 3 screen is divided into five areas, two of which differ between Softkey Style and Pen Style interaction. The anatomy of the screen is described in more detail in Chapter 5.

Softkey Style Screen Layout

- The *Status Bar* (see Figure 2.3a) presents information such as signal strength, battery status and unread messages. It is configured by the phone manufacturer and cannot be used by third-party applications. In Softkey Style, the Status bar is read-only; in Softkey Style Touch, the user can tap on the icons to see further information and access settings.

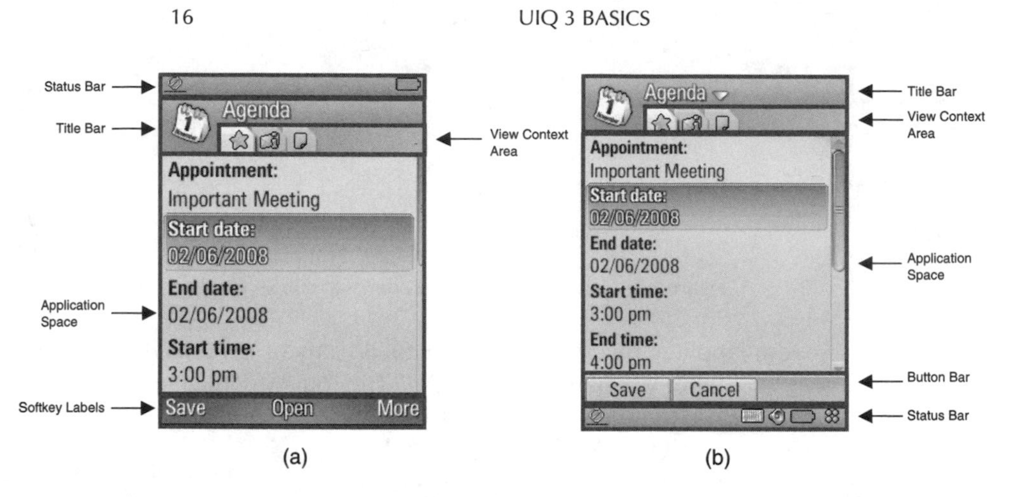

Figure 2.3 Screen areas in a) Softkey Style and Softkey Style Touch and b) Pen Style

- The *Title Bar* displays the application name and icon.
- The *View Context Area* displays status information. Applications typically break down data into a number of pages. The first (default) page shows the most important information, such as the appointment name, data and time, while details such as location and notes are provided on extra pages. Where there is more than a single page, one tab for each page is displayed in the View Context Area. In Softkey style, the left and right keys can be used to change page; in Softkey Style Touch, pages can be selected directly by tapping on the required tab. Additional status information can be provided in the space to the right of any tab.
- The *Application Space* belongs to the application. Building blocks, which we will describe later, are typically used to construct application views.
- In Softkey Style, the labels for the three softkeys are displayed in the *Softkey Labels* area.

Pen Style Screen Layout

In Pen Style (see Figure 2.3b), the Status bar is positioned at the bottom of the screen. The user can tap on its icons to see further information and access settings. Additional icons on the *Title Bar* provide access to the menu and category selection.

Pages in the *View Context Area* can be selected directly by tapping on the required tab. Additional status information can be provided in the space to the right of any tab.

The Softkey labels are replaced by a *Button Bar,* which can be configured to have up to three buttons (rather like the Softkey bar), nine icons or a combination of buttons and icons.

2.1.3 Pre-Defined UI Configurations

A reference configuration and four further configurations make up the five pre-defined UI configurations (see Table 2.2) supported in the application development platform and used in current UIQ 3 phones.

Table 2.2 Pre-defined UI configurations

Name Constant in `qikon.hrh`	Screen Mode	Touchscreen	Interaction Style	Orientation
Softkey Style (reference) `KQikSoftkeyStylePortrait`	Portrait	No	Softkeys	Normal
Softkey Style Touch `KQikSoftkeyStyleTouch Portrait`	Portrait	Yes	Softkeys	Normal
Pen Style `KQikPenStyleTouchPortrait`	Portrait	Yes	Menu Bar	Normal
Pen Style Landscape `KQikPenStyleTouchLandscape`	Landscape	Yes	Menu Bar	Normal
Softkey Style Small `KQikSoftkeyStyleSmall Portrait`	Portrait, small	No	Softkeys	Normal

A phone can only display one UI configuration at a time but it may support a number of different configurations (see Table 2.3). A hardware event may change the configuration; for example, closing the flip on a Sony Ericsson P990i changes from Pen Style to Softkey Style Small. You can access a list of the possible System UI configurations that are available on a phone using the `CQikUiConfigClient` interface.

Table 2.3 Default phone UI configurations

Phone	Name
MOTO Z8	Softkey Style
Sony Ericsson M600i, W950i, P1i, W960i	Softkey Style Touch
Sony Ericsson P990i (flip open)	Pen Style
Sony Ericsson P990i (flip closed)	Softkey Style Small

You need to define which UI configurations your application supports. When it starts, the UIQ 3 application framework selects the most appropriate of your supported UI configurations, meaning the one that fits best to the current UI configuration. This *best fit* ability means that you do not have to care about orientation; your application will work in both normal and inverted orientation, for example. We describe this in more detail in the `ListView1` application (Section 6.1.3).

Alternatively, your application view may request a different screen mode; for example switching from the default Portrait mode to Landscape mode. The UIQ 3 application framework reads this request and, if necessary, changes the current UI configuration before activating the view. The web browser, for example, can switch from portrait mode to landscape mode under user control.

Because of its hardware flip design, the Sony Ericsson P990i (see Figure 2.4) supports more UI configurations than any other UIQ 3 phone to date:

- Pen Style Touch Portrait (default with flip open)
- Pen Style Touch Landscape
- Pen Style Touch Landscape inverted (180°)
- Softkey Style Small Portrait (default with flip closed)
- Softkey Style Small Landscape
- Softkey Style Small Landscape inverted (180°).

Figure 2.4 Sony Ericsson P990i UI configurations

All other Sony Ericsson UIQ 3 phones to date operate in Softkey Style Touch by default and also support Pen Style Touch Landscape in

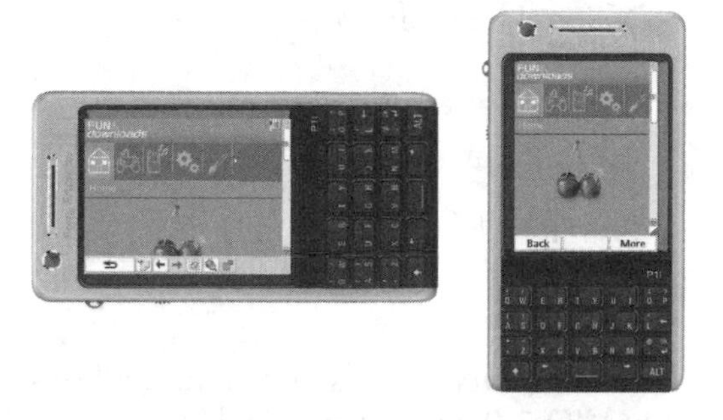

Figure 2.5 Sony Ericsson P1i UI configurations

normal mode (not inverted). Sony Ericsson phones support landscape only in full screen mode, since this the most useful for applications (see Figure 2.5).

Softkey Style (Reference Configuration)

Softkey Style is the *reference configuration*. This is a new UI configuration introduced in UIQ 3. It is intended for phones with smaller displays and keypads and does not require a touchscreen. The UIQ applications are optimized for this UI configuration. As well as the UI configuration parameters for Softkey Style, UIQ makes recommendations for the hardware.

Screen properties:

- 240 × 320 pixels (portrait mode)
- dot pitch: 0.14 mm
- minimum screen size: 34 mm × 45 mm (2.2″)
- color depth: 8 bit (256 colors) to 24 bit (>16 million colors).

Mandatory hardware keys:

- Four-way navigation (up, down, left, right)
- Action key
- Left and right softkeys (LSK and RSK)
- Clear (backspace) key
- Send and End (phone)
- Numeric keypad (0–9, *, #)
- Application launcher key.

Optional hardware keys:

- Cancel (Back) key
- Video call
- QWERTY keyboard
- Done key

See *UIQ Product Description* for further supported keys.

In UIQ 3.0, the Cancel (Back) key is mandatory and the arrangement is called three Softkeys and Back (*3SK&B*, see Figure 2.6a). The Cancel hardware key is used for the Back operation. From UIQ 3.1, the Cancel hardware key is optional. The Back function is instead represented as a Back label placed on the right Softkey. This variant is called three Softkeys (*3SK*, see Figure 2.6b). Manufacturers can choose from two softkey configurations so that the softkeys on phones based on UIQ 3 match other phones in the portfolio.

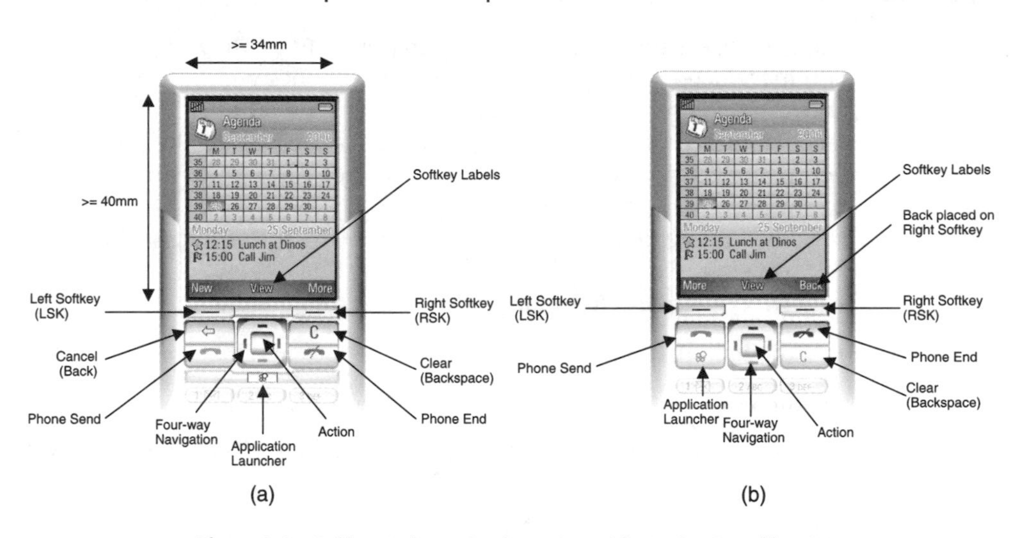

Figure 2.6 Softkey Style UI Configuration with a) 3SK&B and b) 3SK

In this book, we use the 3SK&B model for Softkey Style commands. This is used by current Sony Ericsson phones that run in Softkey Style Touch and the P990i in Softkey Style Small (flip-closed) mode. This mode is presented in the UIQ 3.0 emulator. To see 3SK, use the UIQ 3.1 emulator. We look at command handling in more detail in Section 2.3.

At this point, it's useful if we describe the behavior of the hardware keys in more detail.

The Up and Down keys are used to:

- navigate up and down lists
- navigate inside an activated control.

The Left and Right keys are used to:

- switch page within a view (move across the tabs in the View Context area)
- navigate within a view
- navigate within an activated control.

The Action key is used to:

- select the command displayed on the center Softkey label
- select the highlighted item within an activated control.

The Clear key is used to:

- delete characters in an editor
- delete an item in a list
- cancel and go back when navigating within applications.

The Cancel (Back) key is used to:

- cancel and close an active control or pop-up, such as the menu
- close the view or dialog performing the Back function.

The MOTO Z8 uses the Send key to make a call and the End key to end it. Sony Ericsson's phone application is softkey driven. It does not use Send or End keys.

The Application Launcher key is mandatory for Softkey Style phones, such as the MOTO Z8. Sony Ericsson provides an alternative implementation via the start (idle) screen and status bar.

Softkey Style Touch

Softkey Style Touch can be used in exactly the same way as Softkey Style via the hardware keys (see Figure 2.7). Adding a touchscreen means that applications can also receive input from stylus operations such as tapping

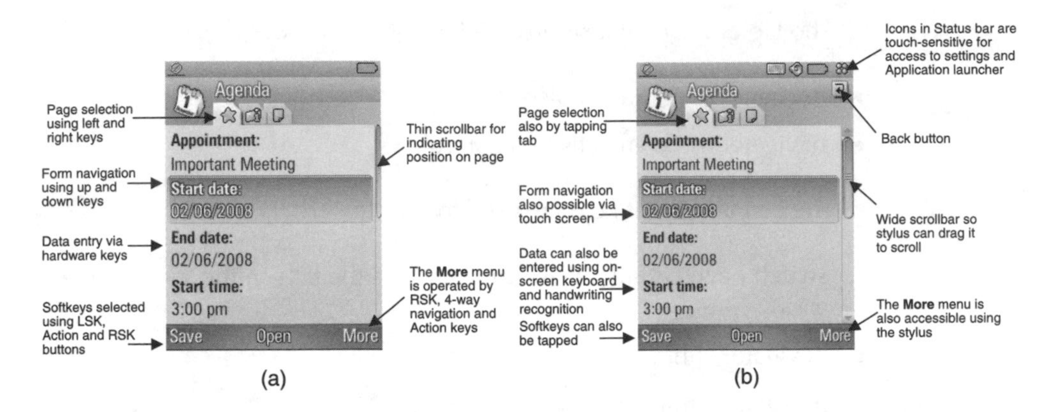

Figure 2.7 Comparison of a) Softkey Style and b) Softkey Style Touch

controls and dragging scrollbars. The UI layout is modified slightly in this mode so that the pen can be used. Softkey Style Touch is a fully-enabled touchscreen interface.

In the UIQ emulator, the Softkey label area has a flat background; there are no graphical dividers between the three labels.

Sony Ericsson phones modify the style of the Softkey labels so that they look and behave like three buttons (see Figure 2.8). This reminds the user that it is possible to touch the screen.

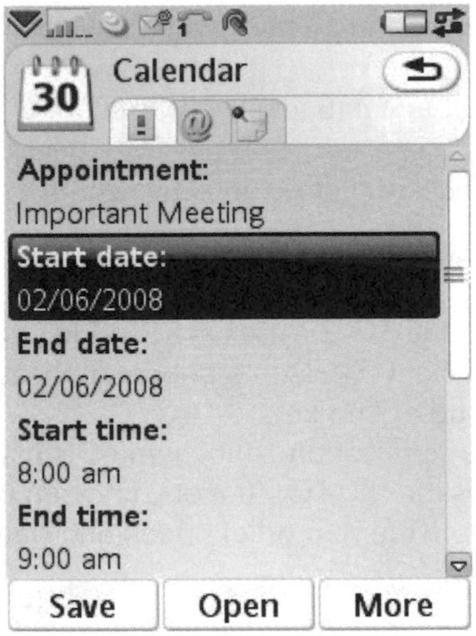

Figure 2.8 Agenda (Calendar) application on Sony Ericsson P1i

From Figure 2.8, you can also see how Sony Ericsson has customized the UI with its own graphics and color scheme.

Pen Style

This configuration (see Figure 2.9) is similar to UIQ versions up to 2.1, but targets displays with a slightly smaller dot pitch. In UIQ 3.0, Pen Style is a reference configuration.

Its interaction style is a pen on a touchscreen with a menu bar.

Screen properties:

- 240 × 320 pixels (portrait mode)
- dot pitch: 0.165 mm
- minimum screen size: 40 mm × 53 mm (2.6″)
- color depth: 8 bit (256 colors) to 24 bit (>16 million colors).

Recommended hardware keys:

- four-way navigation (up, down, left, right)
- Action key
- Send and End (phone) keys
- Done key.

Figure 2.9 Pen Style UI mode

Figure 2.10 shows the UIQ Agenda application in Pen Style mode. While some tasks can be carried out one-handed, it is optimized for two-handed use and some tasks can only be carried out using the stylus.

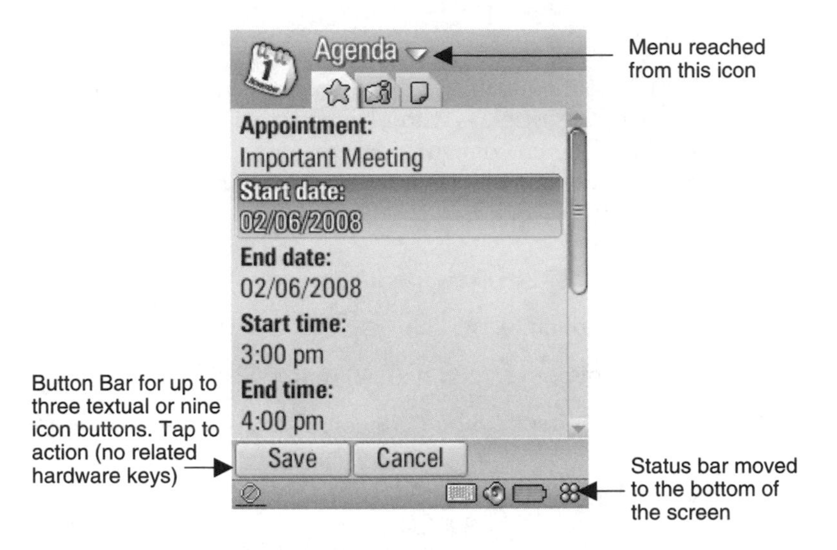

Figure 2.10 Agenda application edit view in Pen Style UI mode

Figure 2.11 Agenda application month view

Figure 2.11 shows the main view of the Agenda application, showing iconic rather than textual buttons in the Button Bar.

Some applications benefit from a different UI approach when a touch-screen is present. You are free to implement a different GUI for each UI configuration where this is beneficial. The Calculator application has been designed in this way so that one-handed use in Softkey Style is optimized (see Figure 2.12a). Numbers can only be entered using the hardware keypad, so no number keys need to be displayed on the screen.

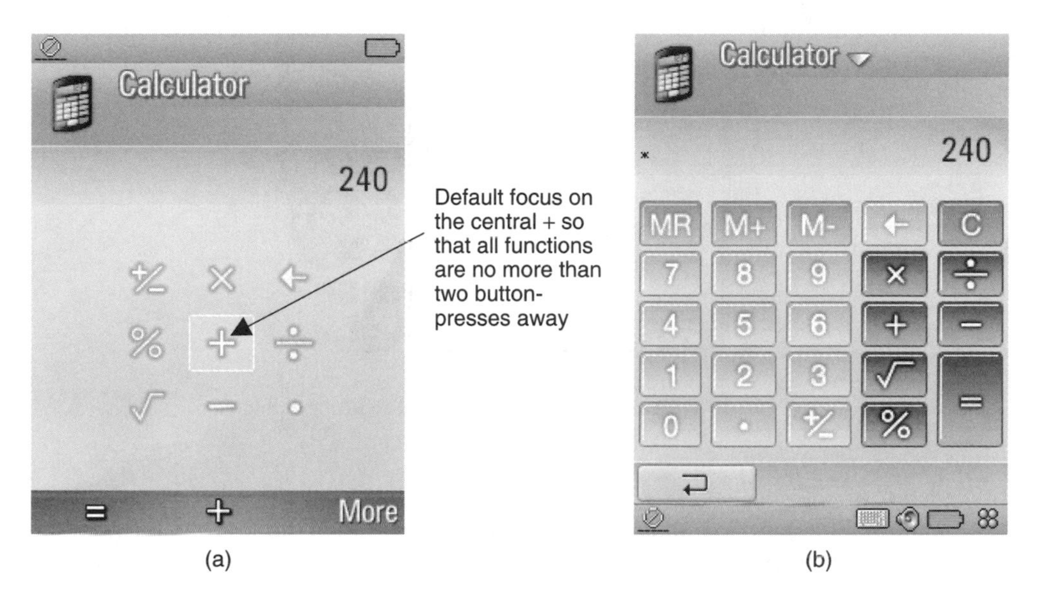

Figure 2.12 Calculator in a) Softkey Style and b) Pen Style

Instead, the calculator functions are laid out in a grid so that they can be easily selected using the four-way navigation and Action keys.

In Pen Style, the calculator can have a traditional layout (see Figure 2.12b). Numbers can be entered via the screen or hardware keypad.

Pen Style Touch Landscape

Some applications require a landscape screen. Pen Style Touch Landscape provides it by re-organizing the screen (see Figure 2.13).

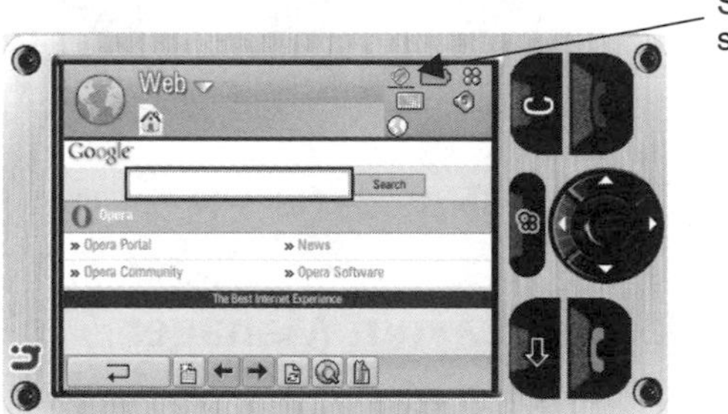

Figure 2.13 Pen Style Touch Landscape

Applications can take the full screen in any UI configuration (see Figure 2.14). This is particularly useful for displaying video, pictures, web content and games, especially in landscape orientation. If your application uses this mode, make sure it is easy for the user to get out of full screen and back to other applications.

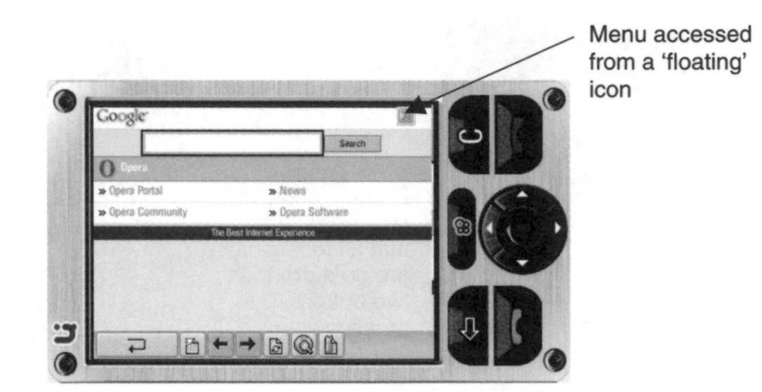

Figure 2.14 Pen Style Touch Landscape (full screen)

Softkey Style Small

The Sony Ericsson P990i uses Softkey Style Small mode when the number-pad flip is closed and part of the screen is obscured. The application space is reduced in size (see Figure 2.15).

Figure 2.15 Softkey Style Small UI configuration

2.2 Building Blocks and Layout Manager

In UIQ 2, the layout of an application was manually created by the application developer. UIQ 3 introduces the concepts of building blocks and a layout manager.

In UIQ 3, you use building blocks to place controls in the application space on the screen. Each building block has a defined layout and handles the placement of the controls to achieve the correct margins and alignment.

2.2.1 System Building Blocks

The 18 standard building blocks can be used in any combination; this makes layout design extremely flexible. You can choose a set of building blocks to compose a layout that suits your application. Technically, building blocks are controls themselves, but they have no visual appearance or state information; they are simply containers for other controls. Each building block has a defined layout and a number of slots where controls are inserted. Using the standard building blocks that UIQ 3 provides makes the behavior of your application predictable and familiar to users. You can also create custom building blocks for specific layout needs.

Keeping the user interface consistent across applications and phone models is easy with building blocks (see Figure 2.16). They manage the layout of the controls as well as managing focus and highlight.

Figure 2.16 Designing a view using building blocks

Prior to choosing which building blocks to use, you need to decide how much information you need to display. Try to use a set of building blocks where all of the blocks are the same height. This way, the highlight will also have the same height no matter which building block has focus. (The building block containing the image in Figure 2.16 does not take focus.) The visual effect of scrolling through the building blocks will then be smooth instead of jumpy, which would distract the user from the task at hand.

2.2.2 Custom Building Blocks

If you require a layout that is not supported by any of the standard building blocks, you can create custom building blocks either programmatically or from resource files (see Figure 2.17). Using custom building blocks means that the appearance and good scrolling behavior of your application is maintained.

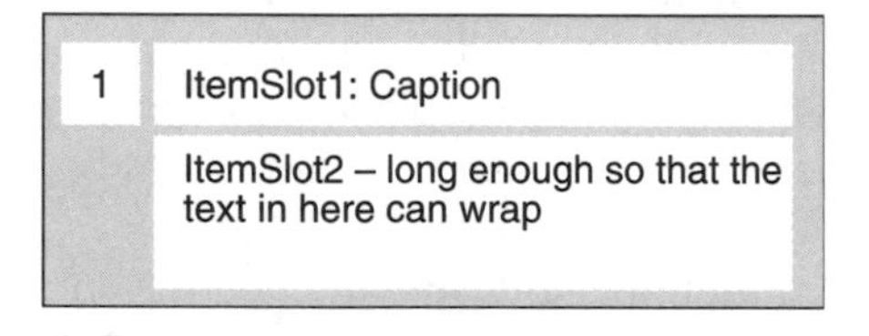

Figure 2.17 Making a custom building block

You can define further parameters of your custom building block such as:

- horizontal and vertical margins
- horizontal and vertical alignment
- number of visible text lines before truncation.

2.2.3 Layout Managers

Layout managers are introduced in UIQ 3 to make it easier to implement the layout of controls. If you are using building blocks, then the role of the layout manager is hidden and you do not need to worry about it. The difference is that building blocks are controls themselves while layout managers are a mechanism for laying out controls. Building blocks are basically compound controls that are also arranged by a layout manager.

RowLayoutManager FlowLayoutManager

GridLayoutManager ColumnLayoutManager

Figure 2.18 Layout manager types

If you create a control which cannot be placed in a building block or for some reason you do not wish to use building blocks, a layout manager is a good alternative for managing layout. The view can be laid out in rows, columns, a grid or a *flow* (see Figure 2.18). Flow means that the controls are laid out along rows, breaking to a new row when there is not enough space for the next control.

Layout managers are covered in more detail in Chapter 8. An example application, `LayoutManager1`, shows row and grid layouts in operation.

2.3 Command Processing Framework

Users interact with a mobile phone and applications by giving commands. An application is typically in a *waiting state* until the user selects an action to be performed.

As we have seen, UIQ 3 introduces a set of UI configurations which use the screen and keys in different ways. It is no longer possible to define a single menu structure that can work effectively across all the available configurations. To make it possible to support all UI configurations from a single codeline, UIQ 3 includes a Command Processing Framework (CPF, see Figure 2.19). Instead of programming commands directly on to buttons and menus, applications present an abstract set of commands and the CPF is responsible for presenting them in the UI, including hardware buttons, softkeys and menus. Each view, dialog and popout has its own command list, while individual controls can also add commands.

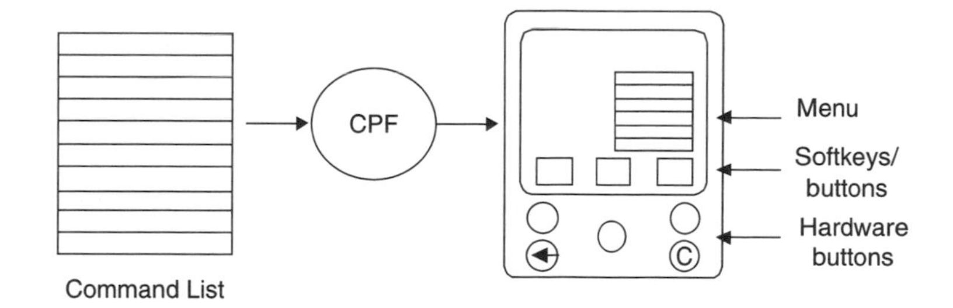

Figure 2.19 Command Processing Framework outline

You give each command in your application parameters, such as `type`, `priority`, `group` and behavior flags, which guide the CPF to present each one in the best way. Some types, such as `Yes`, `No` and `Done`, give very specific direction because softkeys are used for these commands wherever possible. Other types, such as `screen`, make a command generally available so it typically ends up in a menu.

In the same way, you do not explicitly dim a command when it is unavailable; you set the CPF parameter `SetAvailable` to false. In Pen Style, UIQ 3 dims the menu option, however in Softkey Style the option is not displayed so that the user does not have to navigate over it.

Because the CPF does the work, you do not have to worry about arranging commands for each UI mode. It is important that you work with the abstract system rather than trying to force commands to display in a specific way. However, you can specify different commands or parameters for each UI configuration if necessary, so that you can present the most suitable functionality in each case. Chapter 7 provides an example application, `Commands`, and describes in detail how you can add commands to your application.

2.3.1 Softkey Style

In the 3SK&B model, commands are based on the three softkeys and the Cancel (Back) key (see Figure 2.20).

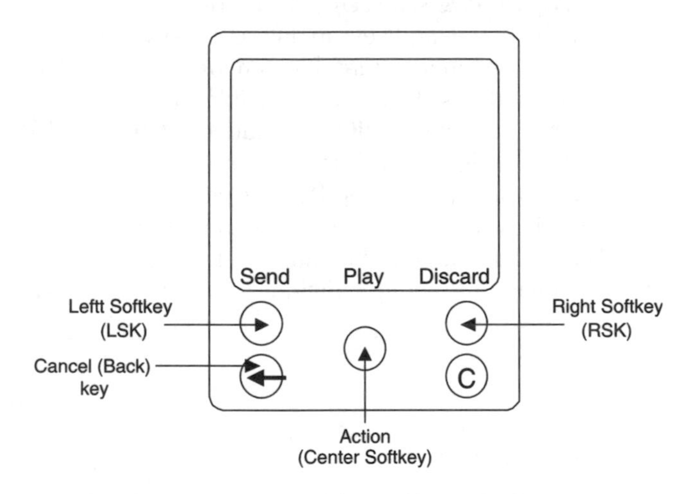

Figure 2.20 Commands in Softkey Style (3SK&B)

The softkey actions are defined by labels across the bottom of the screen.

The Center key is used for the most likely action and is often linked to the item that has focus; these commands have type `item`.

The left softkey is used for:

- Done
- positive and safe actions, such as *Yes* and *Save*
- second-most-likely actions (second item command)
- More – access to a menu (3SK).

The right softkey is used for:

- negative or destructive actions, such as No and Cancel
- third-most-likely actions (third item command)
- More – access to a menu (3SK&B)
- Back (3SK).

'More' is the most common usage of the right softkey (3SK&B) or left softkey (3SK). In this case, the third-most-likely action and other actions are placed in the menu.

The 3SK variant differs from 3SK&B in the following ways:

- The Action key label (Center softkey) duplicates the left softkey when only two actions are available; this means that the Action key is also used to select a positive command. An icon, , is displayed on the Action key when this occurs (see Figure 2.24b).
- The Clear key which is used to delete text in text input mode can also be used to Cancel and go back when navigating between and within applications in non-text-input mode.
- The command on the Action key label (Center softkey) is also available as the first item in the menu when available.
- 3SK places the More menu on the left softkey.

The MOTO Z8 uses the 3SK variant. If you look at this model, you will notice that the 'More' menu is called 'Options' and is located on the left softkey. The right softkey is used as the Back button. Other Motorola phones use this type of interaction model. Keeping a similar interaction model across different mobile phones is important to manufacturers. Customers moving from one model of phone to another expect similar behavior. Consistency also reduces technical support training effort and reduces the number support requests from users.

Examples in this book use the 3SK&B variant, as found on current Sony Ericsson phones that operate in Softkey Style Touch configuration. The Contacts application shows how the hard and softkeys are used in practice. Looking at the detailed view of a contact with the highlight on a telephone number (Figure 2.21), the center softkey is assigned the most obvious action of making a call. The left softkey is assigned the second most likely action of sending a message. Contacts is a rich application with many more commands, therefore the right softkey gives access to a menu. The third most likely action, making a video call, is placed in the menu (see Figure 2.22).

Figure 2.21 Commands in the Contacts application, phone number highlighted

Figure 2.22 Menu in the Contacts application

When the menu is open, the navigation keys move focus, the center softkey is used to select the required menu item and the right softkey closes the menu. It is possible to define cascading menus.

If we move the focus to the email address, the labels change to reflect the actions that are possible with an email address (see Figure 2.23). Sending an email is the most obvious action and is assigned to the center key. A second option is to send an MMS message. The right softkey still provides access to the menu.

Figure 2.23 Commands in the Contacts application, email address highlighted

If we try to delete the contact, the confirmation dialog conforms to this convention by placing Yes on the left softkey and No on the right (see Figure 2.24). Figure 2.24b shows the addition of the symbol in UIQ 3.1, indicating that the Center key duplicates the default action (Yes in this case).

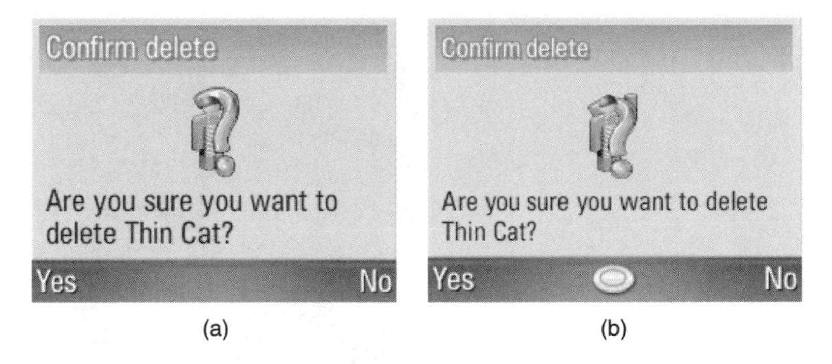

Figure 2.24 Confirmation dialog in the Contacts application in a) UIQ 3.0 and b) UIQ 3.1

If a touchscreen is present, UIQ 3 allows the softkeys and menus to be driven by the pen as well as via the hardware buttons (Softkey Style Touch).

2.3.2 Pen Style

In Pen Style, most commands are actioned via controls displayed on the screen. The action key (clicking in the Jog Dial, on Sony Ericsson phones) is also used, together with the navigation keys.

Looking at the Contacts application detail view in Pen Style UI configuration (see Figure 2.25), we observe that the three most common actions to perform on a telephone number are now available by touching the screen. The most likely action (making a voice call) is also mapped on to the Action key.

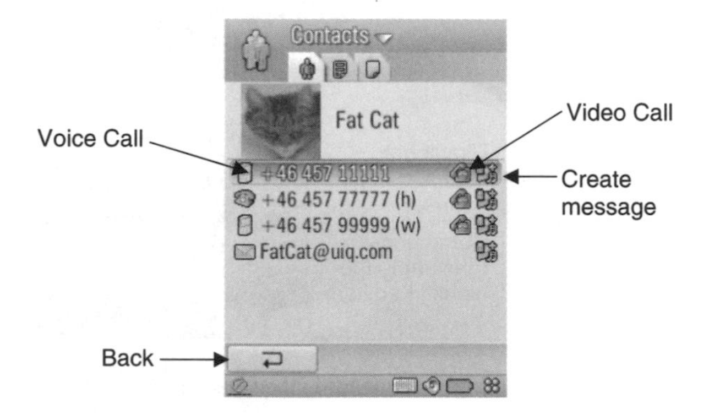

Figure 2.25 Commands in the Contacts application (Pen Style)

The main menu (see Figure 2.26) is accessed by clicking the symbol next to the application name. In this case, the Make video call and Create message commands are repeated.

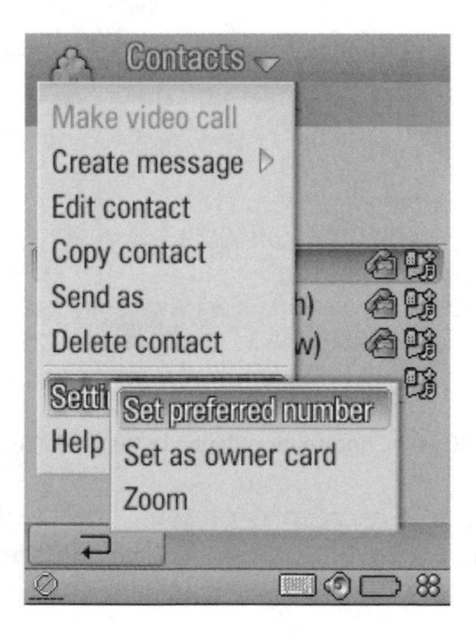

Figure 2.26 Menu in the Contacts application (Pen Style)

The CPF also works in dialog boxes (see Figure 2.27), where up to three textual buttons or nine icon buttons can be displayed. Commands not visible as buttons are placed in a title bar menu, accessed via a

Figure 2.27 Confirmation dialog in the Contacts application (Pen Style)

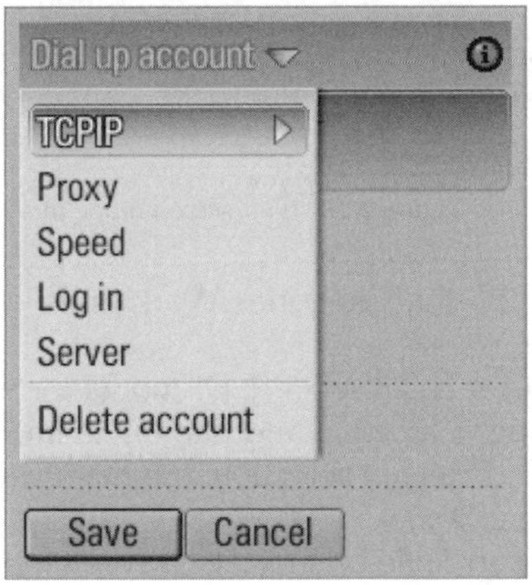

Figure 2.28 Button Bar and Menu in dialog (Pen Style)

symbol (see Figure 2.28). You can specify which commands are preferred on buttons.

2.3.3 Full-Screen Mode

In Softkey Style, it is important to leave the Softkey labels on the screen in full-screen mode, so that the user can still access commands (see Figure 2.29a).

In Pen Style, you can choose whether to display the button bar or not in full-screen mode. The menu is made available from an icon (see Figure 2.29b).

If your application needs to go completely full-screen, it may do so. You must clearly implement a key action or command within your application UI that exits full-screen mode.

2.3.4 Categories

UIQ applications frequently use categories to break up data into more easily manageable subsets. In the Contacts and Agenda applications, for

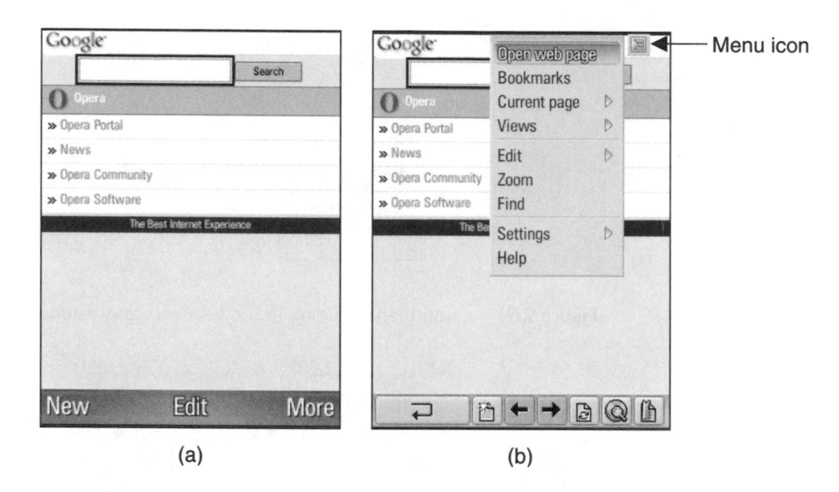

Figure 2.29 Full-screen mode in a) Softkey Style and b) Pen Style

example, the categories Work and Personal can be created to classify entries.

UIQ 3 has a specific UI model for working with categories: creating, deleting, allocating and filtering them. In Softkey Style and Softkey Style Touch, category selection and management is done using the menu (see Figure 2.30a).

In Pen Style, a special category menu is added to the Title Bar and Category operations are selected from a drop-down menu in much the same way as in UIQ 2 (see Figure 2.30b).

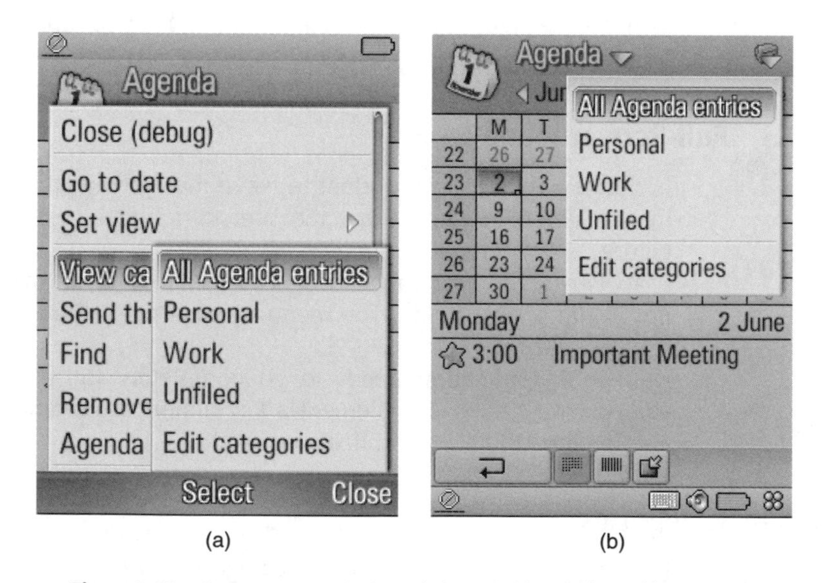

Figure 2.30 Category menu in Agenda in a) Softkey Style and b) Pen Style

Categories are described in more detail in Chapter 7, together with an example application.

2.4 UIQ 3 Operational Model

When a UIQ 3 phone is switched on, it goes to a *start* or *idle* screen which is configured by the phone manufacturer (see Figure 2.31). This screen can also be customized for mobile operators.

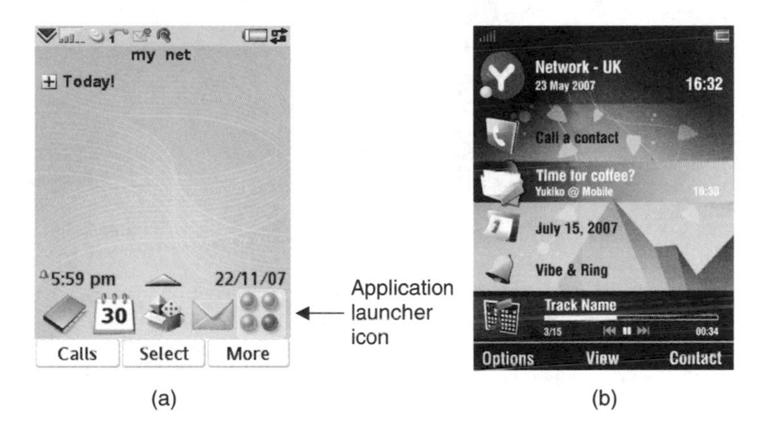

Figure 2.31 Idle screen for a) Sony Ericsson P1i and b) MOTO Z8

Frequently used applications such as Contacts, Agenda and Messaging are typically available directly from the idle screen, which is often user-configurable so that users can access their favorite applications easily. Additional applications are accessed via the application launcher.

The Sony Ericsson P1i provides an icon for the application launcher (see Figure 2.32) which the user can tap with the stylus or access using the Jog-Dial. The MOTO Z8 provides a Home key, , to reach the application launcher.

Newly installed applications appear in the application launcher. UIQ 3 applications typically follow a list-and-details model. When we first open an application, it displays a list view of the data (see Figure 2.33).

We can navigate up and down the list to find the item we want. We can program the list view so that important tasks are possible directly from the list view. UIQ 3 provides extensive support for constructing list views; Chapter 6 describes this in detail.

From the list view, the default operation is typically to view the item in detail. The most frequently needed information is displayed first (see Figure 2.34a). Less frequently used information is available on further pages within the view (see Figure 2.34b).

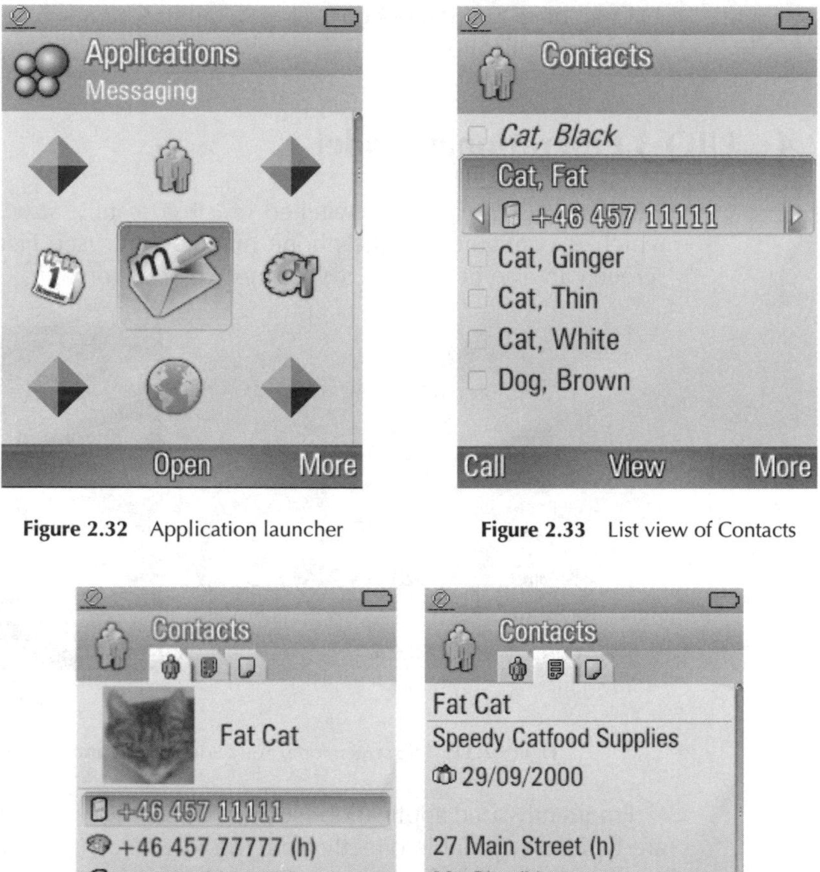

Figure 2.32 Application launcher

Figure 2.33 List view of Contacts

(a) (b)

Figure 2.34 Detail view of Contacts: a) main page and b) Birthday and Address page

Section 2.1.2 has already provided an introduction to the parts of the screen. We provide further information in Chapter 5 and Chapter 9 shows you how to build views and dialogs.

2.5 View Layout Construction

A resource file is a text file in which you can define user interface items. It is compiled into a compressed binary file. UIQ 3 makes extensive use of resource files to define views, dialogs and commands, although they can also be defined in code. The resource structure allows a lot of freedom and can look complicated at first, however, once you understand how a view is constructed you should find it easy.

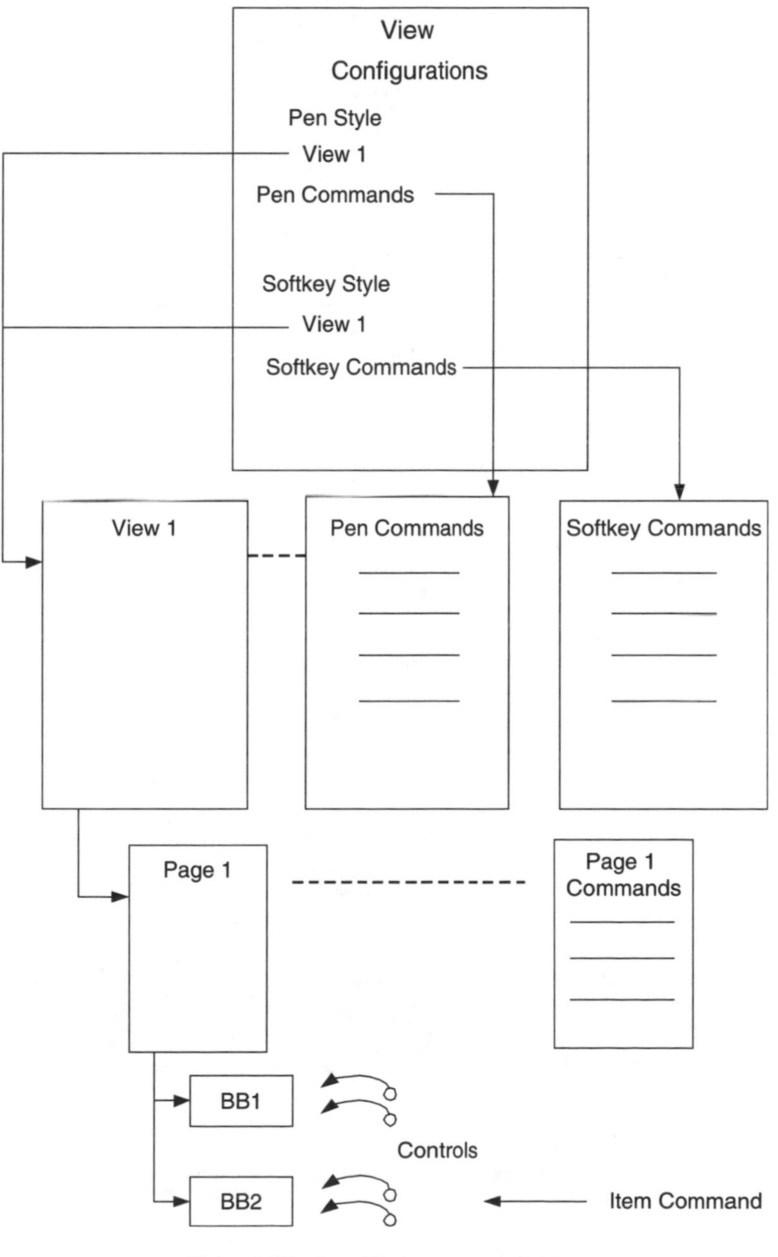

Figure 2.35 Simplified resource definition

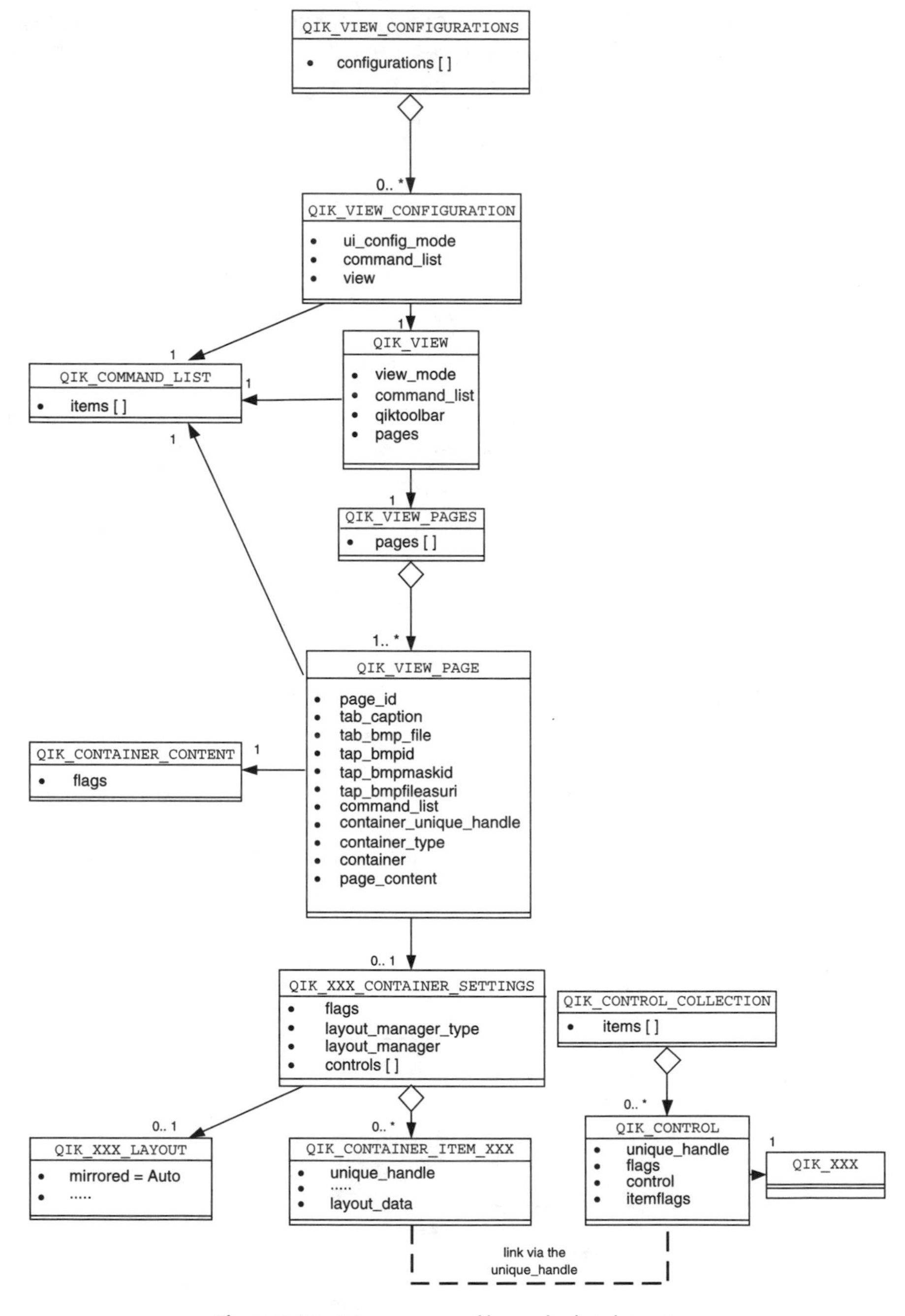

Figure 2.36 View resource files and related structure

In this section, we introduce the structure in simple terms. As you go through the example applications, we show you in detail how to work with resource files.

For each view in your application, the resource file typically defines the following:

- list of UI configurations declared for the view
- view and command lists to be used in each UI configuration
- command definitions for each view
- layout for each view.

In the simplified example in Figure 2.35, we support Pen Style and Softkey Style UI configurations in our view configurations. We use the same layout in each case, View 1, however, we supply a different command definition for each configuration. If the mobile phone is in a UI configuration that is not declared, UIQ 3 selects the declared view configuration that it considers to be the most suitable.

We can define single- or multi-page views and can use building blocks to arrange the layout. Individual controls are placed in our chosen building blocks.

Commands can be defined at various levels including view, page and *item* (linked to controls that gain focus). The CPF takes into account all of these in determining where to place commands.

Figure 2.36 shows the data entities that can be declared and how they are related.

2.6 Changes Between UIQ 2.1 and UIQ 3

If you are familiar with UIQ 2.1, this section summarizes the key changes in UIQ 3 for you. These are the main headline changes:

- UIQ 3 is based on Symbian OS v9; UIQ 2.1 is based on Symbian OS v7.0.
- Platform Security is introduced with Symbian OS v9.
- UIQ 3 provides increased flexibility in UI configuration.
- UIQ 3 enables keypad input as well as touch-screen input, as described in Section 2.1.
- Building blocks and layout managers help you construct your application views, as we describe in Section 2.2 and Chapter 8.

- A Command Processing Framework (CPF) is introduced so that your application can specify its commands without needing to worry about the UI configuration. We describe this in Section 2.3 and Chapter 7.
- UIQ 3 provides enhanced customization features.
- UIQ 3 has improved public APIs.

You can read more in the online documents *UIQ Migration Quick Guide* and *Programmer's Guide to New Features in UIQ 3*.

If you have UIQ 2.1 applications that you wish to port to UIQ 3, you can find examples in the *Special Interest Paper – Porting and Porting DreamConnect to UIQ 3*.

2.6.1 Symbian OS v9

Symbian OS v9 has a new security architecture which introduces two important concepts: *capabilities* and *data caging*.

Capabilities

Functions which are sensitive (they may incur cost to the user or compromise privacy) are protected and your application must have the appropriate *capabilities* to be able to use sensitive APIs.

There are three groups of capabilities:

- User Capabilities cover resources that the user can understand and grant when your application is installed, such as send an SMS or access the phone book.
- System Capabilities cover sensitive operations that could threaten the integrity of the mobile phone.
- Device Manufacturer Capabilities cover unrestricted file system access, access to Digital Rights Management protected content and the ability to install software, load executable code, and so on.

System and Device Manufacturer capabilities are assigned to your finished application by the Symbian Signed process. User capabilities can be granted by the user, but we recommend that you also sign your application for these. During development, you can use Open Signed to test your application on real mobile phones.

Symbian Signed and capabilities are explained in detail in Chapter 14. We also show you how to build an application that is capable of passing Symbian Signed. Chapter 10 introduces `SignedAppPhase1` and `SignedAppPhase2`, while Chapter 11 adds some multimedia functionality to provide `SignedAppPhase3`, which we have successfully signed.

Data Caging

Application data has been separated from application code to provide a simple, trusted path on the platform. An application's stored data can be hidden from other applications.

The file system has the following structure:

- `\sys`. This is the restricted system area which can only be accessed by highly trusted system processes.
- `\sys\bin`. This directory holds all executables such as EXEs, DLLs and plug-ins. The operating system will not run code placed in other locations.
- `\private`. Each application has its own private view of the file system consisting of `\private\<SID>\` where SID is a unique security identifier taken from the program's UID. The application should use this directory tree to hold the data files.
- `\resource`. A public read-only directory that allows files to be publicly shared without compromising integrity. An application should, for example, put its UI resource files and icon files in `\resource\apps`.

Other directories are public and can be read from or written to by any program. An application installation package can install files in, or below, another application's `\private\<SID>\import` directory if such a directory exists.

New Kernel Architecture

Symbian OS v9 introduced a new kernel architecture (known as EPOC kernel architecture 2, *EKA2*), security model and tool chain. There is a binary break from the earlier versions of Symbian OS which means that UIQ 2.1 applications are incompatible with UIQ 3. In some very simple cases, it is enough to recompile the code. However, most applications require code changes.

2.6.2 Application Building

UIQ 3 introduces significant changes to the MMP and PKG files; the AIF file is replaced by two resource files.

MMP File

The target type for applications has been changed from APP to EXE.

UIQ 2.1:

```
TARGET          HelloWorld.app
TARGETTYPE      APP
```

UIQ 3:

```
TARGET          HelloWorld.exe
TARGETTYPE      EXE
```

The Vendor ID (VID) is new in Symbian OS v9 and specifies the vendor of the application executable. In most cases, this will be zero or omitted from the MMP file, which means that the source of the executable is not required for security checks. VIDs are allocated to Symbian licensees, partners, operators and Independent Software Vendors through sign-up programs.

```
VENDORID 0
```

If your application requires capabilities in order to access protected APIs, these are defined in the MMP file:

```
CAPABILITY LocalServices ReadUserData
```

The `LIBRARY` keyword lists the import libraries needed by the application. The build will report link errors (unresolved externals) if you don't specify all the required libraries. The class-level documentation in the Developer Library tells you which library to import for each class. No path needs to be specified and each library statement may contain several libraries, separated by a space. More than one library statement may also be used.

```
// Specifies import libraries
LIBRARY euser.lib
LIBRARY apparc.lib
LIBRARY cone.lib
LIBRARY eikcore.lib
LIBRARY eikcoctl.lib
```

STATICLIBRARY is used to specify statically linked libraries (object code that is built into your application). All UIQ applications should link against a UIQ-specific heap allocator library, which is designed to be more effective in out-of-memory situations. This is done as follows in an MMP file:

```
// New heap allocator, which is more effective in
// out-of-memory situations.
// You only need to include the libs in the MMP file to
// use the new heap allocator.
```

```
STATICLIBRARY  qikalloc.lib
LIBRARY        qikallocdll.lib
```

PKG File

The following changes have been made in the PKG file:

- updated product ID for UIQ 3
- new keywords for mapping media files.

The most important change is the definition of the product ID.
UIQ 2.1:

```
(0x101F617B), 2, 0, 0, {"UIQ20ProductID"}
```

UIQ 3:

```
[0x101F6300], 3, 0, 0, {"UIQ30ProductID"}
```

Resource Files Replace AIF File

The Application Information File (AIF) is no longer used. It is replaced by a pair of resource files:

- the registration resource file, `AppName_reg.rss`, which defines the non-localizable items
- the localizable resource file, `AppName_loc.rss`, which contains localizable information such as caption and icon file definitions.

View Construction

As we explain in Section 2.5, UIQ 3 makes much greater use of resource files than either UIQ 2.1 or S60. Views, dialogs and commands can be defined in resource files.

2.6.3 Application Framework

Application

Since the application target has been changed from `.app`, which was loaded dynamically by the framework as a normal DLL, to an executable (`.exe`), the application entry point has also been changed. The `E32Dll()` entry function, normally placed in the `CQikApplication` derived class, is now replaced by `E32Main()`.

Views

UIQ 3 introduces two new view classes.

The `CQikViewBase` class is the base class for all views and all UIQ views should be derived from it. The `CQikMultiPageViewBase` class is a base class for views with multiple pages. Application views are derived from one of these two base classes instead of from the `CCoeControl` and `MCoeView` as was the case in UIQ 2.1.

The `CQikViewBase` class handles much of the new functionality found in UIQ 3:

- construction of views, controls and layouts from resource files
- basic focus management
- hardware-button navigation between controls
- command handling
- interface for runtime UI configuration switching
- handling the context bar.

Also related to views is the new dialog class `QikViewDialog`. This class implements a view which can be invoked as a dialog. This means that common code can be used for both views and complex dialogs.

A second new class, `CQikSimpleDialog`, is intended for cases where a simple dialog, such as a yes/no query, is needed.

Views and dialogs are covered in detail in Chapter 9.

2.7 Changes Between UIQ 3.0 and UIQ 3.1

UIQ 3.1 is current at the time of writing this book. While moving from UIQ 2.1 to UIQ 3.0 was a very significant change, UIQ 3.1 makes relatively minor changes and additions to UIQ 3.0.

These are the key changes:

- UIQ 3.1 is built on Symbian OS v9.2; UIQ 3.0 is based on Symbian OS v9.1.
- In Softkey Style, the additional softkey arrangement 3SK is introduced for phones that require the Back function to be placed on a softkey rather than a dedicated hardware key (see Section 2.1.3).
- The Scalable Vector Graphics (SVG) Tiny 1.2 player contributes to a richer user experience by, for example, displaying animated SVG icons in the Application Launcher.

- An improved Task list (see Figure 2.37) provides the user with a simple way to navigate between applications. The content can be customized by the phone manufacturer.

Figure 2.37 UIQ 3.1 Task list

2.8 UIQ 3.2

Announced in January 2008, UIQ 3.2 is focused on enhanced support for network operator services. UIQ 3.2 is based on Symbian OS v9.2. Developers do not need a new SDK to develop for UIQ 3.2.

2.8.1 Rich Messaging Suite

The UIQ messaging suite has been updated with an improved design and new features:

- MMS postcard (see Figure 2.38) is an application which allows the user to send a picture as a printed postcard via a network operator service.
- An Instant Messaging (IM) client (see Figure 2.39) allows the user to chat with friends and view their online status information. The IM client supports the OMA IMPS 1.1 specification and enables communication with other IM clients such as MSN and Yahoo.
- Support for the OMA E-Mail Notification 1.0 standard offers operators the possibility of providing their customers with a push email solution.

Figure 2.38 Postcard application

Figure 2.39 Instant Messaging application

2.8.2 New APIs

New API support in UIQ 3.2 includes the following significant enhancements:

- Java MSA Subset (JSR-248); with this set of APIs, developers can create a wide range of media-capable applications
- Push Email APIs, independent of email protocol, which enable integration of third-party email clients with UIQ messaging and make them appear as native applications.

2.8.3 Enhanced Media Support

The media experience in the web browser has been improved by adding support for enjoying and sharing media, for example:

- support for streaming video
- the option to preview content before purchasing
- the ability to view or listen to content before the download is completed.

3
Quick Start

3.1 Introduction

This chapter provides a fast track guide to UIQ 3 development. It describes the development tools and workflow. It highlights the main UIQ architectural concepts that are described in much more detail in later chapters.

We describe how to use both the command line and the Carbide.c++ Integrated Development Environment (IDE). Carbide.c++ is used because a free version is available and because it is positioned as the primary IDE.

An IDE helps you write and debug code. It uses tools provided by the UIQ SDK and the underlying Symbian OS to compile, link and package applications. If you install just the UIQ SDK, without an IDE, you get a command-line development environment which you can use if you prefer. The Carbide.c++ IDE provides a Windows-based environment that uses command-based tools.

In many development environments, you can work entirely within the IDE for most applications. For non-trivial UIQ 3 applications, however, you also need to work outside the IDE. The command line provides a richer set of available operations, for example, to gain greater control over packaging components into an installable SIS file. Hence, you will sometimes find yourself using the command line. Also, as you become more proficient in Symbian OS development, you will probably find it more convenient to use the command line and just use the IDE for editing files and debugging. Due to the variety of available IDEs, specifically Carbide.c++, Carbide.vs and CodeWarrior, the command line also tends to be used in examples and third-party documentation.

3.2 The Development Environment

The development environment is a rich mix of UIQ SDK, Symbian tools, IDE, optional manufacturer-specific extensions and third-party tools. Developers have a choice of several IDEs, each of which has peculiarities and limitations. Running all the tools within a particular PC environment, such as an existing S60 configuration, can also add complexity and cause failures after initial installation. In fact, simple post-installation configuration can solve most problems. Once set up, the development environment is easy to use.

3.2.1 UIQ SDK

The UIQ SDK is a collection of UIQ, Symbian and third-party tools. The SDK is a collection of tools controlled using Perl scripts. For this reason, Activeperl is installed as part of the UIQ SDK. The SDK enables applications to be built from the command line (via Start -> All Programs -> Accessories -> Command Prompt). In simple terms, a Windows-based IDE such as Carbide.c++ submits the command-line commands and redirects or filters output to a window.

The UIQ SDK can be obtained from ***developer.uiq.com***. You will need to register – for free – before you can download the SDK.

Your choice of SDK is a trade-off between having newer features and bug fixes versus compatibility with older phones. Applications written with the UIQ 3.0 SDK will run on phones based on UIQ 3.0 or UIQ 3.1 or any future version of UIQ 3. Writing with the UIQ 3.1 SDK, your code is likely to call system libraries that don't exist on UIQ 3.0 phones. Keeping compatibility in mind, pick the newest SDK you can for the phones that you want to support, so as to gain the benefit of bug fixes and improvements to the documentation and examples.

> **Pro tip**
>
> In order to avoid problems, install the SDK and all additional tools to the same root drive. Avoid using directories other than those provided by default in the setup. Should you change directories, ensure the install path doesn't include spaces.

When you run the extracted setup, you will be asked which components you wish to install (see Figure 3.1).

Enable installation of all the components with the exception of 'Java Developer', unless you also plan to develop for Java ME.

`WinPcap` is a utility that allows you to connect the emulator to the Internet. The GCCE compiler is the one used to compile for the mobile

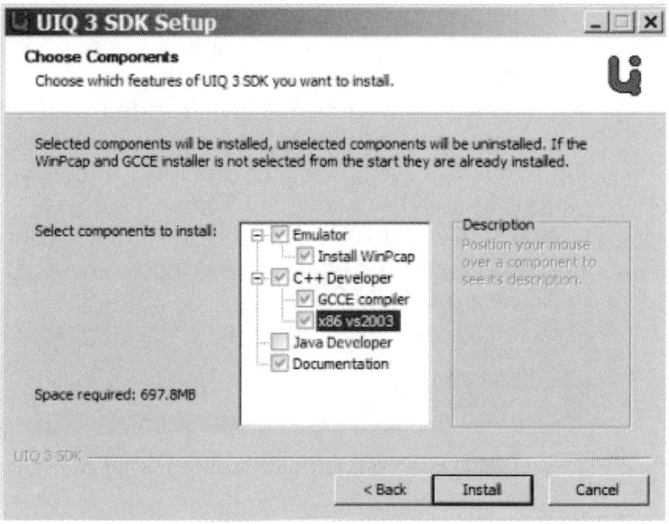

Figure 3.1 UIQ 3 SDK setup window

phone while the x86 vs2003 option installs a Nokia Windows compiler for use when compiling for the emulator.

Part way through the installation you may be asked if you want to set the SDK as the default. The Symbian development environment allows you to switch between various installed S60 and UIQ SDKs. If you confirm selection of the UIQ SDK as the default, this is done automatically for you.

If you encounter problems using the SDK, there are many useful FAQs on ***developer.uiq.com*** and ***developer.symbian.com***.

Pro tip

The default GCCE settings do not enable compiler optimization. This leads to binaries that are slower and much larger than they need to be. It's possible to change the default level of optimization used by GCCE.

Edit the file `epoc32\tools\compilation_config\GCCE.mk` and change the line:

```
REL_OPTIMISATION=
```

to:

```
REL_OPTIMISATION= -O2 -fno-unit-at-a-time
```

If you later run into problems running code on the phone, you should undo this change to determine if it has been caused by optimization of the code.

If you already have an S60 SDK installed, the UIQ installation asks you if you want the UIQ SDK set as the default. The reason for this is because the Symbian development environment allows you to switch between various installed S60 and UIQ SDKs.

You can check or switch between SDKs using the command line `devices` command or using the Windows-based UIQ SDK Configurator.

To list the installed SDKs:

```
C:\>devices
UIQ3:com.symbian.UIQ - default
S60_3rd:com.nokia.s60
```

To change the default SDK to the UIQ 3 SDK:

```
C:\>devices -setdefault @UIQ3:com.symbian.UIQ
```

To use the SDK Configurator, select Start -> All Programs -> UIQ 3 SDK -> SDKConfig. The Devices tab allows you to switch between the SDKs installed on your PC (see Figure 3.2).

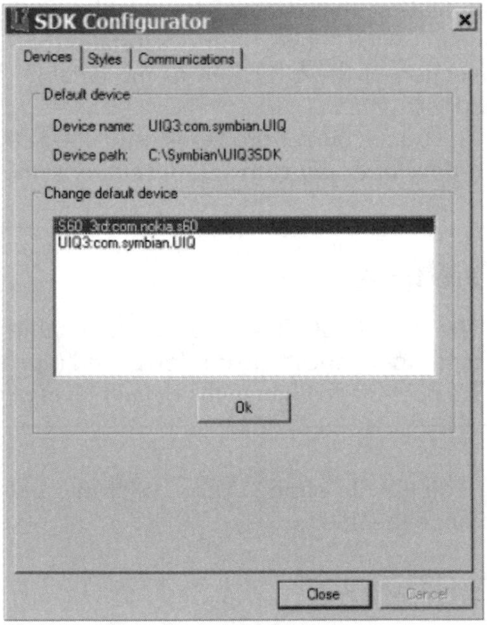

Figure 3.2 SDK Configurator

If you use Carbide.c++ then the IDE automatically takes care of switching between SDKs.

3.2.2 Carbide.c++ IDE

Carbide.c++ is the newest family of Eclipse-based development environments for Symbian OS C++ application development. It's developed and sold by Nokia and is suitable for both UIQ and S60 development.

There are four versions. The free Express edition is suitable for most people doing UIQ development. The Developer edition includes a RAD-style UI builder and on-target (phone) debugging support, neither of which are supported under UIQ as of the time of writing this book; however, they are anticipated to become available in the future. The Professional version is targeted towards advanced developers who need crash debugging and performance investigation tools. The OEM version (which became available during 2007) supports advanced hardware debugging for mobile phone manufacturers.

Carbide.c++ is obtained from ***www.forum.nokia.com/main/resources/tools_and_sdks/carbide_cpp/***. You have to register with Forum Nokia before downloading. As with all Symbian-related tools, install to the same root drive as the proposed workspace and UIQ SDK.

> **Pro tip**
>
> When you start Carbide.c++ you will be prompted for a workspace. Specify a path that doesn't contain spaces otherwise you will have problems importing files and finding headers. Also ensure you use a capitalized drive letter.

3.2.3 Other Development IDEs

CodeWarrior 3.0 and 3.1

CodeWarrior for Symbian OS, also owned by Nokia, is being phased out in favor of Carbide.c++ tools. Therefore, we don't mention it further in this book.

> **Pro tip**
>
> Should you choose CodeWarrior, be sure to obtain version 3.1. Version 3.0 doesn't allow compilation for GCCE.

Carbide.vs and Visual Studio .NET 2003

Carbide.vs (available from ***forum.nokia.com***) and the associated UIQ support package (from ***developer.uiq.com***) are free plug-ins for Visual Studio .NET 2003 that allow developing for UIQ in Visual Studio. While developing under Visual Studio's 'look and feel' might seem attractive, Symbian, UIQ and Nokia are promoting Carbide.c++ as the preferred development IDE. Hence, development based on Visual Studio isn't mentioned further in this book.

> **Pro tip**
>
> Using Carbide.vs requires a licensed version of Visual Studio .NET 2003. It won't work with Visual Studio .NET 2005.

3.3 The `QuickStart` Example

The `QuickStart` example demonstrates features within every UIQ application. For simplicity, all the project files have been placed in the same directory and all of the C++ code in one CPP file. This is unusual and the normal directory and file structure is discussed in Chapter 10.

The `QuickStart` application itself displays a simple list view and processes commands to display an 'about' dialog. It is a simplified version of the `Commands1` application, which you can find in Chapter 7.

3.3.1 Application Architecture

UIQ provides an application framework that relieves you of much of the effort of creating a graphical user interface. The application framework:

- creates a connection to the file server
- creates a connection to the window server
- creates a connection to the memory manager
- does some registration work
- makes sure that we can handle error and out-of-memory situations
- initializes other application services (such as font providers)
- creates the default screen furniture (for example, the status bar, menu bar and softkeys)
- starts the active scheduler (the event loop).

Most applications consist of at least the following four base classes:

- application (`CQikApplication`)
- document (`CQikDocument`)

- application UI, often abbreviated to AppUi (CQikAppUi)
- view (CQikViewBase).

Figure 3.3 shows a diagrammatic representation of what happens during application invocation.

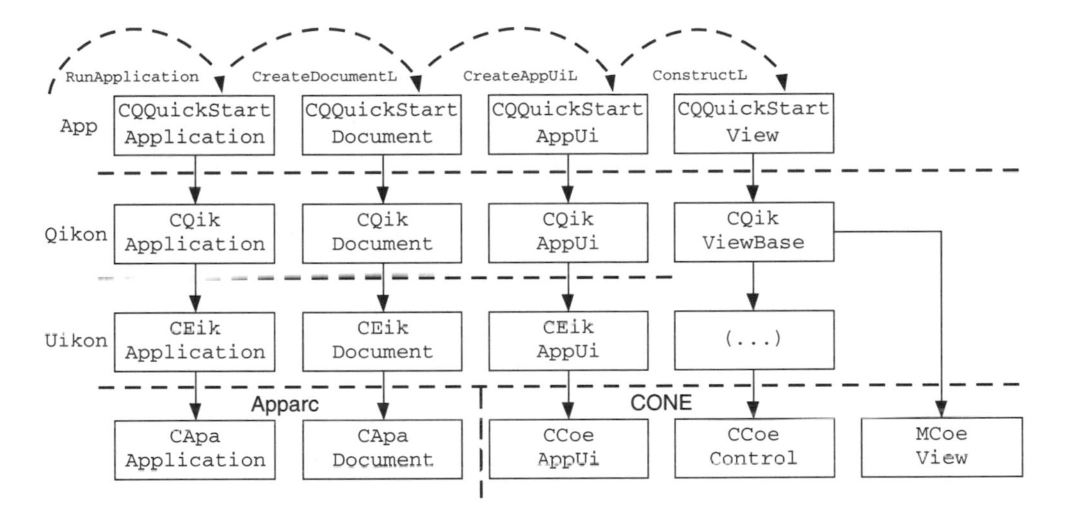

Figure 3.3 Representation of application invocation

The QuickStart example follows a standard format of the application entry point (E32Main) creating a new object derived from CQikApplication, CAppSpecificApplication, which creates a new object derived from CQikDocument, CAppSpecificDocument.

The main responsibility of the application class is to create a document. It is a common software engineering practice to separate the user interface from the engine. The document is basically the engine while the application UI and the view make up the user interface.

```
CApaDocument* CAppSpecificApplication::CreateDocumentL(void)
  {
  return(CAppSpecificDocument::NewL(*this));
  }

LOCAL_C CApaApplication* NewApplication(void)
  {
  return(new CAppSpecificApplication);
  }

GLDEF_C TInt E32Main()
  {
  return(EikStart::RunApplication(NewApplication));
  }
```

The document object is usually responsible for saving and loading objects, for example, to and from files that persist beyond the application lifetime.

Creating the document implicitly causes the CEikAppUi::CreateAppUiL function to be called, which creates the CAppSpecificUi that, in turn, creates the view:

```
void CAppSpecificUi::ConstructL(void)
// Normal primary entry point to a Symbian Application
  {
  CQikAppUi::ConstructL();
  CAppSpecificListView*
               q = new(ELeave)CAppSpecificListView(*this);
  ...
  }
```

The view is constructed using the pre-defined resource (ID):

```
void CAppSpecificListView::ViewConstructL()
  {
  ViewConstructFromResourceL(R_LIST_VIEW_CONFIGURATIONS);
  ...
  }
```

3.3.2 Project Files

The component description file is always called bld.inf. This file defines the targets (e.g., whether the code is built for the Windows emulator and what hardware platforms are supported) and lists the files used to define the project. In the example below, we wish to build for the emulator (WINSCW) and the phone (using the GCCE compiler):

```
PRJ_PLATFORMS
WINSCW GCCE
PRJ_MMPFILES
QuickStart.mmp
```

We have one project definition file called QuickStart.mmp.

The MMP file defines the application attributes such as the executable filename, Platform Security capabilities, compiler and linker options and the files to compile and link. The MMP syntax is covered further in Chapter 10. Platform Security is discussed in Chapter 4 and Symbian Signed is covered in Chapter 14.

3.3.3 Unique Application ID (UID 3)

Every application has a unique ID (see Chapter 4). The QuickStart example is typical of most applications in that the UID (0xE00000AD) can be found in:

- the project definition (MMP) file
- the QuickStart_reg.rss file
- the QuickStart.pkg file
- the code, QuickStart.cpp in this example.

If you change the UID, be sure to change all these files.

3.3.4 Resources, Icons and Controls

Elements of the UIQ user interface, for example, menu panes, views and list boxes, are called resources. They are defined in an editable text file specific to Symbian OS. The text file is compiled into a compressed binary application resource file. Separating resource files from code allows different localized versions of the application to be built without changing the code. Resource files are much more significant in UIQ 3 than in UIQ 2 or S60.

It's also possible to build some resources dynamically in code rather than statically at compile time so as to create dynamic screens and menus based on program data. However, this is beyond the scope of this QuickStart example.

Figure 3.4 shows how an application, in this case the built-in Web application, automatically adapts from being a portrait display with softkey input (note the soft menu options at the bottom of the screen in Figure 3.4a) to landscape with pen input (note the drop-down menu to the right of the application name in Figure 3.4b) when the user requests a switch from portrait to landscape.

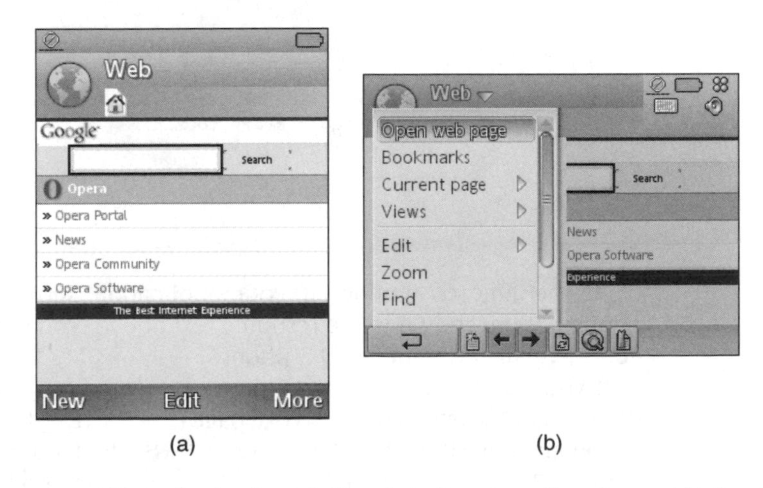

Figure 3.4 Web application in a) Softkey Style Portrait configuration and b) Pen Style Landscape configuration

In the QuickStart example, the screen layout is defined in the QuickStart.rss file. It defines the same view for each supported UI configuration:

```
RESOURCE QIK_VIEW_CONFIGURATIONS r_list_view_configurations
   {
   configurations =
       {
       QIK_VIEW_CONFIGURATION
          {
          ui_config_mode = KQikPenStyleTouchPortrait;
          command_list = r_quickstart_commands;
          view = r_quickstart_view;
          },
       QIK_VIEW_CONFIGURATION
          {
          ui_config_mode = KQikPenStyleTouchLandscape;
          command_list = r_quickstart_commands;
          view = r_quickstart_view;
          },
       QIK_VIEW_CONFIGURATION
          {
          ui_config_mode = KQikSoftkeyStylePortrait;
          command_list = r_quickstart_commands;
          view = r_quickstart_view;
          },
       QIK_VIEW_CONFIGURATION
          {
          ui_config_mode = KQikSoftkeyStyleSmallPortrait;
          command_list = r_quickstart_commands;
          view = r_quickstart_view;
          },
       QIK_VIEW_CONFIGURATION
          {
          ui_config_mode = KQikSoftkeyStyleSmallLandscape;
          command_list = r_quickstart_commands;
          view = r_quickstart_view;
          },
       QIK_VIEW_CONFIGURATION
          {
          ui_config_mode = KQikSoftkeyStyleTouchPortrait;
          command_list = r_quickstart_commands;
          view = r_quickstart_view;
          }
       };
   }
```

At the time of writing, if your application supports the six view configurations shown in the resource text above, you will support all the commercially available UIQ 3 phones.

If you examine the QuickStart.rss file you can see that the view (QIK_VIEW) is defined in terms of a page (QIK_VIEW_PAGE) that contains a container (QIK_CONTAINER_SETTINGS) that contains a container item (QIK_CONTAINER_ITEM_CI_LI) that contains a (QIK_LISTBOX).

QuickStart.rss also defines the about dialog (DIALOG) containing labels (EEikCtLabel).

The exact structure of the RSS file and, specifically, list controls are discussed in detail in Chapter 6.

3.3.5 Commands

The QuickStart.rss file also defines the menu:

```
QIK_COMMAND
  {
  id = EAppCmdAbout;
  type = EQikCommandTypeScreen;
  groupId = EAppCmdMiscGroup;
  priority = EAppCmdAboutPriority;
  text = "About";
  },
```

The "About" string literal would usually be placed within a separate RLS localization file but is hardcoded in this example for simplicity.

The exact syntax of commands is covered in Chapter 7. For now, it is sufficient to know that the resultant ID (EAppCmdAbout) ends up being sent to the command handler:

```
void CAppSpecificListView::HandleCommandL(CQikCommand& aCommand)
//  Handle the commands coming in from the controls
//  that can deliver cmds..
  {
  switch (aCommand.Id())
    {
    case EAppCmdAbout:
      (new(ELeave)CAboutDialog)->ExecuteLD(R_ABOUT_DIALOG);
      break;
    ...
    }
  }
```

The about command causes the dialog defined in the RSS file (R_ABOUT_DIALOG) to be invoked.

3.4 Building from the Command Line

Start a command prompt window and go to the QuickStart example:

```
C:\>cd \symbian\quickstartexample
```

Type:

```
C:\Symbian\QuickStartExample>bldmake bldfiles
```

Pro tip

You may get the warning 'Cannot determine the version of the RVCT Compiler.' You can ignore this warning. The tools are looking for an ARM compiler which you don't have and it doesn't matter as you are using GCCE. Alternatively, you can remove this error by ensuring that your `bld.inf` file contains only `winscw gcce` under `PRJ_PLATFORMS`.

This creates an `abld.bat` file in the current directory that is used for building. To build a debug version for emulation, type:

```
C:\Symbian\QuickStartExample>abld build winscw udeb
```

The executable called `QuickStart.exe` is created in `C:\Symbian\UIQ3SDK\epoc32\release\winscw\udeb`.

To build a release version for the phone, without debug information, type:

```
C:\Symbian\QuickStartExample>abld build gcce urel
```

An executable called `QuickStart.exe` is created in `C:\Symbian\UIQ3SDK\epoc32\release\GCCE\urel`.

Should you need to remove everything built with `abld build`, use:

```
C:\Symbian\QuickStartExample>abld clean
```

To remove everything, including generated makefiles and exported files, use:

```
C:\Symbian\QuickStartExample>abld reallyclean
```

3.5 Running on the Emulator

Start the Symbian OS emulator on the command line by typing:

```
C:\Symbian\QuickStartExample>epoc
```

You may have to wait a few minutes for the emulator to start. Once started, you will see the Applications view with Quick Start listed as an application (see Figure 3.5).

Figure 3.5 Quick Start in the application launcher

Navigate to Quick Start and press return. The application starts (see Figure 3.6).

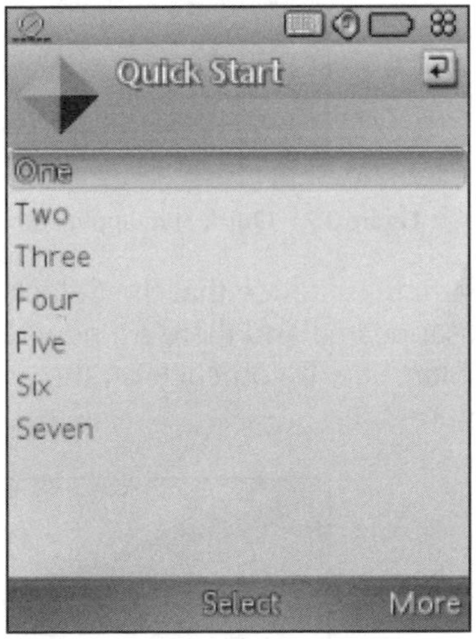

Figure 3.6 Quick Start application in Softkey mode

The emulator itself runs Symbian OS code compiled to run on Windows. Hence, it has a very high fidelity to the device itself and, apart from hardware-specific features, most applications can be developed initially using the emulator for debugging and testing.

It's possible to configure each of the predefined configurations by name: `SoftKey` for Softkey Style, `SoftKeyTouch` for Softkey Style Touch and `Pen` for pure touchscreen-based input in portrait mode:

```
C:\>UiqEnv -ui softkey
C:\>UiqEnv -ui softkeytouch
C:\>UiqEnv -ui pen
```

It's also possible to refine this further by specifying touchscreen on or off, the UI style and the screen orientation. Please see the UIQ SDK documentation for further details on the `UiqEnv -ui` arguments. This is also covered further in Chapter 15.

Running the emulator and the `QuickStart` application again shows that the application framework automatically adapts the application to horizontal and pen mode (see Figure 3.7).

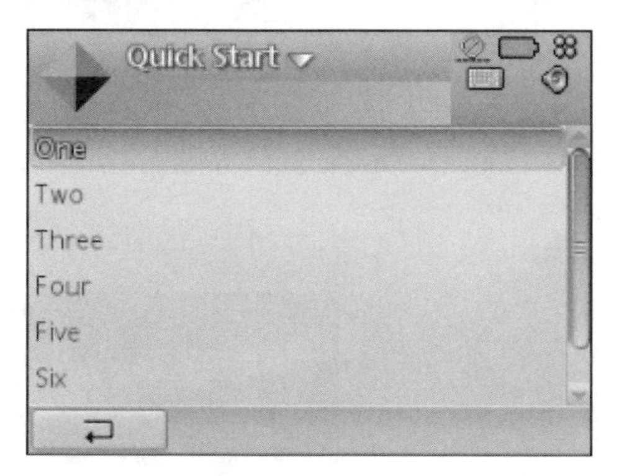

Figure 3.7 Quick Start application in Pen Style Landscape mode

In particular, notice that the Select and More softkeys have automatically disappeared and there's a new drop-down arrow to the right of the Quick Start title. If you click on the arrow, a drop down menu is shown (see Figure 3.8).

Figure 3.8 Quick Start application menu in Pen Style Landscape mode

Selecting About demonstrates how the dialog also automatically adapts to being displayed horizontally (see Figure 3.9).

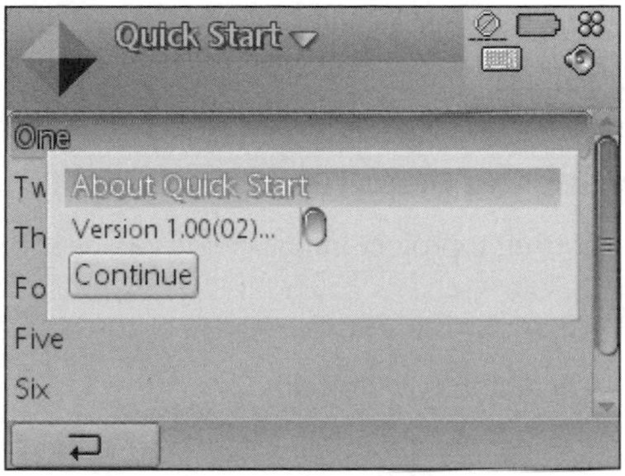

Figure 3.9 Quick Start About dialog

3.6 Packaging for the Phone

The GCCE UREL EXE can't be installed directly on the phone without supporting files such as installation registration files and resources. These files must all be packaged up into an installation SIS file before they can be deployed to the phone.

A PKG file is used to define what files are to be contained within the SIS. The PKG file also declares the software developer's name, the unique ID (UID) for the application, the program version, the versions of the phone on which the application can be installed, language variants and an optional pre-installation message.

The location of the source files depends on the location of your UIQ SDK. The example assumes this is `\Symbian\UIQ3SDK`. If this isn't the case, then you will need to edit the package file to use your UIQ install path.

In our example, the PKG file contains the line:

```
"\Symbian\UIQ3SDK\epoc32\release\GCCE\urel\QuickStart.exe"-
"!:\sys\bin\QuickStart.exe"
```

This copies the application executable from the GCCE release (as opposed to the debug) directory within the SDK to a common `sys\bin` directory used by all applications on the user-specified drive (denoted by the `!`).

To create a SIS file, type:

```
C:\Symbian\QuickStartExample>makesis quickstart.pkg
```

QuickStart.sis is created in the current directory and can be sent to the phone via Bluetooth or a cable connection or by saving on a removable media card which is then placed in the phone.

3.7 Using Carbide.c++

3.7.1 Importing the QuickStart example

Importing a project into Carbide.c++ is straightforward:

1. Start up Carbide.c++.
2. Select File -> Import.
3. Select 'Symbian OS Bld.inf file' and click on Next.
4. Browse to the QuickStart example bld.inf file and click Next.
5. Under Symbian OS SDKs, select the UIQ 3 check box (which selects WINSCW and GCCE) and click Next.
6. QuickStart.mmp should be selected. Click Next.
7. Keep the default project properties and Finish.

The project files are shown in the left-most panel within a tree view (see Figure 3.10).

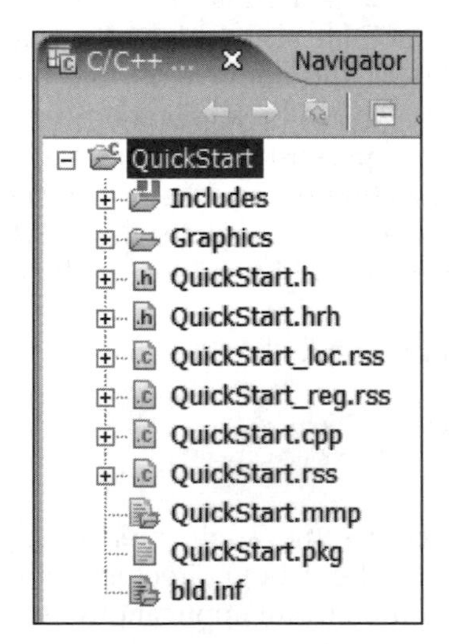

Figure 3.10 Carbide.c++ project files

Double-click on a file to view and edit. Double-clicking on the MMP file shows a wizard-style panel rather than the actual file, which makes it easier to edit project options without having to know the syntax for particular keywords and settings.

3.7.2 Building for Emulation

Carbide.c++ is automatically set up to build for the Windows emulator (WINSCW). However, should you ever need to view and select available builds, use the down arrow next to the hammer icon in the toolbar (see Figure 3.11). Click on the hammer icon to start a build.

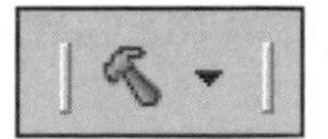

Figure 3.11 Carbide.c++ tools icon

The window at the bottom right of the screen displays the results of the build. The Problems tab gives a quick summary of errors and warnings. Double click on a problem to go directly to the associated source file. The Console tab gives the same information as would usually appear if you were to build from the command prompt.

3.7.3 Running the Emulator from the IDE

In order to run in the Symbian OS emulator from Carbide.c++ it is necessary to perform a one-time configuration to tell the IDE what to run and debug:

1. Select Run -> Debug As.
2. Double click on 'Symbian OS Emulation' and a pane will appear on the right.
3. Use the default settings by pressing the Debug button.

You are now set up for future debug sessions by just selecting the bug icon in the toolbar (see Figure 3.12) or pressing F11.

The emulator starts and operates in the same way as was illustrated for running the emulator from the command line. Carbide.c++ opens a new set of windows called the 'Debug perspective' showing the call stack, variables and console messages.

You can stop debugging either by closing the emulator window or by clicking the red terminate button in the Carbide.c++ toolbar (see Figure 3.13).

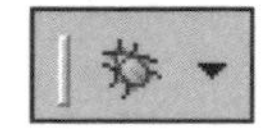

Figure 3.12 Carbide.c++ bug icon

Figure 3.13 Carbide.c++ terminate icon

Pro tip

If the Debug view loses focus, the terminate control will appear disabled. To regain focus, select a thread in the Debug view to update the tool bar.

The Carbide UI stays in the debug perspective. You can change it back to viewing code by selecting Carbide c/c++ via the double right arrow at the very top right of the Carbide IDE (see Figure 3.14).

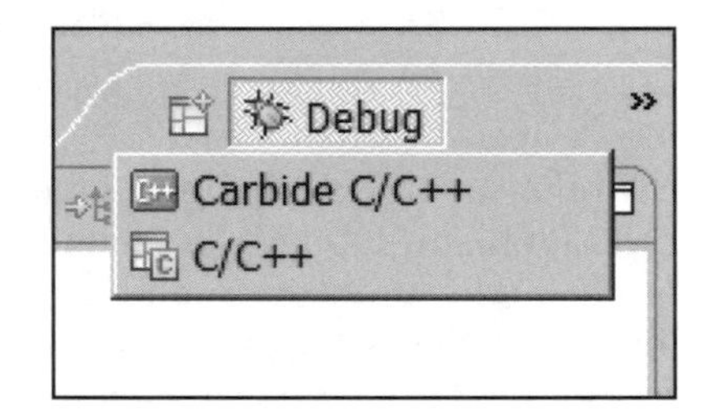

Figure 3.14 Carbide.c++ code/debug selection

3.7.4 Building for the Phone

Use the down arrow next to the hammer icon in the toolbar to select the GCCE configuration. Carbide.c++ will automatically re-build for GCCE.

As with building from the command line, the resulting EXE needs to be packaged up with the installation registration and resource files.

To create a SIS file, right click the `QuickStart.pkg` file and select Build PKG File. `QuickStart.sis` is created in the current directory. It can be sent to the phone via Bluetooth or a cable connection, or by saving on a removable media card which is then placed in the phone.

The console shows that both `MakeSIS` and `SignSIS` are invoked. `SignSIS` self-signs the SIS file, which isn't strictly needed for our simple example application. However, this becomes useful should the application need to be signed. Symbian Signed is covered in Chapter 14.

4

Symbian OS Essentials

4.1 What this Chapter Covers

This chapter covers Symbian OS concepts that you need to understand in order to develop code for UIQ 3 smartphones. It is implicit in the rest of the book that you are familiar with the subjects described here.

There are already a number of Symbian Press books specifically about C++ on Symbian OS, and if this chapter was as comprehensive as those, our book would be very large indeed! Instead, in this chapter, we briefly describe each of the key topics and the aspects of them that commonly cause problems. You can treat this chapter as a summary of the core knowledge required to write good C++ on UIQ 3.

If you are already an expert at working on Symbian OS, you may choose to skip or skim read this chapter, although you may also find it a good resource to confirm that you know the fundamentals. If you are new to C++ programming on Symbian OS, this chapter will allow you to quickly become familiar with the essential information. At the end of each section, we list the resources who cover the subject in more detail, for those who are interested in a deeper understanding, or pursuing formal accreditation.

4.2 Symbian OS Versions

The origins of Symbian OS can be traced back to the 16-bit operating system (known as *SIBO*) that was found in the hugely successful Psion Organizers of the 1980s. When the team who had created that operating system decided to create a 32-bit mobile OS, their experiences inspired them to use C++ and object-oriented design from the kernel upwards. The result was *EPOC*, the operating system first seen in the Psion Series 5, which was launched in 1997.

The following year, Symbian was announced – a private independent company owned jointly by Psion, Nokia, Ericsson and Motorola. This new company used EPOC, which they renamed as Symbian OS, as the foundation of the platform now found in millions of smartphones.

Since 1997, Symbian OS has been updated using numbered releases, each of which introduces new functionality to the platform:

- The original version of EPOC was known as ER3 (EPOC Release 3).
- The first mobile phone release was known as ER5U (the U indicating a Unicode build with native 16-bit character support). This was used in the Ericsson R380 phone, which was released in 2000.
- UIQ 2.0 and UIQ 2.1 were built on Symbian OS v7.0 (the 'ER' was dropped in favor of 'v' for version).

Today, UIQ 3 is built on what is commonly called Symbian OS v9. In fact, Symbian OS v9 is subdivided into a number of minor releases in much the same way as UIQ. So UIQ 3.0 is built on Symbian OS v9.1 and UIQ 3.1 is based on Symbian OS 9.2. Symbian and UIQ are careful to ensure compatibility between the minor releases, so code written for UIQ 3.0 will continue to work on UIQ 3.1, and vice versa.

However, it is important to note the binary break that occurred earlier in the Symbian, and UIQ, release history. Symbian OS v9 introduced a new kernel architecture (known as *EKA2* which stands for EPOC kernel architecture 2), security model and tool chain, compared to that used in previous versions of UIQ. The binary break renders UIQ 3 incompatible with UIQ 2.0 and 2.1, which means that applications and libraries must be rebuilt. In fact, because of the new kernel architecture and security model, the changes required are more extensive than simply requiring recompilation and nearly always require changes to the source code itself. Some of the changes to consider when migrating code for Symbian OS v9 include the following (about which you can find more information in the Symbian Developer Library within the UIQ SDK):

- writable static data is allowed in DLLs, although it is not always recommended because it consumes memory inefficiently
- applications that were previously DLLs running inside `apprun.exe` are now executables (EXEs). This has major advantages because writable static data can be used at will, since it has always been supported by Symbian OS in executables
- the DLL entry point function `E32Dll()` is no longer required in a DLL, and will fail to compile – it can simply be removed and does not need to be replaced with any alternative code

- a number of polymorphic DLL plug-ins to custom frameworks such as recognizers, front-end processors (FEPs), converters and notifiers have been changed to generic ECOM plug-ins
- the system agent is withdrawn and all code that previously used this should migrate to use Publish and Subscribe (described in the Symbian Developer Library documentation)
- the Symbian OS client–server framework has a revised API to secure inter-process communication. If you have need to port client–server code from Symbian OS v7.0 to Symbian OS v9, please consult the Base reference guide, found in the Symbian Developer Library in the UIQ 3 SDK.

One of the most significant differences between Symbian OS v7.0, found in UIQ 2.0 and 2.1, and Symbian OS v9, found in UIQ 3, is the introduction of platform security in the later version of the operating system. As we discuss in Section 4.14, the secure platform places a requirement on anyone creating application code to analyze the system resources it uses and determine the security privileges it needs. Platform security also partitions the file system to protect sensitive data from processes that do not need, or should not have, access to it. This can present a challenge if you are porting code from Symbian OS v7.0. Finally, you must consider whether you need, or want, to submit your application to Symbian Signed for certification. Symbian Signed is discussed in more detail in Chapter 14. In Chapters 10 and 11 we build an application that we have also signed. We show you what you need to do at each stage in the development process in order to make Symbian Signed as pain free as possible.

The list above does not consider the differences between the UIQ 3 and earlier versions. Chapter 2 summarizes the differences between UIQ 2.1 and UIQ 3. Chapter 16 provides information about porting from S60 and from other operating systems such as Windows Mobile or Palm OS.

4.3 Symbian OS Code Conventions

C++ code for Symbian OS uses established coding conventions that were chosen carefully to reflect object cleanup and ownership, and make code more comprehensible. We encourage you to follow the conventions in order for your own code to be understood most easily by other Symbian OS developers. An additional benefit to using the conventions is that your code can then be tested with automatic code analysis tools that can flag potential bugs or areas to review.

If they are unfamiliar, the best way to get used to the conventions, after reading the following section, is to look at code examples in this book and those provided with the UIQ 3 SDK.

4.3.1 Capitalization

The first letter of class names is capitalized:

```
class TColor;
```

The words making up variable, class or function names are adjoining, with the first letter of each word capitalized. Classes and functions have their initial letter capitalized while, in contrast, function parameters, local, global and member variables have a lower case first letter.

Apart from the first letter of each word, the rest of each word is given in lower case, including acronyms. For example:

```
void CalculateScore(TInt aCorrectAnswers,
                    TInt aQuestionsAnswered);
class CActiveScheduler;
TInt localVariable;
CShape* iShape;
class CBbc;    // Acronyms are not usually written in upper case.
```

4.3.2 Prefixes

Member variables are prefixed with a lower case `i` which stands for *instance*.

```
TInt iCount;
CBackground* iBitmap;
```

Parameters are prefixed with a lower case `a` that stands for *argument*. We do not use `an` for arguments that start with a vowel.

```
void ExampleFunction(TBool aExampleBool,
                         const TDesC& aName);
```

(Note `TBool aExampleBool` rather than `TBool anExampleBool`.)

Local variables have no prefix:

```
TInt localVariable;
CMyClass* ptr = NULL;
```

Class names should be prefixed with an appropriate letter (`T`, `C`, `R` or `M` as described fully in Sections 4.4.3–4.4.6):

```
class CActive;
class TParse;
```

Constants are prefixed with K:

```
const TInt KMaxFilenameLength = 256;
#define KMaxFilenameLength 256
```

Enumerations are simple types and so are prefixed with T, according to the convention described in Section 4.4.3. Enumeration members are prefixed with E:

```
enum TWeekdays {EMonday, ETuesday, ...};
```

4.3.3 Suffixes

A trailing L on a function name indicates that the function may leave. This is discussed further in Section 4.5.4:

```
void AllocL();
```

A trailing C on a function name indicates that the function returns a pointer that has been pushed onto the cleanup stack, which is described in more detail in Section 4.5.8:

```
CShapeShifter* NewLC();
```

A trailing D on a function name means that it will result in the deletion of the object referred to by the function.

```
TInt ExecuteLD(TInt aResourceId);
```

4.3.4 Underscores

Underscores are avoided in names except in macros (__ASSERT_DEBUG) or resource files (MENU_ITEM).

4.3.5 Code Layout

You'll notice that the curly bracket layout in Symbian OS code, used throughout this book, is to indent the bracket as well as the following statement. For example:

```
void CNotifyChange::StartFilesystemMonitor()
  { // Only allow one request to be submitted at a time.
    // Caller must call Cancel() before submitting another.
  if (IsActive())
    {
```

```
  _LIT(KAOExamplePanic, "CNotifyChange");
  User::Panic(KAOExamplePanic, KErrInUse);
  }
iFs.NotifyChange(ENotifyAll, iStatus, *iPath);
SetActive(); // Mark this object active.
}
```

4.4 Symbian OS Class Types

4.4.1 Basic Types

Symbian OS defines a set of fundamental types which, for compiler independence, are used instead of the native built-in C++ types for compiler independence. These are provided as a set of `typedefs` as shown in Table 4.1.

Table 4.1 Fundamental data types

<table>
<tr><td><code>TInt</code>, <code>TUint</code>, <code>TIntX</code> and <code>TUintX</code> (where X = 8, 16 and 32)</td><td>For signed and unsigned integers where the X refers to the number of bits each type occupies. In general, the non-specific <code>TInt</code> or <code>TUint</code> types should be used, corresponding to signed and unsigned 32-bit integers, respectively.</td></tr>
<tr><td><code>TInt64</code>, <code>TUInt64</code></td><td>These are <code>typedef</code>'d to <code>long long</code> and <code>unsigned long long</code> respectively on UIQ3, and use the available native 64-bit support.</td></tr>
<tr><td><code>TReal32</code> and <code>TReal64</code> (and <code>TReal</code>, which equates to <code>TReal64</code>)</td><td>For single- and double-precision floating-point numbers, equivalent to float and double respectively. Operations on these types should be avoided unless they are absolutely necessary, since they are likely to be slower than operations on integers.</td></tr>
<tr><td><code>TAny*</code></td><td><code>TAny*</code> is a pointer to something – type unspecified – and is used in preference to <code>void*</code>. But <code>TAny</code> is not used as an equivalent to void.
Thus, we write
<code>void TypicalFunction(TAny* aPointerParameter);</code>
not
<code>void TypicalFunction(void* aPointerParameter);</code>
but not
<code>TAny TypicalFunction(TAny* aPointerParameter);</code></td></tr>
<tr><td><code>TBool</code></td><td>This type should be used for Boolean values. For historical reasons, the Symbian OS <code>TBool</code> type is equivalent to <code>int</code> (<code>ETrue</code> = 1 and <code>EFalse</code> = 0). Since C++ will interpret any non-zero value as true, direct comparisons with <code>ETrue</code> should not be made.</td></tr>
</table>

The fundamental Symbian OS types should always be used instead of the native types (that is, use `TInt` instead of `int`). However, there is one exception, as mentioned above. Always use `void` when a function or method has no return type, instead of `TAny`.

4.4.2 Class Types

Symbian OS also defines several class types, each of which has different characteristics, such as where objects may be created (on the heap, on the stack or on either) and how those objects should later be cleaned up. The conventions make the creation, use and destruction of objects more straightforward. When writing code, the required behavior of a class should be matched to the Symbian OS class characteristics. Later, a user of an unfamiliar class can be confident in how to instantiate an object, use it and then destroy without leaking memory.

4.4.3 T Classes

T classes are simple classes that behave much like the C++ built-in types. For this reason, they are prefixed with the same letter as the `typedefs` described above (the T is for *Type*).

Just like the built-in types, T classes do not have an explicit destructor. In consequence, T classes must not contain any member data which itself has a destructor. T classes contain all their data internally and have no pointers, references or handles to data, unless that data is owned by another object responsible for its cleanup.

T class objects are usually stack-based but they can also be created on the heap – indeed some T classes should only ever be heap-based because the resulting object would otherwise occupy too much stack space. A good rule of thumb is that any object larger than 512 bytes should be heap-based rather than created on the stack.

Although the data contained in these classes is simple, some T classes can themselves have fairly complex APIs, such as the lexical analysis class `TLex` and the descriptor base classes `TDesC` and `TDes`, covered in Section 4.8. In other cases, a T class is simply a C-style `struct` consisting only of public data.

4.4.4 C Classes

C classes are only ever allocated on the heap. Their purpose is to contain and own pointers to other objects. Unlike T classes, it's fine for a C class to have destructors to clean up member variables, for example, those that are heap-based or resource handles.

For Symbian OS memory management to work correctly, C classes must ultimately derive from class `CBase` (defined in the Symbian OS header file `e32base.h`). This class has two characteristics which are inherited by every C class:

- Safe destruction: `CBase` has a virtual destructor, so a `CBase`-derived object is destroyed properly by deletion through a base-class pointer.
- Zero initialization: `CBase` overloads operator new to zero-initialize an object when it is first allocated on the heap. This means that all member data in a `CBase`-derived object will be zero-filled when it is first created, and this does not need to be done explicitly in the constructor.

On instantiation, the member data contained within a C class typically needs to call code which may fail. A good example is instantiation of an object that performs a memory allocation, which fails if there is insufficient memory available. This kind of failure, is called a leave on Symbian OS, and is discussed in more detail shortly. A constructor should never be able to leave, because this can cause memory leaks. To avoid this, C classes are characterized by an idiom called two-phase construction, which also prevents accidental creation of objects of a C class on the stack.

`CBase` also declares a private copy constructor and assignment operator. Their declaration prevents calling code from accidentally performing invalid copy operations on C classes.

4.4.5 R Classes

The `R` which prefixes an R class indicates that it holds an external resource handle, for example a handle to a server session. R classes are diverse, and vary from holding a file server session handle (e.g. class `RFs`) to memory allocated on the heap (e.g. class `RBuf`). However, R classes are most usually used to store client-side handles to server sessions. R classes are often small, and usually contain no other member data besides the resource handle.

R classes may exist as class members, as local stack-based variables or, occasionally, on the heap. If an R class object is created on the stack, it must be made *leave safe*, if used in functions that may leave, by using the cleanup stack as described in Section 4.5.10.

Unlike the C class type, there is no equivalent `RBase` class. A typical R class has a simple constructor and an initialization method, such as `Open()`,`Create()` or `Initialize()`, that must be called after construction to set up the associated resource and store its handle as a member variable of the R class object.

An R class also has a corresponding method, which must be called on cleanup to release the resource. Although, in theory, the cleanup function can be named anything, by convention it is almost always called

`Close()`. It is rare for an R class to have a destructor too – it generally does not need one because cleanup is performed in the `Close()` method.

A common mistake when using R classes is to forget to call `Close()` or to assume that there is a destructor which cleans up the owned resource.

4.4.6 M Classes

The `M` prefix stands for *mixin*, which is a term originating from an early object-oriented programming system, where the mixin class is used for defining interface classes. On Symbian OS, M classes are often used to define callback interfaces or observer classes. The only form of multiple inheritance encouraged on Symbian OS is that which mixes in one or more M classes.

An M class is an abstract interface class which declares pure virtual functions and has no member data. A concrete class deriving from such a class typically inherits from `CBase` (or a `CBase`-derived class) as its first base class and from one or more M class mixin interfaces, and implements the interface functions.

The correct class-derivation order is always to put the `CBase`-derived class first, to emphasize the primary inheritance tree and ensure proper cleanup through the cleanup stack. That is:

```
class CCat : public CBase, public MDomesticAnimal
```

and not

```
class CCat: public MDomesticAnimal, public CBase{...};
```

Since an M class is never instantiated and has no member data, there is no need for an M class to have a constructor. M classes often do not have destructors either. For flexibility, it may be preferable to provide a pure virtual `Release()` method so the owner of an M class pointer can call it to initiate the appropriate cleanup of the object implementing the interface.

An M class should usually have only pure virtual functions, but may occasionally have non-pure virtual functions.

4.4.7 Static Classes

Some Symbian OS utility classes, such as `User`, `Math` and `Mem`, take no prefix letter and contain only static member functions. The classes themselves cannot be instantiated; their functions must instead be called using the scope-resolution operator.

```
User::After(1000); // Suspend the current thread for 1000 microseconds.
```

A static class is sometimes implemented to act as a factory class.

4.4.8 Summary

When writing a Symbian class, it is important to consider the type of member data a class will contain, if any.

There are some cases where a class will not contain any member data, for example, if it is a utility class or an interface class for inheritance only. The latter type of class will usually be defined as an M class but may occasionally be created as a C class which has pure virtual or virtual functions only.

If a class must contain member data, and the member data needs no special cleanup (that is, if it contains only native types, other T classes, or pointers and references to objects owned elsewhere), it will typically be defined as a T class. However, if the size of the object is larger than 512 bytes, it's advisable to prevent the object from being instantiated on the stack, since space is limited. In this case, it's better to make the class a C class since C class objects are always created on the heap.

A C class should also be defined if the class will own data that needs to be cleaned up, such as heap-based buffers, other C class objects or R-class objects.

4.5 Leaves and the Cleanup Stack

Most fundamental to Symbian OS are its lightweight exception handling, which is called leaving, and its use of a cleanup stack, which manages the destruction of memory and other resources in the event of an exceptional condition such as insufficient memory or disk space, or a dropped Internet connection.

On some software platforms, the implications of a memory leak may not seem too bad, because memory is plentiful and any that is leaked is reclaimed when an application is closed or the system rebooted. It's important to remember that Symbian OS was designed to perform well on devices with limited memory that are often not rebooted for days, weeks or even months. Applications are often left running indefinitely rather than closed. Memory leaks, particularly those that occur periodically, have the potential to degrade the behavior of the entire system, and you should take care to avoid them.

Symbian OS was designed before the C++ standard had been formalized, and exceptions were not yet fully supported by the compiler used to build the code to run on hardware. Exception-handling support had the potential to bloat the size of compiled code and add to memory overheads at run time, so standard C++ exception handling was not used by Symbian OS. On early versions, the compilers in the tool chain were explicitly directed to disable C++ exception handling and to flag any use of the `try`, `catch` or `throw` keywords as an error.

Symbian OS v9.1 takes advantage of compiler improvements and now supports C++ standard exceptions. This makes it easier to port existing C++ code to Symbian OS. However, even if your code only uses standard C++ exceptions, leaves are a fundamental part of Symbian error handling and are used throughout the system. It's important to know how to blend leaves and exceptions so hybrid code works correctly, and does not cause panics (which you'll have to find and fix before you release your code) or silent memory leaks, which you may not even be aware of until it is too late.

Like C++ exceptions, leaves are used to propagate errors to where they can be handled. They should not be used to direct the normal flow of program logic. A leave does not terminate the flow of execution on Symbian OS. In contrast, a panic does; it cannot be caught or handled. A panic terminates the thread in which it occurs, and usually the entire application. This results in a poor user experience and panics should only be used in assertion statements to check code logic and flag programming errors during development. Panics and assertion statements are discussed in Section 4.6.

4.5.1 Causes of a Leave

A leave may occur in a function if it:

- explicitly calls one of the system functions in class `User` that cause a leave, such as `User::Leave()` or `User::LeaveIfError()`
- uses the Symbian OS-overloaded form of operator `new` which takes `ELeave` as a parameter
- calls code that may leave (for either of the reasons above) without using a `TRAP` harness.

Let's now examine each of these in turn.

4.5.2 System Functions in Class `User` That Cause a Leave

Class `User` is a static Symbian OS utility class, defined in `e32std.h`. It provides a set of exported system functions, among which are those to cause a leave in the current thread. The methods are defined as follows:

```
IMPORT_C static void Leave(TInt aReason);
IMPORT_C static TInt LeaveIfError(TInt aReason);
IMPORT_C static void LeaveNoMemory();
IMPORT_C static TAny* LeaveIfNull(TAny* aPtr);
```

- `User::Leave()` simply leaves with the integer value passed into it as a leave code.

- `User::LeaveIfError()` tests an integer parameter passed into it and call `User::Leave()` (passing the integer value as a leave code) if the value is less than zero. `User::LeaveIfError()` is useful for turning a non-leaving function which returns a standard Symbian OS error (one of the `KErrXXX` error constants defined in `e32std.h`) into one that leaves with the same value.
- `User::LeaveNoMemory()` simply leaves with `KErrNoMemory` that makes it, in effect, the same as calling `User::Leave(KErrNoMemory)`.
- `User::LeaveIfNull()` takes a pointer value and leaves with `KErrNoMemory` if it is `NULL`.

4.5.3 Heap Allocation Using new(ELeave)

Symbian OS overloads the global operator `new` to leave if there is insufficient heap memory for successful allocation. Use of this overload allows the pointer returned from the allocation to be used without a further test that the allocation was successful, because the allocation would leave if it were not. For example:

```
CBook* InitializeBookL()  // 'L' is explained in the next section.
  {
  CBook* book = new(ELeave) CBook();
  book->Initialize();  // No need to test book against NULL.
  return (book);
  }
```

The code above is preferable to the following, which requires a check to verify that the `book` pointer has been initialized:

```
CBook* InitializeBook()
  {
  CBook* book = new CBook();
  if (book)
    {
    book->Initialize();
    return (book);
    }
  else
    return (NULL);
  }
```

4.5.4 Code That May Leave

How do you know whether a function may leave? As Section 4.3.3 mentioned briefly, there is a naming convention to indicate the potential for a leave within a function – rather like the C++ `throw(...)` exception specification.

> **Pro tip**
>
> If a function may leave, its name must end with a trailing `L` to identify it as such.

Of all Symbian OS naming conventions, this is probably the most important rule: if a leaving function is not named correctly, callers of that function cannot know about it and may leak memory in the event a leave occurs.

Since it is not part of the C++ standard, the trailing `L` cannot be checked during compilation, and can sometimes be forgotten. Symbian OS provides a helpful tool, `LeaveScan` to check code for incorrectly named leaving functions. Other source code scanning tools can also be used for this purpose – a list of those available can be found on the Symbian Developer Network site (***developer.symbian.com/main/tools***).

Another standard practice when defining leaving functions is to avoid returning an error code as well. That is, since leaving functions already highlight a failure (through a *leave code*), they should not also return error values. Any error that occurs in a leaving function should be passed out as a leave; if the function does not leave it is deemed to have succeeded and will return normally. Generally, leaving functions should return `void` unless they return a pointer or reference to a resource that they have allocated.

4.5.5 The `TRAP` and `TRAPD` Macros

The `TRAP` and `TRAPD` macros are provided as harnesses to trap leaves and allow them to be handled. The macros differ only in that `TRAPD` declares the variable in which the leave error code is returned, while code using `TRAP` must declare a `TInt` variable itself first. Thus the following code segments are equivalent:

```
TRAPD(result, MayLeaveL());
if (KErrNone!=result)
  {
  // Handle error.
  }

TInt result;
TRAP(result, MayLeaveL());
if (KErrNone!=result)
  {
  // Handle error.
  }
```

If a leave occurs inside `MayLeaveL()`, control will return immediately to the `TRAP` harness macro. The `result` variable then contains the error code associated with the leave (the parameter passed into `User::Leave()`) or `KErrNone` if no leave occurred.

Any functions called by `MayLeaveL()` are executed within the `TRAP` harness, and so on recursively. Any leave occurring during the execution of `MayLeaveL()` is trapped regardless of many calls deep in the call stack it happens.

`TRAP` macros can also be nested to catch and handle leaves at different levels of code, where they can best be dealt with. However, each `TRAP` has an impact on executable size and execution speed and the number of `TRAP`s should be minimized where possible. For example, suppose that the following function needs to call a number of functions that leave, but must not leave itself. It might seem straightforward simply to put each call in a `TRAP` harness:

```
TInt NonLeavingFunction()
  {
  TRAPD(result, MayLeaveL());
  if (KErrNone==result)
    TRAP(result, MayAlsoLeaveL());
  if (KErrNone==result)
    TRAP(result, MayBeALeaverL());
  // Return any error or KErrNone if successful.
  return (result);
  }
```

The following refactoring is preferable for efficiency:

```
TInt NonLeavingFunction()
  {
  TRAPD(result, LeavingFunctionL());
  // Return any error or KErrNone if successful.
  return (result);
  }

void LeavingFunctionL()
  {
  MayLeaveL();
  MayAlsoLeaveL();
  MayBeALeaverL();
  }
```

4.5.6 Avoiding Memory Leaks

If memory is allocated on the heap and referenced only by a stack-based local variable, what happens if a leave occurs? A leave unwinds the stack back to a `TRAP` and the referencing pointer is destroyed. In effect, the code switches back to (a copy of) the stack frame from the time the `TRAP` harness was called, when the local pointer variable wasn't defined. This means that the allocated heap memory that the pointer referenced becomes unrecoverable, causing a memory leak. The following code illustrates this:

```
void UnsafeL()
  {
  // Heap allocation.
  CBook* book = new(ELeave) CBook();
  book->InitializeL();  // Unsafe - book may leak!
  delete book;
  }
```

The memory allocated on the heap to store `book` will become inaccessible if the call to `InitializeL()` leaves. If that happens, `book` can never be de-allocated (it is said to be *orphaned*) resulting in a memory leak. The Symbian OS cleanup stack must be used prevent this, as the next section explains.

While heap variables referenced only by local variables are orphaned if a leave occurs, member variables do not suffer a similar fate (unless the destructor neglects to delete them when it is called at some later point). The following code is safe:

```
void CTestClass::SafeFunctionL()
  {
  // Heap allocation of member variable.
  iBook = new(ELeave) CBook();
  iBook->InitializeL(); // Safe for iBook.
  }
```

The heap-based `iBook` member is stored safely, and deleted at a later stage with the rest of the `CTestClass` object, through the class destructor.

4.5.7 The Cleanup Stack

The previous section explained that a memory leak can occur if heap objects are accessible only through pointers local to the function that leaves. Let's look at this in terms of the Symbian OS types we discussed in Section 4.4:

- T class objects are leave safe, as long as the convention is followed and none of the member data requires explicit destruction.
- C class objects, which are always created on the heap, are not leave-safe unless they are otherwise accessible (for example, as member variables).
- R class objects are generally not leave-safe, since the resources they own must be freed in the event of a leave (through a call to the appropriate `Close()` or `Release()` function).

- M class objects may or may not be leave-safe, depending on the type of class into which the interface has been mixed.

One way to make objects leave safe is to place a TRAP (or TRAPD) macro around every potential leaving call. However, the use of TRAPs should be limited for reasons of efficiency, and to avoid clumsy code which constantly traps leaves and checks leave codes. In effect, code which uses multiple trap harnesses to catch every possible leave is simply reverting to the basic error-handling approach that leaves were designed to avoid.

Instead, Symbian OS provides a mechanism for storing pointers which are not leave-safe, to ensure they are cleaned up in the event of a leave. This is called the cleanup stack. Each thread has its own cleanup stack and, as long as pointers are stored on it before calling code that may leave, the cleanup stack destroys them in the event of a leave. The cleanup stack is rather like a Symbian OS version of the standard C++ library's smart pointer, `auto_ptr`.

It isn't necessary to create a cleanup stack for a GUI application thread, since the application framework creates one. However, a cleanup stack must be created if writing a server, a simple console-test application or when creating an additional thread which contains code that uses the cleanup stack. This is done as follows:

```
CTrapCleanup* theCleanupStack = CTrapCleanup::New();

... // Code that uses the cleanup stack within a TRAP macro.
delete theCleanupStack;
```

Once created, the cleanup stack is accessed through the static member functions of class `CleanupStack`, defined in `e32base.h`:

```
class CleanupStack
  {
public:
  IMPORT_C static void PushL(TAny* aPtr);
  IMPORT_C static void PushL(CBase* aPtr);
  IMPORT_C static void PushL(TCleanupItem anItem);
  IMPORT_C static void Pop();
  IMPORT_C static void Pop(TInt aCount);
  IMPORT_C static void PopAndDestroy();
  IMPORT_C static void PopAndDestroy(TInt aCount);
  IMPORT_C static void Check(TAny* aExpectedItem);
  inline static void Pop(TAny* aExpectedItem);
  inline static void Pop(TInt aCount, TAny* aLastExpectedItem);
  inline static void PopAndDestroy(TAny* aExpectedItem);
  inline static void PopAndDestroy(TInt aCount,
                      TAny* aLastExpectedItem);
  };
```

4.5.8 Using the Cleanup Stack

The following code illustrates a leave-safe version of the function `UnsafeL()` discussed earlier:

```
void NowSafeL()
  {// Heap allocation.
  CBook* book = new(ELeave) CBook();
  // Push onto the cleanup stack.
  CleanupStack::PushL(book);
  book->InitializeL();// Safe
  // Pop from the cleanup stack.
  CleanupStack::Pop(book);
  delete book;
  }
```

If `InitializeL()` leaves, `book` is destroyed by the cleanup stack. If no leave occurs, the `CBook` pointer is popped from the cleanup stack and the object to which it points is deleted. This code could equally well be replaced by a single call to `CleanupStack::PopAndDestroy(book)` that pops the pointer and makes a call to the destructor in one step.

In the example above, you may have noticed that `PushL()` is a leaving function, which at first sight appears to be self-defeating. If the cleanup stack is supposed to make pointers leave safe, how can it cause a leave when you use it and still be helpful? The reason `PushL()` may leave is that it may need to allocate memory to store the pointers passed to it (and in low-memory conditions this could fail). However, the object passed to `PushL()` will not be orphaned if it leaves. This is because, when the cleanup stack is created, it has at least one spare slot. When `PushL()` is called, the pointer is added to the next vacant slot, and then, if there are no remaining slots available, the cleanup stack attempts to allocate some for future pushes. If this allocation fails, a leave occurs. The pointer passed in was stored safely before this failure could occur, so the object it refers to can be cleaned up.

The cleanup stack is expanded four slots at a time for efficiency and `Pop()` and `PopAndDestroy()` do not release the memory for slots once they have been allocated. This memory is available for future pushes onto the cleanup stack until the cleanup stack is destroyed.

In a function, if a pointer to an object is pushed onto the cleanup stack and remains on it when that function returns, the Symbian OS naming convention is to append a `C` to the function name. For example:

```
void NowSafeLC()
  {// Heap allocation.
  CBook* book = new(ELeave) CBook();
  // Push onto the cleanup stack.
  CleanupStack::PushL(book);
  book->InitializeL();  // Safe.
  return (book);
  }
```

This is used most frequently in static factory functions for C classes, as Section 4.7.1 describes.

4.5.9 Ordering Calls to `PushL()` and `Pop()` or `PopAndDestroy()`

Pointers can be pushed and popped onto the cleanup stack, but since it's a stack, for any series of pushes, the corresponding pops must occur in reverse order. It's a good idea to name the pointer as it is popped, so debug builds can check that it is the correct pointer and flag up any programming errors. For example:

```
void TestPhonesL()
  { // Each object is pushed onto the cleanup stack
    // immediately it is allocated, in case the next
    // allocation leaves.
  CUiq3Phone* phoneP1i = CUiqPhone::NewL(EP1i);
  CleanupStack::PushL(phoneP1i);
  CUiq3Phone* phoneM600i = CUiqPhone::NewL(EM600i);
  CleanupStack::PushL(phoneM600i);
  CUiq3Phone* phoneP990i = CUiqPhone::NewL(EP990i);
  CleanupStack::PushL(phoneP990i);
  CUiq3Phone* phoneW960i = CUiqPhone::NewL(EW960i);
  CleanupStack::PushL(phoneW960i);
  ...  // Leaving functions called here.

  // Various ways to remove the objects from the stack and delete them:
  // (1) All with one anonymous call -  OK
  // CleanupStack::PopAndDestroy(4);

  // (2) Each object individually to verify the code logic
  // Note the reverse order of Pop() to PushL().
  // This is long-winded.
  // CleanupStack::PopAndDestroy(phoneW950i);
  // CleanupStack::PopAndDestroy(phoneP990i);
  // CleanupStack::PopAndDestroy(phoneM600i);
  // CleanupStack::PopAndDestroy(phoneP1i);

  // (3) All at once, naming the last object - best solution.
  CleanupStack::PopAndDestroy(4, phoneP1i);
  }
```

4.5.10 Using the Cleanup Stack with T, R and M Classes

As you can see from the definition of class `CleanupStack` given earlier, there are three overloads of the `PushL()` method. These determine how the item is destroyed when it is cleaned up when a leave occurs or a call is made to `CleanupStack::PopAndDestroy()`.

```
IMPORT_C static void PushL(CBase* aPtr);
```

This overload takes a pointer to a `CBase`-derived object. Cleanup destroys it by invoking `delete`, thus calling the virtual destructor of the `CBase`-derived object.

```
IMPORT_C static void PushL(TAny* aPtr);
```

If the object pushed onto the cleanup stack is a pointer to an object that does not derive from `CBase`, this overload is invoked. On cleanup, the memory referenced by the pointer is simply deallocated by invoking `User::Free()` (`delete` is not called on it, so no destructor is invoked).

This overload of `PushL()` is called, for example, when pointers to heap-based T class objects are pushed onto the cleanup stack. By definition, T classes do not have destructors and thus have no requirement for cleanup beyond deallocation of the heap memory they occupy.

```
IMPORT_C static void PushL(TCleanupItem anItem);
```

This overload takes an object of type `TCleanupItem`, which is designed to accommodate objects requiring customized cleanup processing. The `TCleanupItem` object encapsulates a pointer to the object to be stored on the cleanup stack and a pointer to a function to clean up that object (a local function or a static method of a class). A leave or a call to `PopAndDestroy()` removes the object from the cleanup stack and calls the associated cleanup function.

Symbian OS also provides a set of template utility functions, each of which generates an object of type `TCleanupItem` and pushes it onto the cleanup stack.

- `CleanupClosePushL()` – the cleanup method calls `Close()` on the object in question. This utility method is usually used to make stack-based R class objects leave-safe. For example:

```
void UseFilesystemL()
  {
  RFs theFs;
  User::LeaveIfError(theFs.Connect());
  CleanupClosePushL(theFs);

  ...  // Call functions which may leave,
       // theFs.Close() is called if a leave occurs.
  CleanupStack::PopAndDestroy(&theFs);
  }
```

- `CleanupReleasePushL()` – the cleanup method calls `Release()` on the object in question. This method is typically used to push an object referenced through an M class (interface) pointer that provides a `Release()` method, as discussed in Section 4.4.6.
- `CleanupDeletePushL()` – the cleanup method calls `delete` on the pointer passed into the function. This is also typically used for passing M class pointers.

The following simplistic example illustrates the reason why this is needed. A C class, `CSquare`, uses multiple inheritance from `CBase` and an M class interface, `MPolygon`. The object is instantiated and the M class interface pushed onto the cleanup stack. The object is later destroyed by a call to `PopAndDestroy()`. If the pushing was done with `CleanupStack::PushL()`, `PushL(TAny*)` would be used because the pointer is not a `CBase` pointer. The `PushL(TAny*)` overload means that `User::Free()` is invoked on the pointer on cleanup, but this would result in a `User 42` panic. That's because `User::Free()` can't find the heap information it needs to release the memory to the system when passed an M class pointer to the sub-object of a C class, and even if it could, the virtual destructor of `CSquare` would not be called, so the best-case scenario would be a memory leak! The solution is to use `CleanupDeleteL()` to call delete on the M class pointer and invoke the virtual destructor correctly for the `CSquare` class.

```
class MPolygon
  {
public:
  virtual TInt CalculateNumberOfVerticesL() =0;
  virtual ~MPolygon(){};
  };

class CSquare: public CBase, public MPolygon
  {
public:
  static CSquare* NewL();
  virtual TInt CalculateNumberOfVerticesL();
  virtual ~CSquare(){};
private:
  CSquare(){};
  };

enum TPolygonType { ESquare };

static MPolygon* InstantiatePolygonL(TPolygonType aPolygon);

MPolygon* InstantiatePolygonL(TPolygonType aPolygon)
  {
  if (ESquare==aPolygon)
    return (CSquare::NewL());
  else
    return (NULL);
  }

/*static*/ CSquare* CSquare::NewL()
  { // No two-phase construct needed here.
  CSquare* me = new(ELeave) CSquare();
  return (me);
  }
```

```
TInt CSquare::CalculateNumberOfVerticesL()
  { // Simple implementation, but others may leave.
  return (4);
  }

void DoStuffL()
  {
  MPolygon* square = InstantiatePolygonL(ESquare);
  CleanupDeletePushL(square);

  // ... Leaving calls go here.
  ASSERT(4==square->CalculateNumberOfVerticesL());
  CleanupStack::PopAndDestroy(square);
  }
```

- `CleanupArrayDeletePushL()` – this method is used to push a pointer to a heap-based C++ array of T class objects (or built-in types) on to the cleanup stack. When `PopAndDestroy()` is called, the memory allocated for the array is cleaned up using `delete[]`. No destructor is called on the elements of the array.

4.5.11 Member Variables and the Cleanup Stack

An object on the heap should never be deleted more than once. If a pointer to such an object is stored elsewhere, say as a member variable of another object which is accessible after a leave, it should not also be stored on the cleanup stack. If it were, cleanup after a leave would destroy it, but the other object storing the pointer would also be likely to do so. An attempt to delete an object that has already been released back to the heap will cause a system panic.

For this reason, it is good to follow the general rule that class member variables (prefixed by `i`) should not be pushed onto the cleanup stack. The object may be accessed through the owning object that destroys it when appropriate so does not need to be made leave-safe through use of the cleanup stack.

4.5.12 Mixing Leaves, Exceptions and the Cleanup Stack

Symbian OS `TRAP`s and leaves are implemented internally in terms of C++ exceptions (a leave is an `XLeaveException`). A `TRAP` will panic if it catches any other kind of exception. This means that, if you are mixing C++ exceptions with traditional Symbian OS leaves and `TRAP`s, you must not throw an exception within a leaving function unless also you catch it and handle it internally in that function.

A normal `catch` block will not manage the Symbian OS cleanup stack correctly either, so code that throws exceptions should not use the cleanup stack directly or indirectly. If you are calling leaving functions from exception-style code, do not use the cleanup stack, but make the calls as required and catch and handle any exceptions that arise.

4.6 Panics, Assertions and Leaves Compared

4.6.1 Panics

Panics are used to stop code running. They are intended to ensure robust code logic, by flagging up programming errors in a way that can't be ignored. As a developer, if a system call panics because you are making an invalid assumption (for example, trying to write off the end of a descriptor) the only way you can work around it is to fix the code. Unlike a leave, a panic can't be trapped because it terminates the thread. In fact, if a panic occurs in the main thread of a process, it terminates the process too. If the panic happens in a secondary thread, only that thread is terminated.

On phone hardware, a panic is seen as an **Application closed** message box. When debugging on the emulator builds, the panic breaks the code into the debugger by default, so you can look at the call stack and diagnose the problem.

To cause a panic in the current thread, the `Panic()` method of class `User` can be used. Because of the restrictions of the secure platform, it is not possible to panic other threads in the system except those running in the same process.[1] To panic another thread in the running process, your code should open an `RThread` handle and call `RThread::Panic()`.

`User::Panic()` and `RThread::Panic()` take a string parameter, which is used to specify the reason for the panic, and an integer error code, which can be any value, positive, zero or negative. The panic string should be short and descriptive. They are not intended for a user to see, only for you and other programmers during development.

Symbian OS has a set of well-documented panic categories (for example, `KERN-EXEC`, `E32USER-CBASE`, `ALLOC`, `USER`), the details of which can be found in the reference section of the Symbian Developer Library in the UIQ 3 SDK.

The following example shows the use of a panic to highlight a programming error. Sufficient information is provided so the developer can understand the problem and work out how to fix it:

```
void CTest::GetData(TDes& aBuffer)
  { // Check that the buffer passed in is sufficiently large
    // to contain the data. The caller should have allocated
    // KLengthNeeded.
```

[1] One exception is made for server threads, which may panic badly behaved client threads by using the `RMessagePtr2::Panic()`.

```
if (aBuffer.Length()<=KLengthNeeded)
  {
  _LIT(KCTestGetData, "CTest::GetData");
  User::Panic(KCTestGetData, KErrArgument);
  }
else
  { ... }
}
```

In fact, this kind of checking should be carried out using an assertion statement, as we cover in the next section.

4.6.2 Assertions

Assertions are used to check that code logic is correct. On Symbian OS, an assertion evaluates a statement and halts execution of the code if it is not correct. There is an assertion macro for debug builds only (__ASSERT_DEBUG) and another for both debug and release builds (__ASSERT_ALWAYS). However, you should carefully consider the use of assertions in release builds because the checking will have an impact on the code size and execution speed. If the assertion fails, the code will terminate and the user will see an ugly **Application closed** message box. If you've tested your code in debug builds sufficiently, you shouldn't need to include assertions in release builds.

The assertion macro takes a statement as its first parameter. It evaluates this and, if it proves false, it uses the second parameter passed to flag up the failure. The assertion macros are unfortunately not hard-coded to panic, but they should be. The assertion should always terminate the code at the point of failure to help you, the developer, detect the problem where it occurs.

Here is an example of how to use the debug assertion macro:

```
void CTestClass::SetCount(TInt aCount)
  { // Check aCount >=0 because it shouldn't be negative.
  #ifdef _DEBUG
  _LIT(KPanicDescriptor, "CTestClass::SetCount");
  #endif
  __ASSERT_DEBUG((aCount>=0),
  User::Panic(KPanicDescriptor, KErrArgument));

  ...  // Use aCount.
  }
```

It is common to define a panic function and a set of specific panic enumerators that can be reused by an entire class or application. For example, the following could be added to CTestClass:

```
enum TTestClassPanic
  {
```

```
  ESetCountInvalidArgument, // Invalid argument passed to SetCount().
    ...  // Enum values for assertions in other CTestClass methods.
          };

static void CTestClass::Panic(TTestClassPanic aCategory)
  {
  _LIT(KTestClassPanic, "CTestClass");
  User::Panic(KTestClassPanic, aCategory);
  }
```

The assertion can then be written as follows:

```
void CTestClass::SetCount(TInt aCount)
  {
  __ASSERT_DEBUG((aCount>=0), Panic(ESetCountInvalidArgument));
  ... // Use aCount.
  }
```

By making sure that the panic string is clear and using unique and well named/commented enumeration values for the panic code, developers using an API and receiving a panic can easily track down what the problem is when their code terminates unexpectedly during development.

The `ASSERT` macro is very useful for checking internal code logic. You don't need to pass a descriptor or a error value, but simply pass in the statement for test. In debug builds, this is evaluated and a `USER 0` panic is raised if the result is `EFalse`. It's not helpful for external calling code to receive this panic, since it's difficult to track down the reason, but it is useful when you're writing code and want to check any errors as you test. For example, it can be used in a `switch` statement to ensure that the logic of a state machine is correct:

```
// State machine.
void TLongRunningCalculation::DoTaskStep()
  { // Do a short task step.
  switch (iState)
    {
    case (EWaiting):
      iState = EBeginState; break;
    case (EBeginState):
      iState = EIntermediateState; break;
    case (EIntermediateState):
      iState = EFinalState; break;
    case (EFinalState):
      iState = EWaiting; // Finished
      break;
    default:
      ASSERT(EFalse); // Cause a panic!
                      // Code should never get here.
    }
  }
```

4.6.3 Panics, Assertions and Leaves

Symbian OS leaves occur when unexpected conditions arise that the code should be able to handle gracefully. Good examples are when there is insufficient heap memory for an allocation or when a communications link is dropped. Leaves are the equivalent of standard C++ exceptions, and code implements a recovery strategy, and maybe rollback, by catching the leaves using trap harnesses (`TRAP`s).

Programming errors are *bugs* caused by incorrect assumptions, typos or implementation errors. Examples include writing off the end of an array or making requests on a server in an incorrect sequence (such as trying to write to a file before opening it). Programming errors are persistent and predictable once the code is compiled (they happen deterministically). They should be corrected by the programmer rather than handled. Assertions are used to check programming logic and highlight errors by terminating a thread using a panic. Unlike leaves, panics cannot ever be caught and handled gracefully.

4.7 Construction and Destruction

4.7.1 Two-Phase Construction

Consider the following code, which allocates an object of type `CExample` on the heap and sets the value of `foo` accordingly:

```
CExample* foo = new(ELeave) CExample();
```

The code calls the `new` operator that allocates a `CExample` object on the heap if there is sufficient memory available. Having done so, it calls the constructor of class `CExample` to initialize the object. If the `CExample` constructor itself leaves, the memory already allocated for `foo` and any additional memory the constructor may have allocated will be orphaned. To avoid this, a key rule of Symbian OS memory management is as follows: **No code within a C++ constructor should ever leave**.

However, it may be necessary to write initialization code that leaves, say to allocate memory to store another object or to read from a configuration file that may be missing or corrupt. There are many reasons why initialization may fail, and the way to accommodate this on Symbian OS is to use two-phase construction.

Two-phase construction breaks object construction into two parts or phases:

1. A private constructor which cannot leave. It is this constructor which is called by the `new` operator. It implicitly calls base-class

constructors and may also invoke functions that cannot leave and/or initialize member variables with default values or those supplied as arguments to the constructor.

2. A private class method (typically called `ConstructL()`). This method may be called separately once the object, allocated and constructed by the `new` operator, has been pushed onto the cleanup stack; it will complete construction of the object and may safely perform operations that may leave. If a leave does occur, the cleanup stack calls the destructor to free any resources which have already been successfully allocated and destroys the memory allocated for the object itself.

A class provides a static function that wraps both phases of construction, providing a simple and easily identifiable means to instantiate it. The function is typically called `NewL()` and is static so that it can be called without first having an existing instance of the class. The non-leaving constructors and second-phase `ConstructL()` functions are private so that a caller cannot accidentally call them or instantiate objects of the class except through `NewL()`. For example:

```
class CExample : public CBase
  {
public:
  static CExample* NewL();
  static CExample* NewLC();
  ~CExample(); // Must cope with partial construction
  // Other public methods, e.g. Foo(), Bar().
  ...
private:
  CExample();          // Guaranteed not to leave.
  void ConstructL(); // Second-phase construction code may leave.
  CPointer* iPointer;
  };
```

Note that there is also a `NewLC()` function in class `CExample`. If a pointer to an object is pushed onto the cleanup stack and remains on it when that function returns, the Symbian OS convention is to append a `C` to the function name. This indicates to the caller that, if the function returns successfully, the cleanup stack has additional pointers on it.

Typical implementations of `NewL()` and `NewLC()` are as follows:

```
CExample* CExample::NewLC()
  {
  CExample* me = new (ELeave) CExample(); // First phase construction.
  CleanupStack::PushL(me);
  me->ConstructL();                       // Second phase
  return (me);
  }
CExample* CExample::NewL()
```

```
{
CExample* me = CExample::NewLC();
CleanupStack::Pop(me);
return (me);
}
```

The `NewL()` function is implemented in terms of the `NewLC()` function rather than the other way around (which would be slightly less efficient since this would require an extra `PushL()` call on the cleanup stack).

Each function returns a fully constructed object, or will leave either if there is insufficient memory to allocate the object (that is, if operator `new(ELeave)` leaves) or if the second-phase `ConstructL()` function leaves for any reason. This means that, if an object is initialized entirely by two-phase construction, the class can be implemented without the need to test each member variable to see if it is valid before using it. That is, if an object exists, it has been fully constructed. The result is an efficient class implementation without a test against each member pointer before it is de-referenced.

4.7.2 Object Destruction

A destructor may be called to clean up partially constructed objects if a leave occurs in the second-phase `ConstructL()` function described in Section 4.7.1. The destructor code cannot assume that an object is fully initialized and should not call functions on pointers that may not point to valid objects. For example:

```
CExample::~CExample()
  {
  if (iPointer) // iPointer may be NULL.
    {
    iPointer->DoSomething();
    delete iPointer;
    }
  }
```

A leave should never occur in a destructor or in cleanup code. One reason for this is that a destructor could itself be called as part of cleanup following a leave. A further leave during cleanup would be undesirable, if nothing else because it would mask the initial reason for the leave. More obviously, a leave within a destructor would leave the object destruction incomplete and may leak resources.

It is important to consider the consequences of deleting an object referred to by a class member variable pointer. If the object of the owning class continues to exist, what does the pointer refer to? If the object has been deleted, it points to memory which looks valid but has been released back to the heap. If a destructor call occurs, the destructor attempts to destroy the memory again, and a panic occurs.

To avoid this, when a member variable is deleted, a common programming technique is to set it to zero before instantiating a new object to replace it. If the allocation fails, at least the destructor, when it is called later, will not try to delete the initial object again. For example:

```
void CExample::RenewMemberL()
  {
    delete iPointer; // Destroy old member variable.
    iPointer = NULL;
    iPointer = CDerivedPointer::NewL(); // Create a new one.
    ...
  }
```

However, this causes a problem for the implementation of `CExample`, since if `RenewMemberL()` fails, the `iPointer` member is `NULL`. If `RenewMemberL()` is a public method, the class cannot guarantee that `iPointer` is valid, since it may have been deleted but the replacement allocation may have failed. This means that `iPointer` must be tested before use, which runs contrary to the benefits of two-phase construction described previously, that is, that all member pointers are valid if an object exists.

A solution is to write `RenewMemberL()` to allow rollback:

```
void CExample::RenewMemberL()
  {
  // Create a new pointer.
  CPointer* tempPtr = CDerivedPointer::NewL();
  // Destroy old member variable.
  delete iPointer;
  iPointer = tempPtr;
  ...
  }
```

If the allocation of the new member fails, `RenewMemberL()` leaves, but the original `iPointer` member is retained. The caller of `RenewMemberL()` receives the error and can decide whether to continue to use the object with the original member, or to otherwise handle the error.

(Note that it is never necessary to `NULL` a member variable if the deletion occurs *within* a destructor itself.)

4.7.3 Cleanup of Symbian OS Types

Section 4.4 discussed the Symbian OS class types in some detail, and destruction of each is part of the definition of the characteristic type.

T classes do not have destructors and do not need to be cleaned up explicitly when created on the stack. When created on the heap, the memory allocated to store them is freed by invoking delete on the pointer to the object.

C class objects have a destructor and must always be created on the heap. They are destroyed by deletion of the pointer to the object, unless a specific cleanup method is supplied.

R class objects are cleaned up by calling a particular method, which is nearly always called `Close()`. R class objects are typically stack based but if they are created on the heap, the memory associated with the object should be freed, after the call to `Close()`, by invoking delete on the pointer to the object.

The cleanup of objects referred to through pointers to an M class is discussed in some detail in Section 4.5.10.

4.8 Descriptors: Symbian OS Strings

One of the most common requirements on any programming platform is to handle string data. An issue that frequently arises is how the length of string data is determined, and how programmers can prevent code from writing outside the data area allocated to the string. So-called 'off by one' errors and buffer overruns are common problems. They affected the pace of development of the SIBO operating system that preceded Symbian OS to the extent that when EPOC (later to be known as Symbian OS) was being designed, a set of string classes which imposed a solution to prevent overruns was added from the start. These classes are known as *descriptors* because they are self-describing strings. That is, a descriptor holds the length of the string of data it represents as well as type information, to identify the underlying memory layout of the descriptor data. Descriptors protect against buffer overrun and don't rely on `NULL` terminators to determine the length of the string.

Descriptors are used throughout the operating system, from the very lowest level upwards, and are designed to be very efficient, using the minimum amount of memory necessary to store data, while describing it fully in terms of its length and location. Descriptors are strings and can contain text data. However, they can also be used to manipulate binary data because they don't rely on a `NULL` terminating character to determine their length.

There are probably as many string classes as there are programming platforms, and everyone has their favorite. Many developers find Symbian OS descriptors hard to get used to at first, because there are a lot of different classes to become familiar with. In addition, they can be confusing because they aren't a direct analog to standard C++ strings, Java strings or the MFC `CString` (to take just three examples) since their underlying memory allocation and cleanup must be managed by the programmer.

What this means is that a descriptor object does not dynamically manage the memory used to store its data. Any descriptor method that causes data modification first checks that the maximum length allocated for it can accommodate the modified data. If it can not, it does not re-allocate memory for the operation but instead panics to indicate that a programming error has occurred. As an application developer, before calling a modifiable descriptor method that may expand the descriptor, it is necessary to ensure that there is sufficient memory available for it to succeed.

Some developers new to descriptors try to find ways to avoid using them, such as writing their own alternative string classes. This can lead to inefficient, and potentially leaky, code. What's more, since most native Symbian OS C++ APIs use descriptors, it's impossible to avoid using them completely. This section will discuss the descriptors at a high level to give you the basics required to understand later discussion in this book. For additional information about descriptors, you should consult the Symbian Developer Library in the UIQ 3 SDK, or one of the Symbian Press books dedicated to programming in C++ on Symbian OS, listed in the References and Resources section at the end of this book.

4.8.1 Character Size

In early releases of Symbian OS (up to and including Symbian OS ER5) descriptors had 8-bit native characters. Since that release, Symbian OS supports Unicode character sets with wide (16-bit) characters by default. The character width of descriptor classes can be identified from their names. If the class name ends in 8 (for example, `TPtr8`) it has narrow (8-bit) characters, while a descriptor class name ending with 16 (for example, `TPtr16`) manipulates 16-bit character strings.

There is also a set of neutral classes which have no number in their name (for example, `TPtr`). The neutral classes are `typedef`'d to the character width set by the platform, so are implicitly wide strings on UIQ 3. The neutral descriptor classes were defined for source compatibility purposes to ease a switch between narrow and wide builds and, although today Symbian OS is always built with wide characters, it is good practice to use the neutral descriptor classes where the character width does not need to be stated explicitly. However, to work with binary data, the 8-bit descriptor classes should be used explicitly, and some Symbian OS APIs, for example those to read from and write to a file, all take an eight-bit descriptor, so file handling is agnostic of whether text or binary data is used.

4.8.2 Descriptor Base Classes: `TDesC`

All Symbian OS descriptor classes derive from the base class `TDesC`, apart from the literal descriptors which we discuss in Section 4.8.9. In naming this class, the `T` prefix indicates a simple type class, while the `C` suffix reflects that the class defines a *non-modifiable* type of descriptor, one whose contents are constant.

`TDesC` provides methods for determining the length of the descriptor (`Length()`) and accessing its data (`Ptr()`). Using these methods, it also implements all the standard operations required on a constant string object, such as data access, matching and searching (all of which are documented in full in the Symbian Developer Library within the UIQ 3 SDK). The derived classes all inherit these methods and all constant descriptor manipulation is implemented by `TDesC` regardless of the derived type of the descriptor used.

As the ultimate base class, `TDesC` defines the fundamental layout of every descriptor type. The first four bytes always hold the length of the data the descriptor currently contains. The `Length()` method in `TDesC` returns this value. This method is never overridden by its subclasses since it is equally valid for all types of descriptor.

In fact, only 28 of the available 32 bits are used to hold the length of the descriptor data. This means that the maximum size of a descriptor is limited to 2^{28} bytes (256 MB). The location of the descriptor's data depends on the derived descriptor class in question. The top four bits are used to identify a descriptor object's derived type. The `Ptr()` method of `TDesC` inspects these four bits and uses the associated type to return the correct address for the beginning of the data. The `Ptr()` method is used internally by every descriptor operation that needs to determine the correct address in memory to access the descriptor's data.

This technique is used to avoid the need for each descriptor subclass to implement its own data access method using virtual function overriding. Doing so would add an extra four bytes to each derived descriptor object, for a virtual pointer (*vptr*) to access the virtual function table. Descriptors were designed to be as efficient as possible, and the size overhead to accommodate a C++ vptr was considered undesirable.

`Ptr()` is non-virtual and uses a `switch` statement to check the four-bit descriptor type and return the correct address for the beginning of its data. This requires that the `TDesC` base class has knowledge of the memory layout of its subclasses hard-coded into `Ptr()`.

4.8.3 Descriptor Base Classes: `TDes`

The modifiable descriptor types all derive from the base class `TDes`, which is itself a subclass of `TDesC`. `TDes` has an additional four-byte member variable to store the maximum length of data allowed for the current memory allocated to the descriptor. The `MaxLength()` method of `TDes` returns this value. Like the `Length()` method of `TDesC`, it is not overridden by derived classes. The contents of the descriptor can shrink and expand up to this value.

TDes defines a range of methods to manipulate modifiable string data, including those to append, fill and format the descriptor. All the manipulation code for descriptor modification is implemented by TDes and inherited by its derived classes. The descriptor API methods for both modifiable and non-modifiable descriptor base classes are fully documented in the Symbian Developer Library within the UIQ 3 SDK.

Pro tip

The TDesC and TDes base classes provide the descriptor manipulation methods and must know the type of derived class they are operating on in order to correctly locate the data area.

4.8.4 The Derived Descriptor Classes

As described, the descriptor base classes TDesC and TDes implement all the generic descriptor manipulation code. The derived descriptor classes merely add their own construction and assignment code. However, it is the derived descriptor types that are actually instantiated and used. Although the base classes have member variables to store the length (and maximum length, for TDes) of the descriptor, they do not have storage for descriptor data. There is no valid reason to instantiate them, but to prevent accidental instantiation, the descriptor base classes, TDesC and TDes, have protected default constructors.

The derived descriptor types come in two basic memory layouts: pointer descriptors, in which the descriptor points to data stored elsewhere; and buffer descriptors, where the data forms part of the descriptor. Figure 4.1 shows the inheritance hierarchy of the descriptor classes.

4.8.5 Pointer Descriptors: TPtrC and TPtr

The string data of a pointer descriptor is separate from the descriptor object itself and is stored elsewhere, for example, in ROM, on the heap or on the stack. The memory that holds the data is not 'owned' by the pointer descriptor, which is agnostic about where the memory is located. The class defines a range of constructors to allow TPtrC objects to be constructed from other descriptors, a pointer into memory, or a zero-terminated C string.

TPtrC is the equivalent of using const char* when handling strings in C. The data can be accessed but not modified: that is, the data in the descriptor is constant. All the non-modifying operations defined in the

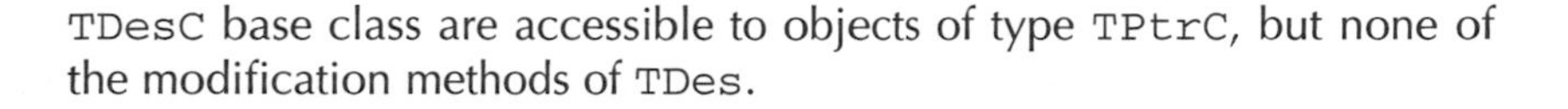

TDesC base class are accessible to objects of type TPtrC, but none of the modification methods of TDes.

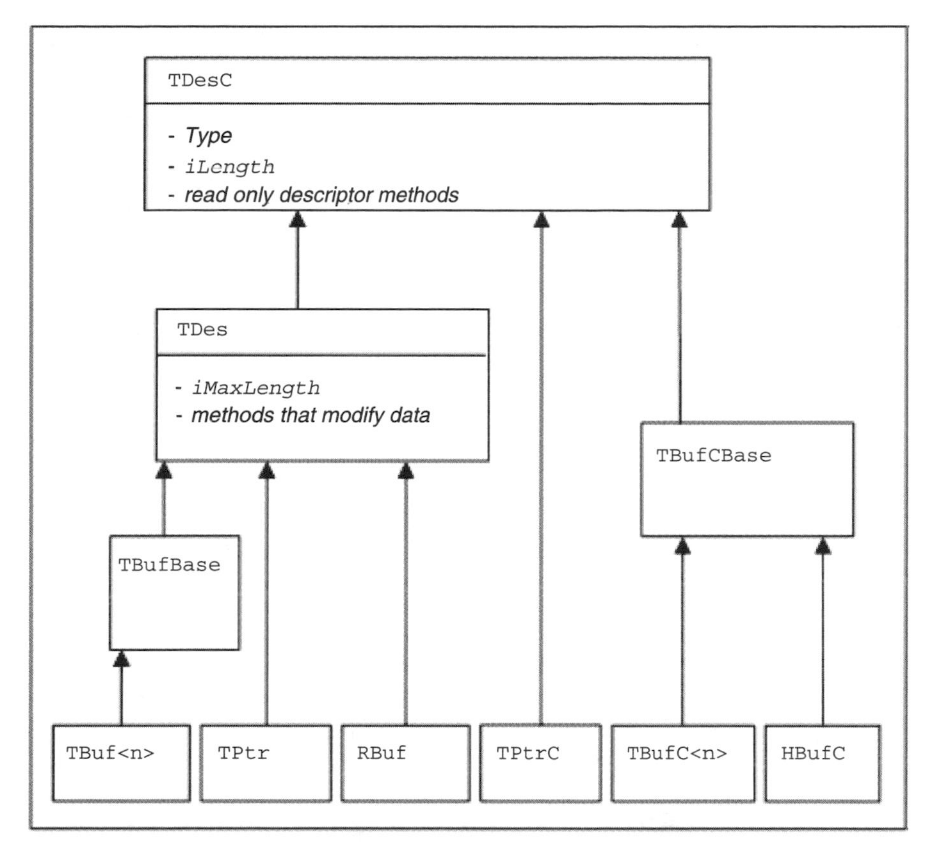

Figure 4.1 Descriptor class inheritance

```
// Literal descriptors are described in Section 4.8.9.
_LIT(KDes, "Sixty zippers were quickly picked from the woven jute bag");
TPtrC pangramPtr(KDes);// Constructed from a literal descriptor.
TPtrC copyPtr(pangramPtr);// Copy constructed from another TPtrC.
TBufC<100> constBuffer(KDes); // Constant buffer descriptor.
TPtrC ptr(constBuffer);        // Constructed from a TBufC.

// TText8 is a single (8-bit) character,
// equivalent to unsigned char.
const TText8* cString = (TText8*)"Waltz, bad nymph, for quick jigs vex";

// Constructed from a zero-terminated C string.
TPtrC8 anotherPtr(cString);
TUint8* memoryLocation; // Pointer into memory initialized elsewhere.
TInt length; // Length of memory to be represented.
...
TPtrC8 memPtr(memoryLocation,length);
```

In comparison, the TPtr class can be used for access to, and modification of, a character string or binary data. All the modifiable and non-modifiable base-class operations of TDes and TDesC, respectively, are accessible to a TPtr. Figure 4.2 compares the memory layouts of TPtr and TPtrC.

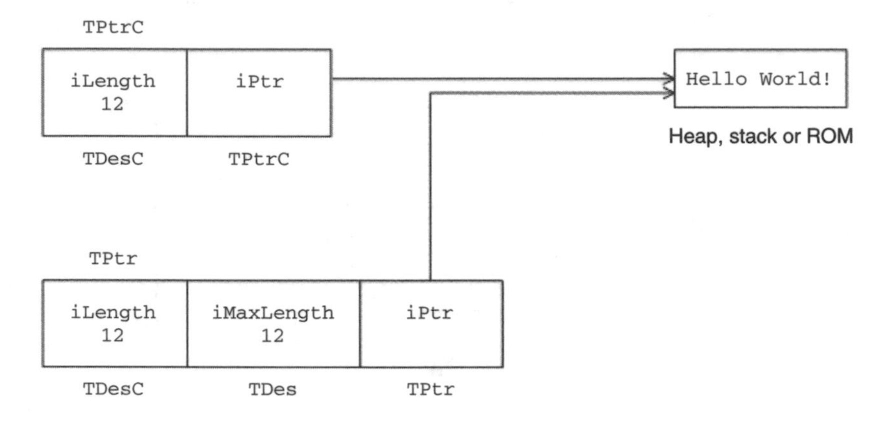

Figure 4.2 Memory layout of pointer descriptors

The class defines constructors to allow objects of type TPtr to be constructed from a pointer into an address in memory, setting the length and maximum length as appropriate. The compiler also generates implicit default and copy constructors and TPtr object may be copy-constructed from another modifiable pointer descriptor, for example, by calling the Des() method on a non-modifiable buffer.

```
_LIT(KLiteralDes1, "Jackdaws love my big sphinx of quartz");
TBufC<60> buf(KLiteralDes1); // TBufC is described later.
TPtr ptr(buf.Des()); // Copy construction can modify the data in buf.
TInt length = ptr.Length(); // Length=37 characters
TInt maxLength = ptr.MaxLength(); // Maximum length=60 characters,
                                  // as for buf.
TUint8* memoryLocation; // Valid pointer into memory.
...
TInt len = 12;     // Length of data to be represented.
TInt maxLen = 32; // Maximum length to be represented.

// Construct a pointer descriptor from
// a pointer into memory.
TPtr8 memPtr(memoryLocation, maxLen); // length=0, max=32.
TPtr8 memPtr2(memoryLocation, len, maxLen); // length=12,max=32
```

4.8.6 Stack-Based Buffer Descriptors TBufC and TBuf

The stack-based buffer descriptors may be modifiable or non-modifiable. The string data forms part of the descriptor object, as shown in Figure 4.3 which compares the memory layouts of TBuf and TBufC.

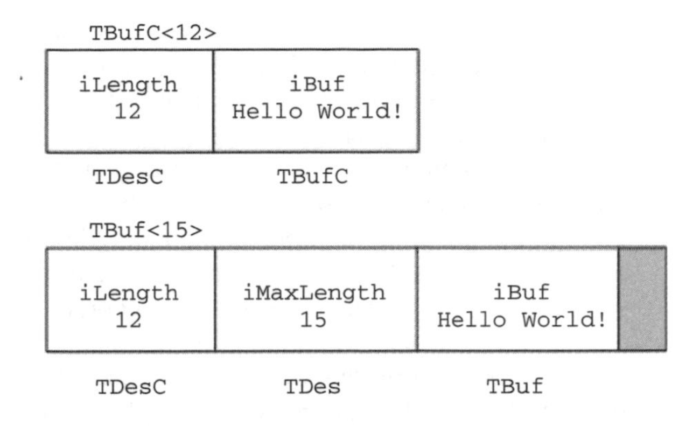

Figure 4.3 Memory layout of buffer descriptors

TBufC<n> is the non-modifiable buffer class, used to hold constant string or binary data. The class derives from TBufCBase (which derives from TDesC and exists only as an inheritance convenience). TBufC<n> is a thin template class that uses an integer value to fix the size of the data area for the buffer descriptor object at compile time. Thin templates are discussed further in Section 4.10.

TBufC defines several constructors that allow non-modifiable buffers to be constructed from a copy of any other descriptor or from a zero-terminated string. They can also be created empty and filled later, since, although the data is non-modifiable, the entire contents of the buffer may be replaced by calling the assignment operator defined by the class. The replacement data may be another non-modifiable descriptor or a zero-terminated string, but in each case the new data length must not exceed the length specified in the template parameter when the buffer was created.

```
_LIT(KPalindrome, "Satan, oscillate my metallic sonatas");
TBufC<50> buf1(KPalindrome); // Constructed from literal
                             // descriptor.
TBufC<50> buf2(buf1);        // Constructed from buf1.

// Constructed from a NULL-terminated C string.
TBufC<30> buf3((TText16*)"Never odd or even");
TBufC<50> buf4; // Constructed empty, length = 0.

// Copy and replace.
buf4 = buf1; // buf4 contains data copied from buf1, length modified.
buf1 = buf3; // buf1 contains data copied from buf3, length modified.
buf3 = buf2; // Panic! Max length of buf3 is insufficient for buf2 data.
```

TBuf<n> is used when a modifiable buffer is required. It derives from TBufBase, which itself derives from TDes, and inherits the full range of descriptor operations in TDes and TDesC. The class defines a number of constructors and assignment operators, similar to those offered by its non-modifiable counterpart, TBufC<n>. The buffer descriptors are useful for fixed-size strings, but should be used conservatively since they are stack-based except when used as members of a class whose objects are instantiated on the heap.

A good example of where stack-based buffers should be used carefully can be made with the TFileName type, which is typedef'd as TBuf<256>, that is, a modifiable buffer descriptor with a maximum of 256 characters. TFileName can be useful when calling various file system functions to parse filenames into complete paths, for example to print out a directory's contents, when the exact length of a filename isn't always known at compile time. However, each character is 16 bits wide, so every time a TFileName object is placed on the stack, it consumes $2 \times 256 = 512$ bytes (plus 12 bytes required for the descriptor object itself). That's just over 0.5 KB and, on Symbian OS, the stack space for each process is limited (it defaults to just 8 KB). This means that a single TFileName object can consume a sizable chunk of the available stack space, and probably wastes a lot of it since most file paths are not 256 characters, or anything approaching that length.

In cases where a descriptor's length is not known at compile time, instead of allocating large TFileName buffers to accommodate all possibilities safely, it's good practice to use one of the dynamic heap descriptor types instead.

4.8.7 Dynamic Descriptors: HBufC and RBuf

The HBufC and RBuf descriptor classes are used for dynamic string data whose size is not known at compile time, and for data that cannot be stack-based because it is too big. These classes are used where malloc'd data would be used in C.

The HBufC8 and HBufC16 classes (and the neutral version HBufC, which is typedef'd to HBufC16) provide a number of static NewL() functions to create the descriptor on the heap. These methods may leave if there is insufficient memory available. All heap buffers must be constructed using one of these methods or from one of the Alloc() or AllocL() methods of the TDesC class that spawn an HBufC copy of any existing descriptor. Once the descriptor has been created to the size required it is not automatically resized if more space is required. Additional memory must be reallocated using the ReAlloc() or ReAllocL() methods. The memory layout of an HBufC * descriptor is shown in Figure 4.4.

As the C suffix of the class name, and the inheritance hierarchy shown in Figure 4.1 indicates, HBufC descriptors are not directly modifiable, although the class provides assignment operators to allow the entire

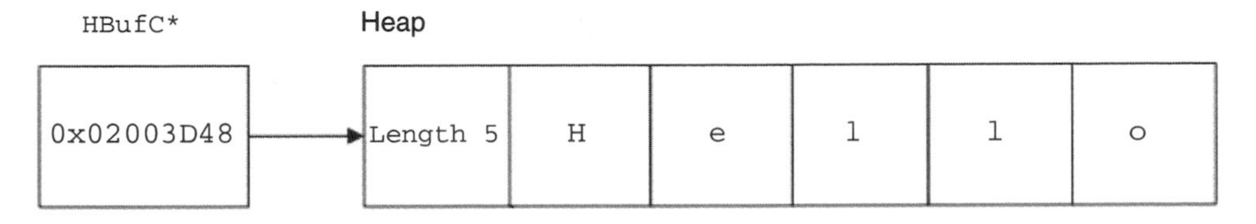

Figure 4.4 Memory layout of HBufC descriptors

contents of the buffer to be replaced. To modify an HBufC object at run time a modifiable pointer descriptor, TPtr, must first be created using the HBufC::Des() method.

```
_LIT(KPalindrome, "Do Geese see God?");
TBufC<20> stackBuf(KPalindrome);

// Allocate an empty heap descriptor of max length 20.
HBufC* heapBuf = HBufC::NewLC(20);

// Modify the heap descriptor through a TPtr.
TPtr ptr(heapBuf->Des());
ptr = stackBuf; // Copies stackBuf contents into heapBuf.

// Allocate a heap descriptor containing stackBuf.
HBufC* heapBuf2 = stackBuf.AllocLC();
_LIT(KPalindrome2, "Palindrome");
*heapBuf2 = KPalindrome2; // Copy and replace data in heapBuf2.
CleanupStack::PopAndDestroy(2,heapBuf);
```

Class RBuf is derived from TDes, so an RBuf object can be modified without the need to create a TPtr around the data first, which often makes it preferable to HBufC. On instantiation, an RBuf object can allocate its own buffer or take ownership of pre-allocated memory or a pre-existing heap descriptor.

To comply with the Symbian OS class name conventions (see Section 4.3), the RBuf class is not named HBuf because, unlike HBufC, it is not directly created on the heap. RBuf descriptors are typically created on the stack, and hold a pointer to a resource on the heap for which it is responsible at cleanup time. This is the definition of an R class.

Internally, RBuf behaves in one of two ways, as shown in Figure 4.5:

- As a TPtr which points directly to the descriptor data stored in memory which the RBuf object allocates or takes ownership of.
- As a pointer to an existing heap descriptor, HBufC*. The RBuf object takes ownership of the HBufC and holds a pointer to memory that contains a complete descriptor object (compared to a pointer to a simple block of data as in the former case).

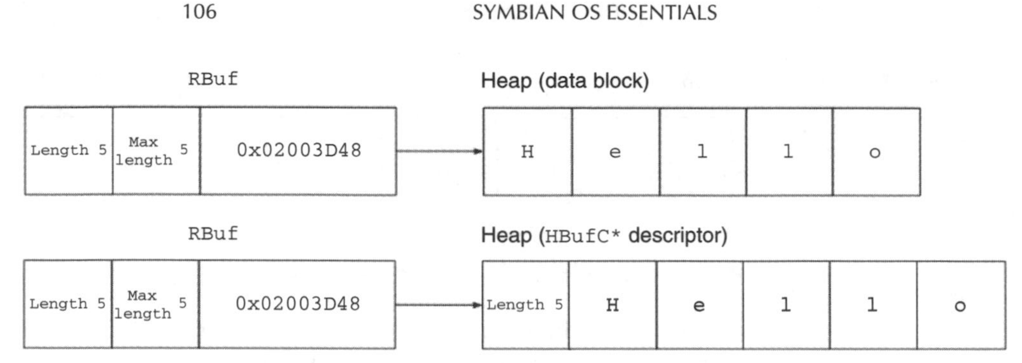

Figure 4.5 Alternative memory layouts of RBuf descriptors

However, this is all transparent, and there is no need to know how a specific RBuf object is represented internally. Using this descriptor class is straightforward too, through the methods inherited from TDes and TDesC.

RBuf is a relatively recent addition to Symbian OS, introduced in Symbian OS v8.0, but first documented in Symbian OS v8.1 and used most extensively in software designed for phones based on Symbian OS v9 and later. A lot of original example code won't use it but it is a simpler class to use than HBufC if you need a dynamically allocated buffer to hold data that changes frequently. The next section will examine how to instantiate and use RBuf in more detail.

HBufC is still ideal when a dynamically allocated descriptor is needed to hold data that doesn't change, that is, if no modifiable access to the data is required.

4.8.8 Using RBuf

RBuf objects can be instantiated using the Create(), CreateMax() or CreateL() methods to specify the maximum length of descriptor data that can be stored. It is also possible to instantiate an RBuf and copy the contents of another descriptor into it, as follows:

```
RBuf myRBuf;
_LIT(KHelloRBuf, "Hello RBuf!"); // Literal descriptor.
myRBuf.CreateL(KHelloRBuf());
```

CreateL() allocates a buffer for the RBuf to reference. If that RBuf previously owned a buffer, CreateL() will not clean it up before assigning the new buffer reference, so this must be done first by calling Close() to free any pre-existing owned memory.

Alternatively, an RBuf can be instantiated and take ownership of a pre-existing section of memory using the Assign() method.

```
// Taking ownership of HBufC.
HBufC* myHBufC = HBufC::NewL(20);
RBuf myRBuf.Assign(myHBufC);
```

Assign() will also orphan any data already owned by the RBuf, so Close() should be called before re-assignment, to avoid memory leaks.

The RBuf class doesn't manage the size of the buffer and reallocate it if more memory is required for a particular operation. If a modification method, such as Append(), is called on an RBuf object for which there is insufficient memory available, a panic will occur. As a programmer, you are responsible for re-allocating memory to the descriptor if it is required, using the ReAllocL() method:

```
// myRBuf is the buffer to be resized e.g. for an Append() operation.
myRBuf.CleanupClosePushL(); // Push onto cleanup stack for leave safety.
myRBuf.ReAllocL(newLength); // Extend to newLength.
CleanupStack::Pop();        // Remove from cleanup stack.
```

Note that the following example uses the CleanupClosePushL() method of the RBuf class to push it onto the cleanup stack. As is usual for other R classes, cleanup is performed by calling Close() (or CleanupStack::PopAndDestroy() if the RBuf was pushed onto the cleanup stack by a call to RBuf::CleanupClosePushL()).

It is easy to migrate code that previously used HBufC to use RBuf, which can be desirable when a dynamic buffer is modifiable. HBufC is rather clumsy to modify, since a TPtr object must first be constructed by calling Des() on it. The following example illustrates this:

```
// Defined elsewhere.
static void FileReader::ReadL(TDes& aModifiable);
TInt KMaxNameLength = 64;

HBufC* socketName = HBufC::NewL(KMaxNameLength);

// Create writable TPtr in order to modify socketName.
TPtr socketNamePtr(socketName->Des());
FileReader::ReadL(socketNamePtr);
```

This can be converted to the following:

```
RBuf socketName;
socketName.CreateL(KMaxNameLength);
FileReader::ReadL(socketName);
```

The first sample requires construction of a separate TPtr around the HBufC, so is slightly less efficient. Because the code using RBuf is simpler, it is also easier to understand and maintain.

4.8.9 Literal Descriptors

Literal descriptors are somewhat different from the other descriptor types. They are equivalent to static const char[] in C and because they are constant, they can be built into ROM to save memory at run time. A set of macros, found in e32def.h, can be used to define Symbian OS literals of two different types, _LIT and _L.

The _LIT macro is preferred for Symbian OS literals, since it is more efficient. It has been used in the sample code throughout this chapter, typically as follows:

```
_LIT(KSymbianOS, "Symbian OS");
```

The _LIT macro builds a named object (KSymbianOS) of type TLitC16 into the program binary, storing the appropriate string (in this case, "Symbian OS"). The explicit macros _LIT8 and _LIT16 behave similarly except that _LIT8 builds a narrow string of type TLitC8.

TLitC8 and TLitC16 do not derive from TDesC8 or TDesC16 but they have the same binary layouts as TBufC8 or TBufC16. This allows objects of these types to be used wherever TDesC is used.

Symbian OS also defines literals to represent a blank string. There are three variants of the *null descriptor*, defined as follows:

```
// Build independent:
_LIT(KNULLDesC,"");
// 8-bit for narrow strings:
_LIT8(KNULLDesC8,"");
// 16-bit for Unicode strings:
_LIT16(KNULLDesC16,"");
```

Use of the _L macro is now deprecated in production code, though it may still be used in test code (where memory use is less critical). The advantage of using _L (or the explicit forms _L8 and _L16) is that it can be used in place of a TPtrC without having to declare it separately from where it is used:

```
User::Panic(_L("testcode.exe"), KErrNotSupported);
```

4.8.10 Descriptor Class Types: Summary

The previous sections discussed the characteristics of each of the various descriptor types and classes; Table 4.2 summarizes them for reference.

Figure 4.6 summarizes how the knowledge in Table 4.2 can be applied when deciding what type of descriptor class to use.

4.8.11 Using the Descriptor APIs

The previous sections discussed the various descriptor classes and how to allocate them. Let's now move on to talk about some of the aspects of their use.

Table 4.2 Descriptor classes

Name	Modifiable	Approximate C equivalent	Type	Notes
TDesC	No	n/a	Not instantiable	Base class for all descriptors (except literals)
TDes	Yes	n/a	Not instantiable	Base class for all modifiable descriptors
TPtrC	No	const char* (doesn't own the data)	Pointer	The data is stored separately from the descriptor object which is agnostic to its location
TPtr	Yes	char* (doesn't own the data)	Pointer	The data is stored separately from the descriptor object which is agnostic to its location
TBufC	Indirectly	const char []	Stack buffer	Thin template – size is fixed at compile time
TBuf	Yes	char []	Stack buffer	Thin template – size is fixed at compile time
HBufC	Indirectly	const char* (owns the data)	Heap buffer	Used for dynamic data storage where modification is infrequent
RBuf	Yes	char* (owns the data)	Heap buffer	Used for modifiable dynamic data storage
TLitC	No	static const char []	Literal	Built into ROM

Length() and Size()

TDesC::Size() returns the size of the descriptor in bytes, while TDesC::Length() returns the number of characters it contains.

For 8-bit descriptors, the length and size of a descriptor are equivalent because the size of a character is a byte. In native 16-bit strings each character occupies two bytes. Thus Size() always returns a value double that of Length() for neutral and explicitly wide descriptors.

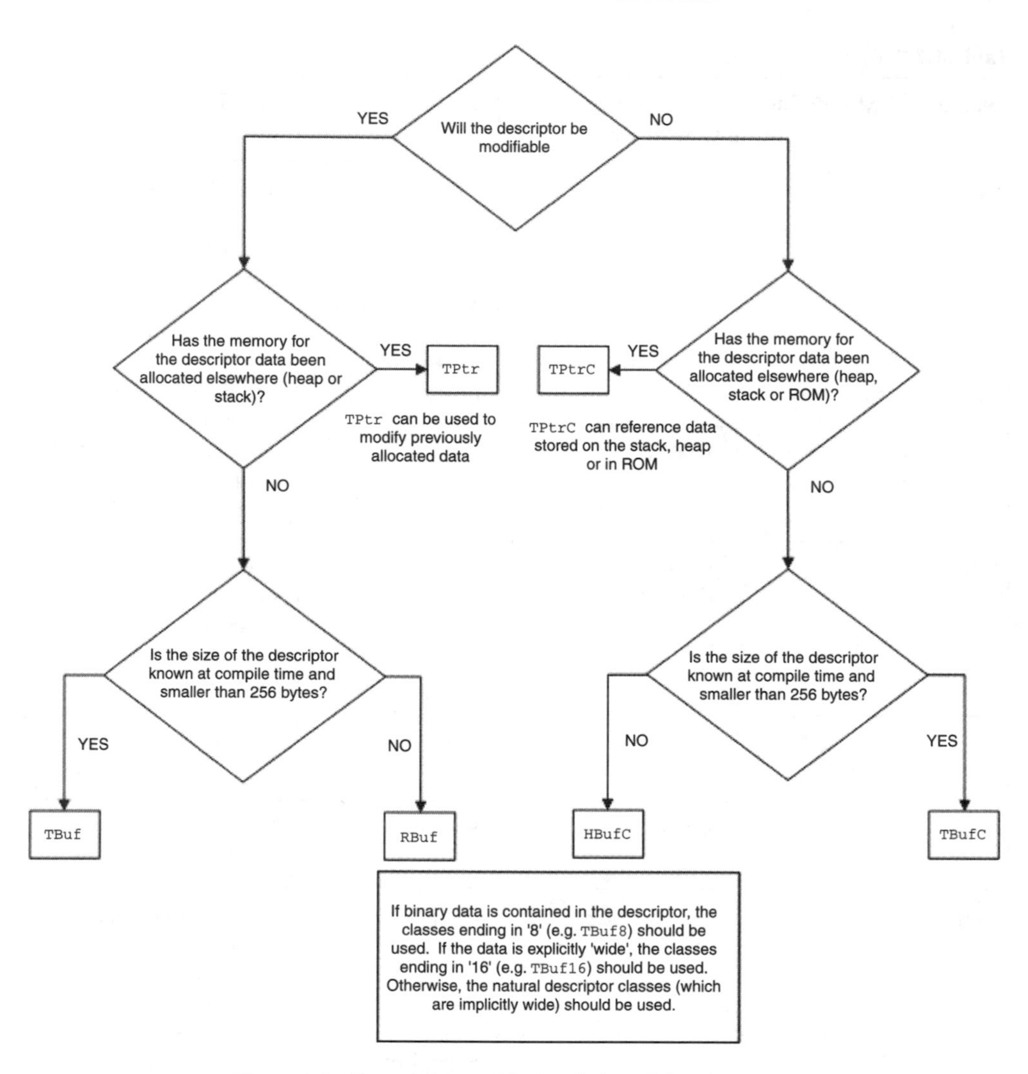

Figure 4.6 Flow chart to guide the choice of descriptor type

`MaxLength()` and length modification methods

`TDes::MaxLength()` returns the maximum length of the modifiable descriptor on which it is invoked.

`TDes::SetMax()` may seem misleadingly named at first. It does not change the maximum length of the descriptor, but instead sets its current length to the maximum it allows.

`TDes::SetLength()` can be used to adjust the descriptor length to any value between zero and its maximum length.

`TDes::Zero()` sets the length of the descriptor object on which it is invoked to zero.

TPtr(C)::Set() and TDes::operator=()

`TPtr` and `TPtrC` both provide `Set()` methods. These can be used to set the pointer to reference different string data. The length and maximum length members of the descriptor object are updated accordingly.

```
// Literal descriptors are described in Section 4.8.9.
_LIT(KDes1, "Sixty zippers were quickly picked from the woven jute bag");
_LIT(KDes2, "Waltz, bad nymph, for quick jigs vex");
TPtrC alpha(KDes1);
TPtrC beta(KDes2);
alpha.Set(KDes2); // alpha points to the data in KDes2.
beta.Set(KDes1);  // beta points to the data in KDes1.
```

`TDes` provides an assignment operator to copy data into the memory already referenced by a modifiable descriptor. The length of the descriptor is updated to that of the new contents, but the maximum length is unchanged. So it is important that the length of new data assigned to the descriptor is no longer than the maximum length, because otherwise the copy will cause a panic.

It is easy to confuse `Set()` with `TDes::operator =()`. Here's an example which illustrates both:

```
_LIT(KLiteralDes1, "Jackdaws love my big sphinx of quartz");
TBufC<60> buf(KLiteralDes1);
TPtr ptr(buf.Des());// Points to the contents of buf.
TUint16* memoryLocation; // Valid pointer into memory.
...
TInt maxLen = 40; // Maximum length to be represented.
TPtr memPtr(memoryLocation, maxLen);  // max length=40.

// Copy and replace.
memPtr = ptr; // memPtr data is KLiteralDes1 (37 bytes),
              // maxLength=40.
_LIT(KLiteralDes2, "The quick brown fox jumps over the lazy dog");
TBufC<100> buf2(KLiteralDes2);
TPtr ptr2(buf2.Des());// Points to the data in buf.
// Replace what ptr points to.
ptr.Set(ptr2); // ptr points to contents of buf2,
               // max length = 100.
memPtr = ptr2; // Attempt to update memPtr panics because
               // the contents of ptr2 (43 bytes) exceeds
               // the maximum length of memPtr (40 bytes).
```

TBufC::Des() and HBufC::Des()

`TBufC` and `HBufC` both provide a `Des()` method which returns a modifiable pointer descriptor to the data held in the buffer. While the content of a non-modifiable buffer descriptor cannot be altered directly, calling `Des()` makes it possible to change the data. The method updates the length members of both the modifiable pointer descriptor and the constant buffer descriptor it points to, if necessary. For example, for HBufC:

```
HBufC* heapBuf = HBufC::NewLC(20);
TPtr ptr(heapBuf->Des());     // Use ptr to modify heapBuf.
```

A common inefficiency when using HBufC, is to use Des() to return a modifiable pointer descriptor object (TPtr), when a non-modifiable descriptor (TDesC) is required. It is not incorrect, but since HBufC itself derives from TDesC, it can simply be de-referenced, which is clearer and more efficient.

```
const TDesC& CExample::Inefficient()
    {
    return (iHeapBuffer->Des());

    // Could be replaced more efficiently with:
    return (*iHeapBuffer);
    }
```

4.8.12 Descriptors as Function Parameters and Return Types

As Section 4.8.2 described, the TDesC and TDes descriptor base classes provide and implement the APIs for all descriptor operations. This means that the base classes can be used as function arguments and return types, allowing descriptors to be passed around in code transparently without forcing a dependency on a particular type.

A method will simply declare whether a descriptor parameter should be modifiable or non-modifiable, and whether it should be 8 or 16 bits in width. It doesn't have to know the type of descriptor passed into it. Unless a function takes or returns ownership, it doesn't even need to specify whether a descriptor is stack or heap-based. It can perform any operation it needs to if the descriptor parameter is modifiable (TDes), and non-modifiable operations otherwise. The client of an API also benefits. It isn't forced to instantiate and pass in a TBuf<64> to a particular method, but can use the most convenient descriptor type for the code in question.

When defining functions, the abstract base classes should always be used as parameters or return values. All descriptor parameters should be passed and returned by reference, either as const TDesC& for constant descriptors or TDes& when modifiable. The only exception comes when transferring ownership of a heap-based descriptor as a return value. This should be specified explicitly so that the caller can clean it up appropriately and avoid a memory leak.

```
// Read only parameter (16 bit).
void SomeFunctionL(const TDesC& aReadOnlyDescriptor);
```

```
// Read only return value (8 bit).
const TDesC8& SomeFunction();

// Read/Write parameter (16 bit).
void SomeFunctionL(TDes& aReadWriteDescriptor);

// Return ownership of HBufC*.
HBufC* SomeFunctionL();
```

The reason for mandating that descriptors should be reference parameters (`TDes&` or `const TDesC&`) and not values (`TDes` or `const TDesC`) is not just one of efficiency. Passing by value uses static binding, and passes the base class in question (`TDesC` and `TDes`) which contain no string data, and should never be used directly.

```
void BadExample(const TDesC aString) // Should be const TDesC&.
  {
  TBufC<10> buffer(aString); // buffer will contain rubbish.
  ...
  }
```

4.9 Arrays

Arrays are used for collections of data, for example, to store pointers to objects on the heap or to hold a set of stack-based objects. There are two families of dynamic array classes on Symbian, both of which can expand as elements are added to them and do not need to be created with a fixed size. Symbian OS also provides a fixed-length array class to wrap the standard C++ `[]` array, which is described briefly later in this section.

Symbian OS arrays are thin template classes, which are discussed in more detail in Section 4.10.

4.9.1 Dynamic Arrays on Symbian OS

You can think of the layout of an array as linear, but on Symbian OS, the implementation of a dynamic array may use:

- A single heap cell as a *flat* buffer to hold the array elements, which is resized as necessary. These are preferred when the speed of element lookup is an important factor or if the array is not expected to be frequently resized.
- A number of segments, using a doubly-linked list to manage them. These are best for large arrays which are expected to be resized frequently, or when elements are regularly inserted into or deleted, since repeated reallocations in a single flat buffer may result in heap thrashing as elements are shuffled around.

The original Symbian OS dynamic array classes are C classes, and there are a number of different types, all of which have names prefixed by `CArray`. We'll refer to them here generically as `CArrayX` classes. The full naming scheme takes the `CArray` prefix and follows it with:

- `Fix` where elements all have the same length and are copied directly into the array buffer (e.g. `TPoint`, `TRect`)
- `Var` where the elements are pointers to objects of variable lengths contained elsewhere on the heap (e.g. `HBufC*` or `TAny*`)
- `Pak` (for 'packed' array) where the elements are of variable length but are copied into the array buffer with each preceded by its length (e.g. T class objects of variable length)
- `Ptr` where the elements are pointers to `CBase`-derived objects.

  The array class name ends with:
  - `Flat` for classes which use an underlying flat buffer layout
  - `Seg` for those that use a segmented buffer.

Figure 4.7 illustrates the various memory layouts available and Table 4.3 summarizes each of the classes and how to destroy the elements it contains at cleanup time.

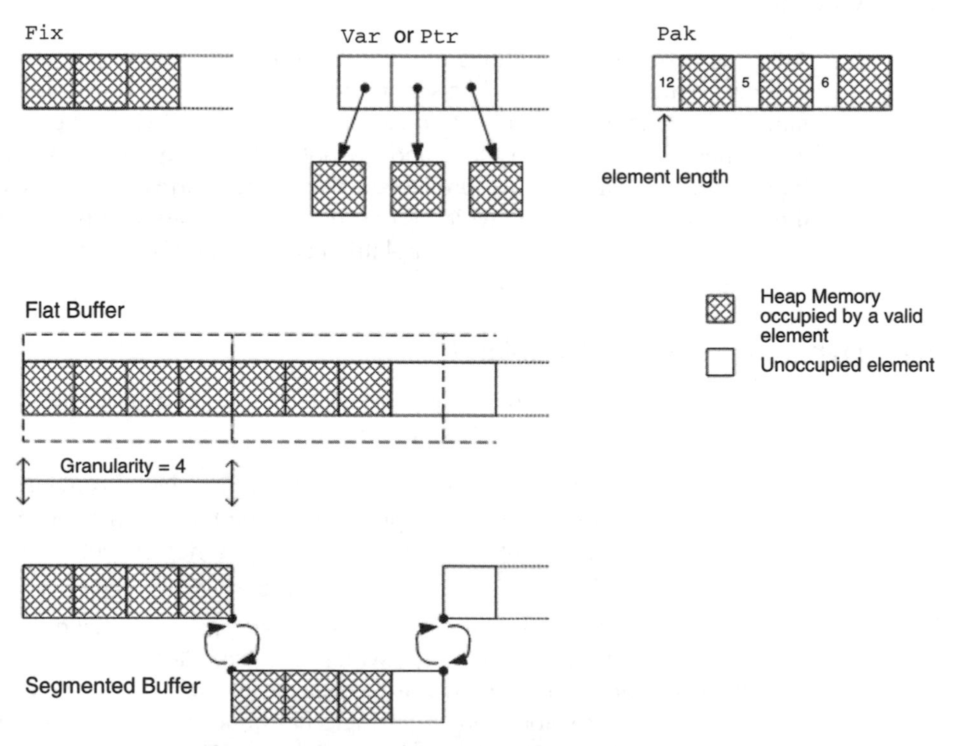

Figure 4.7 Memory layout of Symbian OS dynamic arrays

Table 4.3 Dynamic array classes

Class name	Elements and memory layout	Element cleanup
`CArrayFixFlat`	Fixed size elements contained in the array. Occupies a single memory buffer that is resized as necessary.	Destroyed automatically by the array.
`CarrayFixSeg`	Fixed size elements contained in the array. Occupies segments of memory.	Destroyed automatically by the array.
`CArrayVarFlat`	Elements are pointers to variable size objects contained outside the array. Occupies a single memory buffer that is resized as necessary.	Destroyed automatically by the array.
`CarrayVarSeg`	Elements are pointers to variable size objects contained outside the array. Occupies segments of memory.	Destroyed automatically by the array.
`CArrayPtrFlat`	Elements are pointers to `CBase`-derived objects stored outside the array. Occupies a single memory buffer that is resized as necessary.	Elements must be destroyed by calling `ResetAndDestroy()` on the array before destroying it.
`CarrayPtrSeg`	Elements are pointers to `CBase`-derived objects stored outside the array. Occupies segments of memory.	Elements must be destroyed by calling `ResetAndDestroy()` on the array before destroying it.

(*continued overleaf*)

Table 4.3 (*continued*)

Class name	Elements and memory layout	Element cleanup
`CArrayPakFlat`	Variable size elements occupy the array, each preceded by its length. Occupies a single memory buffer that is resized as necessary.	Destroyed automatically by the array.

As Symbian OS evolved, a second set of dynamic array classes (`RArray` and `RPointerArray`) were added. These classes may be instantiated on the heap or the stack but are usually stack-based. As R classes, they hold a handle to a block of memory used to store the array buffer, which in both cases is a resizable flat buffer rather than a doubly-linked list of segments.

`RArray<class T>` is an array of elements of the same size which are stored within the array buffer. The `Close()` or `Reset()` functions must be called to clean up the array by freeing the memory allocated for the elements.

- `RArray::Close()` frees the memory used to store the array and closes the array object.
- `RArray::Reset()` frees the memory used to store the array but simply resets its state so it can be reused.

`RPointerArray<class T>` is an array of pointer elements addressing objects stored elsewhere. The ownership of these objects must be addressed separately when the array is destroyed. If pointers to the objects are held elsewhere (by other components) then calling `Close()` or `Reset()` to clean up the memory for the array of pointers is sufficient. However, if the array has ownership of the objects its elements point to, it must take responsibility for cleanup. `ResetAndDestroy()` must be called to delete the object associated with each pointer element in the array, and then destroy the array itself.

4.9.2 Granularity and Capacity

All Symbian OS dynamic arrays have the concept of capacity and granularity.

The *capacity* of an array is the number of elements the array can hold within the space currently allocated to its buffer. When the buffer is full, the next insertion will cause the array to resize itself, by reallocating heap memory. The amount of additional space it reserves is set at construction time and is called the *granularity*.

The granularity should be chosen carefully according to the way the array will be used. If it is too small, this means the array must make frequent reallocation, which is inefficient. A common error is to set the granularity to one, which means that every addition to the array causes the underlying memory buffer to be reallocated. This is very inefficient and can potentially cause heap fragmentation. However, on the other side of the scale, if an array granularity is chosen to be too large, the array will waste memory. For example, if the array only ever contains between five and ten items, the granularity should not be set to 20.

4.9.3 Which Dynamic Array Class should be Used?

`RArray` and `RPointerArray` were added to Symbian OS for a reason. Comparing like with like (`RArray` with `CArrayFixFlat`, and `RPointerArray` with `CArrayPtrFlat`), the `RArray` and `RPointerArray` classes have significantly better performance than `CArrayX` classes. Let's consider the reasons behind this.

- The base class to all the `CArrayX` classes requires a `TPtr8` descriptor object to be constructed for every array access, which has a performance overhead, even for a simple flat array containing fixed-length elements.
- Every method that accesses the array has a minimum of two assertions in release and debug builds (through use of the `__ASSERT_ALWAYS` macro described in Section 4.6.2).
- A number of the array-manipulation functions of `CArrayX`, such as `AppendL()`, can leave, for example when there is insufficient memory to resize the array buffer. In some cases, the array must be called within a function which cannot leave, which means that these functions must be called in a `TRAP` macro. As Section 4.5.5 describes, a `TRAP` macro has an associated performance overhead.

The R classes do not require descriptors to be constructed for array access, do not use release build assertion checking and provide both leaving and non-leaving methods (the non-leaving methods may still fail, so the error value they return must always be checked).

For these reasons, the R classes are preferable as efficient and simple flat-memory arrays. The `CArrayX` classes have been retained for legacy code, and because they provide a segmented-memory implementation which may be more appropriate for arrays that are frequently resized. Thus, `CArrayFixSeg` and `CArrayPtrSeg` are useful alternatives to `RArray` and `RPointerArray`, respectively, when segmented memory is required, but `RArray` should be preferred over `CArrayFixFlat`, and `RPointerArray` over `CArrayPtrFlat`.

The examples in this book illustrate the use of both types of array for completeness. You may wish to try both types of array to see which you prefer to work with, since the efficiency advantages of the R classes are likely to be essential to only very few applications. Most code is not sufficiently optimized that the use of a `CArrayX` class instead of an `RArray` class makes a significant difference.

4.9.4 Sorting and Ordering

To sort and search the `CArrayX` classes, an array key is used to define a property of an array. For example, for an array of elements, which are types encapsulating an integer and a string, a key based on the integer value may be used to sort the array. Alternatively, a key may be used to search for an element containing a particular string or sub-string.

The abstract base class for the array key is `TKey`. The following `TKey`-derived classes implement keys for different types of array:

- `TKeyArrayFix` for arrays with fixed-length elements
- `TKeyArrayVar` for arrays with variable-length elements
- `TKeyArrayPak` for arrays with packed elements.

The appropriate `TKeyArrayFix`, `TKeyArrayVar` or `TKeyArrayPak` object is constructed and passed to the `Sort()`, `InsertIsqL()`, `Find()` or `FindIsq()` array-class member function. A search can be sequential, using `Find()`, or binary-chop, using `FindIsq()`. Both functions indicate the success or failure of the search and the position of the element within the array if it is found. An example of the use of `TKeyArrayFix` can be found in the `SignedApp` example of Chapter 10.

The `RArray` and `RPointerArray` classes also provide searching and ordering. Ordering is performed using a comparator function which must be implemented by the class making up the elements themselves, or a function which can access the relevant member data of the elements. The ordering method is passed to the `InsertInOrder()` or `Sort()` methods, by wrapping it in a `TLinearOrder<class T>` package.

Search operations on the `RArray` and `RPointerArray` classes are similar. The classes have several `Find()` methods that take a method, usually provided by the element class, that determines whether an element objects matches the item to be found.

4.9.5 Static Arrays on Symbian OS: `TFixedArray`

The `TFixedArray` class can be used when the number of elements it will contain is known at compile time. It wraps a standard fixed-length C++ `[]` array to add range checking, and prevents out of bounds access by panicking to flag the programming error. The range checking assertions are called in debug builds only, or in both debug and release builds, if you require it, by calling different methods on the class to access the elements. The class also provides some additional functions to navigate to the beginning and end, return the number of elements in the array, and clean up the elements it contains. Further information can be found in the Symbian Developer Library that accompanies each UIQ 3 SDK.

4.10 Templates

C++ templates allow for code reuse, for example when implementing container classes such as dynamic arrays that may contain different types. The use of templates makes the code generic and re-usable.

```
template<class T>
class CDynamicArray : public CBase
  {
public:
  // Functions omitted for clarity.
  ...
  void Add(const T& aEntry);
  T& operator[](TInt aIndex);
  };
```

The use of templates is preferable to an implementation using `void*` arguments and casting because templated code can be checked for type safety at compile time. However, template code can lead to increased code size because separate code is generated for each templated function for each type used. For example, if an application uses the `CDynamicArray` class to store an array of `HBufC*` and a separate array of `TUid` values, there would be two copies of the `Add()` function generated, and two copies of `operator[]`. On a limited resource system such as Symbian OS, templates can have a significant and undesirable impact on its size, and you should avoid using them except as Symbian OS does, through the thin template pattern.

The thin template pattern gives the benefits of type-safety, while reducing the amount of duplicated object code at compile time. Symbian OS uses this idiom in all its container classes, collections and buffers. The thin template idiom implements the container using a generic base class with `TAny*` pointers. A templated class is then defined that uses private inheritance of the generic implementation.[2]

[2] In case your C++ is a bit rusty: private inheritance allows implementation to be inherited, but the methods of the base class become private members. The deriving class does not inherit the interface of the base class so the generic implementation can only be used internally and not externally, which avoids accidental use of the non-type-safe methods.

Instead of the generic interface, the derived class presents a templated interface to its clients and implements it inline by calling the private base class methods. The use of templates means that the class can be used with any type required and that it is type-safe at compile time. There are no additional runtime costs because the interfaces are defined inline, so it will be as if the caller had used the base class directly. Only a single copy of the base class code is generated because it is not templated, so the code size remains small, regardless of the number of types that are used. This is ideal, if a bit confusing. Let's look at an example to make it clearer.

A snippet of the Symbian OS `RArrayBase` class and its deriving class `RArray` are shown below (the `RArray` class was discussed in Section 4.9.1). `RArrayBase` is the generic base class that implements the code logic for the array, but the code cannot be used directly because all its methods are protected.

```
class RArrayBase
  {
protected:
  IMPORT_C RArrayBase(TInt aEntrySize);
  IMPORT_C RArrayBase(TInt aEntrySize,TAny* aEntries,TInt aCount);
  IMPORT_C TAny* At(TInt aIndex) const;
  IMPORT_C TInt Append(const TAny* aEntry);
  IMPORT_C TInt Insert(const TAny* aEntry, TInt aPos);
  ...
  };
```

The derived `RArray` class is templated, defining a clear, usable API for clients. The API is defined inline and uses the base class implementation which it privately inherits from the base class. Elements of the array are instances of the template class.

```
template <class T>
class RArray : private RArrayBase
  {
public:
  ...
  inline RArray();
  inline const T& operator[](TInt aIndex) const;
  inline T& operator[](TInt aIndex);
  inline TInt Append(const T& aEntry);
  inline TInt Insert(const T& aEntry, TInt aPos);
  ...
  };

template <class T>
inline RArray<T>::RArray()
: RArrayBase(sizeof(T))
  {}
template <class T>
inline const T& RArray<T>::operator[](TInt aIndex) const
  {return *(const T*)At(aIndex); }
```

```
template <class T>
inline T& RArray<T>::operator[](TInt aIndex)
  {return *(T*)At(aIndex); }

template <class T>
inline TInt RArray<T>::Append(const T& aEntry)
  {return RArrayBase::Append(&aEntry);}

template <class T>
inline TInt RArray<T>::Insert(const T& aEntry, TInt aPos)
  {return RArrayBase::Insert(&aEntry,aPos);}
```

4.11 Active Objects and Threads

Active objects are used for lightweight event-driven multi-tasking on Symbian OS, and are fundamental for responsive and efficient event handling. They are used in preference to threads to minimize the number of thread context switches that occur, and make efficient use of system resources. This section describes what active objects are, how to use them, and how to avoid the most common programming errors that can occur. Section 4.11.9 discusses the RThread API provided by Symbian OS for occasions when multi-threaded code is unavoidable.

4.11.1 What is Event-Driven Code?

Symbian OS uses event-driven code extensively for user interaction (which has a non-deterministic completion time), and in system code, for example for asynchronous communications. To explain further – a function can be either synchronous or asynchronous:

- A *synchronous* function performs the service it offers and only then completes, returning control to the caller. A typical synchronous function is one that appends an item to a collection, such as the `RArray` class described in Section 4.9.1.
- An *asynchronous* function submits a request and immediately returns control to the caller. The service is performed and the request completes some time later, signaling the caller to notify it. This signal is known as an *event*, and the code can be said to be *event-driven*. While the caller waits for the completion event, it can perform other processing if necessary, and so be responsive, or it can wait in a low-power state. Asynchronous functions are commonly found where a service is provided across a network or otherwise takes some time to complete (including timers).

Events are managed by an *event handler*, which, as its name suggests, waits for an event and then handles it, and may re-submit a request for the asynchronous service. An operating system must have an efficient event-handling model to respond to an event as soon as possible after it occurs. It is also important that, if more than one event occurs simultaneously, the response occurs in the most appropriate order (user-driven events should be handled rapidly for a good user experience).

An event-driven system is important on a battery-powered device. The alternative to responding to events is for the submitter of an asynchronous event to keep polling to check if a request has completed. On a mobile operating system this can lead to significant power drain because a tight polling loop prevents the OS from powering down all but the most essential resources. Symbian OS is designed to avoid this, and provides the active object framework to make it easy for event handling, thus allowing it to manage its power consumption carefully by moving to an idle mode while events are pending, if no other processing is required.

4.11.2 The Active Object Framework

The *active object framework* is used on Symbian OS to simplify asynchronous programming. It is used to handle multiple asynchronous tasks in the same thread and provides a consistent way to write code to submit asynchronous requests and handle completion events.

A Symbian OS application or server will usually consist of a single main event-handling thread with an associated active scheduler that runs a loop waiting on completion events. The events are generated by asynchronous services encapsulated in active objects, each with an event-handling function that the active scheduler calls if the request associated with the active object completes.[3] So, when the asynchronous service associated with an active object completes, it generates an event, which is detected by the active scheduler. The active scheduler determines which active object is associated with the event and calls the appropriate active object to handle the event.

Pro tip

An active object encapsulates a task; it requests an asynchronous service from a service provider and handles the completion event later when the active scheduler calls it.

Within a single application thread, active objects run independently of each other, similar to the way that threads are independent of each other

[3] Windows programmers may recognize the pattern of message loop and message dispatch which drives a Win32 application. The active scheduler takes the place of the Windows message loop and the event-handling function of an active object acts as the message handler. This is discussed further in Chapter 16.

in a process. A switch between active objects in a single thread incurs a lower overhead than a thread context switch, described shortly, which is ideal for lightweight event-driven multi-tasking on Symbian OS.

4.11.3 Pre-Emption

Within a single thread, the active object framework uses non-pre-emptive multi-tasking. Once invoked, an event handler must run to completion before any other active object's event handler can run – it cannot be pre-empted.

Some events strictly require a response within a guaranteed time, regardless of any other activity in the system (e.g. low level telephony). This is called *real-time* event handling. Active objects are not suitable for real-time tasks and on Symbian OS real-time tasks should be implemented using high-priority threads. Symbian OS threads are scheduled *pre-emptively* by the kernel, which runs the highest-priority thread eligible. The kernel controls thread scheduling, allowing the threads to share system resources by time-slice division, pre-empting the running of a thread if another, higher-priority thread becomes eligible to run.

A *context switch* occurs when the current thread is suspended (for example, if it becomes blocked, has reached the end of its time-slice, or a higher priority thread becomes ready to run) and another thread is made current by the kernel scheduler. The context switch incurs a runtime overhead in terms of the kernel scheduler and, if the original and replacing threads are executing in different processes, the memory management unit and hardware caches.

4.11.4 Class `CActive`

An active object class must derive directly or indirectly from class `CActive`, defined in e32base.h. `CActive` is an abstract class with two pure virtual functions, `RunL()` and `DoCancel()`.

Construction

On construction, classes deriving from `CActive` must call the protected constructor of the base class, passing in a parameter to set the priority of the active object. Like threads, all active objects have a priority value to determine how they are scheduled. As the previous section described, once an active object is handling an event, it cannot be pre-empted until the event-handler function has returned back to the active scheduler. During this time, a number of completion events may occur. When the active scheduler next gets to run, it must resolve which active object gets to run next. It would not be desirable for a low priority active object to handle its event if a higher-priority active object was also waiting, so events are handled sequentially in order of the highest priority rather than in order of completion.[4]

A set of priority values is defined in the `TPriority` enumeration of class `CActive`. In general, the priority value `CActive::EPriorityStandard` (=0) should be used unless there is good reason to do otherwise.

In its constructor, the active object should also add itself to the active scheduler by calling `CActiveScheduler::Add()`.

Making a request and handling its completion in `RunL()`

Figure 4.8 illustrates the basic sequence of actions performed when an active object submits a request to an asynchronous service provider. An active object class supplies public methods that submit such a request, for which the standard behavior is as follows:

1. Check for outstanding requests.

   An active object must never have more than one outstanding request, so before attempting to submit a request, the active object must check to see if it is already waiting on completion. If it is, there are various ways to proceed:

   - Panic – if this scenario could only occur because of a programming error.
   - Refuse to submit another request – if it is legitimate to attempt to make more than one request but that successive requests should fail until the original one completes.
   - Cancel the outstanding request and submit the new one – if it is legitimate to attempt to make more than one request and successive requests supplant earlier pending requests. Cancellation of outstanding asynchronous requests is discussed in more detail in the following section.

2. Submit the request.

   The active object submits a request to the service provider, passing in the `TRequestStatus` member variable (`iStatus`). The service provider must set this value to `KRequestPending` before initiating the asynchronous request.

3. Call `SetActive()` to mark the object as *waiting*.
   A call to `CActive::SetActive()` indicates that a request has been submitted and is currently outstanding. This call should not be made until after the request has been submitted.

---

[4] Note that if an event associated with a high-priority active object occurs while a lower priority active object handler is executing, no pre-emption occurs. The priority value is only used to determine the order in which event handlers are run, not to re-schedule them.

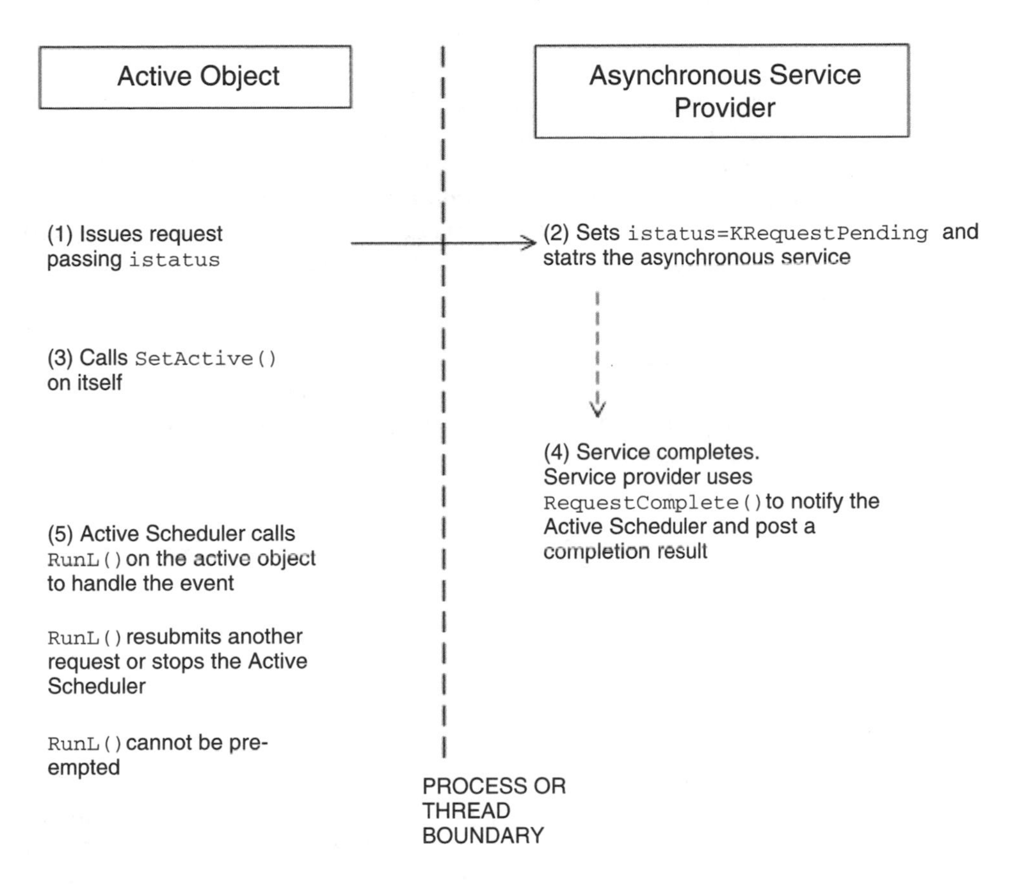

Figure 4.8 A request to an asynchronous service provider that generates an event on completion

Each active object class must implement the pure virtual `RunL()`[5] method inherited from the `CActive` base class. This is the event handler method called by the active scheduler when a completion event occurs.

`RunL()` should check whether the asynchronous request succeeded by inspecting its completion code, which is the 32-bit integer value stored in the `TRequestStatus` object (`iStatus`) of the active object. Depending on the result, `RunL()` may submit another request or perform other processing such as writing to a log file. The complexity of `RunL()` code can vary considerably.

Since `RunL()` cannot be pre-empted by other active objects' event handlers while it is running, it should complete as quickly as possible so that other events can be handled without delay.

---

[5] `RunL()` should probably be called `HandleEventL()` or `HandleCompletionL()` to make it more obvious that it is the event handler method.

DoCancel()

An active object must be able to cancel an outstanding asynchronous request. An active object class must implement the pure virtual DoCancel() method of the base class to terminate a request by calling the appropriate cancellation method on the asynchronous service provider. DoCancel() must not leave or allocate resources and should not carry out any lengthy operations, but simply cancel the request and perform any associated cleanup.

CActive::Cancel() calls DoCancel() and waits for notification that the request has terminated. Cancel() must be called whenever a request is to be terminated, not DoCancel(), since the base class method checks whether a request is outstanding and performs the necessary wait until it has terminated.

RunError()

The CActive base class provides a virtual RunError() method which the active scheduler calls if RunL() leaves. If the leave can be handled, this should be done by overriding the default implementation of CActive::RunError() to handle the exception.

Destruction

The destructor of a CActive-derived class should always call Cancel() to cancel any outstanding requests. The CActive base-class destructor checks that the active object is not currently active. It panics with E32USER-CBASE 40 if any request is outstanding, that is, if Cancel() has not been called. This catches any programming errors where a call to Cancel() has been forgotten. Having verified that the active object has no issued requests outstanding, the CActive destructor removes the active object from the active scheduler.

The reason it is so important to cancel requests before destroying an active object is that otherwise the request would complete after the active object had been destroyed. This would cause a *stray signal* because the active scheduler is unable to find a handler for the event. This results in a panic (E32USER-CBASE 46). Other reasons for receiving a stray signal are described in Section 4.11.7.

4.11.5 An Active Object Example: File System Monitoring

The following example illustrates the use of an active object class to wrap an asynchronous service: a file system change monitor that generates a completion event when the file system location specified is modified (for example, by addition of a file, modification or deletion). An asynchronous request is submitted by StartFilesystemMonitor(), that simply makes a call to the RFs::NotifyChange() method, the documentation for which you can find in the Symbian Developer Library. Section 4.12.1 discusses the basics of the Symbian OS file system in more detail.

When a change occurs in the location specified, the file system completes the `NotifyChange()` request, which in turn causes the `RunL()` event handler of class `CNotifyChange` to be invoked. This method checks the active object's `iStatus` result and leaves if it contains a value other than `KErrNone`, so that the `RunError()` method can handle the problem. In this case, the error handling is very simple: the error returned from the request is logged to debug output. This could have been performed in the `RunL()` method, but has been separated into the `RunError()` method to demonstrate how to use the active object framework to split error handling from the main logic of the event handler.

If no error occurred, the `RunL()` event handler resubmits the `RFs::NotifyChange()` request without further ado, to ensure no future changes to the file system are missed. In effect, once the initial `StartFilesystemMonitor()` request has been submitted, it continues monitoring until it is stopped by an error or a call is made to `Cancel()`.

```
class CNotifyChange: public CActive
  {
public:
  ~CNotifyChange();
  static CNotifyChange* NewL(const TDesC& aPath);
  void StartFilesystemMonitor();
protected:
  CNotifyChange();
  void ConstructL(const TDesC& aPath);
protected:
  virtual void RunL(); // Inherited from CActive.
  virtual void DoCancel();
  virtual TInt RunError(TInt aError);
private:
  RFs iFs;
  HBufC* iPath;
  };

CNotifyChange::CNotifyChange() : CActive(EPriorityStandard)
{ CActiveScheduler::Add(this); }

void CNotifyChange::ConstructL(const TDesC& aPath)
  { // Open a fileserver session.
  User::LeaveIfError(iFs.Connect());
  iPath = aPath.AllocL();
  }

CNotifyChange* CNotifyChange::NewL(const TDesC& aPath)
  {} // Standard two-phase construction code omitted
     // for clarity. See Section 4.7.1 for details.

CNotifyChange::~CNotifyChange()
  {// Stop any outstanding requests.
  Cancel();
```

```
  delete iPath;
  iFs.Close();
  }

void CNotifyChange::StartFilesystemMonitor()
  { // Only allow one request to be submitted at a time.
    // Caller must call Cancel() before submitting another.
  if (IsActive())
    {
    _LIT(KAOExamplePanic, "CNotifyChange");
    User::Panic(KAOExamplePanic, KErrInUse);
    }

  iFs.NotifyChange(ENotifyAll, iStatus, *iPath);
  SetActive(); // Mark this object active.
  }

// Event handler method.
void CNotifyChange::RunL()
  {
  // If an error occurred handle it in RunError().
  User::LeaveIfError(iStatus.Int());

  // Resubmit the request immediately so as not
  // to miss any future changes.
  iFs.NotifyChange(ENotifyAll, iStatus, *iPath);
  SetActive();

  // Now process the event as required.
  ...
  }

void CNotifyChange::DoCancel()
  { // Cancel the outstanding file system request.
  iFs.NotifyChangeCancel(iStatus);
  }

TInt CNotifyChange::RunError(TInt aError)
  { // Called if RunL() leaves
    // aError contains the leave code.
  _LIT(KErrorLog, "CNotifyChange::RunError %d");
  RDebug::Print(KErrorLog, aError); // Logs the error.
  return (KErrNone); // Error has been handled.
  }
```

The file system monitor is a useful illustration of a request that does not complete in a deterministic time, and must be asynchronous (i.e. there is no way otherwise to know when a change will occur in the location being watched). It is useful for applications that, for example, maintain a display of the files in a particular directory. If active objects were not used, the application would have to perform periodic polling on the file system to get the contents of the directory and compare it against the previous listing (which is inefficient and could unduly affect battery life if the polling were too frequent). This is a good illustration of the utility of a lightweight active object implementation, which avoids the need for polling, and only executes code when an event (in this case, a file system change) occurs.

It is also clear that the amount of code required is minimal, and easy to refactor for similar monitoring code, thus promoting code reuse. It is simple to understand and maintain, and is easier than creating a separate monitor thread to watch the location. The application and the active object code all run in a single thread so synchronization and being able to re-enter are not issues that need to be considered.

4.11.6 The Active Scheduler

Most threads running on Symbian OS have an active scheduler that is usually created and started implicitly by a framework (for example, CONE for the GUI framework). There is only one active scheduler created per thread, although it can be nested in advanced cases.

Console-based test code must create an active scheduler in its main thread if it depends on components which use active objects. The code to do this is as follows:

```
CActiveScheduler* scheduler = new(ELeave) CActiveScheduler;
CleanupStack::PushL(scheduler);
CActiveScheduler::Install(scheduler);
```

Once the active scheduler has been created and installed, its event-processing wait loop must be started by calling `CActiveScheduler::Start()`. The event-processing loop starts and does not return until a call is made to `CActiveScheduler::Stop()`. For implementation reasons, there must be at least one asynchronous request issued before the loop starts, otherwise the thread simply enters the wait loop indefinitely. Let's look at why this happens in more detail by examining the active scheduler wait loop. If you are not too bothered by the actual mechanics, you can skip on to the following section which is essential reading for anyone using active objects because it describes the common mistakes and problems that occur!

When it is not handling other completion events, the active scheduler suspends a thread by calling `User::WaitForAnyRequest()`, which waits for a signal to the thread's request semaphore. If no events are outstanding in the system, the operating system can power down to sleep.

When an asynchronous server has finished with a request, it indicates completion by calling `User::RequestComplete()` (if the service provider and requestor are in the same thread) or `RThread::RequestComplete()` otherwise. The `TRequestStatus` associated with the request is passed into the `RequestComplete()` method along with a completion result, typically one of the standard error codes. The `RequestComplete()` method sets the value of `TRequestStatus` to the given error code and generates a completion event in the requesting thread by signaling a semaphore.

When a signal is received and the thread is next scheduled, the active scheduler determines which active object should handle it. It checks the priority-ordered list of active objects for those with outstanding requests (these have their `iActive` Boolean set to `ETrue` as a result of calling `CActive::SetActive()`). If an object has an outstanding request, the active scheduler checks its `iStatus` member variable to see if it is set to a value other than `KRequestPending`. If so, this indicates that the active object is associated with the completion event and that the event handler code should be called. The active scheduler clears the active object's `iActive` Boolean and calls its `RunL()` event handler.

Once the `RunL()` call has finished, the active scheduler re-enters the event processing wait loop by issuing another `User::WaitForAnyRequest()` call. This checks the thread's request semaphore and either suspends it (if no other events need handling) or returns immediately to lookup and event handling.

The following pseudo-code illustrates the event-processing loop.

```
EventProcessingLoop()
  {
  // Suspend the thread until an event occurs.
  User::WaitForAnyRequest();
  // Thread wakes when the request semaphore is signaled.
  // Inspect each active object added to the scheduler,
  // in order of decreasing priority.
  // Call the event handler of the first which is active
  // and completed.
  FOREVER
    {
    // Get the next active object in the priority queue.
    if (activeObject->IsActive())&&
                    (activeObject->iStatus!=KRequestPending)
      { // Found an active object ready to handle an event.
        // Reset the iActive status to indicate not active.
      activeObject->iActive = EFalse;

      // Call the active object's event handler in a TRAP.
      TRAPD(r, activeObject->RunL());
      if (KErrNone!=r)
        { // Event handler left,
          // call RunError() on active object.
        r = activeObject->RunError();
        if (KErrNone!=r) // RunError() didn't handle error
          Error(r);       // call CActiveScheduler::Error()
        }
      break; // Event handled, break out of lookup loop and resume.
      }
    } // End of FOREVER loop
  }
```

4.11.7 Common Problems with Active Objects

Stray signal panics

The most common problem when writing active objects is when you encounter a stray signal panic (`E32USER-CBASE 46`). These occur when an active scheduler receives a completion event but cannot find an active object to handle it. Stray signals can occur because:

- `CActiveScheduler::Add()` was not called when the active object was constructed
- `SetActive()` was not called after submitting a request to an asynchronous service provider
- the asynchronous service provider completed the request more than.

If you receive a stray signal panic, the first thing to do is to work out which active object is responsible for submitting the request that later generates the stray event. One of the best ways to do this is to use file logging in every active object and, if necessary, eliminate them from your code one by one until the culprit is tracked down.

Unresponsive UI

In an application thread, in particular, event-handler methods must be kept short to allow the UI to remain responsive to the user. No single active object should have a monopoly on the active scheduler since that prevents other active objects from handling events. Active objects must cooperate and should not:

- have lengthy `RunL()` or `DoCancel()` methods
- repeatedly resubmit requests that complete rapidly, particularly if the active object has a high priority, because the event handler will be invoked at the expense of lower-priority active objects waiting to be handled
- have a higher priority than is necessary.

Other causes of an unresponsive UI are temporary or permanent thread blocks that result because of:

- a call to `User::After()`, which stops the thread executing for the length of time specified
- incorrect use of the active scheduler. There must be at least one asynchronous request issued, via an active object, before the active

scheduler starts. If no request is outstanding, the thread simply enters the wait loop and sleeps indefinitely

- incorrect use of `User::WaitForRequest()` to wait on an asynchronous request, rather than correct use of the active object framework.

4.11.8 Background Tasks

Active objects can also be used to implement tasks that can run in chunks at low priority in the background. This avoids the need to create a separate thread. The task is divided into multiple increments, for example performing background recalculations. The increments are performed in the event handler of a low-priority active object, which is why they must be short, since `RunL()` cannot be pre-empted once it is running. The active object must track the progress of the task, using a series of states, if necessary, to implement a state machine.

The active object must be assigned a very low priority such as `CActive::TPriority::EPriorityIdle(=-100)`, to ensure that the task only runs when there are no essential events, such as user input, to handle.

The following simple example code illustrates a basic background task (only the relevant code is shown; two-phase construction, destruction and most error handling are omitted for simplicity).

A basic T class performs the background task, in this case some kind of calculation (for clarity, this code is omitted). The class provides an API to start the task, perform the various discrete task steps using a state machine to track the stages and a cancellation method to stop the calculation.

The active object class, `CBackgroundTask`, drives the task by generating events in `SelfComplete()` to invoke the `RunL()` event handler when the active scheduler has no higher priority active object to run. It does this by calling `User::RequestComplete()` on its own `iStatus` object.

In the event handler, `CBackgroundTask` checks whether there are any further steps to run by calling `TLongRunningCalculation::ContinueTask()`. If there are, it performs a step and sends another self completion event, continuing to do so until the task is complete. When the calculation is complete, the caller is notified by generating a completion event.

```
class TLongRunningCalculation
  {
public:
  TLongRunningCalculation() {iState=EWaiting;};
  void StartCalculation(); // Initialization before starting the task.
  void DoTaskStep();     // Perform a short task step.
  TBool ContinueTask(); // Returns ETrue if more left to do.
  void CancelCalculation() {iState=EWaiting;};
```

```
private:
  enum TCalcState
  { EWaiting, EBeginState, EIntermediateState, EFinalState };
  TCalcState iState;  // Keeps track of the current calculation state.
  };

_LIT(KExPanic, "Example Panic");

void TLongRunningCalculation::StartCalculation()
  {
  // Flag a programming error if it's already running.
  __ASSERT_DEBUG(iState==EWaiting, User::Panic(KExPanic, KErrInUse));
  iState=EBeginState;
  }

// State machine.
void TLongRunningCalculation::DoTaskStep()
  {
  // Do a short task step.
  switch (iState)
    {
  case (EWaiting):
    iState = EBeginState; break;
  case (EBeginState):
    iState = EIntermediateState; break;
  case (EIntermediateState):
    iState = EFinalState; break;
  case (EFinalState):
    iState = EWaiting; // Finished
    break;
  default:
    ASSERT(EFalse);   // Cause a panic! Should never get here.
    }
  }

// Return ETrue if there are further steps to run.
TBool TLongRunningCalculation::ContinueTask()
  {
  return (iState!=EWaiting);
  }

class CBackgroundTask : public CActive
  {
public:
  // Public method to kick off a background calculation.
  void PerformCalculation(TRequestStatus& aCallerStatus);
protected:
  CBackgroundTask();
  void SelfComplete();
  virtual void RunL();
  virtual void DoCancel();
private:
  TLongRunningCalculation iCalc;
  TBool iMoreToDo;
  TRequestStatus* iCallerStatus; // To notify caller on completion.
  };

CBackgroundTask::CBackgroundTask()
```

```
  : CActive(EPriorityIdle) // Low priority task
  { CActiveScheduler::Add(this); }

// Issue a request to initiate a lengthy task.
void CBackgroundTask::PerformCalculation(TRequestStatus& aStatus)
  {
  // Save the parameter to notify when complete.
  iCallerStatus = &aStatus;
  *iCallerStatus = KRequestPending;
  __ASSERT_DEBUG(!IsActive(), User::Panic(KExPanic, KErrInUse));
  iCalc.StartCalculation(); // Start the task.
  SelfComplete(); // Self-completion to generate an event.
  }

void CBackgroundTask::SelfComplete()
  {
  // Generate an event by completing on iStatus.
  TRequestStatus* status = &iStatus;
  User::RequestComplete(status, KErrNone);
  SetActive();
  }

// Perform the background task in increments.
void CBackgroundTask::RunL()
  {
  // Resubmit request for next increment of the task
  // or stop.
  if(iCalc.ContinueTask())
    {
    iCalc.DoTaskStep();
    SelfComplete();
    }
  else
    User::RequestComplete(iCallerStatus, iStatus.Int());
  }

void CBackgroundTask::DoCancel()
  {
  // Call iCalc to cancel the task.
  iCalc.CancelCalculation();

  // Notify the caller of completion (through cancellation).
  User::RequestComplete(iCallerStatus, KErrCancel);
  }
```

4.11.9 Threads

In many cases on Symbian OS, it is preferable to use active objects rather than threads, since these are optimized for event-driven multi-tasking on the platform. However, when you are porting code written for other platforms or writing code with real-time requirements, it is often necessary to write multi-threaded code.

The Symbian OS class used to manipulate threads is `RThread` that represents a handle to a thread; the thread itself is a kernel object. The base class of `RThread` is `RHandleBase`, that encapsulates the behavior of a generic handle and is used as a base class throughout Symbian OS to identify a handle to another object, often a kernel object.

Class `RThread` defines several functions for thread creation, each of which takes a descriptor for the thread's name, a pointer to a function in which thread execution starts, a pointer to data to be passed to that function (if any), and the stack size of the thread, which defaults to 8 KB. `RThread::Create()` is overloaded to allow for different heap behavior, such as definition of the thread's maximum and minimum heap size or whether it shares the creating thread's heap or uses a separate heap.

When the thread is created, it is assigned a unique thread identity, which is returned by the `Id()` function of `RThread`. If the identity of an existing thread is known, it can be passed to `RThread::Open()` to open a handle to that thread, or alternatively the unique name of a thread can be passed to open a handle to it.

A thread is created in the suspended state and its execution initiated by calling `RThread::Resume()`. On Symbian OS, threads are pre-emptively scheduled and the currently running thread is the highest-priority thread ready to run. If there are two or more threads with equal priority, they are time-sliced on a round-robin basis. A running thread can be removed from the scheduler's ready-to-run queue by a call to `RThread::Suspend()`. It can be scheduled by calling `Resume()`. To terminate a thread permanently, you should call `Kill()` or `Terminate()` to stop it normally and `Panic()` to highlight a programming error. If any of the termination methods are called on the main thread of a process, it is also terminated. For this reason, and because of the secure platform, it is not possible for a thread to suspend or terminate another thread in a different process. A thread can only call the functions to halt itself or other threads in the same process.

Symbian OS provides several kernel object classes for thread synchronization:

- A semaphore can be used either for sending a signal from one thread to another or for protecting a shared resource from being accessed by multiple threads at the same time. On Symbian OS, a semaphore is created and accessed with a handle class called `RSemaphore`. A global semaphore can be created, opened and used by any process in the system, while a local semaphore can be restricted to all threads within a single process. Semaphores can be used to limit concurrent access to a shared resource, either to a single thread at a time, or allowing multiple accesses up to a specified limit.
- A mutex is used to protect a shared resource so that it can only be accessed by one thread at a time. On Symbian OS, the `RMutex` class is used to create and access global and local mutexes.

- A critical section is a region of code that should not be entered simultaneously by multiple threads. An example is code that manipulates global static data, since it could cause problems if multiple threads change the data simultaneously. Symbian OS provides the `RCriticalSection` class that allows only one thread within the process into the controlled section, forcing other threads attempting to gain access to that critical section to wait until the first thread has exited from the critical section. `RCriticalSection` objects are always local to a process, and a critical section cannot be used to control access to a resource shared by threads across different processes – a mutex or semaphore should be used instead.

Please see the Symbian Developer Library for more information about working with Symbian OS threads. Example code is also provided in the SDK to show the use of `RThread` and other Symbian OS system classes.

4.12 The Client–Server Framework

Most of the system services on Symbian OS are provided using a client–server framework. This means that a client uses a library to interact with a server and use the services it provides. The client sends messages to the server which handles them, either synchronously or asynchronously.

Servers are used to manage shared access to system resources and services. Symbian OS system servers used in this way include the file system server (which coordinates access to the file system) and the serial communications server (for access to the serial ports). The server can service multiple client sessions and be accessed concurrently by clients running in separate threads.

Servers also protect the integrity of the system, acting as a buffer between clients and the resource in question. The server ensures that the resource is shared properly between clients and that those clients use the resource securely and correctly. A Symbian OS server running in a separate process can guarantee that malicious clients in other processes cannot corrupt the resource it manages or crash the server by overwriting data, because the hardware architecture provides memory protection. That is, the client and server cannot access each other's virtual address spaces.

Even if it does not run in a separate process to its clients, a server will always run in a separate thread, and all client–server communication takes place by messages that are mediated by the kernel. The active object framework is used on both sides of the framework. The server has an active object that receives notification when a request is made by a client. A client making an asynchronous request uses an active object to receive notification when the request has completed. As Section 4.11 described, the use of active objects allows the threads to suspend themselves when they are not active. This helps to improve power management by removing the need for a tight polling loop, which is important for optimizing the battery life of the smartphone.

The communication channel for passing messages between the client and server is known as a session. Sessions are initiated by a client and when a session is open, the client holds a handle to identify it, typically in an R class object. The client's messages to the server are known as requests; each request can contain parameter data packaged up and passed to the server. Simple data can be passed with the request, while more complex data is accessed by the server using inter-thread data transfer

To hide the details of the message passing and data packaging from application code, a typical server has associated client-side code to format requests and pass them to the server, via the kernel. This is usually provided in a separate library. For example, `mytest.exe`, an application which is a *client* of the Symbian OS file server (`efile.exe`) accesses the file system by linking against (`efsrv.dll`) which is the file server's client-side implementation library. This is illustrated in Figure 4.9. The client-side access library is an efficient way of providing access to a server by code re-use, rather than expecting every caller to implement their own message passing code.

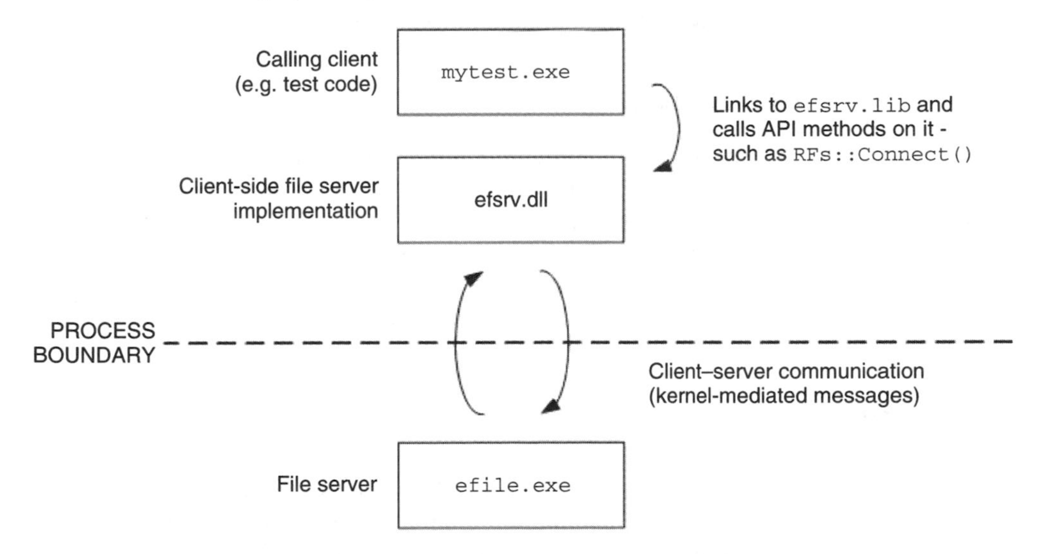

Figure 4.9 The file system server, client library and calling code

4.12.1 The Symbian OS File System Server

The file system server, or simply the *file server*, provides access to files and directories on the phone's internal and removable storage, providing a consistent interface regardless of the physical location. In order to use the file server, you must first open a file server session by using class `RFs`, as shown in the following example.

```
RFs fs;
User::LeaveIfError(fs.Connect()); // Connect the session.

// Ensure the handle is safe if a leave occurs.
CleanupClosePushL(fs);

// Use the handle to access files, directories, etc.
...

// Call RFs::Close on fs via the cleanup stack.
CleanupStack::PopAndDestroy(&fs);
```

An open `RFs`session can be used to access any number of files or directories, or to perform any other file-related operations such as the file system change monitoring demonstrated in Section 4.11.5. A file server session can be kept open indefinitely, and since connections to the file server can take a significant amount of time to set up, if the file system is used throughout an application's lifetime, it makes sense to store and re-use an open session where possible. Rather than opening multiple `RFs`sessions throughout the application, for efficiency, it is good practice to pass a previously opened handle between functions and classes, as long as it is cleaned up when no longer needed.

4.13 System Information

This section contains information about various aspects of C++ development on Symbian OS, including a description of the two common types of libraries used to share and re-use code and how to use and acquire unique identifiers (UIDs) for your binaries.

On Symbian OS, there are two main types of dynamic link library (DLL) for implementing code that may be re-used by other components: shared library DLLs and polymorphic interface DLLs.

4.13.1 Shared Library (Static-Interface) DLLs

A shared library exports any number of API functions, each of which is an entry point into the DLL. The library releases a header file (`.h`) for other components to compile against and an import library (`.lib`) to link against. The filename extension of a shared library is `.dll` – examples include the user library (`EUser.dll`) and the file system library (`EFile.dll`).

4.13.2 Polymorphic Interface DLLs

A polymorphic interface DLL implements an interface that is usually defined separately, for example by a framework. This type of DLL is often used to extend the framework when a range of different implementations of an interface are possible, each of which is known as a plug-in. The plug-ins are loaded dynamically by the framework.

Polymorphic DLLs have a single entry-point function, usually called a *gate* or *factory* function. This has responsibility for creating an instance of the concrete class that implements the interface. It may have a DLL filename extension, but a different extension is often used to identify it as a particular type of plug-in, for example `.fsy` for a file system plug-in or `.prt` for a protocol module plug-in.

In Symbian OS v9, the most common types of plug-in are ECOM plug-ins. ECOM is a generic framework for locating and loading plug-ins, and many Symbian OS frameworks require their plug-ins to be written as ECOM plug-ins, thus avoiding the need for them to write custom code for plug-in identification and loading.

4.13.3 Unique Identifiers (UIDs)

On Symbian OS, a UID is a 32-bit value used to identify a file type, and a combination of up to three UIDs are used to identify a binary executable file uniquely.

UID1

This is the system-level identifier which distinguishes between EXEs and DLLs. The value is built into code by the build tools depending on the keyword used with the `targettype` identifier in the MMP file.

For example, the `targettype` specified for a shared library is `DLL`. This means that UID1 is set by the build tools to be `KDynamicLibraryUid` (`0x10000079`). An application has UID1 set to be `KExecutableImageUid` (`0x1000007a`) by specifying a `targettype` of `EXE`.

UID2

This UID is used to differentiate between shared library and polymorphic interface DLLs. For shared libraries, UID2 is `KSharedLibraryUid` (`0x1000008d`) but the value for polymorphic DLLs varies depending on their plug-in type (for example, the socket server protocol module UID2 value is `0x1000004A`).

UID2 is not relevant for `targettype` EXE and can simply be set to 0 or not specified at all in the MMP file. In some examples, particularly those that have been ported from previous versions of UIQ, UID2 is often set to `0x100039CE` for an application. This is because, prior to Symbian OS v9, applications were polymorphic DLLs that were loaded dynamically by a framework that used the UID2 value to identify them. Now that applications are executables, UID2 is no longer relevant. It does not cause a problem to use the legacy value, but it is better simply not to specify it or use 0.

UID3

The third UID value identifies a file uniquely. To ensure that each binary has a different value, Symbian manages UID allocation through a central database. You must register with Symbian Signed to request UIDs for use as UID3 values in production code, although you can assign yourself a value from the development range for test code. The choice of UID range to use for your application is discussed in Section 10.1.2 but we list the full range in Table 4.4 for completeness. The shaded ranges are allocated on request by Symbian Signed (*www.symbiansigned.com*).

Table 4.4 Ranges of UID3

UID Range	Intended Use	Type
0x00000000	`KNullUID`	Protected
0x00000001 – 0x0FFFFFFF	Reserved	Protected
0x10000000 – 0x1FFFFFFF	Legacy allocated UIDs, not for Symbian OS v9	Protected
0x20000000 – 0x2FFFFFFF	UID3/SID range	Protected
0x30000000 – 0x6FFFFFFF	Reserved	Protected
0x70000000 – 0x7FFFFFFF	Vendor IDs	Protected
0x80000000 – 0x9FFFFFFF	Reserved	Unprotected
0xA0000000 – 0xAFFFFFFF	UID3/SID range	Unprotected
0xB0000000 – 0xE0FFFFFF	Reserved	Unprotected
0xE1000000 – 0xEFFFFFFF	Development/testing range	Unprotected
0xF0000000 – 0xFFFFFFFF	Legacy UID compatibility range	Unprotected

4.13.4 Memory Management Macros

Symbian OS provides a set of debug-only macros that can be added directly to code to check that memory is not leaked.

There are a number of macros available, but the most commonly used are defined as follows:

```
#define __UHEAP_MARK User::__DbgMarkStart(RHeap::EUser)
#define __UHEAP_MARKEND User::__DbgMarkEnd(RHeap::EUser,0)
```

The macros verify that the default user heap is consistent. The check starts after a call to `__UHEAP_MARK` and a later call to `__UHEAP_MARKEND` verifies that the heap is in the same state as at the start. If heap cells allocated in the interim have not been freed, a panic is raised to indicate a potential leak. The panic is `ALLOC nnnnnnnn`, where `nnnnnnnn` is a hexadecimal pointer to the heap cell in question.

The heap-checking macros can be nested used anywhere in code. They are not compiled into release builds so do not have any impact on the code size or speed of production code. Sometimes you may find that you get *false positives* where memory is allocated by the system and not freed within a `MARK/MARKEND` pair. A good example is the cleanup stack which is expanded to store pointer values where necessary, but when they are popped off, the allocated memory is retained and re-used later or destroyed when the cleanup stack is deleted.

In addition, an application must gracefully handle any out of memory leaves that occur when memory is not available for allocation. Symbian OS provides a set of macros to simulate heap failure, either on the next allocation, randomly or by allowing you to specify the number of allocations that can succeed before failure. These are extremely useful: for example, a test suite can call a function repeatedly, increasing the heap failure point by one, to check that each allocation fails gracefully. Please consult the Symbian Developer Library documentation for more information about `__UHEAP_FAILNEXT` and `__UHEAP_SETFAIL` for more information about how to use them.

4.14 Platform Security

Platform security prevents applications from unauthorized access to hardware, software and system or user data on Symbian OS v9. The intention is to prevent malware, or even just badly written code, from compromising the operation of the phone, corrupting or stealing confidential user data, or adversely affecting the phone network.

4.14.1 Trust

Symbian OS defines the process as the smallest unit of trust, because it is the unit of memory protection on Symbian OS. The phone hardware raises a processor fault if access is made in one process to the virtual address space of another process. Thus, Symbian OS, assisted by the hardware architecture, can control what a process can and cannot do, and will not allow it to access a service if it is not deemed trustworthy enough.

4.14.2 Trusted Computing Base (TCB)

The TCB is the most trusted part of Symbian OS, and controls the security mechanisms themselves, with responsibility for maintaining the operating system's integrity. The TCB includes the kernel, file server (because it is used to load program code into a process) and the software installer (SWInstall) on phones that are *open* to after-market installable software.

4.14.3 The Trusted Computing Environment (TCE)

The TCE consists of trusted software built into the mobile phone by Symbian, the UI platform provider (UIQ) and phone manufacturers like Sony Ericsson and Motorola. The TCE code is considered trustworthy but does not have the same level of privilege as the TCB. The TCE provides a layer by which applications can access lower-level services and each component is trusted to perform the set of services it provides, but only those. For example, the window server and the telephony server are part of the TCE. The window server has privileged access to the screen hardware but has no need to access the phone network. Conversely, the telephony server (ETEL) has privileged access to the communications device driver but has no need to access the screen hardware.

4.14.4 Signed Software

Most application software lies outside the TCE, but needs certain privileges to use the services the TCE provides. For example, an application that uses a network socket uses the Symbian OS socket server, ESOCK, which is part of the TCE, to do so. ESOCK will check that its client is trusted, before granting a request – trust is conferred when software is digitally signed.

Signing does not require full code inspection, but it does require that the general behavior of an application is tested, that the developer makes certain guarantees about how they use sensitive APIs and that the developer is identified by means of a Certificate Authority identifier (Publisher ID). This is the basis of Symbian Signed scheme, which is discussed in detail in Chapter 14.

4.14.5 Untrusted Software

The trustworthiness of software that is not signed by a trusted Certificate Authority cannot be determined. It does not mean that the software is necessarily malicious or worthless, but simply that it can only access certain levels of Symbian OS, and can have no impact on the integrity of the phone, network or user information. Untrusted software can be installed and run on the phone but it is effectively restricted in what it can do. This may sound limited, but there are many useful operations that can be performed because security checks protect only about 40 % of all Symbian OS APIs.

4.14.6 Capabilities

On Symbian OS, every process is assigned a level of privilege, through a set of capabilities which are like tokens. Holding a capability indicates that the process is trusted not to abuse the services associated with that privilege. The kernel holds a list of capabilities for every running process and checks it before allowing a process to access a protected service. The software installer verifies the digital signature of software at install time to ensure that it is authorized to use the capabilities it was built with, and the digital signature can only come from a trusted authority. This prevents applications from arbitrarily assigning themselves the capabilities they want prior to installation.

As Chapter 14 describes, there are three types of capabilities: User capabilities, System and Device Manufacturer capabilities.

User capabilities

User capabilities (sometimes referred to as user-grantable capabilities) are intended to be meaningful to a phone user. For example, a user can be comfortable about deciding whether to install software with the capability to make network connections, or access their contacts data.

Installable software that needs only user-grantable capabilities does not need to be signed by a trusted authority, because the user can make the decision instead. A user should not, however, make decisions about capabilities which affect whether the phone works properly, that is, capabilities that affect system services.

System capabilities

A user should not make decisions about granting trust to software if this could affect whether the phone works properly. System-level capabilities are for this, and allow a process to access sensitive operations, misuse of which could threaten the integrity of the phone. System capabilities are hidden from the user, so applications that need these capabilities must be tested and signed by a trusted authority.

Device Manufacturer capabilities

Some capabilities are so sensitive that they are granted by the mobile phone manufacturer.

4.14.7 Capability Rules

> **Pro tip**
>
> The capabilities of a process cannot be changed at run time.

The Symbian OS loader starts a process by reading the executable and checking the capabilities it has been assigned. Once the process starts running, it cannot change its capabilities, nor can the loader or any other process or DLL that loads into it affect its capability set.

> **Pro tip**
>
> A process can only load a DLL if that DLL is trusted with at least the same capabilities as the process.

A process can load a DLL which has been assigned more capabilities than itself, but that DLL will only be allowed to use the capabilities of the process in which it runs. Since this is possible, DLL code cannot assume that it will always be running with the capabilities it needs, since these are dependent on the process it has been loaded into. If a DLL requires a particular capability, this should be clearly specified in its documentation.

A process cannot load a DLL that has been assigned fewer capabilities than itself. This makes sense because, if it does not have them, it indicates that the DLL cannot be trusted to run securely within the process, and may prevent it from functioning.

This is illustrated in Figure 4.10 (capabilities of the binaries are indicated as 'Cn').

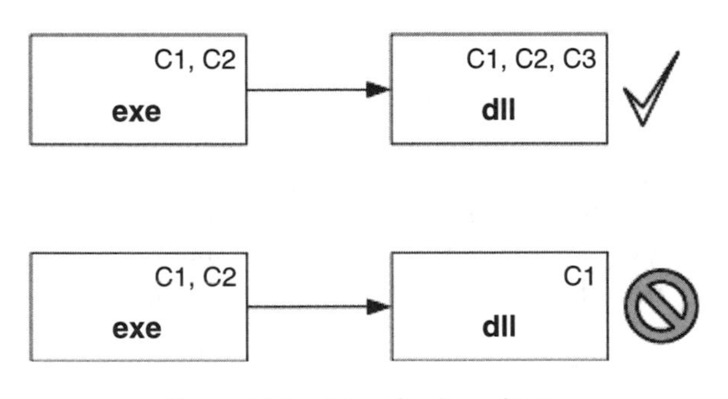

Figure 4.10 Direct loading of DLLs

4.14.8 Data Caging

The Symbian OS file system is partitioned to protect system files (critical to the operation of the phone), application data (to prevent other applications from stealing copyrighted content or accidentally corrupting data) and data files personal to the user (which should remain confidential). This partitioning is called data caging. It is not used on the entire file system; there are some public areas for which no capabilities are required.

The private areas are to be found under three top-level directories: `\sys`, `\resource` and `\private` and access restrictions apply to these directories and their subdirectories.

\sys

The `\sys\bin` directory is where all executables are stored (ROM files are in `z:\sys\bin`, while installed software is written into the equivalent directory on the c: drive or removable media). Binary code stored elsewhere cannot be executed, so all applications and DLLs are stored here. As Chapter 10 describes, this has implications for name clashing; you should not give your binaries such generic names that they may already exist in `\sys\bin` because if you do, your code will fail to install.

Only code within the TCB has read/write access to `\sys` and code with `AllFiles` capabilities has read access to `\sys`.[6] Only the most trusted code can therefore access the executables, preventing accidental or deliberate corruption.

\resource

The `\resource` stores read-only resources such as bitmaps, fonts and help files. Only the TCB can write into this directory, to prevent corruption. No capabilities are required to read from this directory.

\private

Each process has its own caged file system area as a subdirectory of `\private` on every drive. The subdirectory is identified by the SID of the EXE, which is discussed shortly. Only processes with the matching SID or those with `AllFiles` capability can read from or write to the subdirectory. There is one exception to this, which is for installation where, for example, a plug-in needs to put data into the private directory of the framework in which it will run. This is possible by building it into a special `\import` subdirectory, as long as the owner of the directory has already created it.

Note that a DLL does not have its own `\private` directory but uses that of the loading process, which can be discovered at run time by calling `RFs::PrivatePath()`.

---

[6] Section 14.3 lists the Symbian OS capabilities and describes them in more detail.

The capabilities required for access to the protected areas of the file system are summarized in Table 4.5.

`AllFiles` allows read access to the whole file system and write access to all subdirectories of `\private`, but no write access to `\sys` or `\resource`. For this, `Tcb` capability is required.

Table 4.5 File system capabilities

Directories and subdirectories (all drives)	**Reading**	**Writing**
`\sys`	`AllFiles`	`Tcb`
`\resource`	None needed	`Tcb`
`\private\<ownSID>`	None needed	None needed
`\private\<otherSID>`	`AllFiles`	`AllFiles`
`\other` (e.g. `c:` or `c:\temp`)	None needed	None needed

4.14.9 Secure Identifier (SID)

A Secure Identifier (SID) is required to identify each EXE on the phone and is used to create its private directory. The SID is similar to the UID3 identifier and the default value of the SID is the UID3 value if the SID is not explicitly specified by use of the SECUREID keyword in the MMP file.

> **Pro tip**
>
> It is usually recommended not to specify a SID but simply allow it to default to the UID3 value as assigned by Symbian Signed.

4.15 Acknowledgements

Sections of this chapter include material that is reproduced from the following titles:

Symbian OS Explained: Effective C++ Programming for Smartphones (2004) John Wiley & Sons

The Accredited Symbian Developer Primer: Fundamentals of Symbian OS (2006) John Wiley & Sons

Symbian OS C++ for Mobile Phones, Volume 3 (2007) John Wiley & Sons.

5

Understanding User Interface Components

In Chapter 3, we built a simple application, `QuickStart`, and introduced the application framework. In this chapter we present some important concepts in building a user interface: controls, windows, user input events and views. We also examine the parts of the UIQ 3 screen in more detail. In the next chapter, we will put this all together and start programming list boxes with UIQ-specific controls.

5.1 Controls and Windows

5.1.1 What Are Controls?

Controls are the basic elements used by applications to interact with users. They include such items as buttons, menus, list boxes and views. In Symbian OS, all controls are derived from the `CCoeControl` base class. This ensures that all controls present a common set of features and capabilities; for example, position and size.

Controls occupy a rectangular area on screen (see Figure 5.1). They usually display some text or an image to the user and perhaps respond to user, application or system-generated events such as key presses or pointer events.

Control Types

Simple controls perform very basic tasks, such as displaying lines, text and images.

We need much more complicated controls, but rather than writing each control afresh, we can combine simple controls together to make

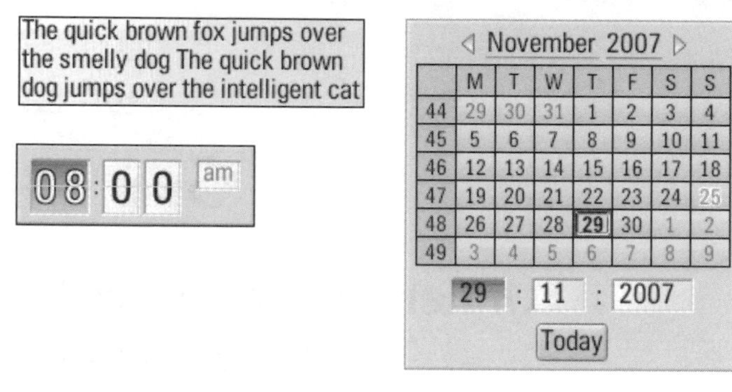

Figure 5.1 Simple text display control, time and date editor controls

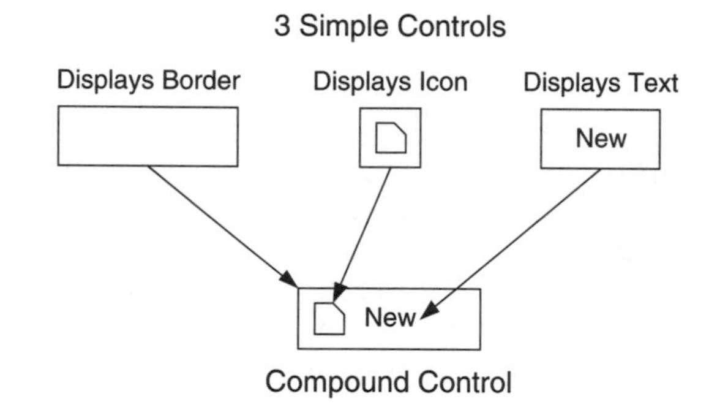

Figure 5.2 Building a compound control from simple controls

a compound control. For example, we can build a compound button control from three simple controls (see Figure 5.2).

In a sophisticated GUI such as UIQ 3, more complicated effects are needed to provide an attractive display. A button needs a graphical background and needs to look different when it is clicked (see Figure 5.3).

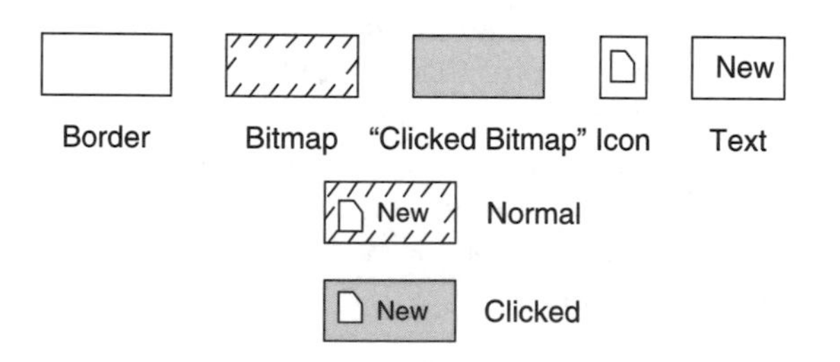

Figure 5.3 Building a button control

Compound controls can themselves be incorporated into other controls. This enables a tree-like structure of controls to be constructed (see Figure 5.4). A compound control (for example, C3) can be made up from other compound controls (C1 and C2) and some simple controls (S2 and

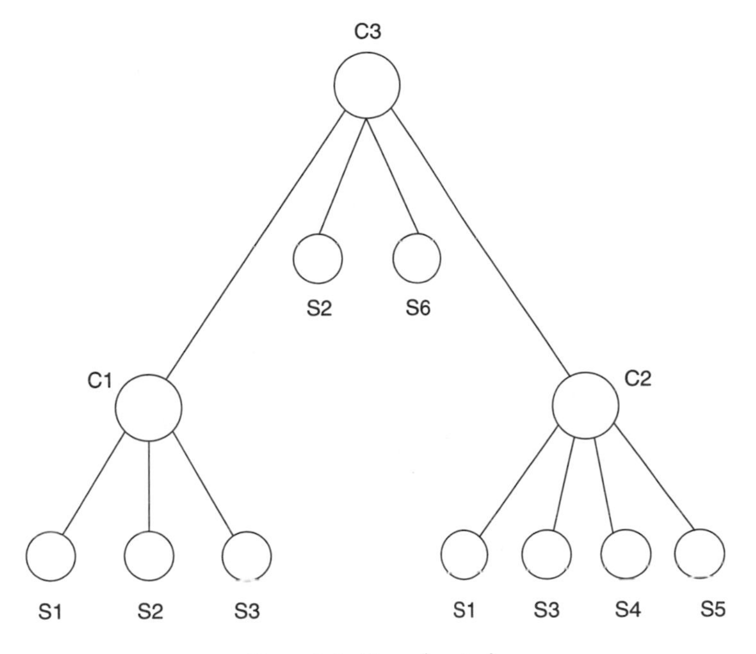

Figure 5.4 Tree of controls

S6). Following the tree, we find simple controls at the tips. For example, S1 could be a simple text display control that is utilized in C1 and C2.

The primary advantage to this approach is component re-use. Symbian OS only needs a single text control or a single image control irrespective of how it is incorporated into more substantial user interface controls.

As a second example, a menu could be considered to be a bordered control. Within the menu border, the textual menu items and possibly a graphic image is displayed. The exact same controls used to display text and graphics in a button can be reused to display a menu (see Figure 5.5).

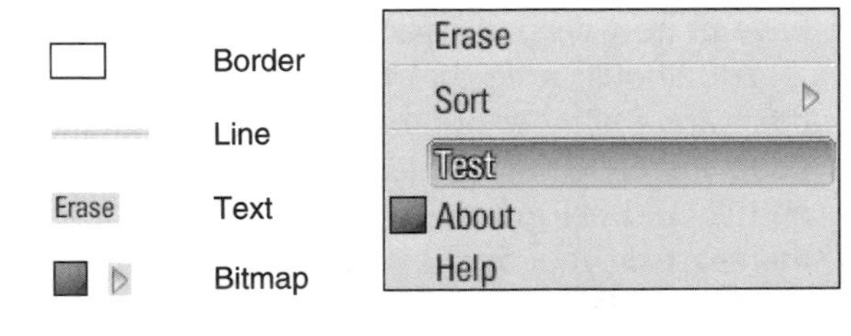

Figure 5.5 Constructing a menu

Drawing Controls

For controls to be visible on the screen, they need to be drawn. The following pseudo-code shows approximately how drawing is handled within `CCoeControl` when an application calls the `DrawNow()` method. In particular, it demonstrates how compound controls are drawn first, when their `Draw()` methods are called, then any component controls are drawn.

```
void CCoeControl::Draw(const TRect& aRect) const
  {
  // Implement the control specific drawing here.
  }

void CCoeControl::DrawComponents(const TRect& aRect) const
  {
  TInt count=CountComponentControls();
  for (TInt i=0;i<count;i++)
    {
    CCoeControl* ctrl=ComponentControl(i);
    TRect rect(ctrl->Rect());
    rect.Intersection(aRect);
    if (!rect.IsEmpty())
      {
      ctrl->Draw(rect);
      // Recursively call this method.
      ctrl->DrawComponents(rect);
      }
    }
  }
void CCoeControl::DrawNow() const
  {
  TRect rect(Rect());
  ActivateGc();
  Window().Invalidate(rect);
  Window().BeginRedraw(rect);
  Draw(rect);
  DrawComponents(rect);
  Window().EndRedraw();
  DeactivateGc();
  }
```

The `DrawComponents()` method demonstrates that there are no specific limits as to how many components an individual control may have or how deep the nesting of controls can be.

You should note that this is not the exact code that is executed as there are a number of other conditions that need to be checked for. The code merely demonstrates the recursive nature of compound controls and should help in the understanding of the tree-like control structure that exists in your application.

5.1.2 What Are Windows?

The Window Server

Symbian OS is a multi-tasking operating system. This means that more than one application can be running at the same time. A system process called the window server exists to manage shared access to the screen and input hardware, so that user interaction events such as key presses or pen events are delivered to the correct application.

Individual applications are said to be clients of the window server (see Figure 5.6). The window server communicates with its clients by sending them events via the system standard inter-process communication infrastructure. Clients listen out for these events using an active object.

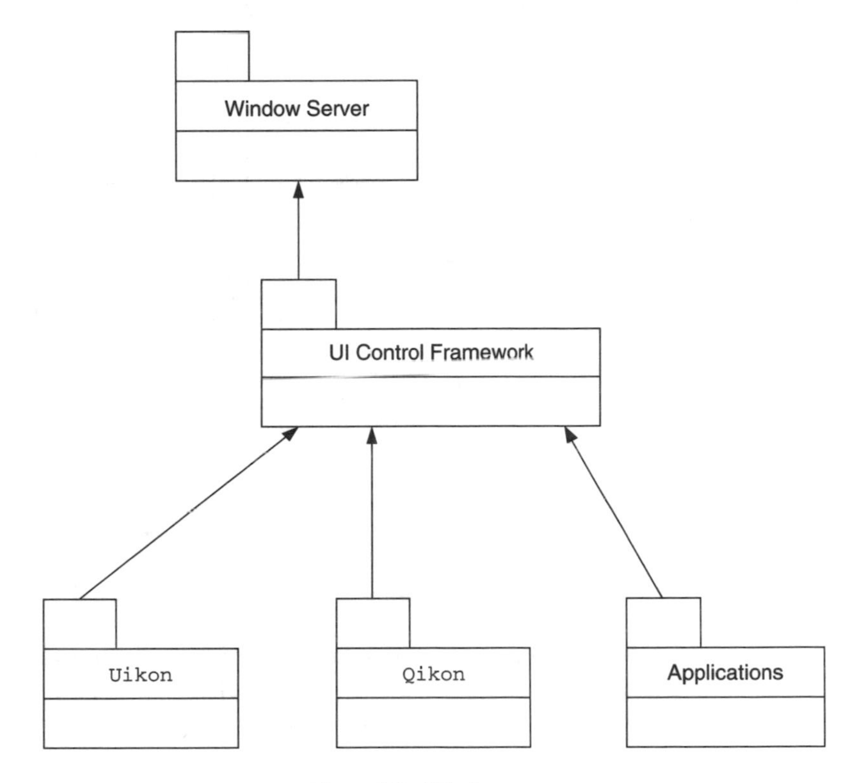

Figure 5.6 Window server

Controls and Windows

A window is a resource owned by the window server (`WSERV`) and represents a rectangular area of the screen. Controls exist inside windows (see Figure 5.7). While every control could have its own window, such an arrangement is very resource intensive and, in general, many controls share a single window. Each control must be able to reference the window in which it exists.

Controls that own a window are called window-owning controls and use one of the `CreateWindowL()` methods during their construction. In comparison, a non-window-owning control requires you to call `SetContainerWindowL()` to specify the window this control shares. In terms of functionality, there is little difference between whether a control

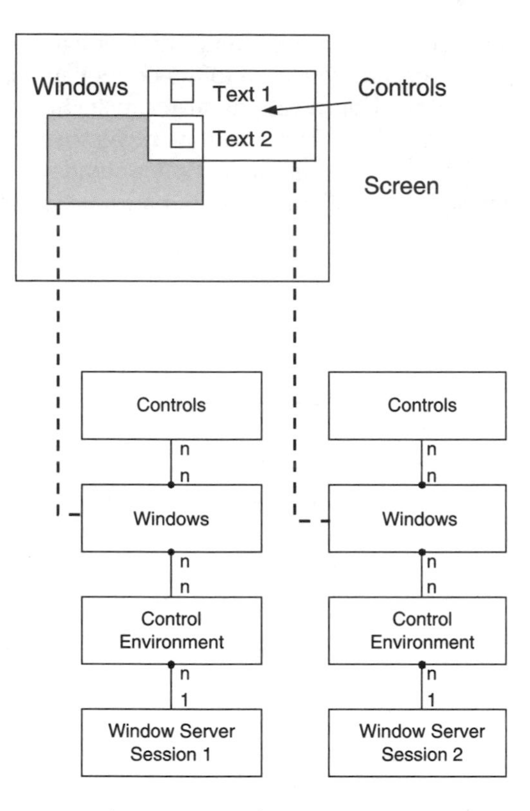

Figure 5.7 Window server sessions

owns a window or not. Indeed a control is usually unaware of whether it is window-owning or not. In a single application, the same control could be window-owning when used in one way and non-window-owning when used in another way.

5.2 The Control Environment

Controls do not exist in isolation; they exist within an environment called the control environment, also known as *CONE*. This environment is provided by the `CCoeEnv` class and objects such as `CEikonEnv` which subclass `CCoeEnv` to extend its functionality.

The `CCoeEnv` class is an active object. In particular, it is responsible for receiving the events sent by the window server. Once it has received these events, it performs a certain amount of low-level generic processing, enabling controls to focus on specific tasks such as redrawing themselves or handling a pointer event.

It is possible to write applications without using any of the control environment and UI framework code; for example, software that does not have a UI, such as a server, does not need the UI framework. For applications with a UI, you are not forced to use the provided framework, but avoiding it is a substantial task and is outside the scope of this book.

5.2.1 User Interaction with Controls

The window server delivers a range of events, but the three events of most interest to us at this time are redraw, pointer and keyboard events.

Redraw events

The original design principle of the window server is that it does not update windows by itself. Instead, it requests applications to perform the task. UIQ 3 incorporates an advanced technology called the `Redraw Store` that is capable of caching most drawing operations and replaying them on demand. This technology enables some of the advanced transition and image blending effects that are available on UIQ 3 phones. It does not, however, completely eliminate the need for applications to be able to redraw their screens on demand.

Redraw events are sent to an application when part of a window owned by the application needs to be updated; for example, due to a dialog box being removed (see Figure 5.8).

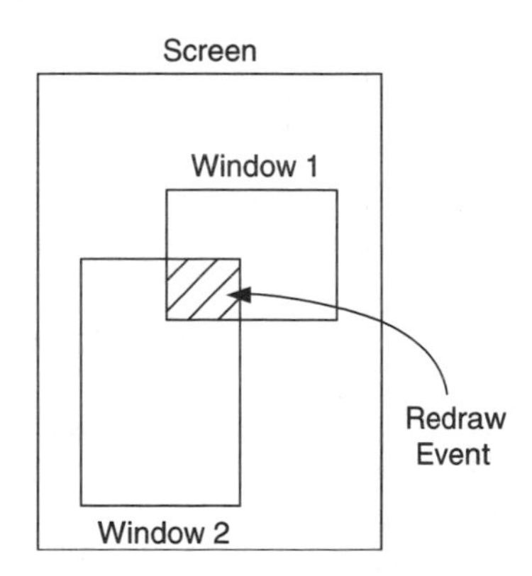

Figure 5.8 Window redraw event

The window server does not have any concept of controls, only windows. An application is responsible for determining which specific controls intersect with the window region that needs redrawing and cause those controls to redraw themselves. The client-side representation of a window is presented through objects derived from the `RWindowBase` class such as `RWindow`. When you construct such an object, you can specify a unique id or *handle*, which the window server will keep associated with the window.

When a window-owning `CCoeControl` is constructed, it creates the `RWindow` and sets this handle to be the `CCoeControl` object pointer. When a redraw event is received, the framework simply has to obtain the handle, which is conveniently passed with the redraw event, and interpret it as a `CCoeControl` object. Once the framework has identified the `CCoeControl` it can proceed with drawing all controls and component controls by using the `Draw()` and `DrawComponents()` methods outlined previously.

Pointer events

As with redraw events, pointer events occur within a window. The application is responsible for determining which control happens to be positioned at the location within the window where the pointer event occurred and therefore which control should process it (see Figure 5.9).

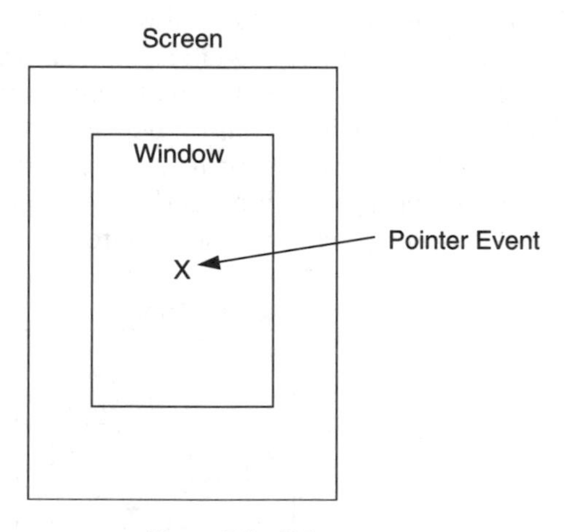

Figure 5.9 Pointer event

As with redraw events, the handle given to the window construction method is passed with the pointer event. The framework interprets this handle as a `CCoeControl` and begins its search for the component that controls it, performing something like:

```
void CCoeControl::HandlePointerEventL(
  const TPointerEvent& aPointerEvent) const
  {
  TInt count=CountComponentControls();
  for (TInt i=0; i<count; i++)
    {
    CCoeControl* ctrl=ComponentControl(i);
    TRect rect(ctrl->Rect());
    if (rect.Contains(aPointerEvent.iPosition))
      {
      ctrl->ProcessPointerEventL(aPointerEvent);
```

```
        break;
        }
    }
}
```

You should note that pointer events are slightly harder to deal with fully than the above code fragment implies. However, it shows how the framework traverses the set of compound controls to deal with pointer events in a very similar fashion to dealing with redraw events.

Key events

Unlike redraw and pointer events, key events are not associated with a single window but with a window group. Strictly speaking, the window server sends key events to the window group that currently has focus. Within the control environment, there is only a single window group per application, so the two are considered synonymous.

On phones that share the screen between multiple applications, the foreground application is the one that has focus (see Figure 5.10). On mobile phones based on Symbian OS, applications effectively run maximized all the time, therefore the displayed application is in the foreground and has focus; only one application can be displayed at a time. Key events are therefore sent to the current application (see Figure 5.11).

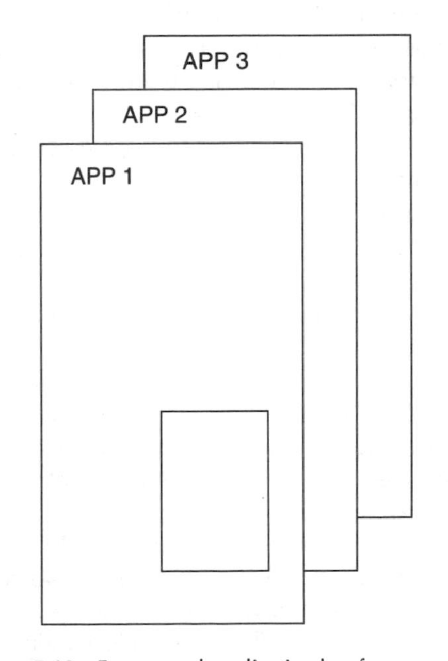

Figure 5.10 Foreground application has focus

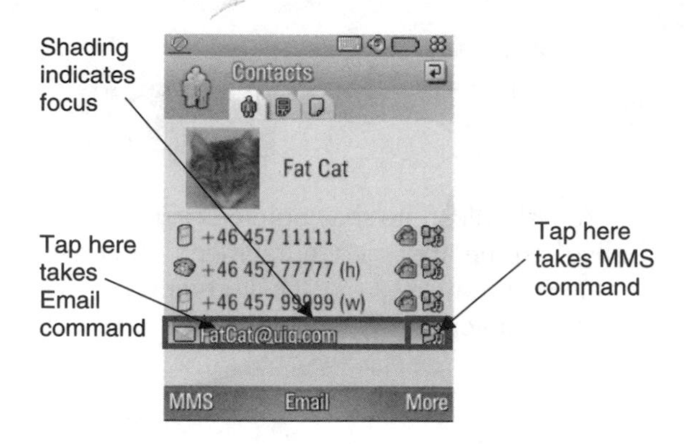

Figure 5.11 Focus inside the Contacts application

Pro tip

The use of focus correlating to the foreground application should not be confused with the use of focus within an individual application. Focus within an individual application is often displayed visually and is used to attract the user's attention to a specific control.

Once the control environment has received a key event, it has to determine how to deliver that event to the correct control. Unlike redraw and pointer events, it cannot simply associate the key with a window. Instead, the control environment uses a structure called the control stack. The control stack is a priority-ordered list of all controls that are interested in receiving keyboard events. Typically, only a small number of controls are present on the control stack at any point in time; compound controls only need to register the parent with the control stack for all the components to be registered. If required, the parent can pass the key event to its components.

When a key event is received, the control environment searches the control stack from highest to lowest priority to locate a control prepared to handle the event. You can see this in the method name `OfferKeyEventL()` and its return values. The key event is first offered to a high-priority control. If it reports `EKeyWasNotConsumed`, then it is offered to a lower priority control, and so on until a control returns `EKeyWasConsumed`.

The control environment adds a number of controls to the control stack, for example, if the phone has a front end processor (FEP) for data entry, a control with priority `ECoeStackPriorityFep` is added to the control stack. The application menu control is added at `ECoeStackPriorityMenu`. In general, views within your application are added to the control stack at `ECoeStackPriorityDefault`.

The full set of defined priorities is:

```
enum
  {
  ECoeStackPriorityDefault=0,
  ECoeStackPriorityMenu=10,
  ECoeStackPriorityDialog=50,
  ECoeStackPriorityCba=60,
  ECoeStackPriorityAlert=200,
  ECoeStackPriorityFep=250,
  ECoeStackPriorityEnvironmentFilter=300
  };
```

You may have noticed that there are gaps between the priority values. This enables applications to insert controls with intermediate priorities if there is a requirement to do so.

5.3 Views and the View Server

We say that applications display a *view* of their data. Some applications may only present a single view. Others, such as a diary application, have multiple views (see Figure 5.12).

Figure 5.12 Agenda application views of day, week and month

Views may also display temporary data; for example, they can display the progress of an action.

Users frequently perform tasks that need to use the data and functionality of several applications. For example, the user may wish to phone a colleague. The phone application is responsible for actually dialing numbers and establishing a voice call. The user does not want to enter the phone number of the colleague every time and copying and pasting the number from the contacts application has too many steps involved. The best solution is to provide a 'call' function in the Contacts application.

We could provide the phone call functionality in a shared code library, with both applications using that library (see Figure 5.13).

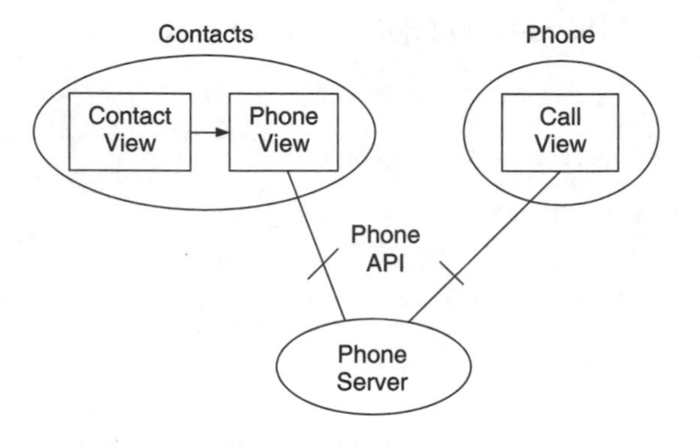

Figure 5.13 Phone server solution

Alternatively, we could provide a separate dialer application to implement the phone call functionality (see Figure 5.14). The Phone and Contacts applications launch the Dialer when the user makes a phone call.

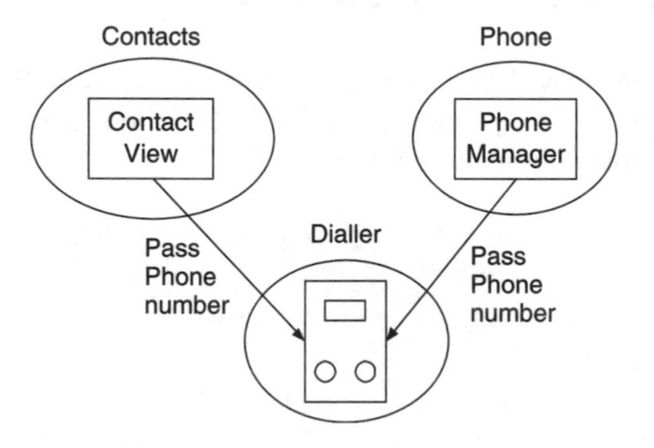

Figure 5.14 Dialer application solution

We might also want our Agenda application to send a 'Happy Birthday' SMS when a friend in the Contacts database has a birthday. There are many opportunities to share data and use the communications facilities of a mobile phone, so it is important to provide a flexible but simple way for applications to perform these cross-functional tasks.

Symbian OS provides a substantial framework to support views, called the view architecture. It was developed to provide a flexible environment within which individual applications exist, but the user can easily perform tasks that happen to be implemented across multiple applications.

The view architecture has two key components: a set of views and the view server. Views are UI controls, ultimately derived from `CCoeControl`, which implement the abstract view interface defined by `MCoeView`. All views need to inherit `CQikViewBase`. Applications supply the views.

The view server is a system-wide component responsible for coordinating the changing of views displayed on a phone. It is implemented as a separate process, hence its name. The view server facilitates the switching of views depending on the task a user is performing rather than navigating by application. When switching from a view on one application to a view in another, it is called a direct navigation link (DNL). These are described in detail in Chapter 9. Although implemented as a separate process, all the inter-process communications with the view server are handled by the framework (see Figure 5.15). Your application needs to interact with the UIQ 3 framework. This may seem complicated but we provide example applications to help you.

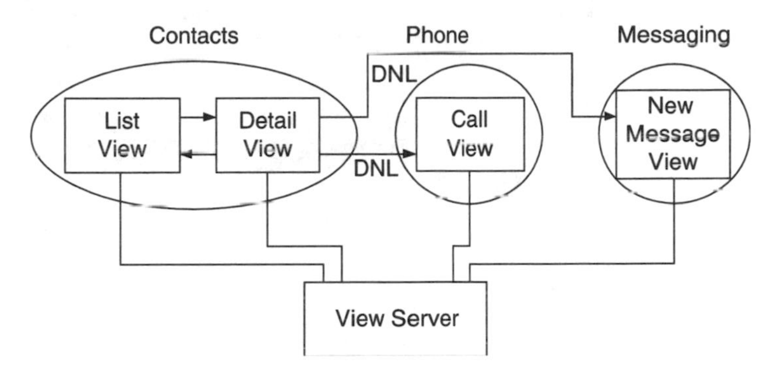

Figure 5.15 UIQ 3 applications, views and DNLs between views

5.4 Anatomy of the Screen

We now take a look at the UIQ 3 screen in detail and describe the various parts of the UI (see Figure 5.16).

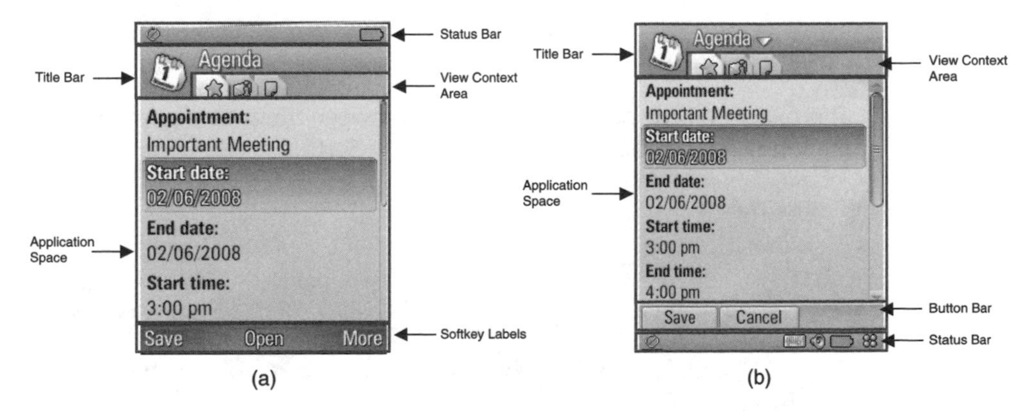

Figure 5.16 Screen areas in a) Softkey Style and b) Pen Style

5.4.1 Status Bar

The status bar contains system indicators such as the battery level, Bluetooth state and mobile network signal strength. The content of the status bar is defined by the phone manufacturer and is not available to application developers. In Softkey Style Touch and Pen Style, the status bar items can be tapped to access further information and settings. In Figure 5.17, you can see the task menu and the battery status indicator that Sony Ericsson has implemented.

Your application can perform some basic actions such as hiding the status bar:

```
MEikAppUiFactory* q=iEikonEnv->AppUiFactory();
q->StatusPane()->MakeVisible(EFalse);
```

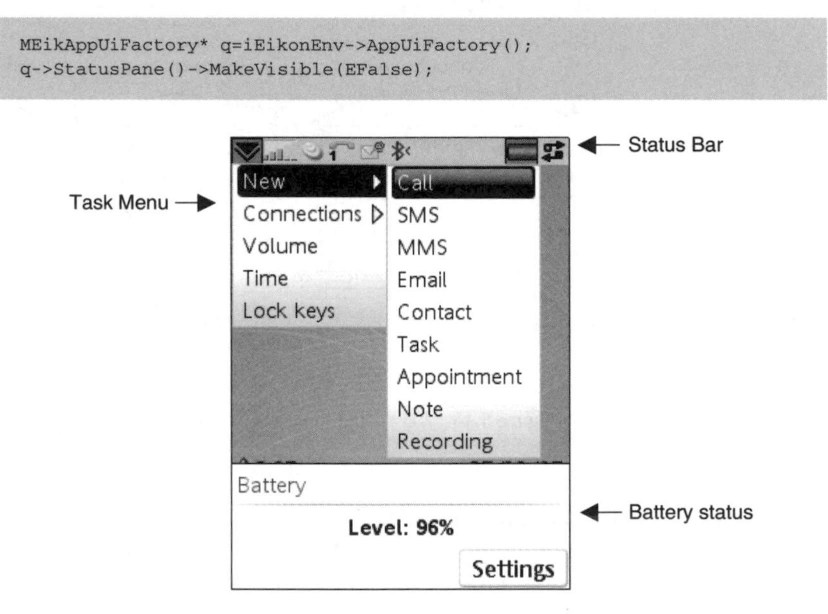

Figure 5.17 Status bar on Sony Ericsson P1i

5.4.2 Title Bar

The exact content of a title bar varies between view configurations but the following components exist (see Figure 5.18).

Figure 5.18 Title bar (Softkey Style Touch)

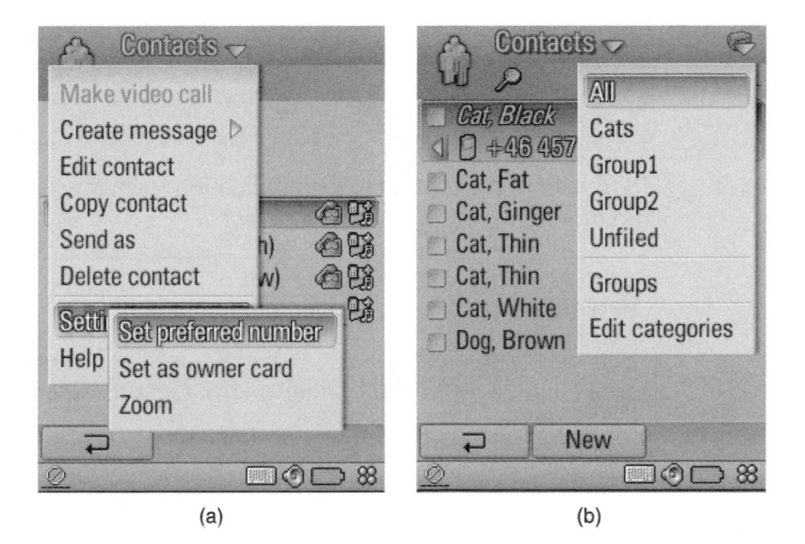

Figure 5.19 a) Menu and b) category selection in the Contacts application (Pen Style)

Title Text

This initially contains the application name but you can change the text, for example:

```
void CAppSpecificListView::ViewConstructL()
  {
  _LIT(KNewTitle,"New title");
  SetAppTitleNameL(KNewTitle);
  }
```

In Pen Style configurations, an icon is displayed immediately to the right of the title indicating you can tap on it to display a menu (see Figure 5.19a). If your application supports categories, then on Pen Style phones a second icon is displayed to the far right of the title bar (see Figure 5.19b).

To restore the content to the application name you can simply pass `KNullDesc` to the `SetAppTitleNameL()` method.

Icon

This initially contains the application icon but you can change the icon, for example:

```
void CAppSpecificListView::ViewConstructL()
  {
  _LIT(KMbmFileName,"*");
  CFbsBitmap* bit=iEikonEnv->CreateBitmapL(
                KMbmFileName,EMbmAppIcon);
```

```
CFbsBitmap* mask=iEikonEnv->CreateBitmapL(
                KMbmFileName,EMbmAppMask);
SetAppTitleIconL(bit,mask);
}
```

To restore the content to the application icon, pass `NULL` for the image and mask icons.

Pro tip

You may wish to note that `"*"` can be used as a shortcut to mean 'my application's MBM file', rather than having to determine the full file name yourself.

View Context Area

This area is normally used to display tabs if your view is a multi-page view or to display some application-specific context information, such as a view title, in single-page views.

For multi-page views, the tab information is normally defined in a `QIK_VIEW_PAGE` application resource:

```
QIK_VIEW_PAGE
  {
  page_id = EAppSpecificListViewPageId1;
  tab_bmpid = EMbmListview2Tab1;
  tab_bmpmaskid = EMbmListview2Tab1mask;
  page_content = r_list_view_page1_control;
  },
QIK_VIEW_PAGE
  {
  page_id = EAppSpecificListViewPageId3;
  tab_caption = "Tab3";
  page_content = r_list_view_page3_control;
  }
```

For single-page views, you can set icons, text or progress information controls as decorators in the View context area (see Figure 5.20).

Here is the code:

```
TBuf<64>bb;
iEikonEnv->ReadResource(bb,R_STR_VIEW_TITLE);
ViewContext()->AddTextL(1,bb);
_LIT(KMbmFile,"*");
CFbsBitmap* bit=iEikonEnv->CreateBitmapL(KMbmFile,
                                    EMbmAppIcon);
CFbsBitmap* mask=iEikonEnv->CreateBitmapL(KMbmFile,
                                    EMbmAppMask);
ViewContext()->AddIconL(2,bit,mask);
```

```
ViewContext()->AddProgressInfoL(
                          EEikProgressTextPercentage,100);
ViewContext()->SetAndDrawProgressInfo(50);
```

Figure 5.20 Text, icon and progress indicator controls in View context area

> **Pro tip**
>
> Although the above example does not contain any error-handling code, you should note that the `ViewContext()` takes ownership of the `CFbsBitmap` objects.

5.4.3 Button Bar

On Pen Style phones, applications usually have a button bar containing a Back button (see Figure 5.21). Applications can specify any additional commands they wish to be displayed on the button bar, for example:

```
RESOURCE QIK_CONTENT_MBM r_function1_button
  {
  bmpid=EMbmAppIcon1;
  bmpmask=EMbmAppIcon1mask;
  }
QIK_COMMAND
  {
  id = EAppCmdFunction1;
  type = EQikCommandTypeScreen;
  cpfFlags = EQikCpfFlagPreferToBePlacedInButtonbar;
  icon = r_function1_button;
  },
```

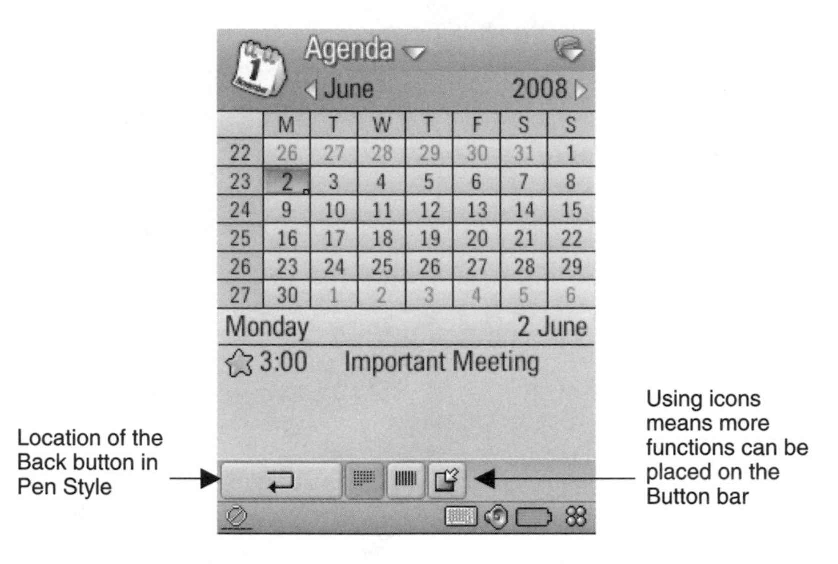

Figure 5.21 Button bar in Agenda application

Using icons in the button bar means that you can present more commands that the user can select with a single tap.

5.4.4 Softkey Bar

In contrast to a button bar, the softkey bar on a Softkey Style phone usually only contains labels indicating which action will occur if a softkey is depressed (see Figure 5.22).

Figure 5.22 Softkey bar in Contacts

While applications don't have direct access to the button bar or softkey bar region, they can strongly influence what is displayed in those regions by appropriate usage of commands.

> **Pro tip**
>
> The button bar and softkey bar are treated quite differently within UIQ 3 and should not be considered as alternatives dependent on phone type. For example, the Command Processing Framework flag `EQikCpfFlagPreferToBePlacedInButtonbar` is only applicable to button bars.

Sony Ericsson phones operating in Softkey Style Touch have a modified design for the softkey bar which looks like a button bar (see Figure 5.23). By using button-style labels, the user is reminded that the softkeys can be touched on screen.

Figure 5.23 Softkey bar on Sony Ericsson P1i

5.4.5 Application Space

Lastly, there is the application space region where your views are displayed (see Figure 5.24).

In general, applications should focus on only using the application-specific region but there are options to hide the other components therefore expanding this region. This code shows you how:

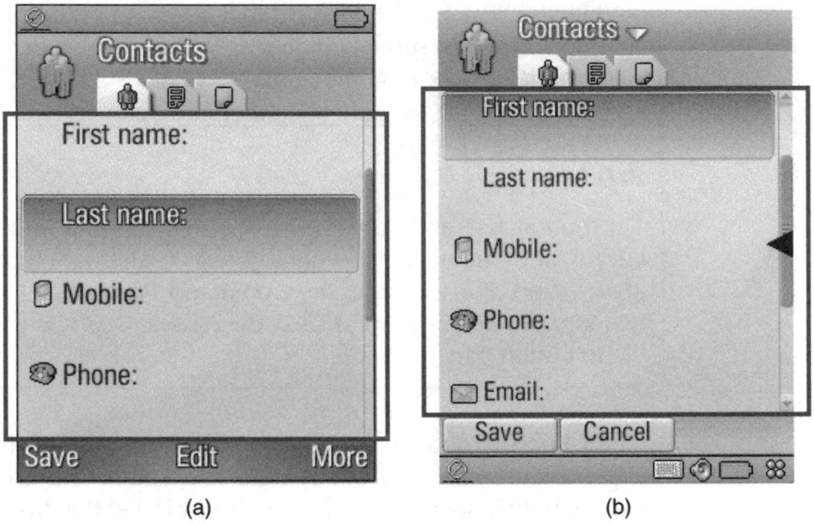

Figure 5.24 Application space in a) Softkey Style and b) Pen Style

```
void CAppSpecificListView::ViewConstructL()
  {
  TQikViewMode mode = ViewMode();
  // Normally only 1 of the following used at a time.
  // Hides status bar.
  mode.SetStatusBar(EFalse);
  // Hides the title bar.
  mode.SetAppTitleBar(EFalse);
  // Hides button bar/soft key bar and status bar.
  mode.SetButtonOrSoftkeyBar(EFalse);
  // Applies the requested view mode.
  SetViewModeL(mode);
  }
```

There is also the option to use the entire screen:

```
void CAppSpecificListView::ViewConstructL()
  {
  TQikViewMode mode = ViewMode();
  mode.SetFullscreen();
  SetViewModeL(mode);
  }
```

You should exercise care if you hide any of the standard components. Ease of use is often derived from a consistent user interface. Varying the interface presented by your application compared to standard applications, such as by hiding the status bar, may be detrimental to your application. On some phones, a *Task* switch and shortcut icon is present within the status bar. If you choose to hide the status bar, the user loses the part of the UI that is normally there. Ensuring that you integrate with the features and functions on a phone can make the difference between happy customers and frustrated ones.

6

List Boxes

UIQ 3 applications frequently use a list–detail model. This means that when the application starts, data is presented in a list view. The user selects an item from the list and opens a detail view to see and do more. Agenda, Contacts and Jotter use this method of presentation.

Previous versions of UIQ supported list boxes but the old controls have been replaced by a new highly flexible, highly functional list box class `CQikListBox`.

This new class fulfils a core UIQ 3 design goal of allowing applications to use a single control, while the control adapts itself to be appropriate for the type of phone on which it is being displayed. For example, on a phone using Pen Style, the list box has a wide pen-operated scrollbar (see Figure 6.1a); on a phone using Softkey Style, scrolling uses the Up and Down keys but retains a narrow scrollbar to provide visual feedback (Figure 6.1b).

This chapter presents the functionality of list boxes through two example applications. The first example presents a simple, traditional list box application. The second explores the wide variety of functionality that can be included within a UIQ 3 list box control.

As part of the first example application, we step through the framework code that each application needs to provide. The general structure of the framework code is common across most UIQ 3 applications and so we do not repeat it in later examples.

6.1 `ListView1` Application

This application presents a list box control containing a fixed number of entries. The list box is presented in a single view (see Figure 6.2). Key,

(a) (b)

Figure 6.1 Scrollbars in a) Softkey Style Touch and b) Softkey Style

Figure 6.2 `ListView1` application

pointer and redraw events are all delivered to the application although this is not obvious since these are all handled automatically by the framework. The application supports the six primary view configurations available within the emulator. As you switch between configurations, so the list box is updated to display correctly within those configurations.

The application is made up of the following files:

- `Listview1.mmp`: The Symbian makefile for the application.

- Listview1_reg.rss: The application registration file. The meaning and use of registration files is discussed fully in Chapter 10. For the purpose of the examples in this chapter you can ignore these files.
- ListView1_loc.rss: The localization information required by the application registration file. As with registration files, these are described fully in Chapter 10.
- ListView1.rss: The application resource file. UIQ 3 applications are resource intensive so you should be able to read and understand the structure and content of resource files.
- ListView1.hrh: The application's shared resource and source code header file.
- ListView1.h: The public object definitions. For the example applications, these contain minimal information.
- ListView1.cpp: The entire example application source code.

The application contains two objects of interest:

- CAppSpecificListView: The application-specific list view, which subclasses the CQikViewBase class and, hence, ultimately the CCoeControl class. It presents the view as a list box as opposed to any other type of control.
- CAppSpecificUi: The application user interface object, which subclasses CQikAppUi and includes the application entry point. Amongst other tasks, CQikAppUi creates the control environment for you.

In a graphical application, code execution effectively starts at the ConstructL() method of the application-specific UI object. Our application contains the following code:

```
void CAppSpecificUi::ConstructL()
  {
  CQikAppUi::ConstructL();
  // Create and set up an engine.
  iEngine=new(ELeave)CAppEngine(EQikCmdZoomLevel2);
  TBuf<KMaxListItemText>bb;
  const TInt KListView1Items=7;
  for (TInt i=0;i<KListView1Items;i++)
    {
    iEikonEnv->ReadResourceL(bb,R_STR_LIST_CONTENT_1+i);
    iEngine->SetListItem(bb);
    }
  // Create and set up the single view.
  CAppSpecificListView* q = new(ELeave)
                    CAppSpecificListView(*this,iEngine);
  CleanupStack::PushL(q);
```

```
q->ConstructL();
AddViewL(*q);
CleanupStack::Pop(q);
}
```

The call to `CQikAppUi::ConstructL()` is very important as it is responsible for setting up various parts of the control environment such as creating the control stack.

The first application-specific task is creating and initializing an engine object. The second is creating and initializing the single view. Of all the above lines of code, the `AddViewL()` function is one to focus on; the rest should be self explanatory.

6.1.1 `AddViewL()`

This method takes ownership of the passed object, registers it with the view server and adds it to the control stack. Unlike in previous versions of UIQ, the `CAppSpecificUi` object is no longer required to track or delete the view object; the framework now performs that task on your behalf.

For a view to be registered with the view server, it needs to implement the `MCoeView` interface. A primary task of our view is to implement that interface.

As we explained in Chapter 5, the control stack is a priority-ordered list of controls that are interested in receiving key events from the window server. Our view has been registered as such a control and is therefore delivered key presses if no other higher priority control consumes them.

Code execution reaches the end of the `ConstructL()` method and returns to the framework. A short period of time later, the `View ConstructL()` method of our view is called.

6.1.2 The `CAppSpecificListView` object

This object implements a view on the application data. As part of implementing the `MCoeView` interface, all views need to be uniquely identified. This is achieved using a `TVwsViewId` object. A `TVwsViewId` object comprises two UIDs: the first must be your application UID and the second is a value which is unique within your application.

A `TVwsViewId` is defined as:

```
const TUid KAppSpecificUid={0xEDEAD001};
const TUid KUidListView={0x00000001};
#define KViewIdListView
                    TVwsViewId(KAppSpecificUid,KUidListView)
```

It is returned by the `ViewId()` method:

```
TVwsViewId CAppSpecificListView::ViewId() const
  {
  return(KViewIdListView);
  }
```

To reiterate, the first UID needs to be your application UID, as opposed to any other globally unique ID you may have been allocated. This is how the view server identifies the application that contains the view and how you can display views belonging to other applications.

After the `ViewId()` method, `ViewConstructL()` is called:

```
void CAppSpecificListView::ViewConstructL()
  {
  ViewConstructFromResourceL(R_LIST_VIEW_CONFIGURATIONS);
  CQikViewBase::SetZoomFactorL(CQikAppUi::ZoomFactorL(
            iEngine->ListViewZoomState(),*iEikonEnv));
  LocateControlByUniqueHandle<CQikListBox>(
            EListViewListId)->SetListBoxObserver(this);
  // Add a view title.
  TBuf<64>bb;
  iEikonEnv->ReadResourceL(bb,R_STR_LIST_TITLE);
  ViewContext()->AddTextL(1,bb);
  }
```

UIQ 3 substantially expands the view functionality so that applications can support a wide variety of phones within a single executable. A single phone may support multiple UI configurations, such as portrait and landscape or Softkey Style and Pen Style. Such configurations may vary so that functionality available in one configuration cannot reasonably be performed in another or the way in which certain functionality is presented has to change to ensure optimal usage. The UIQ 3 Calculator application is such a case (see Figure 6.3).

UIQ 3 deals with these configuration issues via extensive use of resource files. The first line of our `ViewConstructL()` method passes the resource ID of the view configuration information specific to our view.

The remainder of the method sets our default zoom size, specifies that the view wants to hear about list box events and sets the text displayed below the application name.

The framework code has performed a substantial number of tasks on our behalf. For example, it has read and interpreted the configuration resource, created the controls that resource has defined, laid them out and presented a type-safe mechanism by which we can access the controls.

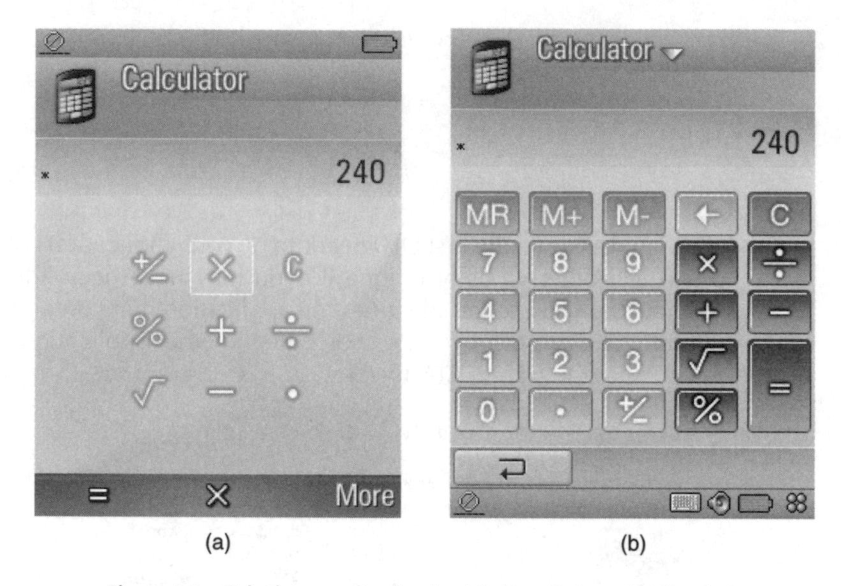

Figure 6.3 Calculator application in a) Softkey Style and b) Pen Style

To complete the implementation of the `MCoeView` interface, we need to provide the `ViewActivatedL()` and `ViewDeactivated()` methods.

```
void CAppSpecificListView::ViewActivatedL(
  const TVwsViewId& aPrevViewId,
  const TUid aCustomMessageId,
  const TDesC8& aCustomMessage)
  {
  AddItemsL();
  }

void CAppSpecificListView::ViewDeactivated()
  {
  CQikListBox* listbox= LocateControlByUniqueHandle<
           CQikListBox>(EAppSpecificListViewListId);
  listbox->RemoveAllItemsL();
  }
```

The `ViewActivatedL()` method of a view is called when the view server determines a particular view should become the foreground view. If completed without error, the `ViewDeactivated()` method of the previous view is called.

In our example, we have chosen to generate the content of the list box when the view is activated, freeing any resources when the view is deactivated. In a resource-constrained environment, it is good practice to allocate resources only when they are needed and to free them up when not required. It should be noted however that there is often a trade-off between the time taken to build any data structure, how much memory it uses and how appropriate such actions are for your specific application.

You can test the `ViewActivatedL()` and `ViewDeactivated()` code by switching back and forth between `ListView1` and the application launcher.

6.1.3 View Configuration Resource Definition

Most resource structures used by a view are defined in `qikon.rh`. If you are unsure about how resource files are structured, we recommend that you review the resource files information in the UIQ 3 SDK or read Chapter 13 of *Symbian OS C++ for Mobile Phones, Volume 3*.

When our view is constructed, we call the `ViewConstructFromResourceL()` method, passing it a resource ID, `R_LIST_VIEW_CONFIGURATIONS`. This resource is defined in `ListView1.rss` as:

```
RESOURCE QIK_VIEW_CONFIGURATIONS r_list_view_configurations
  {
  configurations =
    {
    QIK_VIEW_CONFIGURATION
      {
      ui_config_mode = KQikPenStyleTouchPortrait;
      command_list = r_list_view_pen_style_commands;
      view = r_list_view_pen_style_view;
      },
    QIK_VIEW_CONFIGURATION
      {
      ui_config_mode = KQikPenStyleTouchLandscape;
      command_list = r_list_view_pen_style_commands;
      view = r_list_view_pen_style_view;
      },
    QIK_VIEW_CONFIGURATION
      {
      ui_config_mode = KQikSoftkeyStylePortrait;
      command_list = r_list_view_key_style_commands;
      view = r_list_view_key_style_view;
      },
    QIK_VIEW_CONFIGURATION
      {
      ui_config_mode = KQikSoftkeyStyleSmallPortrait;
      command_list = r_list_view_key_style_commands;
      view = r_list_view_key_style_view;
      },
    QIK_VIEW_CONFIGURATION
      {
      ui_config_mode = KQikSoftkeyStyleSmallLandscape;
      command_list = r_list_view_key_style_commands;
      view = r_list_view_key_style_view;
      },
    QIK_VIEW_CONFIGURATION
      {
      ui_config_mode = KQikSoftkeyStyleTouchPortrait;
```

```
    command_list = r_list_view_key_style_commands;
    view = r_list_view_key_style_view;
    }
  };
}
```

The `r_list_view_configurations` resource comprises a set of six `QIK_VIEW_CONFIGURATION` structures, one for each of the configurations (see Chapter 2) supported by our application.

The framework code looks at the set of view configurations and compares them with the phone's current UI configuration to choose the most appropriate entry. For example, if a phone has a portrait-oriented touch screen and is in Pen Style UI configuration, the `KQikPenStyleTouchPortrait` entry is chosen. If the UI configuration changes, for example, if the flip is closed on a Sony Ericsson P990i, the framework code chooses the entry for `KQikSoftkeyStyleSmallPortrait`. Once the framework has chosen an entry, it reads the list of views and commands associated with that entry to configure dynamically the application.

We have determined that the example application can use a single command list and view for both of the Pen Style configurations. Similarly, a single command list and view can be used for all four of the Softkey Style configurations. In more sophisticated applications you may choose to have separate commands or views for each of the configurations.

The full list of configurations as defined in `Qikon.hrh`, is:

```
KQikSoftkeyStylePortrait
KQikSoftkeyStyleLandscape
KQikSoftkeyStyleLandscape180
KQikSoftkeyStyleSmallPortrait
KQikSoftkeyStyleSmallLandscape
KQikSoftkeyStyleSmallLandscape180
KQikSoftkeyStyleTouchPortrait
KQikSoftkeyStyleTouchLandscape
KQikSoftkeyStyleTouchLandscape180
KQikPenStyleTouchPortrait
KQikPenStyleTouchLandscape
KQikPenStyleTouchLandscape180
```

There are a number of factors that you need to consider when determining what configurations an application should support. We have chosen to support six UI configurations in our example applications. At the time of writing, if your application supports these six, you will support all the commercially available UIQ 3 devices. As an absolute minimum you should support the default configurations of the UIQ 3 phones that you wish to support.

If the phone is in a UI configuration that is not explicitly listed in `QIK_VIEW_CONFIGURATIONS`, then the framework selects the `QIK_VIEW_CONFIGURATION` that it considers to best match the phone's configuration. Your application will run; however, in some instances the view layout and command handling may not be optimal.

Pro tip

Your application may choose not to support certain view configurations, hence classes of phones, for technical or commercial reasons. The UIQ 3 framework does not require you to support a specific number of configurations. However, you need to understand the consequences of only supporting certain configurations and running in 'best match' mode, particularly if you choose not to support a configuration used on commercially available phones.

In our example application, we have identified two views, one for Pen Style configurations and one for Softkey Style configurations, r_list_view_pen_style_view and r_list_view_key_style_view respectively. The resource referenced by the view field is expected to be of type QIK_VIEW.

At this point it is worth recalling that the UIQ 3 resource definitions are required to support a wide range of application usages. Specific applications use only a subset of the fields defined within a particular resource. The first example of this is the QIK_VIEW resource, defined as:

```
STRUCT QIK_VIEW
  {
  BYTE version = 2;
  LLINK view_mode = 0;
  LLINK command_list = 0;
  LLINK qiktoolbar = 0;
  LLINK pages = 0;
  }
```

If you look at the r_list_view_pen_style_view and r_list_view_key_style_view resources, you will see we only use the pages field.

```
RESOURCE QIK_VIEW r_list_view_pen_style_view
  {
  pages = r_list_view_pen_style_pages;
  }
```

In general, an application view is made up of a number of pages of information. In its simplest form, the number of pages is one. The r_list_view_pen_style_pages link is expected to reference a resource of type QIK_VIEW_PAGES.

```
RESOURCE QIK_VIEW_PAGES r_list_view_pen_style_pages
  {
  pages =
    {
    QIK_VIEW_PAGE
      {
      page_id = EAppSpecificListViewPageId;
      page_content = r_list_view_page_control;
      }
    };
  }
```

Since our application only has a single page of controls we only need a single `QIK_VIEW_PAGE` structure.

Applications often need to reference controls created from resources within the application code. To achieve this, controls are allocated some type of unique ID; in the case of a `QIK_VIEW_PAGE`, it is the `page_id`. For this application, the `page_id` is set to `EAppSpecificListViewPageId`, an application-specific value defined in `ListView1.hrh`.

A page needs to contain some content. In UIQ 3, a page usually contains a set of controls within a containing control. The `r_list_view_page_control` references a resource of type `QIK_CONTAINER_SETTINGS` which defines the container and the controls within that container.

```
RESOURCE QIK_CONTAINER_SETTINGS r_list_view_page_control
  {
  layout_manager_type = EQikRowLayoutManager;
  layout_manager = r_row_layout_manager_default;
  controls =
    {
    QIK_CONTAINER_ITEM_CI_LI
      {
      unique_handle = EAppSpecificListViewListId;
      type = EQikCtListBox;
      control = r_app_listview_listbox;
      layout_data = r_row_layout_data_fill;
      }
    };
  }
```

The UIQ 3 application framework provides a set of sophisticated controls called layout managers that will, within certain bounds, lay out controls automatically for you irrespective of the current configuration. We discuss this in detail in Chapter 8. For now, all you need to know is that our list box is laid out in rows using `EQikRowLayoutManager` (see Figure 6.4) with default parameters.

```
RESOURCE QIK_ROW_LAYOUT_MANAGER r_row_layout_manager_default
  {
  default_layout_data = QIK_ROW_LAYOUT_DATA {};
  }
```

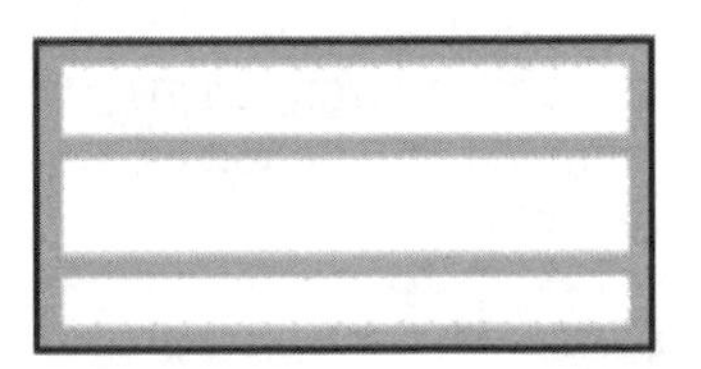

Figure 6.4 Row layout manager

Within the container we have a set of controls, each defined by a QIK_CONTAINER_ITEM or derivative, such as:

```
QIK_CONTAINER_ITEM_CI_LI
QIK_CONTAINER_ITEM_CD_LI
QIK_CONTAINER_ITEM_LD
QIK_CONTAINER_ITEM_CI_LD
QIK_CONTAINER_ITEM_CD_LD
QIK_CONTAINER_ITEM_NESTED_CONTAINER
QIK_CONTAINER_ITEM_NESTED_CONTAINER_CI_LI
```

For the inquisitive amongst you, the way the resource-reading code knows how to differentiate between the above structures is to query the first byte or the struct_type field. You may note from resource definitions this field varies from 0 to 7 depending on which of the above resources is used.

Each of the QIK_CONTAINER_ITEM resource definitions has a different set of fields to complete depending on exactly what you are attempting to achieve, and some individual preferences about how to achieve the desired results. So how do we go about determining which variant is applicable to our application?

A container item requires two fundamental pieces of information: a description of the control to put in the container and some associated layout information. Resources can be defined to expect a reference to another resource or to contain that resource directly. A reference to another resource is said to be indirect.

In general, the QIK_CONTAINER_ITEM resource definitions follow a naming convention:

- C = control information
- L = layout information
- D = directly defined
- I = indirectly defined, via a reference to another resource.

Thus a QIK_CONTAINER_ITEM_CD_LI resource contains the **C**ontrol defined **D**irectly within the structure, while the **L**ayout information is defined **I**ndirectly.

Our application has chosen to reference both the control and layout information using a resource name. Therefore our example application uses a QIK_CONTAINER_ITEM_CI_LI structure.

Had we made other choices, we could have used the following code to achieve the same result:

```
QIK_CONTAINER_ITEM_CI_LD
  {
  unique_handle = EAppSpecificListViewListId;
  type = EQikCtListBox;
  control = r_app_listview_listbox;
  layout_data = QIK_ROW_LAYOUT_DATA
    {
    vertical_alignment = EQikLayoutVAlignFill;
    vertical_excess_grab_weight = 1;
    };
  }
```

There are two ways in which controls can be referenced indirectly within a resource file: via resource name or by a unique ID. Controls referenced by their unique ID need to exist within a QIK_CONTROL_COLLECTION. For simplicity, the first example does not use a QIK_CONTROL_COLLECTION.

Apart from enabling the usage of a QIK_CONTROL_COLLECTION, the unique_handle field allows us to uniquely identify a control within our application. As with the page_id earlier, the EAppSpecificListViewListId is defined in ListView1.hrh. The layout_data field in our QIK_CONTAINER_ITEM_CI_LI specifies how the row layout manager needs to handle some layout attributes. In this example, with a single list box control, little information is required. Finally, we get to define what type of control is to be created, in this instance an EQikCtListBox, then the reference to the list box configuration information.

```
RESOURCE QIK_LISTBOX r_app_listview_listbox
  {
  view = r_app_listbox_view_default;
  layouts = { r_app_normal_layout_pair };
  }
RESOURCE QIK_LISTBOX_ROW_VIEW r_app_listbox_view_default
  {
  }
RESOURCE QIK_LISTBOX_LAYOUT_PAIR r_app_normal_layout_pair
  {
  standard_normal_layout = EQikListBoxLine;
  }
```

The primary field of interest above is the standard_normal_layout field. List boxes come in a variety of styles depending on what information

South Africa

Figure 6.5 List box type EQikListBoxLine

you are attempting to present and how you want to present it. For Listbox1 we have chosen the simplest type of list box, EQikListBoxLine, where each line contains a single text item (see Figure 6.5).

Wow, all that just to display a list box? Well, yes, however you have a UI control that supports multiple phone configurations. You can observe this by switching the emulator into different view configurations, which you can do in a number of ways:

- Set the configuration using UiqEnv -ui at the command prompt (see Chapter 15).
- Use the UiConfigTest application in the application launcher.
- Switch through configurations using Ctrl+Alt+Shift+U.

We now have a starting point from which we can extend the functionality. For example we might change the type of list box, add more pages and use more controls in our application.

The next example builds on this simple list box example to demonstrate usage of all the standard list box types and how to start customizing list boxes.

6.2 ListView2 Application

This application (see Figure 6.6) extends the ListView1 example application to show:

- support for multi-page applications
- support for all list box types
- an alternative approach to creating and initializing list boxes.

Unlike on many platforms, a significant proportion of a UIQ 3 application user interface can be defined in resource files. The software that interprets the resource information is provided within the application framework code. If there is a problem with interpreting any resource, your application will typically panic. One of the difficulties with UIQ 3 is being able to debug why the panic occurred and which resource caused the problem.

Figure 6.6 `ListView2` example application

> **Pro tip**
>
> Due to the difficulty in debugging the framework resource interpretation code, it is often easier to take a known working example, get that to work within your application and then change the code and resources to suit your specific application. This application provides the working examples; in particular, showing how to set up each of the list box types to display content.

This application presents at least one implementation of every type of standard list box as listed earlier. It also shows how to construct a custom list box should you need something more than UIQ 3 provides as standard.

Structurally the application is identical to the `ListView1` example. The primary difference is that the application view subclasses `CQikMultiPageViewBase` instead of `CQikViewBase`, since this application wants to display more than one page to the user. The two pieces of code are shown below:

```
class CAppSpecificListView : public CQikMultiPageViewBase
  {
  };

class CAppSpecificListView : public CQikViewBase
  {
  };
```

Since we have multiple pages of content, multiple QIK_VIEW_PAGE entries are defined in the r_list_view_pages resource, one for each page. We actually have 12 pages but only the first three are shown here.

```
RESOURCE QIK_VIEW_PAGES r_list_view_pages
  {
  pages =
    {
    QIK_VIEW_PAGE
      {
      page_id = EAppSpecificListViewPageId1;
      tab_bmpid = EMbmListview2Tab1;
      tab_bmpmaskid = EMbmListview2Tab1mask;
      page_content = r_list_view_page1_control;
      },
    QIK_VIEW_PAGE
      {
      page_id = EAppSpecificListViewPageId2;
      tab_bmpid = EMbmListview2Tab2;
      tab_bmpmaskid = EMbmListview2Tab2mask;
      page_content = r_list_view_page2_control;
      },
    QIK_VIEW_PAGE
      {
      page_id = EAppSpecificListViewPageId3;
      tab_caption = "Tab3";
      page_content = r_list_view_page3_control;
      },
    // Remaining pages repeat the above info.
    };
  }
```

Pages are presented as a tabbed list across the view context area of the title bar (see Figure 6.7). These tabs can contain icons or text. In our code fragment, three tabs contain icons and the fourth contains some text.

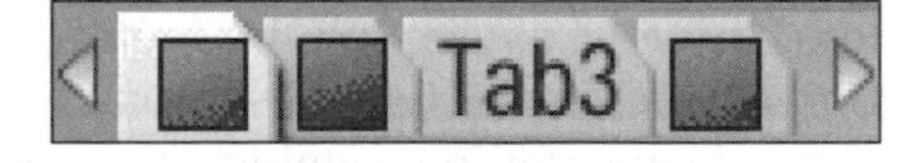

Figure 6.7 Page tabs

Each page displays a different type of list box. Page 1 displays a list box similar to ListView1, demonstrating how little needs to change to incorporate that entity within a more complex application. The list box content creation code is modified slightly to demonstrate the use of separators.

```
// Add a captioned separator.
lbData=model.NewDataLC(MQikListBoxModel::EDataSeparator);
iEikonEnv->ReadResourceL(bb,R_STR_LIST_SEPARATOR_1);
lbData->AddTextL(bb,MQikListBoxData::EDefaultSlot);
```

```
CleanupStack::PopAndDestroy(lbData);

// Add a plain separator.
lbData=model.NewDataLC(MQikListBoxModel::EDataSeparator);
CleanupStack::PopAndDestroy(lbData);
```

Page 2 displays a text only list box. The example has added some left and right margins to illustrate the potential usage of QIK_LISTBOX_ROW_VIEW fields.

```
RESOURCE QIK_LISTBOX_ROW_VIEW r_app_listview_listbox2_view
  {
  left_margin=8;                    // Add 8 pixel margins.
  right_margin=8;
  }
```

We also introduce the usage of two layouts within the same list box; the first layout is used for regular entries, the second for the highlighted entry (see Figure 6.8).

```
RESOURCE QIK_LISTBOX_LAYOUT_PAIR r_app_listbox2_layout_pair
  {
  standard_normal_layout = EQikListBoxLine;
  standard_highlight_layout = EQikListBoxTwoLines;
  }
```

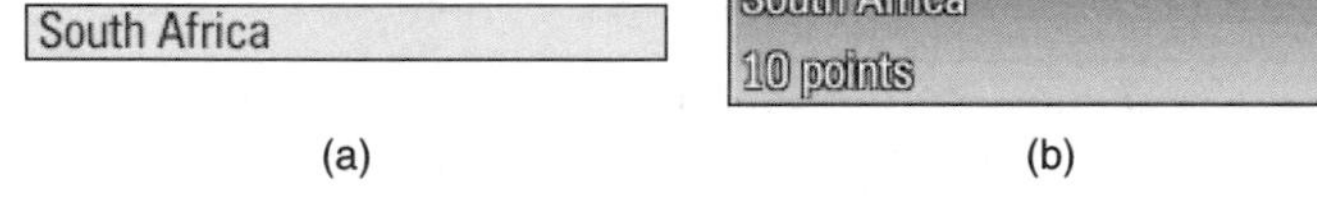

Figure 6.8 List box with different layout for a) normal state and b) highlighted state

Here, the row containing the highlight is displayed using two lines of content. This facilitates the display of additional content about the highlighted entry without significantly reducing the number of items that are visible within a screen full of information.

Pages 3 to 10 of our application present examples of the other standard list box types. The application code demonstrates how you go about setting up the icons and text fields displayed by each of the different list boxes. Table 6.1 shows all list box styles and in which tab of ListView2 you will find each style (for example, 1 means Tab 1; 2H means the highlighted row in Tab 2). We have populated all text and icons in the top row; in other rows we show that you can leave text or icons blank.

Pages 11 and 12 present examples of custom list boxes (discussed in Section 6.2.2).

Table 6.1 Types of list box layout in `ListView2`

List Box Style	Tab	Name and Contents
South Africa	1	`EQikListBoxLine` Text1
South Africa 10 points	2H	`EQikListBoxTwoLines` Text1 Text2
South Africa	3	`EQikListBoxIconLine` Icon and text
South Africa	2	`EQikListBoxLineIcon` Text1, Icon1
South Africa	4	`EQikListBoxIconLineIcon` Icon1, Text1, Icon2
South Africa	4H	`EQikListBoxIconIconLine` Icon1, Icon2, Text
South Africa	5	`EQikListBoxLineIconIcon` Text, Icon1, Icon2
South Africa	5H	`EQikListBoxIconLineIconIcon` Icon, text, icon, icon
South Africa Africa	6	`EQikListBoxIconHalfLineHalfLine` Icon, Text1, Text2
South Africa Africa 10 points	6H	`EQikListBoxIconHalfLineHalfLineLine` Icon, Text1, Text2 Text 3
South Africa 10 points	7	`EQikListBoxIconTwoLines` Icon, Text1 Text2
South Africa 10 points	7H	`EQikListBoxIconIconTwoLines` Icon1, Icon2, Text1 Text2
South Africa 10 points	3H	`EQikListBoxMediumIconTwoLines` MediumIcon1, Text1 Text2
South Africa 10 points	8	`EQikListBoxMediumThumbTwoLines` Thumbnail1, Text1 Text2

South Africa
Africa
	9	`EQikListBoxCheckLine`
		`SwappingLine`
		Checkbox, Text1
		Line with swappable content

South Africa 10 points	10	`EQikListBoxCheckLineIconLine` Checkbox, Text1 Icon, Text2

6.2.1 Creating and Initializing List Boxes

This example creates and initializes the list boxes within the `View-ConstructL()` method instead of the `ViewActivatedL()` method. This clearly requires more resources to be used when our view is not displayed, that is, when the view is `ViewDeactivated()`, but some applications need to maintain contextual information when you switch between views. For example, if you display the details of a list item then switch back to the list, it is desirable that the list box highlights the item for which you displayed the details. If you remove all items from the list box when the view is `ViewDeactivated()` and re-construct the list box in `ViewActivatedL()`, as we did in `ListView1`, you need to supply an external mechanism to track the originally selected item. This example shows that the list box can perform that task if required:

```
void CAppSpecificListView::ViewConstructL()
  {
  ViewConstructFromResourceL(R_LIST_VIEW_CONFIGURATIONS);
  CQikViewBase::SetZoomFactorL(CQikAppUi::ZoomFactorL(
           iEngine->ListViewZoomState(),*iEikonEnv));
  AddItemsToList1L();
  AddItemsToList2L();
  AddItemsToList3L();
  AddItemsToList4L();
  AddItemsToList5L();
  AddItemsToList6L();
  AddItemsToList7L();
  AddItemsToList8L();
  AddItemsToList9L();
  AddItemsToList10L();
  AddItemsToList11L();
  AddItemsToList12L();
  }
void CAppSpecificListView::ViewDeactivated()
  {
  // No code required.
  }
void CAppSpecificListView::ViewActivatedL(
  const TVwsViewId& aPrevViewId,
  const TUid aCustomMessageId,
  const TDesC8& aCustomMessage)
  {
  // No code required.
  }
```

By setting up the list boxes in the `ViewConstructL()` method, they are constructed once during the lifetime of the application. Consequently, it is quite reasonable to move the loading of the text resource directly into the list box creation code, compared to an engine supplying the text. This is particularly true of the UIQ 3 platform since the list boxes own a copy of the content.

Pro tip

In general, it is desirable to push functionality into an engine, even if that functionality is associated with displaying content, hence `ListView1` prefers to store and supply the list view content from an engine object. The `ListView1` and `ListView2` examples merely demonstrate the various choices that need to be made appropriate to your application requirements.

ItemIds

Entries added to a list box for display are usually derived from a data structure maintained elsewhere within an application. In that respect, our example applications are somewhat unusual.

Since items within a list box can be re-ordered; for example, if the list box is sorted, any one-to-one correspondence between list box entries and alternate data structure may not be permanently maintained. By allowing the list box entries to be assigned an ID, in this case a 32-bit number; a link between the two structures can be maintained. It is up to the individual application to generate these unique IDs. Although this sounds complex, it can be as simple as assigning the index value used when creating the list box entries, as demonstrated by the following code:

```
void CAppSpecificListView::AddItemsToList2L()
  {
  CQikListBox* listbox = LocateControlByUniqueHandle<
           CQikListBox>(EAppSpecificListViewListId2);
  MQikListBoxModel& model(listbox->Model());
  model.ModelBeginUpdateLC();
  TBuf<KMaxListItemText>bb;
  for (TInt i=0;i<KListView2Items;i++)
    {
    MQikListBoxData* lbData = model.NewDataL(
             MQikListBoxModel::EDataNormal);
    CleanupClosePushL(*lbData);
    iEikonEnv->ReadResourceL(bb,R_STR_LIST_CONTENT_1+i);
    lbData->AddTextL(bb,EQikListBoxSlotText1);
    iEikonEnv->ReadResourceL(bb,R_STR_LIST_DETAILS_1+i);
    lbData->AddTextL(bb,EQikListBoxSlotText2);
    lbData->SetItemId(i);
    CleanupStack::PopAndDestroy(lbData);
    }
  model.ModelEndUpdateL();
  }
```

Pro tip

If you plan to have 'add and delete' type functionality then the simple scheme mentioned above is unlikely to be adequate. In this case you will usually have to generate and store a unique ID in your engine code so you can translate between items being displayed on screen and the corresponding engine entries.

Slots

We use a set of standard enums to specify the content that is displayed in a list box, rather than dealing with any list-box-specific structure. The full list is defined by TListBoxStdLayoutSlots. The ListView1 example only displays a single line of text placed in EQikListBoxSlotText1. In ListView2, the list boxes on pages 2 and 3 display up to two lines of text. Irrespective of whether the list box displays icons or not the text is assigned to EQikListBoxSlotText1 and EQikListBoxSlotText2. The text in slot two is only displayed when the second text line is visible.

On page 3, the small icon is placed in the EQikListBoxSlotLeft-SmallIcon1 while the larger icon is placed in the EQikListBoxS-lotLeftMediumIcon1 (see Figure 6.9). The list box determines which content to display at run time.

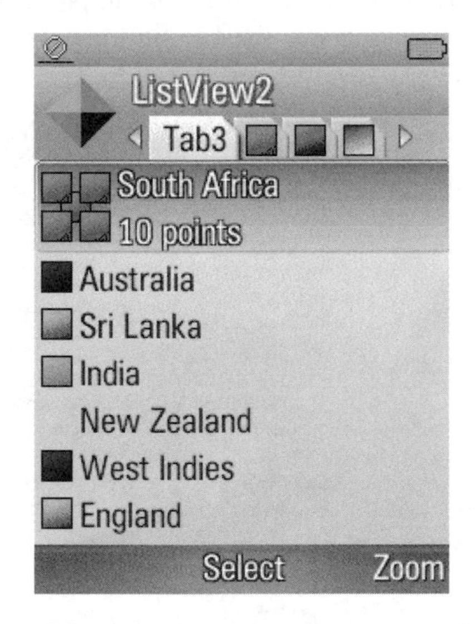

Figure 6.9 EQikListBoxMediumIconTwoLines

The following code, simplified slightly from the actual example code, shows text and icons being added to various slots within a list box.

```
void CAppSpecificListView::AddItemsToList3L()
  {
  CQikListBox* listbox = LocateControlByUniqueHandle<
          CQikListBox>(EAppSpecificListViewListId3);
  MQikListBoxModel& model(listbox->Model());
  model.ModelBeginUpdateLC();
  TBuf<KMaxListItemText>bb;
  for (TInt i=0;i<KListView2Items;i++)
    {
    MQikListBoxData* lbData = model.NewDataL(
            MQikListBoxModel::EDataNormal);
    CleanupClosePushL(*lbData);
    // Add text to slots.
    iEikonEnv->ReadResourceL(bb,R_STR_LIST_CONTENT_1+i);
```

```
    lbData->AddTextL(bb,EQikListBoxSlotText1);
    iEikonEnv->ReadResourceL(bb,R_STR_LIST_DETAILS_1+i);
    lbData->AddTextL(bb,EQikListBoxSlotText2);
    // Add small icons to slots.
    CQikContent* icon = CQikContent::NewL(NULL, KMbmFile,
            EMbmListview2Icon0, EMbmListview2Icon0mask);
    CleanupStack::PushL(icon);
    lbData->AddIconL(icon,EQikListBoxSlotLeftSmallIcon1);
    CleanupStack::Pop(icon);
    // Add medium icons to slots.
    icon = CQikContent::NewL(NULL, KMbmFile,
                  EMbmListview2Largeicon0,
             EMbmListview2Largeicon0mask);
    CleanupStack::PushL(icon);
    lbData->AddIconL(icon,
     EQikListBoxSlotLeftMediumIcon1);
    CleanupStack::Pop(icon);      // lbData taken ownership.
    CleanupStack::PopAndDestroy(lbData);
    }
  model.ModelEndUpdateL();
  }
```

Slots are defined by the creator of the resource structure describing a particular list box. For the pre-defined list box types, you only need to know which data is associated with which slot and where that is displayed. The example application provides that information. If you create custom list boxes (see Section 6.2.2), you become the resource creator. You therefore decide which slots are defined and where they are displayed.

Each slot may have multiple elements associated with it. By default, the first element added to a slot will be displayed. Usually, you would assign multiple elements to a slot if you need to display different content depending on the internal state of your application. As an alternative, it is possible to construct a list box where the user can switch between the different elements, a so-called 'content-swapping' list box (see Figure 6.10). Examples of such list boxes are presented by `r_list_view_page9_control` and `r_list_view_page12_control`, the latter being a custom list box.

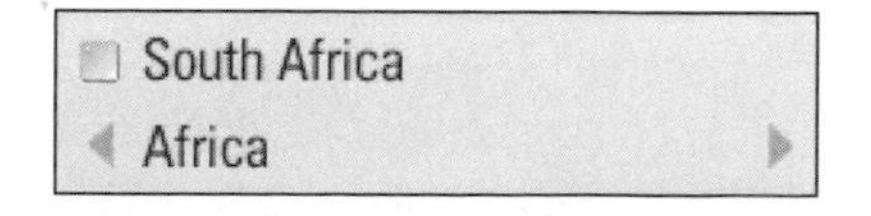

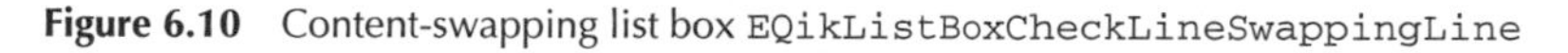

Figure 6.10 Content-swapping list box `EQikListBoxCheckLineSwappingLine`

The code to add items to page 9 is:

```
void CAppSpecificListView::AddItemsToList9L()
  {
```

```
CQikListBox* listbox=LocateControlByUniqueHandle<
       CQikListBox>(EAppSpecificListViewListId9);
MQikListBoxModel& model(listbox->Model());
model.ModelBeginUpdateLC();
TBuf<KMaxListItemText>bb;
for (TInt i=0;i<KListView2Items;i++)
  {
  MQikListBoxData* lbData = model.NewDataL(
            MQikListBoxModel::EDataNormal);
  CleanupClosePushL(*lbData);
  // This listbox has two lines of data.
  iEikonEnv->ReadResourceL(bb,R_STR_LIST_CONTENT_1+i);
  lbData->AddTextL(bb,EQikListBoxSlotText1);
  // We swap between items in the second line.
  iEikonEnv->ReadResourceL(bb,R_STR_LIST_ALT_TEXT_1+i);
  lbData->AddTextL(bb,EQikListBoxSlotText2);
  iEikonEnv->ReadResourceL(bb,R_STR_LIST_DETAILS_1+i);
  lbData->AddTextL(bb,EQikListBoxSlotText2);
  CleanupStack::PopAndDestroy(lbData);
  }
model.ModelEndUpdateL();
}
```

You should be able to see two items being added to `EQikListBox SlotText2`. Figure 6.11 shows how the content would be presented to the user.

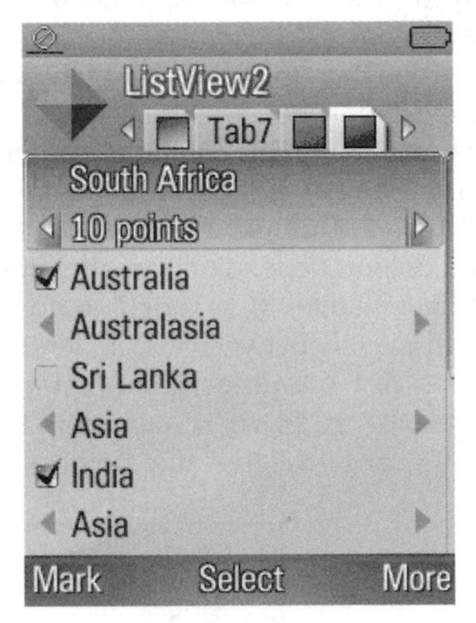

Figure 6.11 Content-swapping list box in a view

Since `EQikListBoxSlotText2` allows content swapping, the user can change the display between the content associated with the slot.

TabActivatedL(TInt aTabId)

Often there is a requirement to know when the user navigates between the pages presented by an application. The `TabActivatedL()` method is called when the current page is changed; the parameter indicates the page that is made active.

```
void CAppSpecificListView::TabActivatedL(TInt aTabId)
  {
  // Display info message saying which tab is activated.
  TBuf<128>bb;
  if (aTabId==EAppSpecificListViewPageId1)
    iEikonEnv->Format128(bb,R_STR_TAB_ACTIVATED_INFO,1);
  else if (aTabId==EAppSpecificListViewPageId2)
    iEikonEnv->Format128(bb,R_STR_TAB_ACTIVATED_INFO,2);
  else if (aTabId==EAppSpecificListViewPageId3)
    iEikonEnv->Format128(bb,R_STR_TAB_ACTIVATED_INFO,3);
  else
    ...
  if (bb.Length())
    iEikonEnv->InfoMsg(bb);
  // Let base class perform its work.
  CQikMultiPageViewBase::TabActivatedL(aTabId);
  }
```

6.2.2 Custom List Boxes

`ListView2` introduces the ability to define custom list boxes. At the simplest level you can choose between numerous combinations of `standard_normal_layout` and `standard_highlight_layout` to produce various effects. The list boxes displayed on pages 11 and 12 of our example demonstrate how to construct highly customized list boxes.

Anatomy of a List Box

From a high-level perspective, a list box starts with a single column of content. The column can then be broken into one or more rows of content. Each of those rows can be further broken down into columns. This process can continue to form a tree-like structure of arbitrary depth but, in practice, it is unlikely to go more than a few levels deep.

If you consider the `EQikListBoxIconLine` list box (see Figure 6.12a), at the highest level there is a single column of content, the list box. Each element in the column comprises a single row, the rows in the list box. The single row contains two columns, one for the icon and the other for the text component. The icon and text components complete the tree so no further items are required.

In contrast, `EQikListBoxCheckLineSwappingLine` has two rows, each broken into a different number of columns (Figure 6.12b).

Grid List Box

A grid list box is created using a custom layout. While there is a slightly different interpretation for grid-like list boxes, the general principles are the same.

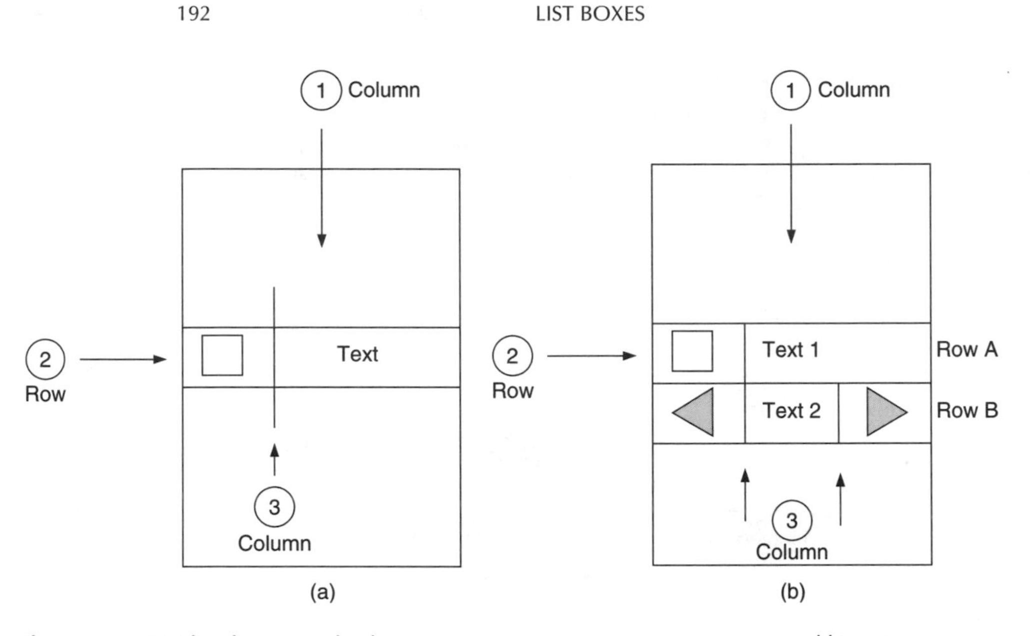

Figure 6.12 Dividing by row and column to create a) `EQikListBoxIconLine` and b) `EQikListBox-CheckLineSwappingLine`

A grid list box starts with a single column containing the content. In the case of the grid, the column is interpreted as an individual grid cell. The column can be broken down into a set of rows. If required, each row can be further broken into columns, and so on (see Figure 6.13).

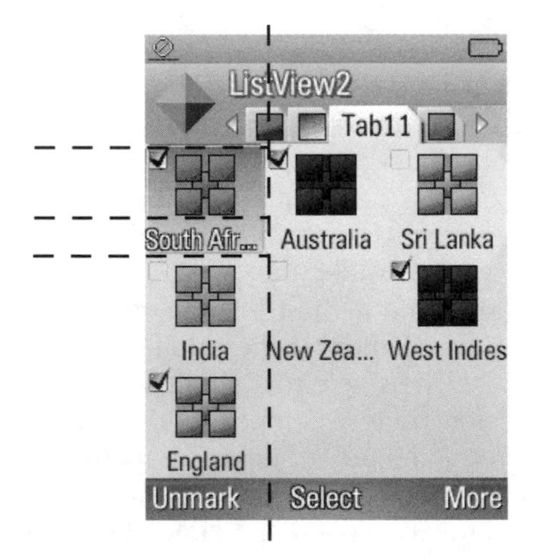

Figure 6.13 Dividing up a grid list box

Page 11 of `ListView2` displays a grid list box. Each cell in the grid consists of an icon displayed above some text. Mapping that onto the previous description, you should see that we have a single column or grid cell. Each element in that *column* comprises two rows. No further sub-division is required to display our content so we have reached the end of the hierarchy.

Our grid is described by the following resource code:

```
RESOURCE QIK_LISTBOX_LAYOUT r_app_list11_custom_layout
  {
  flags = 0;
  columns =
    {
    QIK_LISTBOX_COLUMN
      {
      type = QIK_LISTBOX_PARENT_TYPE;
      width_type = EQikListBoxColWidthGrab;
      slot_id = EQikListBoxSlotParent1;
      rows =
        {
        QIK_LISTBOX_ROW
          {
          type = QIK_LISTBOX_ICON_TYPE
            {
            size=EQikListBoxSizeMediumIcon;
            };
          height_type=EQikListBoxRowHeightGrab;
          height_value=1;
          slot_id=EQikListBoxSlotLeftMediumIcon1;
          },
        QIK_LISTBOX_ROW
          {
          type = QIK_LISTBOX_TEXT_TYPE
            {
            alignment=EQikListBoxTextAlignCenter;
            font_size=EQikListBoxFontSmall;
            };
          height_type =
           EQikListBoxRowHeightFromContentType;
          height_value=1;
          slot_id=EQikListBoxSlotText1;
          }
        };
      }
    };
  }
```

You should be able to identify the two rows; one containing the icon and one containing the text. The two rows belong to a parent or containing column element. You should be able to see where the slot IDs are being defined. When you come to populate your list box, the slot ID values you have assigned to the individual components are used to set the content.

In order for the list box to correctly display as a grid there is one other important field to configure: the *flags*. By default, list boxes are said to be row aware. We need to disable this for grids. This is achieved by setting the flags to zero.

Content-Swapping List Box

A second example of custom list box design is displayed on page 12 of the `ListView2` example (see Figure 6.14). The list box is a single column. Each element in the column is a single row. The single row has four columns, left and right arrows, an icon and some text.

Figure 6.14 Content-swapping list box

The resource defining the content-swapping list box is:

```
RESOURCE QIK_LISTBOX_LAYOUT r_app_list12_custom_layout
  {
  columns =
    {
    QIK_LISTBOX_COLUMN
      {
      type = QIK_LISTBOX_PARENT_TYPE;
      slot_id = EQikListBoxSlotParent1;
      rows =
        {
        QIK_LISTBOX_ROW
          {
          type = QIK_LISTBOX_PARENT_TYPE;
          slot_id = EQikListBoxSlotParent2;
          columns=
            {
            QIK_LISTBOX_COLUMN
              {
              type = QIK_LISTBOX_LEFT_ARROW_TYPE
                {
                slots_to_swap =
                  {
                  EQikListBoxSlotLeftSmallIcon1,
                  EQikListBoxSlotText1
                  };
```

```
                        };
                      width_type =
                          EQikListBoxColWidthFromContentType;
                          slot_id = EQikListBoxSlotLeftArrow1;
                      },
                    QIK_LISTBOX_COLUMN
                      {
                      type = QIK_LISTBOX_ICON_TYPE
                        {
                        size=EQikListBoxSizeSmallIcon;
                        };
                      width_type =
                          EQikListBoxColWidthFromContentType;
                          slot_id = EQikListBoxSlotLeftSmallIcon1;
                      },
                    QIK_LISTBOX_COLUMN
                      {
                      type = QIK_LISTBOX_TEXT_TYPE;
                      width_type = EQikListBoxColWidthGrab;
                      width_value = 1;
                      slot_id = EQikListBoxSlotText1;
                      },
                    QIK_LISTBOX_COLUMN
                      {
                      type = QIK_LISTBOX_RIGHT_ARROW_TYPE
                        {
                        slots_to_swap =
                          {
                          EQikListBoxSlotLeftSmallIcon1,
                          EQikListBoxSlotText1
                          };
                        };
                      width_type=
                          EQikListBoxColWidthFromContentType;
                      slot_id = EQikListBoxSlotRightArrow1;
                      }
                    };
                  }
                };
              }
            };
          }
```

This example demonstrates the ability for a single row to swap between multiple pieces of information. When data is added to the list box, up to two icons and two pieces of text are added to each slot. By default the first item added to a slot, that is, the content at index 0 of the slot, is displayed. Tapping the left or right arrows changes the current slot index, causing alternative content to be displayed (see Figure 6.15).

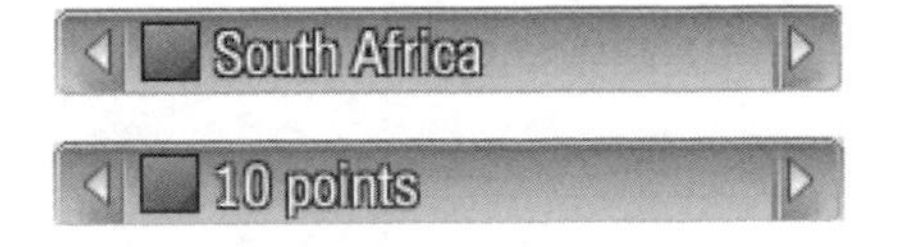

Figure 6.15 Content-swapping in ListView2 page 12

The Contacts application makes good use of the content-swapping capability of `EQikListBoxCheckLineSwappingLine` (see Figure 6.16). The user can select the required telephone number directly in the list view and then dial it.

Figure 6.16 Content-swapping in Contacts list view

When constructing grid list boxes or list boxes containing items that can be swapped it is important to understand what happens to left and right arrow keys, particularly in the context of multi-page views.

In pen-enabled configurations, the user can freely decide whether to tap a page tab and select a page, or tap the arrow in a content-swapping list box to select alternative content.

In softkey-only configurations, the left and right keys are used for both actions. Where the content-swapping list box has focus, it consumes the arrow key command and changes the content. The user cannot change page. We therefore recommend that content-swapping list boxes should not be part of a multi-page design in commercial applications.

7

Commands and Categories

The UIQ 3 Command Processing Framework (CPF) maps your application's commands to hardware keys, softkeys and menus. In this chapter, we show you how to add commands to your application. The `Commands1` example application demonstrates how to use the CPF in your application.

Categories are used to subdivide data such as Work and Personal appointments in the Agenda application. Categories are linked to commands because category selection and management are handled via the menu. The `Categories` example application shows how you can start to use categories in your application.

7.1 Commands Overview

The functional specification for your application should define the set of commands that the user can perform, for example, the commands 'Add entry' and 'Delete entry'. Your specification may go further and define how the user interacts with the application to perform these commands. You might specify that a menu option entitled 'Add entry' should be placed within the Edit menu.

Mobile phones have very particular interaction styles based on the hardware, operating system and manufacturer preferences. This has an especially large impact on command processing because commands are actioned by pressing buttons and, if available, by interaction with a touchscreen.

UIQ 3 supports a variety of interaction styles and allows applications to support the whole range of styles from a single executable. You must therefore avoid trying to define a single way for the user to interact with your application. UIQ 3 has a CPF that is designed to map application commands on to the available hard and softkeys and the menu. The concepts behind the CPF are described in Chapter 2.

We can break command handling into two parts:

- presenting a command to the user via some kind of UI component
- processing the command.

When processing a command, we do not need to worry about how the command was invoked but only that it was invoked. A command identifier, usually in the form of an integer value, is delivered from the command presentation code to the command processing code.

On the other hand, presenting commands to a user has become somewhat harder since what is correct for one UI style is not always correct or even possible for a different UI style.

When we present commands to the user, we must also consider the following factors:

- Is the command valid, given the current application state?
- How does the user cause the command to occur?
- How frequently is the command likely to be required?
- How important is the command?

For the purpose of presenting commands, UIQ 3 phones are classified as:

- **Pen Style:** there is a menu available and, optionally, a button bar that can contain commands. A limited set of other keys may also be available. The screen is touch sensitive.
- **Softkey Style:** there is a menu available and a softkey area is normally displayed directly above a set of two or three keys, usually at the bottom of the screen. A limited set of other keys may also be available. The screen is not touch sensitive.
- **Softkey Style Touch:** commands are presented in the same way as in Softkey Style, but they can also be selected using the touchscreen. In practice, this is the same as Softkey Style when you are defining commands in your application.

For an optimal user interface, a mechanism of dynamically allocating commands to the menu, buttons, softkeys and other keys available on a particular phone is required, dependent on the phone type. On the other hand, a professional quality application needs to maintain a degree of control over where commands are positioned, how they are grouped together, how they are separated and in what order they are presented to the user to aid in application usability. For example, if your application supports the Cut, Copy, Paste and Exit commands but they are presented in the order Copy, Exit, Paste and Cut then users familiar with such commands would find the application more difficult to operate.

The UIQ 3 solution to these conflicting requirements is to associate a set of commands with each view configuration you define. Each command is given a set of attributes that inform the framework how to present the command. Different attributes can be associated with a particular command dependent on the configuration.

Section 7.2 describes how UIQ 3 implements commands by way of an example application.

7.2 Commands1 Example Application

Commands1 is a two-page application in which each page contains a single list box. The list boxes are single line, text only. To demonstrate using commands, we implement the ability to add, delete and sort entries.

The example application also demonstrates numerous aspects of the CPF including:

- defining commands on a view-wide basis with resources
- defining commands on a page-specific basis with resources
- adding and removing commands in software instead of using resources
- processing commands within an application
- configuring the command state depending on the application state.

7.2.1 Page-Independent Commands

Commands1.rss shows a set of view configurations. Here is one of them:

```
QIK_VIEW_CONFIGURATION
  {
  ui_config_mode = KQikPenStyleTouchPortrait;
  command_list = r_list_view_generic_commands;
  view = r_list_view_view;
  }
```

We specify r_list_view_generic_commands to define a set of commands that are available to the user irrespective of the page that is displayed within a particular view configuration.

The r_list_view_generic_commands resource looks like this:

```
RESOURCE QIK_COMMAND_LIST r_list_view_generic_commands
  {
```

```
items=
  {
  QIK_COMMAND
    {
    id = EAppCmdHelp;
    type = EQikCommandTypeHelp;
    groupId = EAppCmdMiscGroup;
    priority = EAppCmdHelpPriority;
    text = "Help";
    },
  QIK_COMMAND
    {
    id = EAppCmdAbout;
    type = EQikCommandTypeScreen;
    groupId = EAppCmdMiscGroup;
    priority = EAppCmdAboutPriority;
    text = "About";
    },
  QIK_COMMAND
    {
    id = EAppCmdDebugTest;
    type = EQikCommandTypeScreen;
    stateFlags = EQikCmdFlagDebugOnly;
    groupId = EAppCmdMiscGroup;
    priority = EAppCmdDebugTestPriority;
    text = "Test";
    },
  QIK_COMMAND
    {
    id = EAppCmdDelete;
    type = EQikCommandTypeDelete;
    text = "";
    }
  };
}
```

When we run `Commands1`, we see that the Help, About and Test commands have been added to the menu in both Softkey Style Touch and Pen Style UI configurations (in Figure 7.1).

The **Delete** command, `EAppCmdDelete`, is allocated to the Delete key.

7.2.2 Page-Specific Commands

When we look at a page definition resource, we can see the following piece of code:

```
QIK_VIEW_PAGE
  {
  page_id = EAppSpecificListViewPageId2;
  tab_bmpid = EMbmCommands1Tab2;
  tab_bmpmaskid = EMbmCommands1Tab2mask;
  page_content = r_list_view_page2_control;
  command_list = r_page2_specific_commands;
  }
```

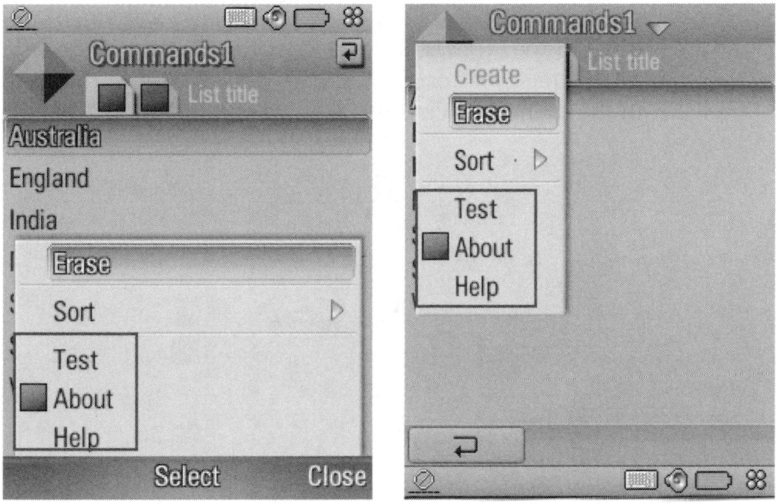

(a) (b)

Figure 7.1 Page 1 of Commands1 in a) Softkey Style Touch and b) Pen Style UI configurations

As suggested by the name, r_page2_specific_commands, we can define a set of commands that are only available when a particular page is selected; no external software is required, only a resource element.

```
RESOURCE QIK_COMMAND_LIST r_page2_specific_commands
  {
  items=
    {
    QIK_COMMAND
      {
      id = EAppCmdAdd;
      type = EQikCommandTypeScreen;
      groupId = EAppCmdEditGroup;
      priority = EAppCmdAddPriority;
      text = "Add";
      },
    QIK_COMMAND
      {
      id = EAppCmdDelete2;
      type = EQikCommandTypeScreen;
      groupId = EAppCmdEditGroup;
      priority = EAppCmdDeletePriority;
      text = "Delete";
      }
    };
  }
```

If you run the application, you can observe how the generic and page-specific commands are combined to produce the application command set for each page. No application-specific software is required to achieve this. All the work is performed by the framework code.

In page 2 of `Commands1`, we can see the Add and Delete commands in the menu (see Figure 7.2).

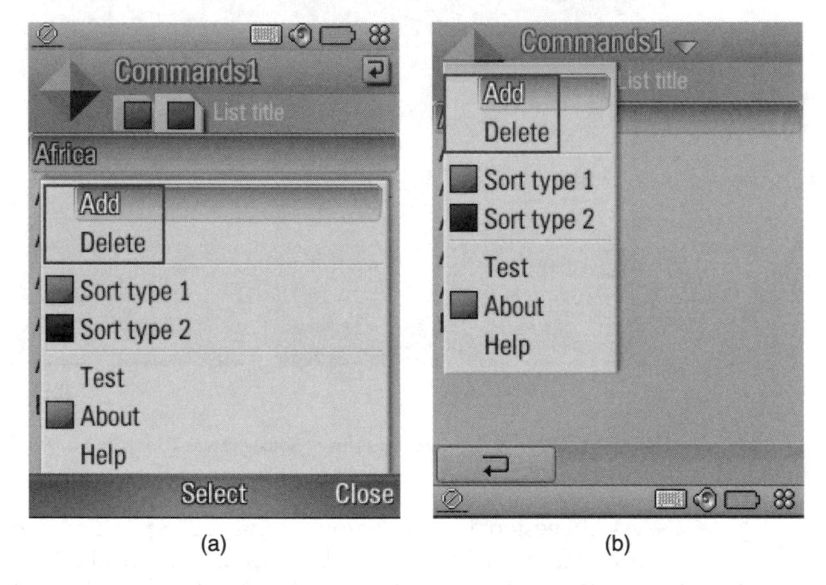

Figure 7.2 Page 2 of `Commands1` in a) Softkey Style Touch and b) Pen Style configurations

The command `EAppCmdDelete2` adds a second Delete command delivered by a menu (see page 2). In contrast, `EAppCmdDelete` is delivered by the Delete key press.

Your application may not need to distinguish between how a user selects a specific command. Our example application simply demonstrates how you can achieve this if required.

7.2.3 Item-Specific Commands

This type of command is associated with the control that currently has focus. For example, if a phone number is highlighted, the first choice action may be 'Make a call' and a second action could be 'Send an SMS'.

In 3SK&B mode, the CPF attempts to place the commands like this:

- Center softkey
- Left softkey (if no Done command)
- Right softkey
- in the menu (no more softkeys available).

7.2.4 Command Definition

A QIK_COMMAND_LIST comprises a set of QIK_COMMAND elements. An individual command is defined by the QIK_COMMAND resource:

```
STRUCT QIK_COMMAND
  {
  BYTE version=3;
  LONG id=EQikCmdUnassigned;
  LONG type;
  LONG groupId=0;
  LONG namedGroupLinkId=0;
  LONG namedGroupId=0;
  LONG priority=0;
  LTEXT text;
  LTEXT shortText="";
  LLINK icon=0;
  LONG stateFlags=0;
  LONG cpfFlags=0;
  }
```

Version

This is internal to UIQ 3 and should not be changed by the developer.

ID

Set this field to contain the command identifier that you wish to deliver to the command-processing code. Command identifiers are usually defined as a set of enums within the application-specific HRH file. For the Commands1 example, the following enums are defined:

```
enum TAppCommandIds
  {
  EAppCmdAbout=0x1000,  // Below 0x1000 reserved for UIQ 3.
  EAppCmdHelp,
  EAppCmdContinue,
  EAppCmdAdd,
  EAppCmdAdd2,
  EAppCmdDelete,
  EAppCmdDelete2,
  EAppCmdDelete3,
  EAppCmdSortCascade,
  EAppCmdSortType1,
  EAppCmdSortType2,
  EAppCmdSortType3,
  EAppCmdSeparator,
  EAppCmdSortOrder,
  EAppCmdSortAltType1,
  EAppCmdSortAltType2,
  EAppCmdDebugTest,
  EAppCmdLastItem
  };
```

You may note that some commands appear to be duplicated, such as EAppCmdAdd and EAppCmdAdd2. It is possible to define more than one way in which a user may request a particular operation; for example, a command might exist both in a menu and on a key press. Even though the same functionality is requested by the user, the CPF needs to distinguish between each QIK_COMMAND. As an alternative to duplicating commands, the CPF flag, EQikCpfFlagDuplicateInMenuPane, described later, may be more appropriate.

Type

This is perhaps the most important field and is the most influential one in determining where the command is displayed within a particular UI style. This field categorizes a command into one of the following types:

```
enum TQikCommandType
  {
  EQikCommandTypeScreen,
  EQikCommandTypeItem,
  EQikCommandTypeYes,
  EQikCommandTypeNo,
  EQikCommandTypeDone,
  EQikCommandTypeDelete,
  EQikCommandTypeCancel,
  EQikCommandTypeHelp,
  EQikCommandTypeSystem,
  EQikCommandTypeOperator,
  EQikCommandTypeFep,
  EQikCommandTypeCategory
  };
```

Since the objective is to allow individual phones to assign commands to the range of controls available on a specific phone, you specify your commands in abstract terms rather than stating where they appear in the UI. The descriptions below apply to the generic UIQ 3.0 3SK&B as presented by the emulator. Real mobile phones may display different behavior: however, as long as you follow the guidelines and describe your commands using the UIQ command definitions, you do not need to worry too much about where they are displayed because the CPF will take care of that.

EQikCommandTypeScreen

Perhaps the most common type, this is used to define commands that are applicable to the entire view or dialog. This is used when the command does not belong to any other command type. Such commands are typically displayed in a menu.

EQikCommandTypeItem

This defines commands that are only available for the control that has focus. For example, if the focus is on a text string that represents a phone number, then the item commands Call and SMS could be displayed. If the next item represents an email address, then a Send Mail command could be displayed.

Controls add and remove their commands in their PrepareForFocusGainL() and PrepareForFocusLoss() methods. The CQikCommandManager uses CCoeControl::Parent() to find out to which list to add the control's commands.

EQikCommandTypeYes

This indicates that the command is a positive response to a question presented by the UI. This command type is normally used in conjunction with EQikCommandTypeNo. When both command types are present, the CPF attempts to present them within the same container in the UI, with the EQikCommandTypeYes to the left and EQikCommandTypeNo to the right (see Figure 7.3).

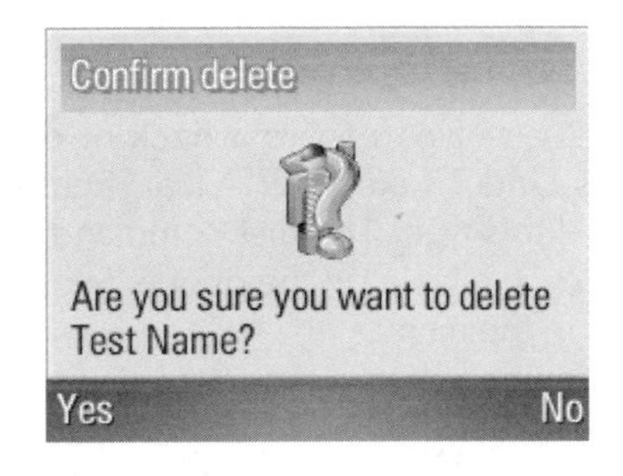

Figure 7.3 Confirmation dialog example

EQikCommandTypeNo

This indicates that the command is a negative response to a question presented by the UI. This command type is normally used in conjunction with EQikCommandTypeYes as described above.

EQikCommandTypeDone

This command type represents the positive completion of an action. Positive actions are usually placed towards the left. For example, in a typical softkey style phone, the command with type EQikCommandTypeDone is placed on the Left softkey. If the Center softkey is empty, the command is moved there. You can use the EQikCpfFlagDoneCommandPrefer-

ToBePlacedOnPositiveSoftkey flag, described later, to indicate you prefer the command to be placed on the Left softkey.

Commands of the types EQikCommandTypeDone and EQikCommand TypeCancel are frequently used as a pair. Similarly, commands of type EQikCommandTypeYes and EQikCommandTypeNo are frequently used as a pair. You should not mix the two command pairings.

EQikCommandTypeCancel

This indicates that the command represents the canceling of an action. There is typically only a single command of type EQikCommandType-Cancel per view or dialog. If a specific phone has a Back key, the first command of type EQikCommandTypeCancel is assigned to that key. On Pen Style phones, this command is usually placed to the right of any EQikCommandTypeDone.

One feature of the CPF is that if a view or dialog does not define a command of type EQikCommandTypeCancel then the CPF automatically adds one, irrespective of whether the application wants a Cancel command. Applications need to explicitly remove the command type if it is not required. We show how to do this later on.

EQikCommandTypeDelete

The Delete type causes a back or delete action to occur. Commands such as Undo, Take back, Clear, Remove and Delete would usually be assigned this type. The first command of this type is associated with any Clear key available on the phone. Subsequent commands of this type are placed in the menu.

```
QIK_COMMAND
  {
  id = EAppCmdDelete;
  type = EQikCommandTypeDelete;
  text = "";
  }
```

We have defined a single command of type EQikCommandType Delete in our command set. This is assigned to the Clear key in the emulator and currently available phones but, in general, it is not advisable to assume this would happen on every UIQ 3 phone. In a production-quality application, you should ensure there is at least one other mechanism that performs this command. You can see this in Commands1: page 1 has a Delete command assigned to the Clear key and page 2 adds an alternative Delete command which is placed in the menu.

EQikCommandTypeHelp

This indicates that the command is used to display some context-sensitive help. Commands of this type only differ from EQikCommandType Screen commands in the position the framework assigns the command within a menu. The framework places commands of this type in a consistent position across all applications.

```
QIK_COMMAND
  {
  id = EAppCmdHelp;
  type = EQikCommandTypeHelp;
  groupId = EAppCmdMiscGroup;
  priority = EAppCmdHelpPriority;
  text = "Help";
  }
```

You should note that command types act as a first-order sort key for commands. The priority acts as a second-order sort key.

EQikCommandTypeSystem

Occasionally commands need to be added by the system to perform certain actions. For example, on some Sony Ericsson phones, a task manager command is added to the set of commands presented by the menu bar. In general, applications should not use commands of this type.

EQikCommandTypeOperator

Commands can be added when an application is interacting with various operator services. Applications should not use commands of this type.

EQikCommandTypeFep

Commands can be added by the current front end processor (FEP) to allow for configuration of the FEP. Applications should not use commands of this type.

EQikCommandTypeCategory

Commands of this type should be associated with the category management presented by the application. On Softkey Style phones, these commands are presented in a cascade menu item; on Pen Style phones, these commands are placed in the category drop-down menu (see Figure 7.4). We discuss categories in more detail in Section 7.3.

Command types offer a high degree of flexibility when constructing your UI. With flexibility comes responsibility. The framework does not

(a) (b)

Figure 7.4 Category menu in a) Softkey Style Touch and b) Pen Style UI configurations

prevent you defining all your commands as one particular type should you wish, however such an action is unlikely to result in an acceptable UI for your application across all phones.

> **Pro tip**
>
> We recommend that you determine which commands should be assigned to the single controls such as the Back and Clear keys, then which commands should be associated with any softkey area or button bar. The remaining commands form the menu and are therefore assigned the `EQikCommandTypeScreen` type.

`groupId`

Often a set of commands is related to each other. To indicate this you can assign commands the same group ID from an application-specific `enum` list.

In our example we have defined the following groups:

```
enum TAppCommandGroupIds
  {
  EAppCmdEditGroup,
  EAppCmdSortGroup,
  EAppCmdSortCmdsLink,
  EAppCmdMiscGroup,
  EAppCmdLastGroupId
  };
```

We have identified three logical groups of commands: those associated with editing, those associated with sorting and a miscellaneous group containing the remaining commands.

The CPF displays those items with the same `group` together. If there are other commands in the menu, the group is surrounded by dividers. A `groupId` value defines the order in which groups of commands are presented. We have not yet specified any order for the commands within the group; we use the priority field to do that.

There are two types of group, unnamed (or anonymous) and named groups. The `groupId` field represents unnamed groups. In `Commands1` page 2, we define three groups of commands, which are displayed in a menu as shown in Figure 7.5:

- `EAppCmdEditGroup`: Add, Delete
- `EAppCmdSortGroup`: Sort type 1, Sort type 2
- `EAppCmdMiscGroup`: About

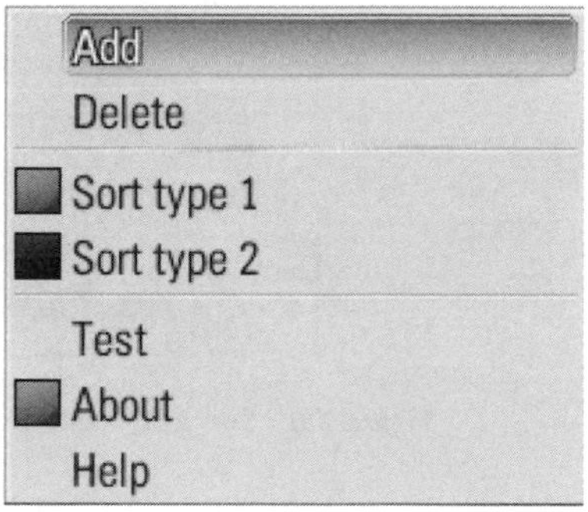

Figure 7.5 Menu for Page 2 of `Commands1`

`namedGroupId`

We use `namedGroupId` to create a named group. This field contains the named group ID value. In contrast to unnamed groups, which are only displayed within menu panes, named groups can be presented in sub-menus, on the softkeys or on button bars depending on other command attributes. In our example, we use the `namedGroupId` to display the sort commands within a cascade sub-menu (see Figure 7.6).

`namedGroupLinkId`

Commands belonging to a named group need to be linked back to a command from an unnamed group. For example, if a named group is displayed as a sub-menu, the `namedGroupLinkId` field of the command is set to the same value as the namedGroupId. The following resource fragment helps explain this further:

```
QIK_COMMAND
  {
  namedGroupLinkId = EAppCmdSortCmdsLink;
  priority = EAppCmdSortCascadePriority;
  text = "Sort";
  },
  QIK_COMMAND
  {
  namedGroupId = EAppCmdSortCmdsLink;
  priority = EAppCmdSortType1Priority;
  text = "Sort by name";
  },
  QIK_COMMAND
  {
  namedGroupId = EAppCmdSortCmdsLink;
  priority = EAppCmdSortType2Priority;
  text = "Sort by 2nd letter";
  },
  // Remaining commands removed for clarity.
```

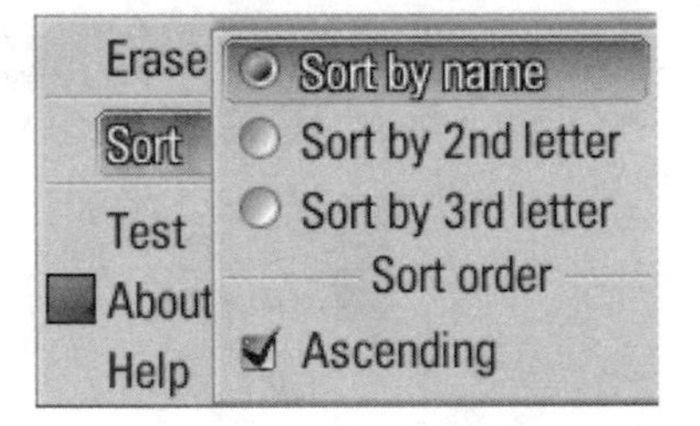

Figure 7.6 Sort sub-menu for page 1 of Commands1

The Sort command belongs to an unnamed group. The Sort by name and Sort by 2nd letter commands belong to a named group, EAppCmd SortCmdsLink, which forms the cascade sub-menu (see Figure 7.7).

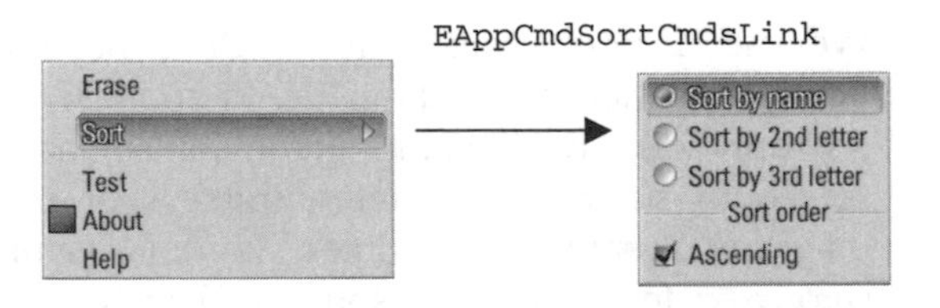

Figure 7.7 Using namedGroupId to link a sub-menu

Since the Sort command contains a namedGroupLinkId value equal to the namedGroupId (EAppCmdSortCmdsLink), a link between the two sets of commands is established. When the link is followed, a cascade style sub-menu is presented.

While this may seem complicated, it means that you can code once and your menu tree is displayed correctly in each UI configuration.

> **Pro tip**
>
> You should note that the order in which commands are defined within a resource and any indentation are purely cosmetic. We only lay out the commands in this way so as to document our intentions for the end result. It is the values assigned to the group and priority fields that influence the presentation order.

`priority`

When you have commands of the same type, it is often important to ensure that they are grouped together and displayed in a specific order. For example, Cut, Copy and Paste commands should always be displayed in that order for consistency with other applications. We have already seen how we can group commands using unnamed or named groups. Within a group, the priority field defines the order of the commands.

Our example defines the current `enum` list:

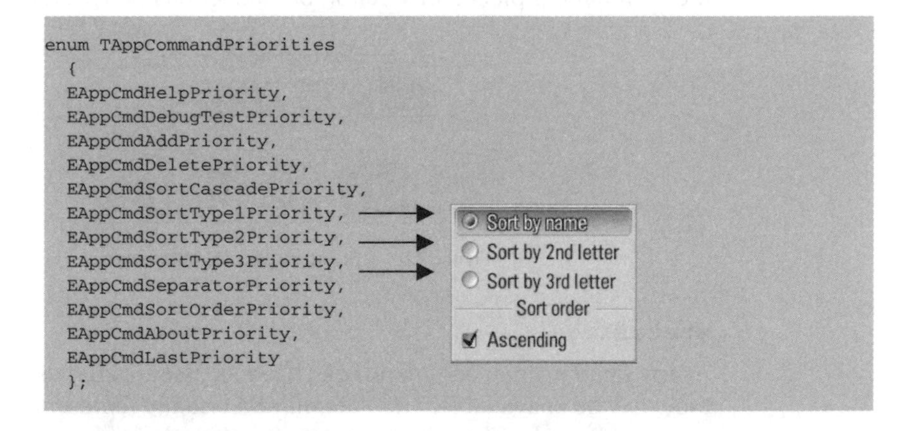

```
enum TAppCommandPriorities
  {
  EAppCmdHelpPriority,
  EAppCmdDebugTestPriority,
  EAppCmdAddPriority,
  EAppCmdDeletePriority,
  EAppCmdSortCascadePriority,
  EAppCmdSortType1Priority,
  EAppCmdSortType2Priority,
  EAppCmdSortType3Priority,
  EAppCmdSeparatorPriority,
  EAppCmdSortOrderPriority,
  EAppCmdAboutPriority,
  EAppCmdLastPriority
  };
```


If you compare the priority list with the sort sub-menu image, you should be able to see that `EAppCmdSortType1Priority` has a lower value and hence higher priority than `EAppCmdSortType2Priority`. Therefore the Sort by name command is displayed before Sort by 2nd letter within the named group `EAppCmdSortCmdsLink`.

Pro tip

You should also consult the *UIQ 3 Style Guide* document in the UIQ 3 SDK to ensure applications use the group and priority attributes to present commands in the recommended groups and order within menus.

`text`

If you want the command to be displayed as text, supply the text content here.

`shortText`

If a text command is to be displayed on a UI component that has restricted display space, such as a softkey or on the button bar, the CPF chooses the `shortText` if the `text` field is too long.

Icons

As an alternative to displaying commands with text, commands can be displayed as icons or as a combination of icon and text (see Figure 7.8). If the command is placed within a menu, the icon is displayed beside the text. If the command is placed in a button bar the icon is preferred over any text.

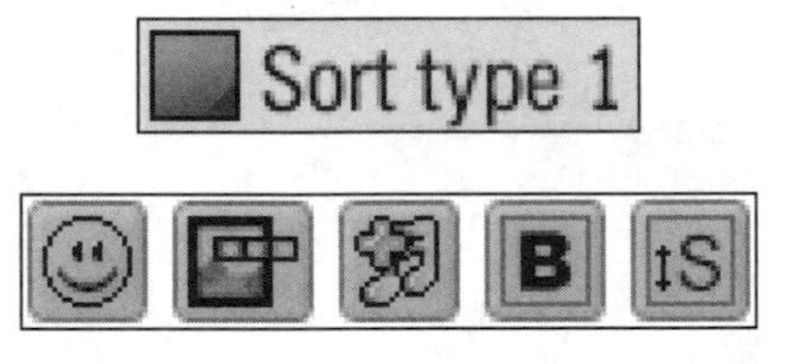

Figure 7.8 Icons in menu and button bar

`stateFlags`

At any point in time, commands can have an associated state, for example they may be unavailable. The state may be visually represented differently in each UI mode. The following state flags are defined.

`EQikCmdFlagDimmed`

The command is currently dimmed. A dimmed command is displayed to the user but it cannot be activated. Your application should not use this flag directly. Rather it would make the command unavailable. The phone-specific configuration would then apply its rules as to whether the configuration dims or hides unavailable commands.

EQikCmdFlagInvisible

The command is not displayed. As with `EQikCmdFlagDimmed`, an application should not use this flag directly.

EQikCmdFlagUnavailable

This indicates that a command is unavailable. On Softkey Style phones this usually makes the command invisible; in Pen Style phones it usually dims the command (see Figure 7.9). In our example, we have chosen to restrict the number of items displayed within two bounds. To achieve this we make the commands to create or delete entries unavailable at those boundaries.

```
CQikCommandManager& cm=CQikCommandManager::Static();
CQikListBox* listbox = LocateControlByUniqueHandle<
         CQikListBox>(EAppSpecificListViewListId1);
TBool avail=EFalse;
if (listbox->ItemCount()>=KCommand1Items)
  avail=ETrue;
cm.SetAvailable(*this,EAppCmdDelete3,avail);
avail=EFalse;
if (listbox->ItemCount()<KCommand1Items)
  avail=ETrue;
cm.SetAvailable(*this,EAppCmdAdd2,avail);
```

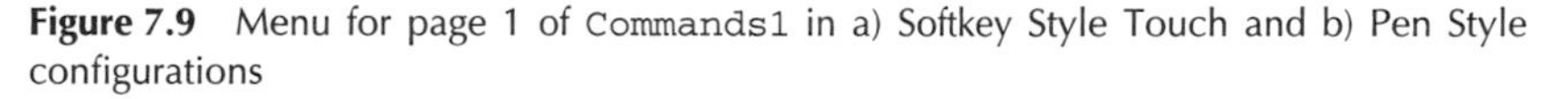

Figure 7.9 Menu for page 1 of `Commands1` in a) Softkey Style Touch and b) Pen Style configurations

EQikCmdFlagCheckBox

This indicates that the command is shown with an associated check box (see Figure 7.10).

```
QIK_COMMAND
  {
  id = EAppCmdSortOrder;
  type = EQikCommandTypeScreen;
  stateFlags = EQikCmdFlagCheckBox;
  namedGroupId = EAppCmdSortCmdsLink;
  priority = EAppCmdSortOrderPriority;
  text = "Ascending";
  }
```

Figure 7.10 Menu command with checkbox

This state flag is usually set within a `QIK_COMMAND` resource and not manipulated by the application. In contrast the application normally determines the actual value to be presented at run time.

EQikCmdFlagRadioStart

This indicates that the command is the first in a group of radio buttons (see Figure 7.11). This state flag is usually set within a `QIK_COMMAND` resource and not manipulated by the application. The application usually determines the value to be presented at run time.

```
QIK_COMMAND
  {
  id = EAppCmdSortType1;
  type = EQikCommandTypeScreen;
  stateFlags = EQikCmdFlagRadioStart;
  namedGroupId = EAppCmdSortCmdsLink;
  priority = EAppCmdSortType1Priority;
  text = "Sort by name";
  },
```

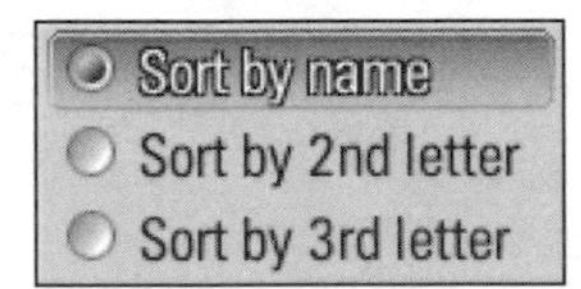

Figure 7.11 Menu commands with radio buttons

EQikCmdFlagRadioMiddle

This indicates that the command belongs to a group of radio buttons; however, it is neither the first or last in the list. This state flag is usually set within a `QIK_COMMAND` resource such as `EQikCmdFlagRadioStart`.

```
QIK_COMMAND
  {
  id = EAppCmdSortType2;
  type = EQikCommandTypeScreen;
  stateFlags = EQikCmdFlagRadioMiddle;
  namedGroupId = EAppCmdSortCmdsLink;
  priority = EAppCmdSortType2Priority;
  text = "Sort by 2nd letter";
  },
```

EQikCmdFlagRadioEnd

This indicates that the command is the last in a group of radio buttons. This state flag is usually set within a QIK_COMMAND resource such as EQikCmdFlagRadioStart.

```
QIK_COMMAND
  {
  id = EAppCmdSortType3;
  type = EQikCommandTypeScreen;
  stateFlags = EQikCmdFlagRadioEnd;
  namedGroupId = EAppCmdSortCmdsLink;
  priority = EAppCmdSortType3Priority;
  text = "Sort by 3rd letter";
  },
```

EQikCmdFlagSymbolOn

For a checkbox command, the box is checked. For a group of radio buttons, this indicates the currently chosen option. While the application can set this within a QIK_COMMAND resource, the application software is more likely to need to set or clear this flag to match any internal meaning. When commands are added to a view, the DynInitOrDelete CommandL() method is called for each command. In that method, you would perform actions such as the following:

```
CQikCommand* CAppSpecificListView::DynInitOrDeleteCommandL(
 CQikCommand* aCommand,
 const CCoeControl& aControlAddingCommands)
  {
  TBool val=EFalse;
  switch (aCommand->Id())
    {
    // Determine which radio buttons should be checked.
  case EAppCmdSortType1:
    if (iEngine->ListViewSortType()==ESortType1)
      val=ETrue;
    aCommand->SetChecked(val);
    break;
    // Determine whether the check box should be checked.
  case EAppCmdSortOrder:
    if (iEngine->ListViewSortOrder() == EAscending)
      val=ETrue;
    aCommand->SetChecked(val);
    break;
  default:
    break;
    }
  return(aCommand);
  }
```

EQikCmdFlagInlinePane

The command is displayed as a section divider containing text (see Figure 7.12).

```
QIK_COMMAND
  {
  id = EAppCmdSeparator;
  type = EQikCommandTypeScreen;
  stateFlags = EQikCmdFlagInlinePane;
  namedGroupId = EAppCmdSortCmdsLink;
  priority = EAppCmdSeparatorPriority;
  text = "Sort order";
  },
```

The section divider is not selectable and does not deliver a command. To display section dividers without text, use the group facility.

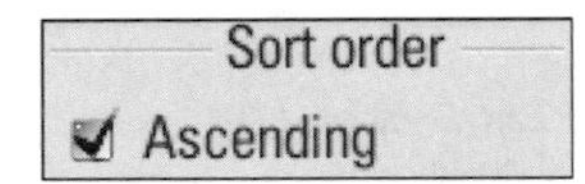

Figure 7.12 Section divider with text

EQikCmdFlagSortAlphabetic

If set, this indicates that commands are sorted alphabetically instead of by command priority.

EQikCmdFlagDebugOnly

This indicates that the command should only be presented in debug builds. The developers of UIQ 3 recognized that there is a requirement to allow application developers to have commands within a UI that would not normally be present in the final application. Prior to UIQ 3, application developers needed to perform some kind of automatic or manual pre-processing of a resource file to include or exclude such commands. UIQ 3 formalizes this process.

cpfFlags

A second set of flags, known as the CPF flags, is associated with a command. In general, these flags give hints to the framework as to how to prioritize the positioning of commands. The following CPF flags are defined.

EQikCpfFlagReplaceContainerPopoutDone

Setting this flag prevents any `CQikContainerPopouts` automatically adding a Done command. This flag is used by custom controls that wish to replace a Done command in any container or pop-out in which they may be placed.

EQikCpfFlagPreferToBePlacedInButtonbar

This informs the framework that the command should be placed in any available button bar rather than in the menu.

EQikCpfFlagDuplicateInMenuPane

This informs the framework that the command should be added in the menu pane as well as being present in the softkey or button bar region. It may be preferable to use this flag rather than manually duplicate a command through the use of separate `QIK_COMMAND`s.

EQikCpfFlagOkToExclude

On mobile phones, such as the Sony Ericsson P990i, that support two different screen sizes, this flag informs the CPF that it is acceptable to remove this command from the smaller screens. You need to carefully consider the subset of features that you make available. While this flag allows a single command for multiple view configurations, there is a trade-off between the complexity and maintainability introduced by attempting to run a single command list for all view configurations, compared to a separate command list, one for each view configuration.

EQikCpfFlagIsDefault

This informs the framework that the command is the primary action for the dialog or view and should be mapped to any Confirm key. You should ensure that only a single command has this attribute. When the command is displayed within a button bar, that item is emphasized.

EQikCpfFlagDoneCommandPreferToBePlacedOnPositiveSoftkey

This informs the CPF that a command should be placed on the Left softkey as opposed to the Center softkey should the Left softkey be unused. This applies only to commands of type `EQikCommandTypeDone`.

EQikCpfFlagHardwarekeyOnly

This informs the CPF that the command should only be placed on hardware buttons.

EQikCpfFlagTouchscreenOnly

This informs the CPF that the command should be made available only if the current configuration has a touchscreen. As with `EQikCpfFlagOkToExclude` there is a trade-off between maintaining a single command list and having a separate list for each view configuration.

EQikCpfFlagNoTouchscreenOnly

This informs the CPF that the command should only be made available if the current configuration does not have a touchscreen.

EQikCpfFlagInteractionMenubarOnly

This informs the CPF that the command should only be made available if the current configuration has a menu bar. This is only applicable when the mobile phone is in a Pen Style UI configuration.

EQikCpfFlagInteractionSoftkeysOnly

This informs the CPF that the command should only be made available if the current configuration has a softkey region. This is the opposite of `EQikCpfFlagInteractionMenubarOnly`, and only applies for mobile phones in a Softkey Style UI configuration.

EQikCpfFlagPortraitOnly

This informs the CPF that the command should only be made available if the current configuration is in portrait mode.

EQikCpfFlagLandscapeOnly

This informs the CPF that the command should only be made available if the current configuration is in landscape mode.

EQikCpfFlagExecuteRepeat

This requests the CPF to issue the command ID repeatedly while the key to which command is allocated remains depressed.

Summary

The CPF is rich in functionality, and command handling can be influenced by using the large number of available parameters. Getting to know these parameters and learning to read `QIK_COMMAND` resources are the keys to utilizing the functionality.

```
QIK_COMMAND
  {
  id = EAppCmdDebugTest;
  type = EQikCommandTypeScreen;
  stateFlags = EQikCmdFlagDebugOnly;
  groupId = EAppCmdMiscGroup;
  priority = EAppCmdDebugTestPriority;
  text = "Test";
  },
```

In this resource definition, we have created a command that:

- Delivers a command with ID `EAppCmdDebugTest` to the command-processing code.

- Is most likely displayed in a menu (has a type of EQikCommandTypeScreen).
- Is only visible in debug builds (due to the EQikCmdFlagDebugOnly state flag).
- Belongs to the unnamed EAppCmdMiscGroup and is therefore displayed with any other commands belonging to that group. In this application, the Help and About options also belong to that group of commands.
- Has been given a priority of EAppCmdDebugTestPriority.
- Displays the text 'Test' to the user.

Pro tip

If you get a QIKON-PANIC : 39 when attempting to display a view or running your application, the most likely cause is that you have defined more than one QIK_COMMAND with the same ID value.

7.2.5 Processing Commands

Commands are delivered to the HandleCommandL() method of a view. To process them, you simply override that method and perform the actions required by the indicated command. A typical HandleCommandL() looks like this:

```
void CAppSpecificListView::HandleCommandL(CQikCommand& aCommand)
  {
  switch (aCommand.Id())
    {
    case EAppCmdAbout:
      (new(ELeave)CAboutDialog)->ExecuteLD(R_ABOUT_DIALOG);
      break;
    case EAppCmdHelp:
      iEikonEnv->InfoWinL(R_HELP_TITLE,R_HELP_TEXT);
      break;
    case EAppCmdSortType1:
      iEngine->SetListViewSortType(ESortType1);
      SortListViewL();
      break;
    case EAppCmdSortType2:
      iEngine->SetListViewSortType(ESortType2);
      SortListViewL();
      break;
    default:           // For example, the back button...
      CQikViewBase::HandleCommandL(aCommand);
      break;
    }
  }
```

As can be seen from the function prototype, an object of type CQik-Command is delivered to the HandleCommandL() method. This object is the C++ equivalent of the QIK_COMMAND resource. If you view the class definition in QikCommand.h, you can see the object property that maps directly onto the resource fields.

Command Lifetime

Commands defined in a QIK_VIEW_CONFIGURATION are created at view construction time and remain loaded until the view is deleted, usually when the application terminates. In contrast, commands defined within a QIK_VIEW or QIK_VIEW_PAGE are loaded when the view or page comes to the foreground and unloaded when it switches to the background. Item type commands are loaded and unloaded as focus moves to and from a control. Any state information stored with a command will be lost when commands are unloaded.

Applications should present commands that are consistent with their internal data. By defining commands within the QIK_VIEW_CONFIG-URATION, it is possible to allow the command framework to maintain any state information. For example, if we define the sort commands as part of the QIK_VIEW_CONFIGURATION command set, we may be able to let the command framework store the sort order information for the lifetime of the application. However, we have chosen to define the sort commands within a QIK_VIEW_PAGE. When we swap pages, the commands are unloaded and the state information is lost. If we return to the page of sorted items, it is desirable that the menu options are consistent with sort order of the displayed list box items. To achieve this, we record the sort order state information when it is changed:

```
void CAppSpecificListView::HandleCommandL(
 CQikCommand& aCommand)
  {
  switch (aCommand.Id())
    {
    case EAppCmdSortType1:
     iEngine->SetListViewSortType(ESortType1);
     SortList1L();
     break;
    default:             // For example  the back button...
     CQikViewBase::HandleCommandL(aCommand);
     break;
    }
  }
```

The command state is then set when commands are loaded:

```
CQikCommand* CAppSpecificListView::
             DynInitOrDeleteCommandL(CQikCommand* aCommand,
```

```
                const CCoeControl& aControlAddingCommands)
{
TBool val=EFalse;
switch (aCommand->Id())
  {
  // Determine which radio buttons should be checked.
  case EAppCmdSortType1:
    if (iEngine->ListViewSortType()==ESortType1)
      val=ETrue;
    aCommand->SetChecked(val);
    break;
  default:
    break;
  }
return(aCommand);
}
```

Commands and Menus

If you compare UIQ 3 to other systems, it does not offer the opportunity for applications to modify the command state immediately prior to menus being displayed. The main reason for this is that applications do not explicitly define menus, only a set of commands for the UI to present to a user. Within a particular UI configuration, some commands may be presented within a drop-down menu; within other phone configurations, it is possible they are displayed permanently on screen. If commands are displayed permanently on screen, the command state should accurately represent the underlying data state.

Command State and `HandleCommandL()`

As previously noted, commands in UIQ 3 maintain their own state information within a `CQikCommand` object. When an application attempts to process a command, it may fail. The `HandleCommandL()` method is designed to be able to handle error conditions; in particular, the method leaving. However, simply leaving is often not sufficient, particularly if state information is maintained in more than one place. In general, applications should perform some kind of rollback when errors occur. A common requirement is to ensure that the application state is the same after the failed operation as it was before the operation was attempted.

Consider the following piece of code:

```
iEngine->SetListViewSortType(ESortType1);
SortListViewL();
```

You can see that the engine state is updated before the sort method, which can leave, is called. If the `SortListViewL()` leaves, we must assume that the sort has failed. The sort order is not the value we have now stored in the engine. The software should roll back to the previous value for the engine and list to remain synchronized.

You could try to solve the issue with this code change:

```
SortListViewL(ESortType1);
iEngine->SetListViewSortType(ESortType1);
```

However, we have a third component to keep in synchronization and that component is the command state.

If you observe the `Commands1` application closely, you may note that the radio button controls associated with the sort commands are updated to the new state before the command is delivered to the `HandleCommandL()` method. If the sort operation fails and we simply leave, the radio buttons display one state and the list box a different state. We need to roll back the command state such that all components remain synchronized.

The following code from our example ensures all components remain synchronized:

```
case EAppCmdSortType3:
  TAppSortTypes oldType=iEngine->ListViewSortType();
  iEngine->SetListViewSortType(ESortType3);
  TRAPD(err,SortListViewL(););
  if (err!=KErrNone)
    {
    // Set engine back to its previous state.
    iEngine->SetListViewSortType(oldType);
    // Now roll back the UI.
    CQikCommandManager& cm = CQikCommandManager::Static();
    TInt commandId=EAppCmdSortType1;
    if (oldType==ESortType2)
      commandId=EAppCmdSortType2;
    else if (oldType==ESortType3)
      commandId=EAppCmdSortType3;
    cm.SetChecked(*this,EAppCmdSortType3,EFalse);
    cm.SetChecked(*this,commandId,ETrue);
    User::Leave(err); // inform user
  }
  break;
```

It also shows how to implement rollback such that, if the sort operation fails, all constituent parts of the system remain in synchronization.

CQikCommandManager

The example code above introduces the `CQikCommandManager` object. Each application has a single `CQikCommandManager` created on its behalf by the UI framework, which is responsible for maintaining the command set presented by your application. As shown above, the `CQikCommandManager` also allows you to access individual commands to update their state. In an application, processing one command sometimes affects the availability of another command. For example, if you have a Delete command, it may be that no items remain after the deletion, so you would disable the Delete command and probably others too. The opposite may be true with View or Edit commands.

Adding or Removing Commands Programmatically

Our previous discussion has centered on the use of resources to define automatically what commands are available within a given view of your application. If you look at the `CQikCommandManager` object definition, you should observe various methods for creating, adding and deleting commands. Should the functionality provided automatically by resources and the framework be inadequate for your particular application, you can add or remove commands programmatically.

In our example, we add the commands defined by the `R_ALTERNATE_COMMANDS` resource when page 1 is activated and remove them otherwise.

```
void CAppSpecificListView::TabActivatedL(TInt aTabId)
  {
  CQikCommandManager& cm=CQikCommandManager::Static();
  if (aTabId==EAppSpecificListViewPageId1)
    {

    // Page one adds a Create and erase command.
    cm.InsertIntoCommandListL(*this,*this,
                    R_ALTERNATE_COMMANDS);
    cm.DeleteCommand(*this,EAppCmdSortAltType1);
    cm.DeleteCommand(*this,EAppCmdSortAltType2);
    }
  else
    {
    // Remove any create and erase commands
    // added in page one.
    cm.DeleteFromCommandList(*this,R_ALTERNATE_COMMANDS);
    // Manually add a couple of sort commands.
    TBuf<64>bb;
    CQikCommand* q = CQikCommand::NewLC(EAppCmdSortAltType1);
    q->SetType(EQikCommandTypeScreen);
    q->SetPriority(EAppCmdSortType1Priority);
    q->SetGroupId(EAppCmdSortGroup);
    q->SetIcon(KMbmFile, EMbmCommands1Icon0, EMbmCommands1Icon0mask);
    iEikonEnv->ReadResourceL(bb,R_STR_SORT_TYPE1);
    q->SetTextL(bb);
    q->SetHandler(this);
    cm.InsertCommandL(*this,q);
    CleanupStack::Pop(q);
    q=CQikCommand::NewLC(EAppCmdSortAltType2);
    q->SetType(EQikCommandTypeScreen);
    q->SetPriority(EAppCmdSortType1Priority);
    q->SetGroupId(EAppCmdSortGroup);
```

```
    q->SetIcon(KMbmFile, EMbmCommands1Icon1, EMbmCommands1Icon1mask);
    iEikonEnv->ReadResourceL(bb,R_STR_SORT_TYPE2);
    q->SetTextL(bb);
    q->SetHandler(this);      // We are MQikCommandHandler
    cm.InsertCommandL(*this,q);
    CleanupStack::Pop(q);
    }
  CQikMultiPageViewBase::TabActivatedL(aTabId);
  }
```

If we do not want to use resources at all, it is possible to construct `CQikCommand`, or sub-classes thereof, by hand and add them to the available command set. The following code performs such a task:

```
CQikCommand* q=CQikCommand::NewLC(EAppCmdSortAltType1);
q->SetType(EQikCommandTypeScreen);
q->SetPriority(EAppCmdSortType1Priority);
q->SetGroupId(EAppCmdSortGroup);
q->SetIcon(KMbmFile, EMbmCommands1Icon0, EMbmCommands1Icon0mask);
iEikonEnv->ReadResourceL(bb,R_STR_SORT_TYPE1);
q->SetTextL(bb);
q->SetHandler(this);
cm.InsertCommandL(*this,q);
CleanupStack::Pop(q);
```

It has performed the equivalent of:

```
QIK_COMMAND
  {
  id = EAppCmdSortAltType1;
  type = EQikCommandTypeScreen;
  priority = EAppCmdSortType1Priority;
  groupId = EAppCmdSortGroup;
  icon = r_command_icon;
  text = "Sort by name";
  },
```

7.3 Categories

7.3.1 Overview

UIQ 3 applications often use a list view to present data and a detail view to read and work with specific content. To make it easier for users to work with list views, UIQ 3 applications can use the *category* paradigm. Each data record is associated with a category, often the Unfiled category by default. The list view normally displays a list of all data records using category All. The list can be filtered by category to make it easier to work with subsets of data; this is a useful technique to help the user organize and find data on a mobile phone.

For example, in the Agenda application the user can classify appointments using categories. In Figure 7.13, we have created 'Personal' and 'Work' categories using the New Category menu option. The 3pm appointment is set as a Work meeting. In the monthly (list) view, we can change from the default of All Agenda Categories to view only Work appointments.

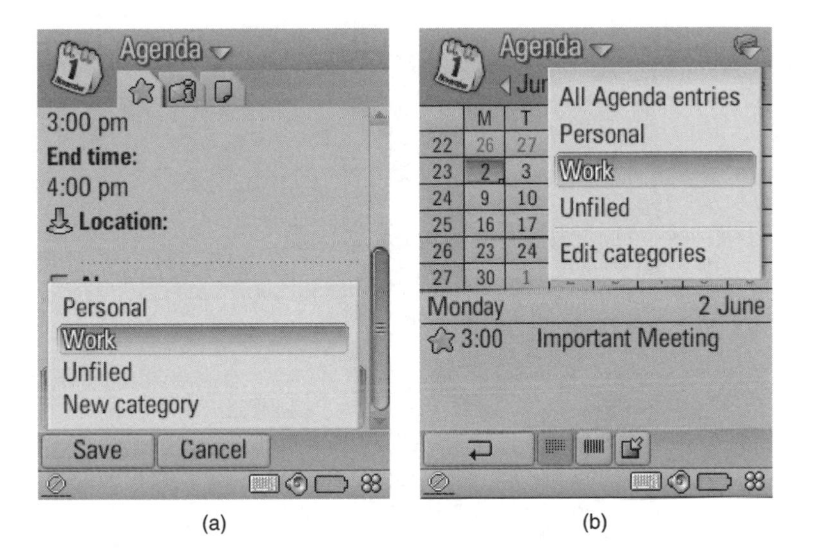

Figure 7.13 Category allocation and filtering in Agenda

Category support is embedded within the view architecture; our `Categories` example application shows you how to implement category support in your application.

7.3.2 `Categories` Example Application

The `Categories` example demonstrates how to:

- add categories to your application
- support the creation of new categories
- handle deletion of existing categories
- support renaming of categories
- change the currently active category and ensure any displayed list is kept up to date.

Firstly, we define a default set of categories using a resource:

```
RESOURCE ARRAY r_app_categories
  {
  items=
    {
    QIK_CATEGORY
      {
      name = "All";
      flags = EQikCategoryCantBeRenamed |
              EQikCategoryCantBeDeleted |
                         EQikCategoryAll;
      handle = EAppCategoryAll;
      },
    QIK_CATEGORY
      {
      name = "Unfiled";
      flags = EQikCategoryCantBeRenamed |
              EQikCategoryCantBeDeleted |
                     EQikCategoryUnfiled;
      handle = EAppCategoryUnfiled;
      },
    // This app chooses to define the following category
    // as one that cannot be deleted.
    QIK_CATEGORY
      {
      name = "Australasian";
      flags = EQikCategoryCantBeDeleted;
      handle = EAppCategoryAustralasian;
      },
    // This app chooses to define the following category
    // as one that cannot be renamed.
    QIK_CATEGORY
      {
      name = "Asian";
      flags = EQikCategoryCantBeRenamed;
      handle = EAppCategoryAsian;
      },
    QIK_CATEGORY
      {
      name = "European";
      flags = 0;
      handle = EAppCategoryEuropean;
      }
    };
  }
```

When you have categories within your application, you should define two standard categories: the All and Unfiled categories.

Usually the All category is displayed by default and, as its name suggests, when the All category is selected, all entries should be listed.

The Unfiled category is for entries that have not been assigned any other category. The All and Unfiled categories need to be available at all times and must not be renamed or deleted.

Our application has chosen to define three other categories, corresponding to the continents in which the countries are located. They have been given different properties (flags) merely to demonstrate that this is possible. You should note that only one entry should be defined for each of the `EQikCategoryAll` and `EQikCategoryUnfiled` flags.

The handle field is set from an application specific `enum` set:

```
enum TAppCategoryIds
  {
  EAppCategoryAll,
  EAppCategoryUnfiled,
  // We have three application-specific pre-defined categories.
  EAppCategoryAustralasian,
  EAppCategoryAsian,
  EAppCategoryEuropean,
  EAppCategoryLastItem
  };
```

Adding the categories to an application is usually performed within the `ViewConstructL()` method of the view with which they are associated:

```
void CAppSpecificListView::ViewConstructL()
  {
  ViewConstructFromResourceL(R_LIST_VIEW_CONFIGURATIONS);
  // Create the category list
  CQikCategoryModel* categories = QikCategoryUtils::
           ConstructCategoriesLC(R_APP_CATEGORIES);
  SetCategoryModel(categories);
  CleanupStack::Pop(categories);
  // By default we view the 'All' category.
  SelectCategoryL(EAppCategoryAll);
  // Cause the category picker to be visible.
  SetCategoryModelAsCommandsL();
  }
```

In our demonstration example, we do not save any user changes when the application exits or restore the changes when the application is run next time. Though unrealistic for a commercial-quality application, this example is intended to demonstrate the principles involved. Chapter 10 will add the required load and save support for categories.

From the code fragment, you should be able to observe a `CQikCategoryModel` object being created and initialized with the default category set; this handle is given to the view via the `SetCategoryModel()` method and an initial category is selected via `SelectCategoryL()`.

As a final step, we request a set of commands to be generated with the `SetCategoryModelAsCommandsL()` method, so the category information can be displayed on screen as a set of commands. On Pen Style phones, a drop-down icon is displayed top right; tapping on the icon displays the category commands (see Figure 7.14a). On a Softkey Style phone, a named group is created and added to the menu (Figure 7.14b).

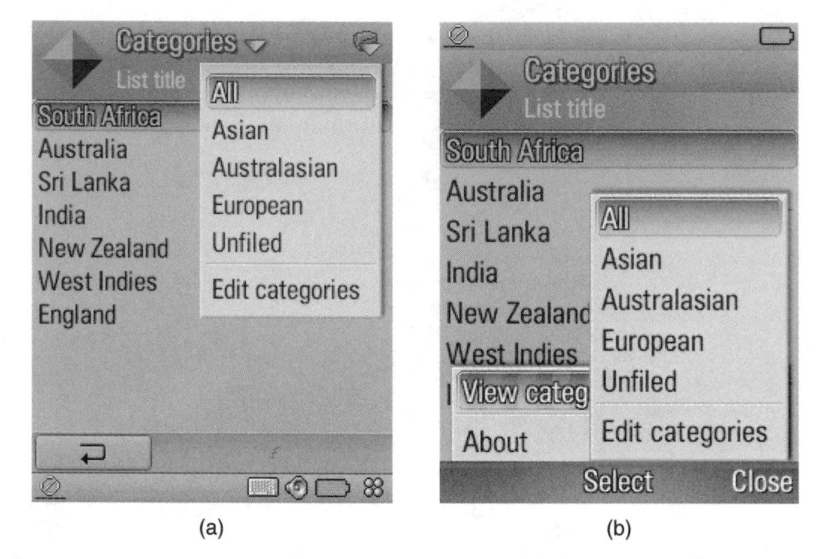

Figure 7.14 `Categories` application in a) Pen Style and b) Softkey Style UI configuration

The Edit categories command appears due to the content of our `QIK_COMMAND_LIST` for this view:

```
RESOURCE QIK_COMMAND_LIST r_list_view_commands
  {
    items=
    {
    QIK_COMMAND
      {
      id = EAppCmdAbout;
      type = EQikCommandTypeScreen;
      groupId = EAppCmdMiscGroup;
      priority = EAppCmdAboutPriority;
      text = "About";
      },
    QIK_COMMAND
      {
      id = EAppCmdEditCategories;
      type = EQikCommandTypeCategory;
      groupId =EQikCommandGroupIdAfterCategoryCommands;
      text = "Edit categories";
      }
    };
  }
```

Category commands are delivered to our `HandleCommandL()` method in the same way as any other command. However, rather than only processing by command ID, we have to handle category commands slightly differently:

```
void CAppSpecificListView::HandleCommandL(
                    CQikCommand& aCommand)
```

```
{
if (aCommand.Type()==EQikCommandTypeCategory)
  {
  if (aCommand.Id()==EAppCmdEditCategories)
    CQikEditCategoriesDialog::RunDlgLD(CategoryModel(),this);
  else
    {
    SelectCategoryL(aCommand.CategoryHandle());
    UpdateListBoxL();
    }
  }
else
  {
  switch (aCommand.Id())
    {
    case EAppCmdAbout:
      (new(ELeave)CAboutDialog)->ExecuteLD(R_ABOUT_DIALOG);
      break;

   default:       // For example, the back button...
      CQikViewBase::HandleCommandL(aCommand);
      break;
    }
  }
}
```

As you can see, we first check to see if the command type is a category command. If it is, then we either run the system category editor or update the list box due to selection of a different category.

The system-supplied category editor requires us to pass an object of type `MQikEditCategoryObserver` to process various user actions. In our example, the `CAppSpecificListView` implements that interface.

The `MQikEditCategoryObserver` interface comprises nine methods, one of which is shown below:

```
TBool CAppSpecificListView::DoDeleteCategoryL(TInt aHandle)
  {
  iEngine->DeleteCategory(aHandle);

  if (aHandle==CurrentCategoryHandle())
    {
    SelectCategoryL(EAppCategoryAll);
    UpdateListBoxL();
    }
  return(ETrue);
  }
```

This method is called back to implement the deletion of a category from the underlying model; in our example, the engine runs that model. The user presentation is updated by the framework code. The code checks to see if the category that is being deleted is the one currently being displayed by the list box. If so, an alternative category is chosen, usually the All category, and the list box is updated.

Updating the List Box

In our example, we present a very basic approach to updating the list box when a category changes. We simply remove all the entries and rebuild from the beginning. An alternative approach would be to run though all the existing entries, checking whether the entries within the list box should remain there and adding any missing entries.

Whichever approach you adopt, do not forget that some kind of rollback should be incorporated, in order to ensure that any selected category and displayed list remain synchronized. We cover rollback in more detail in Chapter 10.

Changing Entry Categories

The Categories application does not allow you to change the category that is allocated to each data record. You cannot change the Australasia category that is allocated to Australia, for example.

Setting the category is normally done in a detail view; UIQ 3 applications such as Agenda and Contacts typically place a Category drop-down below the data presented in the detail view (see Figure 7.15). Clicking on the category drop-down opens a menu in which the user can select an existing category or create a new one.

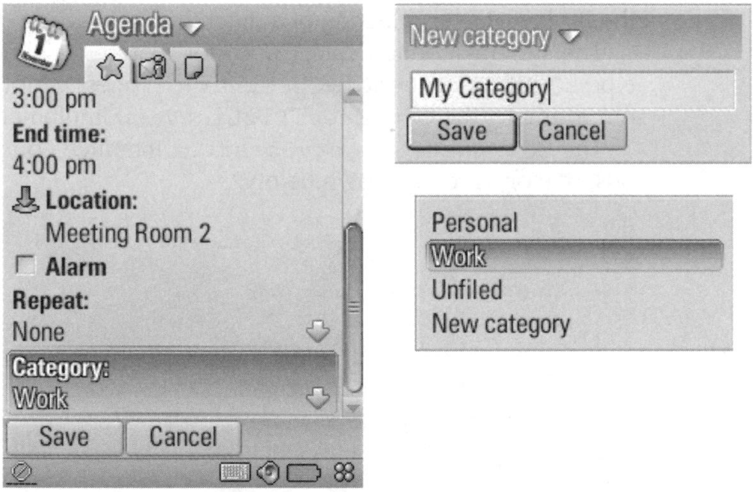

Figure 7.15 Allocating and creating categories in Agenda (Pen Style UI configuration)

Before we can add this functionality to our application, we must first describe how to create views. We return to categories and add the ability to allocate a category to a data record in the `SignedAppPhase2` example in Chapter 10.

7.4 Further Information

You can find more information on the Command Processing Framework in the *Programmer's Guide to New Features* and the *UIQ Style Guide*.

8

Layout Managers and Building Blocks

8.1 Layout Managers

Layout managers were introduced in UIQ 3 to assist in the layout of controls. We have already used layout managers in our example applications. In ListView2, we specify the type and details for a row layout manager in the resource file:

```
RESOURCE QIK_CONTAINER_SETTINGS r_list_view_page1_control
  {
  layout_manager_type = EQikRowLayoutManager;
  layout_manager = r_row_layout_manager_default;
  controls =
    {
    QIK_CONTAINER_ITEM_CI_LI
      {
      unique_handle = EAppSpecificListViewListId1;
      type = EQikCtListBox;
      control = r_app_listview_listbox;
      layout_data = r_row_layout_data_fill;
      }
    };
  }
```

UIQ 3 defines four layout manager types:

```
enum TQikStockLayoutManagers
  {
  EQikGridLayoutManager,
  EQikRowLayoutManager,
  EQikFlowLayoutManager,
  EQikColumnLayoutManager
  };
```

Visually, the layout managers present content as shown in Figure 8.1.

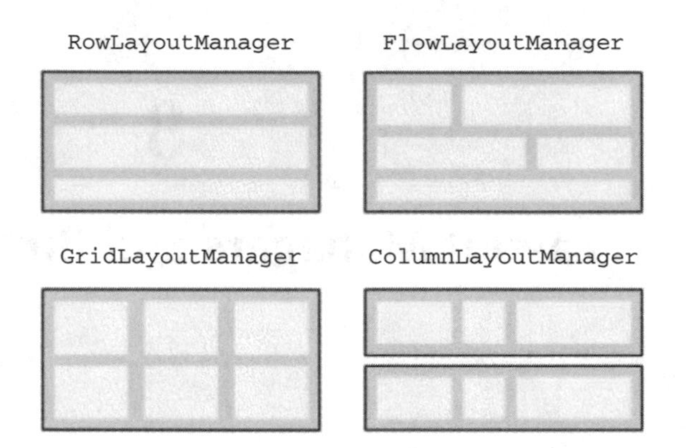

Figure 8.1 Types of layout manager

The `layout_manager` field references a resource whose type depends on the `layout_manager_type`, described in Table 8.1.

Table 8.1 Types of layout manager

Layout Manager Types	Properties
EQikGridLayoutManager	Places controls in a grid; each cell in the grid is the same size.
EQikRowLayoutManager	Places controls in a row, one control per row.
EQikFlowLayoutManager	Places controls end to end in rows, breaking the rows when a control cannot fit in the remaining space within a row.
EQikColumnLayoutManager	Positions controls on a single row, then aligns additional controls to create the appearance of columns.

The `LayoutManager1` example application shows a simple example of row and grid layout managers in use (see Figure 8.2). The example application has two pages:

```
RESOURCE QIK_CONTAINER_SETTINGS r_page1_control
  {
  layout_manager_type = EQikGridLayoutManager;
  layout_manager = r_grid_layout_manager;
```

```
  controls =
    {
    .. list of controls
    };
  }
RESOURCE QIK_CONTAINER_SETTINGS r_page2_control
  {
  layout_manager_type = EQikRowLayoutManager;
  layout_manager = r_row_layout_manager;
  controls =
    {
    .. list of controls
    };
  }
```

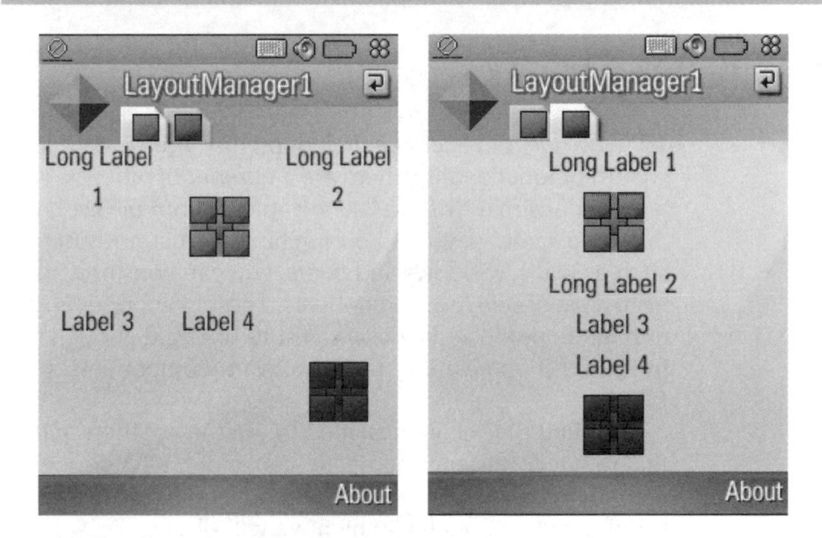

Figure 8.2 Page 1 (grid manager) and page 2 (row manager) of `LayoutManager1`

The grid layout references a resource:

```
RESOURCE QIK_GRID_LAYOUT_MANAGER r_grid_layout_manager
  {
  // This app has decided to have 3 columns.
  columns = 3;
  default_layout_data = QIK_GRID_LAYOUT_DATA
    {
    horizontal_alignment = EQikLayoutHAlignCenter;
    };
  }
```

The row layout references:

```
RESOURCE QIK_ROW_LAYOUT_MANAGER r_row_layout_manager
  {
  default_layout_data = QIK_ROW_LAYOUT_DATA
```

```
    {
    // Align items to centre of the display region.
    horizontal_alignment = EQikLayoutHAlignCenter;
    // 8 pixels to left, item not on display edge.
    left_margin = 8;
    };
  }
```

The layout of the entire application is handled by the resource definitions; there is no application-specific software required.

Most applications do not need to use layout managers directly; instead they use building blocks that hide the usage of layout managers.

8.2 Building Blocks

To date we have concentrated on list boxes. While they provide a wealth of functionality, there are a number of other standard controls that applications may wish to use to capture input or display content.

In the same way that you might construct a house with components such as walls, windows and doors, you can construct an application view from a set of software components. These components need to be flexible enough to produce the desired results but rigid enough to still fit together. In a UIQ 3 application, these software components are called *building blocks*.

Building blocks are designed to aid the creation and display of views that retain consistency across the range of UIQ 3 phones. Building blocks are, in many ways, like compound controls, however they also add significant functionality, which we will discuss.

Building blocks differ from standard compound controls in that they have attributes such as margins, alignment and size. They also integrate closely with the UIQ 3 layout managers such that much of the work associated with presenting controls on different mobile phones is handled automatically by the framework. Layout managers were introduced in UIQ 3 to assist in the layout of controls. Prior to UIQ 3, an application would present a variety of controls to the user and manage the layout of those controls in a rather ad hoc, application-specific manner. You can find more on layout managers in the UIQ 3 SDK.

Figure 8.3 shows you how a view may be constructed using building blocks (it may also be constructed using controls directly).

Note that:

- the view is made up building blocks
- each building block contains a number of slots
- each slot can hold one control.

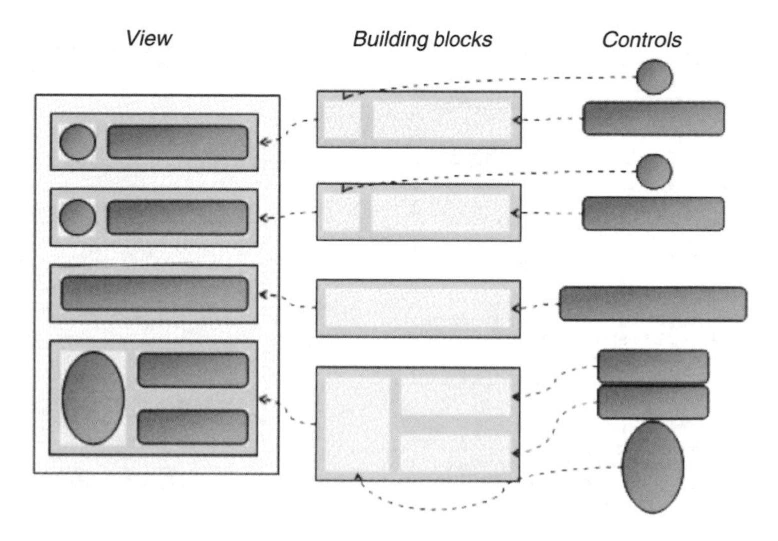

Figure 8.3 Constructing a view using building blocks

You may place any control you like within a slot but you need to ensure the application presents data sensibly. For example, placing an icon into a long thin slot or a date editor into a tall, narrow, multi-line slot would be inappropriate.

A large number of building blocks are supplied as standard but if they are insufficient for your application you can create your own, either using resources or programmatically.

The standard building blocks are defined by the following extract from the standard UIQ 3 control set, found in `qikstockcontrols.hrh`:

```
enum TQikStockControls
  {
  EQikCtOnelineBuildingBlock,
  EQikCtCaptionedTwolineBuildingBlock,
  EQikCtTwolineBuildingBlock,
  EQikCtManylinesBuildingBlock,
  EQikCtIconOnelineBuildingBlock,
  EQikCtIconCaptionedTwolineBuildingBlock,
  EQikCtIconTwolineBuildingBlock,
  EQikCtOnelineIconBuildingBlock,
  EQikCtIconOnelineIconBuildingBlock,
  EQikCtIconTwolineIconBuildingBlock,
  EQikCtMediumThumbnailDoubleOnelineBuildingBlock,
  EQikCtLargeThumbnailThreelineBuildingBlock,
  EQikCtCaptionedOnelineBuildingBlock,
  EQikCtIconCaptionedOnelineBuildingBlock,
  EQikCtTwolineIconBuildingBlock,
  EQikCtIconIconOnelineBuildingBlock,
  EQikCtHalflineHalflineBuildingBlock,
  EQikCtCaptionedHalflineBuildingBlock,
  };
```

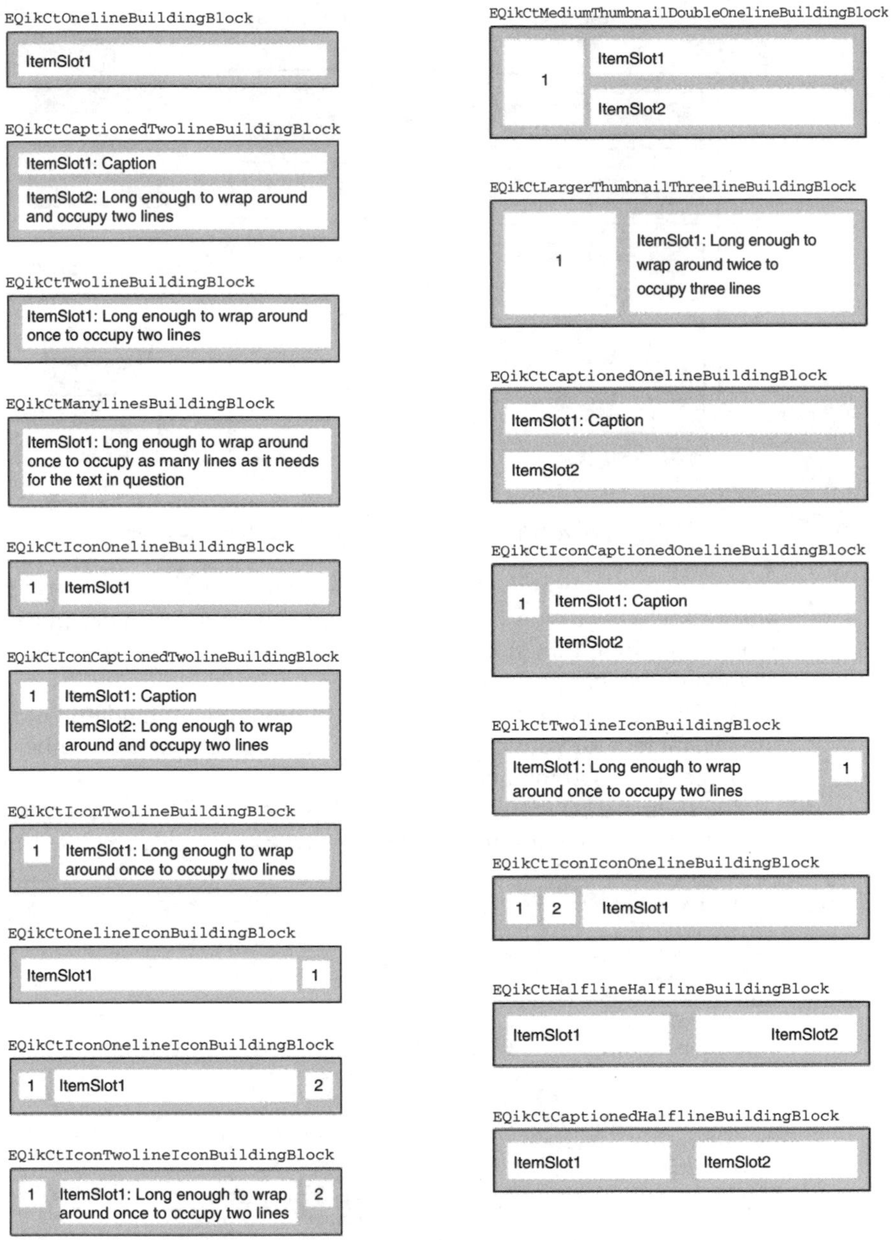

EQikCtOnelineBuildingBlock

ItemSlot1

EQikCtCaptionedTwolineBuildingBlock

ItemSlot1: Caption
ItemSlot2: Long enough to wrap around and occupy two lines

EQikCtTwolineBuildingBlock

ItemSlot1: Long enough to wrap around once to occupy two lines

EQikCtManylinesBuildingBlock

ItemSlot1: Long enough to wrap around once to occupy as many lines as it needs for the text in question

EQikCtIconOnelineBuildingBlock

1 ItemSlot1

EQikCtIconCaptionedTwolineBuildingBlock

1 ItemSlot1: Caption
ItemSlot2: Long enough to wrap around and occupy two lines

EQikCtIconTwolineBuildingBlock

1 ItemSlot1: Long enough to wrap around once to occupy two lines

EQikCtOnelineIconBuildingBlock

ItemSlot1 1

EQikCtIconOnelineIconBuildingBlock

1 ItemSlot1 2

EQikCtIconTwolineIconBuildingBlock

1 ItemSlot1: Long enough to wrap around once to occupy two lines 2

EQikCtMediumThumbnailDoubleOnelineBuildingBlock

1 ItemSlot1
ItemSlot2

EQikCtLargerThumbnailThreelineBuildingBlock

1 ItemSlot1: Long enough to wrap around twice to occupy three lines

EQikCtCaptionedOnelineBuildingBlock

ItemSlot1: Caption
ItemSlot2

EQikCtIconCaptionedOnelineBuildingBlock

1 ItemSlot1: Caption
ItemSlot2

EQikCtTwolineIconBuildingBlock

ItemSlot1: Long enough to wrap around once to occupy two lines 1

EQikCtIconIconOnelineBuildingBlock

1 2 ItemSlot1

EQikCtHalflineHalflineBuildingBlock

ItemSlot1 ItemSlot2

EQikCtCaptionedHalflineBuildingBlock

ItemSlot1 ItemSlot2

Figure 8.4 System building blocks

Figure 8.4 shows you each building block in more detail. A building block can take one or two controls. These are placed in ItemSlot1 and, where available, ItemSlot2. Some building blocks also have one or two slots for icons, shown as numbered blocks in Figure 8.4.

8.2.1 `BuildingBlocks1` Example Application

In the `BuildingBlocks1` application, we construct a view that has three controls, each of which is a building block. Almost the entire application is defined by its resources. The key resource to understand is as follows:

```
RESOURCE QIK_CONTAINER_SETTINGS r_page_control
  {
  controls =
    {
    QIK_CONTAINER_ITEM_CD_LI
      {
      type = EQikCtOnelineBuildingBlock;
      control = QIK_SYSTEM_BUILDING_BLOCK
        {
        flags = EQikBuildingBlockDividerBelow;
        content =
          {
          QIK_SLOT_CONTENT
            {
            slot_id = EQikItemSlot1;
            caption = "One line building block.";
            }
          };
        };
      },
    QIK_CONTAINER_ITEM_CD_LI
      {
      type = EQikCtTwolineBuildingBlock;
      control = QIK_SYSTEM_BUILDING_BLOCK
        {
        content =
          {
          QIK_SLOT_CONTENT_INDIRECT
            {
            slot_id = EQikItemSlot1;
            type = EEikCtLabel;
            itemflags = EQikCtrlFlagIsFocusing;
            control= r_indirect_two_line_building_block;
            }
          };
        };
      },
    QIK_CONTAINER_ITEM_CD_LI
      {
      type = EQikCtManylinesBuildingBlock;
      control = QIK_SYSTEM_BUILDING_BLOCK
        {
        content =
          {
```

```
        QIK_SLOT_CONTENT_DIRECT
          {
          slot_id = EQikItemSlot1;
          type = EEikCtLabel;
          itemflags = EQikCtrlFlagIsFocusing;
          control = LABEL
            {
            standard_font = EEikLabelFontAnnotation;
            txt = "Many line building blocks wrap text
                   across as many lines as is required
                                to display the text.";
            };
          }
        };
      };
    }
  };
}
```

You can easily see each control from the `QIK_CONTAINER_ITEM_CD_LI` elements in the code. The control data is defined inline and the layout information is defined by reference. In this example, the control data is defined by a `QIK_SYSTEM_BUILDING_BLOCK`. There is no additional layout information; that field is omitted. The application is shown in Figure 8.5.

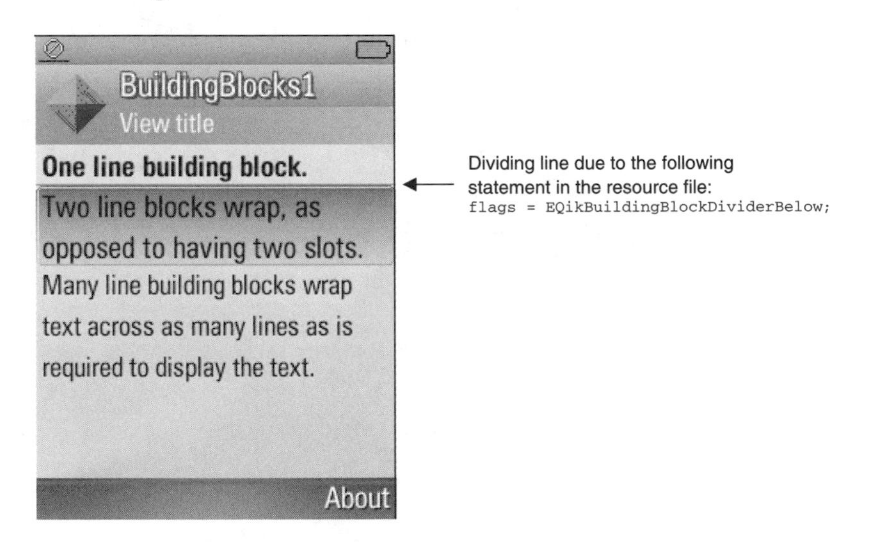

Figure 8.5 `BuildingBlocks1` application

Each of the `QIK_CONTAINER_ITEM_CD_LI` items comprises a different type of building block. This example uses the building blocks listed in Table 8.2.

Table 8.2 Types of building block

Types of building block	Description
EQikCtOnelineBuildingBlock	ItemSlot1
EQikCtTwolineBuildingBlock	ItemSlot1: Long enough to wrap around once to occupy two lines
EQikCtManylinesBuildingBlock	ItemSlot1: Long enough to wrap around as many times as necessary for the text that is displayed

Each building block is described by a `QIK_SYSTEM_BUILDING_BLOCK` structure:

```
STRUCT QIK_SYSTEM_BUILDING_BLOCK
  {
  BYTE version = 2;
  LONG flags = 0;
  LTEXT default_caption = "";
  STRUCT content[];
  }
```

The values that can be assigned to the flags field are:

- `EQikBuildingBlockDividerBelow`. This causes a dividing line to be displayed below a building block. Our example displays a dividing line between the first and subsequent building blocks (see Figure 8.5).
- `EQikBuildingBlockMirror`. This flag indicates that the building block slots should be mirrored, that is, instead of running left-to-right, they run right-to-left. For example, for a building block having an icon on the left and text on the right, the icon would be displayed to the right and the text to the left.
- `EQikBuildingBlockNoMirroring`. When applications are run on mobile phones that present content in a right-to-left order, compared to typical European left-to-right order, this flag indicates whether the building block should obey the system mirroring requirements.
- `EQikBuildingBlockDebugMode`. When set, the building block draws a black outline around itself. This makes it easier to see if your controls are laid out correctly when developing your application.

The content of each building block is described by one of three possible resource structures:

```
STRUCT QIK_SLOT_CONTENT
  {
  BYTE struct_type = 0;
  LONG slot_id = -1;
  LONG unique_handle = -1;
  LTEXT caption = "";
  }
STRUCT QIK_SLOT_CONTENT_INDIRECT
  {
  BYTE struct_type = 1;
  LONG slot_id = -1;
  LONG unique_handle = -1;
  LONG itemflags = 0;
  LONG type = -1;
  LLINK control = 0;
  }
STRUCT QIK_SLOT_CONTENT_DIRECT
  {
  BYTE struct_type = 2;
  LONG slot_id = -1;
  LONG unique_handle = -1;
  LONG itemflags = 0;
  LONG type = -1;
  STRUCT control;
  }
```

Which of the three structures you choose depends on what information you have and which feature set you wish to use. We use one of each type in `BuildingBlocks1` to give you an example of all three.

`struct_type`

This defines which structure type is being used. It should be left as is.

`slot_id`

This field informs the software with which slot the resource description should be associated. For our first example, the system building blocks only have a single slot identified by `EQikItemSlot1`, defined in `qikon.hrh`. When you use the system building blocks that are provided with UIQ 3, the number and type of slots, and how they are referenced, is predefined.

`unique_handle`

A control has a unique value, or handle, associated with it. If the application is using a collection of controls, as opposed to defining the controls inline, this value is used to determine which `QIK_CONTROL` from the collection is to be used. The value only needs to be unique within a view but, in practice, its value is taken from your application-specific `enum` lists and is often unique within your application.

The `unique_handle` is also used if you need to reference the control from your application, for example, to set up its initial value or collect the user data.

`caption`

This field is only defined in the `QIK_SLOT_CONTENT` and is a shorthand way of defining text content to be associated with the slot. If no caption is defined, the `unique_handle` field should reference a control from a control collection.

`itemFlags`

This field contains a set of flags from the following list:

- `EQikCtrlFlagIsFocusing`. The control can take focus. Controls that take focus usually do so to indicate that the user can change their content. In our example, we have used the flag on a text-only item, which is somewhat unusual. Focus is often indicated with a highlight, the highlight normally moves between those controls that can take focus.
- `EQikCtrlFlagIsNonFocusing`. The control does not take focus. In general, controls that don't take focus cannot have their content changed. A highlight does not normally move to items with this flag set.
- `EQikCtrlFlagIsEditInPlace`. For some controls, such as a text editor, a pop-out control is typically used to perform content editing. This flag indicates that the inline control performs that task instead.
- `EQikCtrlFlagIsNonEditInPlace`. For some controls, the actual control is normally used to perform content editing. This flag indicates that a pop-out type control should be used instead.

You should take care when using the edit-in-place and non-edit-in-place flags. In particular, you should ensure that a user is able to navigate around controls that are presented to them, irrespective of whether they are using a soft key or touchscreen type mobile phone.

`type`

This field indicates the control type from the set of stock controls that the control field references and is used to indicate how the control field should be interpreted. The type values for stock controls are contained in the `TEikStockControls` and `TQikStockControls` `enum` lists.

`control`

This contains either a reference to or the actual resource associated with the control indicated by the type field. In our example, both the type fields contain EEikCtLabel. This control needs to describe a resource of type LABEL. For a QIK_SLOT_CONTENT_INDIRECT, it is a reference to a resource of that type. For a QIK_SLOT_CONTENT_DIRECT, it is the LABEL resource itself.

8.2.2 BuildingBlocks2 Example Application

For reference, we have created a second example application, BuildingBlocks2 (see Figure 8.6). We do not discuss it further here since it simply extends the previous example to illustrate a multi-page application that demonstrates the use of each of the standard building blocks listed by the TQikStockControls enum.

As in BuildingBlocks1, the controls are restricted to simple icons and text to minimize the complexity and help focus on the usage of each of the standard building block types. This is a good example to help you investigate the suitability of each of the building blocks for your particular application.

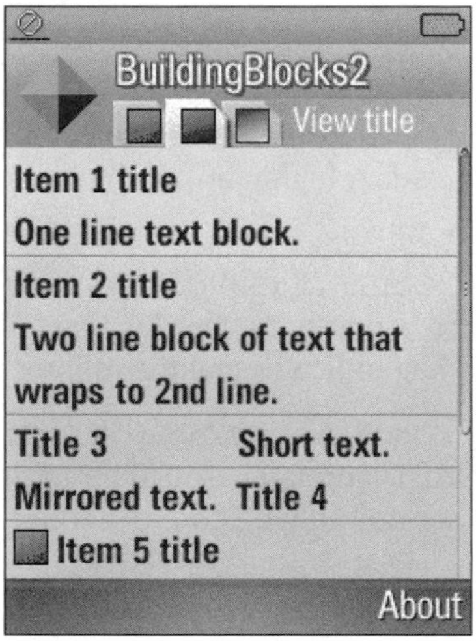

Figure 8.6 BuildingBlocks2 application

8.2.3 BuildingBlocks3 Example Application

The two previous examples have demonstrated how to define and use, at a basic level, the standard building blocks in UIQ 3. While the ability to display icons and text is important, there are a number of other controls that are equally useful in real-world applications. You may recall from the earlier description of building blocks that you can place any control within a slot. This example shows how you can go about placing a wide variety of system controls within building block slots to construct highly functional applications with minimal effort.

Figure 8.7.shows three choice lists: a check box, a slider and a button control.

Figure 8.7 BuildingBlocks3 application

Controls are used not only to display information but also to capture user input. To achieve this, an application needs to interact with the controls, for example, to set them to a specific value when first viewed and to collect information from them when it is entered by a user.

Page 1 of BuildingBlocks3 presents three choice lists. The first fully defines the entries within the resource:

```
QIK_SLOT_CONTENT_DIRECT
  {
  slot_id = EQikItemSlot1;
  type = EEikCtChoiceList;
  itemflags = EQikCtrlFlagIsFocusing;
  control = CHOICELIST
    {
    array_id = r_page1_choicelist_array;
    };
  }
RESOURCE ARRAY r_page1_choicelist_array
  {
  items=
    {
    LBUF {txt="Choice 1";},
    LBUF {txt="Choice 2";},
    LBUF {txt="Choice 3";}
    };
  }
```

The second and third require application-specific software to set up the list box items.

The following resource fragment shows the declaration of an empty choice list. It also shows how the example application obtains a reference to the underlying control, in this case a `CEikChoiceList`, by using the same value used by the `unique_handle` field. You should note the importance of the `unique_handle` field:

```
QIK_SLOT_CONTENT_DIRECT
  {
  slot_id = EQikItemSlot2;
  type = EEikCtChoiceList;
  itemflags = EQikCtrlFlagIsFocusing;
  unique_handle = EAppChoiceList1;
  control = CHOICELIST
    {
    };
  }
```

The code fragment below shows how to reference the choice list object created by the UIQ 3 framework:

```
// Create and initialize a choice list.
iAppChoiceList=new(ELeave)CChoiceListArray;
// Tell list box control about choice list.
CEikChoiceList* cl = LocateControlByUniqueHandle<
                 CEikChoiceList>(EAppChoiceList1);
cl->SetArrayL(iAppChoiceList);
```

You may notice the use of a templated entity `LocateControlByUniqueHandle`. When controls are created by interpreting the resource information or, more precisely, the `type` field of the `QIK_SLOT_CONTENT`, a control of the required type is created. You may recall that all controls derive from the `CCoeControl` base class. When the framework stores the created object handles, it does so as a set of `CCoeControl` objects. When specific controls need to be manipulated, it is often the case that you need to convert back from the stored `CCoeControl` type into the real object type. This is what `LocateControlByUniqueHandle<>()` is used for.

Table 8.3 brings together the control types, their resource structures and the equivalent C++ object.

The following code fragment taken from the example application's `ViewConstructL()` method demonstrates how we set controls to contain default values:

```
// Locate the check box and ensure it is set.
CEikCheckBox* cb = LocateControlByUniqueHandle<
                   CEikCheckBox>(EAppCheckbox1);
cb->SetState(CEikButtonBase::ESet);
// Locate the slider and set a default value.
CQikSlider* sl = LocateControlByUniqueHandle<CQikSlider>(EAppSlider1);
sl->SetValue(21);
// Locate the edwins and set a default value.
```

```
_LIT(KSomeText,"Some text");
TBuf<16> text(KSomeText);
CEikEdwin* edwin = LocateControlByUniqueHandle<CEikEdwin>(EAppEdwin1);
edwin->SetTextL(&text);
_LIT(KOtherText,"Other text");
text=KOtherText;
edwin=LocateControlByUniqueHandle<CEikEdwin>(EAppEdwin2);
edwin->SetTextL(&text);
// Locate the number editor and set a default value.
CQikNumberEditor* numEd = LocateControlByUniqueHandle<
                        CQikNumberEditor>(EAppEdwin4);
numEd->SetValueL(21);
// Locate the floating point editor and set a default value.
CQikFloatingPointEditor* fltEd = LocateControlByUniqueHandle<
                        CQikFloatingPointEditor>(EAppEdwin5);
fltEd->SetValueL(21.21);
// Locate and set the IP editor to a default value.
CQikIpEditor* ipEd = LocateControlByUniqueHandle<
                        CQikIpEditor>(EAppEdwin6);
_LIT(K10203040," 10.20.30.40");
ipEd->SetIpAddress(K10203040);
// Locate and set the time & date editor to a default value.
CQikTimeAndDateEditor* dtEd = LocateControlByUniqueHandle<
                    CQikTimeAndDateEditor>(EAppDateTime3);
TTime time;
time.HomeTime();                        // Default value is now.
dtEd->SetTimeL(time);
// Locate and set the duration editor to a default value.
CQikDurationEditor* durEd = LocateControlByUniqueHandle<
                      CQikDurationEditor>(EAppDateTime4);
durEd->SetDurationL(1200); // 1200 secs = 1200/60 mins = 20 mins.
```

Table 8.3 Control types, resources and C++ objects

Control Type	Resource Type	C++ Object Type
EEikCtCommandButton	CMBUT	CEikCommandButton
EEikCtEdwin	EDWIN	CEikEdwin
EEikCtRichTextEditor	RTXTED	CEikRichTextEditor
EEikCtSecretEd	SECRETED	CEikSecretEditor
EEikCtCheckBox	CHECKBOX	CEikCheckBox
EEikCtChoiceList	CHOICELIST	CEikChoiceList
EEikCtOptionButton	OPBUT	CEikOptionButton
EEikCtHorOptionButList	HOROPBUT	CEikHorOptionButtonList

(*continued overleaf*)

Table 8.3 (*continued*)

Control Type	Resource Type	C++ Object Type
EEikCtListBox	LISTBOX	CEikListBox
EEikCtImage	IMAGE	CEikImage
EEikCtLabel	LABEL	CEikLabel
EEikCtComboBox	COMBOBOX	CEikComboBox
EEikCtProgInfo	PROGRESSINFO	CEikProgressInfo
EEikCtGlobalText Editor	GTXTED	CEikGlobalText Editor
EEikCtClock	CLOCK	CEikClock
EEikCtCalendar	CALENDAR	CEikCalendar
EEikCtTextButton	TXTBUT	CEikTextButton
EEikCtBitmapButton	BMPBUT	CEikBitmapButton
EEikCtTwoPicture CommandButton	PICMBUT	CEikTwoPicture CommandButton
EEikCtLabeledCheckBox	LABELEDCHECKBOX	CEikLabeledCheckBox
EQikCtVertOption ButtonList	QIK_VERTOPBUT	CQikVertOption ButtonList
EQikCtSoundSelector	QIK_SOUND_SELECTOR	CQikSoundSelector
EQikCtTimeEditor	QIK_TIME_EDITOR	CQikTimeEditor
EQikCtDateEditor	QIK_DATE_EDITOR	CQikDateEditor
EQikCtTimeAndDate Editor	QIK_TIME_AND_DATE_ EDITOR	CQikTimeAndDate Editor
EQikCtDurationEditor	QIK_DURATION_ EDITOR	CQikDurationEditor
EQikCtColorSelector	QIK_COLOR_SEL	CQikColorSelector
EQikCtSlider	QIK_SLIDER	CQikSlider
EQikCtNumberEditor	QIK_NUMBER_EDITOR	CQikNumberEditor
EQikCtFloating PointEditor	QIK_FLOATING_POINT _EDITOR	CQikFloating PointEditor
EQikCtIpEditor	QIK_IP_EDITOR	CQikIpEditor

8.2.4 Edit in Place and Control Stand-ins

When we place controls on a page, it is important to ensure that the user can still navigate around using the standard mechanism provided by the mobile phone, especially in Softkey Style UI configuration. For example, if a phone contains a four-way navigation key, a user would expect to be able to move through pages using left and right keys and move vertically through the list of controls with the up and down keys.

Some controls, such as a choice list, require use of the Up and Down keys to enable the user to choose an item. Other controls, such as a text editor, require the use of Left and Right keys to move a cursor around the text. It is very important that controls do not consume the navigation keys. If this happens, the user is unlikely to be able to navigate around the application.

To support these conflicting requirements, controls tend not to be set as 'edit in place'. To select a choice list item or change the text within an editor, you need to pop out the control (see Figures 8.8 and 8.9).

Figure 8.8 Choice list stand-in and control

This is how the framework operates: even though you have specified that you wish to have an `EEikCtChoiceList` or `EEikCtEdwin` control, a replacement or *stand-in* control is created and is displayed in its place. It is only when a user chooses to change the content that a control of the correct type is displayed. The user interacts with the control, which can now consume any navigation keys, until the user completes the operation. The stand-in and real control co-operate so that any changes are displayed correctly when the operation completes.

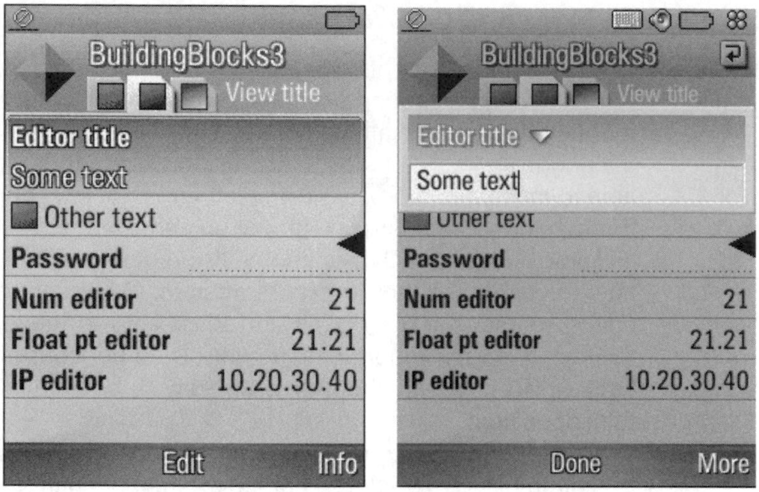

Figure 8.9 Text editor stand-in and control

Pro tip

You should take care when distinguishing between the PC keyboard arrow keys and the four-way navigation keys displayed on the emulator skin. When testing applications on the emulator, you should always click on the four-way navigation key images to determine what will happen on a real mobile phone.

9

Views and Dialogs

9.1 Overview

Many applications follow a list and details model. Using this model, it is common for the list and details components to be in separate views. The list view is typically a single page while detail views are frequently multi-page.

The UIQ 3 Contacts application follows this model. The list view has a single page (see Figure 9.1).

Figure 9.1 Contacts application list view

When opening a contact (see Figure 9.2), the most commonly used information is displayed on page 1. Other information is presented in

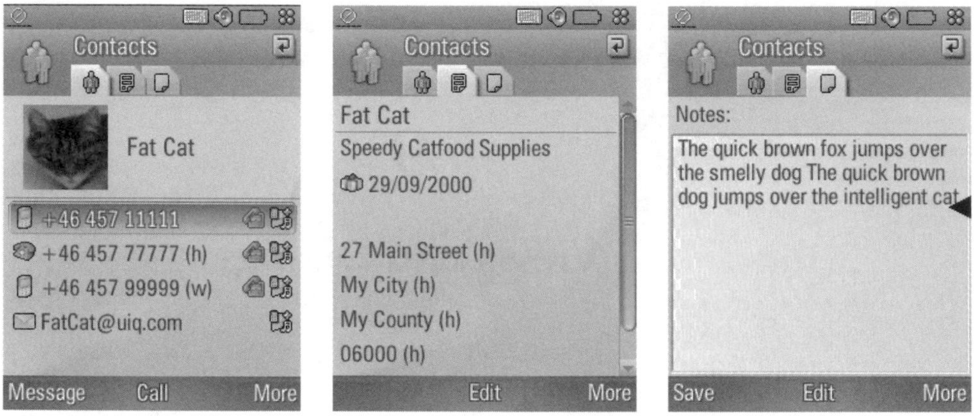

Figure 9.2 Contacts application detail views

pages 2 (address) and 3 (notes). Information that requires a lot of screen space is often best separated out. The user can easily switch pages to find the required information.

If all the contact information were presented in a single page, the user would have to scroll down a long way to get to the notes.

The Contacts detail view is optimized for reading information and taking actions such as making phone calls. Empty fields are not displayed. To add or edit data, a separate view, called the *edit* view (see Figure 9.3), is used. The edit view is laid out in the same way as the details view, spreading the data over the same three pages. Empty fields are displayed so that data can be entered.

Figure 9.3 Contacts application edit views

9.2 Working with Views

9.2.1 Views and Pages

Until now our example applications have only used a single `QIK_VIEW_CONFIGURATIONS` resource to display a view of the application data. While this may have contained multiple `QIK_VIEW` resources, one per supported configuration, the application is said to have a single view. The example applications have demonstrated both single and multi-page displays within a single view.

For multi-page views, the tab information is normally defined in a `QIK_VIEW_PAGE` application resource:

```
QIK_VIEW_PAGE
  {
  page_id = EAppSpecificListViewPageId1;
  tab_bmpid = EMbmListview2Tab1;
  tab_bmpmaskid = EMbmListview2Tab1mask;
  page_content = r_list_view_page1_control;
  },

QIK_VIEW_PAGE
  {
  page_id = EAppSpecificListViewPageId3;
  tab_caption = "Tab3";
  page_content = r_list_view_page3_control;
  }
```

The pages are represented as tabs in the application title bar's view context area (see Figure 9.4).

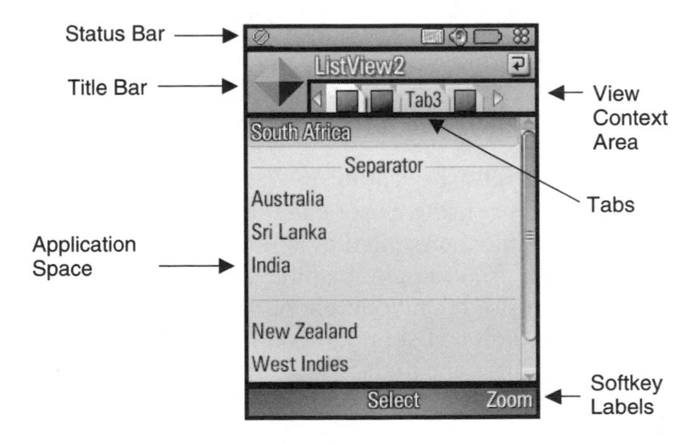

Figure 9.4 `ListView2` application showing screen layout

Within the application space, you display one or more views of your application data. Within a single application there are no specific rules about when to use multiple views compared to multiple pages. It is possible to use any combination. Most applications display a single data record per view. When there is a large amount of data associated with a single entity, this data is generally split across two or more pages.

If necessary, your application can extend outside the application space by hiding the status bar, title bar or button bar/soft key bar:

```
void CAppSpecificListView::ViewConstructL()
  {
  TQikViewMode mode = ViewMode();

  // Normally only 1 of the following
  // would be used at a time.

  // Hides status bar.
  mode.SetStatusBar(EFalse);

  // Hides the title bar.
  mode.SetAppTitleBar(EFalse);

  // Hides button bar/soft key bar and status bar.
  mode.SetButtonOrSoftkeyBar(EFalse);

  // Applies the requested view mode.
  SetViewModeL(mode);
  }
```

You can also use the entire screen:

```
void CAppSpecificListView::ViewConstructL()
  {
  TQikViewMode mode = ViewMode();
  mode.SetFullscreen();
  SetViewModeL(mode);
  }
```

You must be careful if you extend outside the normal application space. A consistent user interface is generally much easier to use. Varying the interface presented by your application compared to standard applications, for example, by hiding the status bar, may make it hard for the user to use the mobile phone as expected. Some UIQ 3 phones have a task switch and shortcut icon in the status bar. Careful integration with the features and functions on a phone ensures that your application does not upset your customers.

9.2.2 Defining Multiple Views: `MultiView1`

If you want to design an application with multiple views, it is straightforward. As you might expect, there is one `QIK_VIEW_CONFIGURATIONS` and a corresponding CQikViewBase or CQikMultiPageViewBase derived class for each view.

The MultiView1 example application presents a basic multi-view application. It implements the classic list and details model where one view is the list and a second view presents the details.

There are two QIK_VIEW_CONFIGURATIONS defined in the resource file; their corresponding CQikViewBase-derived objects are defined in MultiView1.cpp:

```
RESOURCE QIK_VIEW_CONFIGURATIONS r_list_view_configurations
  {
  configurations =
    {
    };
  }

class CAppSpecificListView : public CQikViewBase,
                        public MQikListBoxObserver
  {
  };

// and ...

RESOURCE QIK_VIEW_CONFIGURATIONS r_details_view_configurations
  {
  configurations =
    {
    };
  }

class CAppSpecificDetailsView : public CQikViewBase
  {
  };
```

9.2.3 View Navigation

A very important aspect of views is the ability to navigate between them. Figure 9.5 displays the typical view navigation on a UIQ 3 phone.

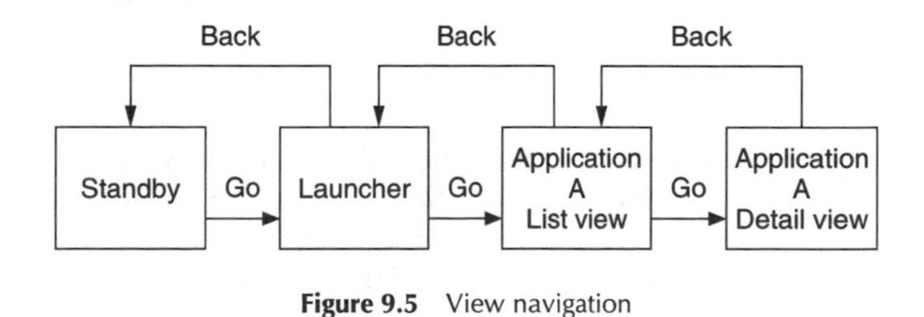

Figure 9.5 View navigation

From the diagram you can see there is a parent–child relationship between the views. For example, the parent to the application list view is the launcher view; its child view is the application details view.

So far, all example applications have had a single application view. Entering and exiting the view is handled automatically by the framework; our application did not appear to have to do anything for this to happen but our previous applications do perform two tasks. Firstly, the view constructor code passes `KNullViewId` to the base class:

```
CAppSpecificListView::CAppSpecificListView(CAppSpecificUi& aAppUi) :
                           CQikMultiPageViewBase(aAppUi,KNullViewId)
  {
  }
```

Secondly, our `HandleCommandL()` method performs an action on commands it does not otherwise handle. When the user presses the Back button, the `EQikCmdGoBack` command is sent to the currently active view in the application.

```
void CAppSpecificListView::HandleCommandL(CQikCommand& aCommand)
  {
  switch (aCommand.Id())
    {
    default:              // For example the back button...
      CQikViewBase::HandleCommandL(aCommand);
      break;
    }
  }
```

The application framework processes the `EQikCmdGoBack` command by looking up which view has been defined as the parent to the currently active view. In `MultiView1`, as in previous examples, we have defined the parent as `KNullViewId`.

For the detail view, the object constructor code passes `KViewIdListView` to the base class:

```
CAppSpecificDetailsView::CAppSpecificDetailsView(CAppSpecificUi& aAppUi)
                                   :CQikViewBase(aAppUi,KViewIdListView)
  {
  }
```

This defines the parent of the details view to be the list view, which is uniquely identified by `KViewIdListView` in our case.

So when the `EQikCmdGoBack` command is processed by the details view, it navigates to the list view. When the list view processes the command, it uses the value `KNullViewId` as a shortcut to mean the application launcher.

Applications navigate between views by calling the `ActivateViewL()` method. In the `MultiView1` example, we display the details view when an item in the list view is selected:

```
void CAppSpecificListView::HandleListBoxEventL(
  CQikListBox* aListBox,
  TQikListBoxEvent aEventType,
  TInt aItemIndex,
  TInt aSlotId)
  {
  switch (aEventType)
    {
    case EEventItemConfirmed:
    case EEventItemTapped:
      {
      iQikAppUi.ActivateViewL(KViewIdDetailsView);
      break;
      }
    default:
      break;
    }
  }
```

9.2.4 View Activation

When an application attempts to switch between views, it is important to understand what actually happens. In simple terms, the following occurs:

```
void CCoeAppUi::ActivateViewL(const TVwsViewId& aViewId)
  {

  // Locate current view.
  MCoeView* oldView=GetCurrentView();

  // Locate view to activate.
  MCoeView* newView=GetViewToActivate(aViewId);

  // Activate new view and tell the view it is activated.
  newView ->ViewActivatedL();

  // Only now the new view has been activated will the
  // previous view be deactivated.
  oldView->ViewDeactivated()
  }
```

UIQ adds a layer of functionality to the view server. The `CQikViewBase` class hides and handles the real view-server-invoked `ViewActivatedL()` call in `CQikViewBaseProxy`. It then does an asynchronous call and returns directly. The asynchronous call invokes the `CQikViewBase::ViewActivatedL` function which is overridden by your application. This means, from a view server perspective, that view activations are always successful and hence the previous view is always deactivated via its `ViewDeactivated()` method. This occurs *before* your new view receives its `ViewActivatedL()` request.

If your application leaves in the `CQikViewBase::ViewActivatedL` override, the UIQ framework calls the `CQikViewBase::HandleErrorL()` method of your view which you should override and handle the `EViewActivatedL` case. You should note that your previous view will have been deactivated, so rolling back to displaying that view may be difficult, for example, if you have run out of memory and cannot re-activate the previous view.

If your application leaves whilst in the `CQikViewBase::HandleErrorL()` method, the application will exit and bring the previous application to the foreground.

You should note that this is different to the usage of the view server in other environments where the previous view is *not* deactivated until the new view is successfully activated.

Pro tip

This is a very good example of why you need to understand the consequences of functions that can leave. Performing error handling and cleanup is only part of the task; the state your application is left in after a leave occurs is equally as important.

There are several ways you may wish to handle this particular problem:

- In the `MultiView1` application, the `ViewActivatedL()` method does not perform any actions that can leave, all set up is performed in the `ViewConstructL()` method.
- You can `TRAP` any leave and perform some sensible actions such that the `ViewActivatedL()` method itself does not leave but the user is informed of a problem occurring. For example, a list view may be reset to contain zero entries and the user informed of the problem via an `iEikonEnv->AlertWin()`.
- Alternatively, you can simply allow the `ViewActivatedL()` to leave but ensure that the application state is consistent should a leave occur.

9.2.5 Direct Navigation Links

In the `MultiView1` application, we switch from the list view to the details view by calling the `iQikAppUi.ActivateViewL(KViewIdDetailsView)`. While this successfully changes views, an important piece of information required by the details view, which item we are required to display, is not communicated. There are several ways to achieve this; all have advantages and disadvantages.

The details view could have addressability to the list view's list box, and thus it could ask the list box which entry is the current item. Unfortunately, obtaining the list box handle is not that easy. Even if it were, this approach relies on the list view maintaining the list box even when the view is deactivated. This solution is probably the least desirable.

Alternatively, the list view's current item index could be stored in a global variable (available in applications since the advent of Symbian OS 9) or, preferably, in an object common to both the list and the details view, such as an engine.

Finally, the current item index can be passed from the list view to the details view by packaging the parameter into a descriptor and using the `ActivateViewL(TVwsViewId&, TUid, TDesC8&)` variant of the `ActivateViewL()` method.

The process of communicating parameters between views and causing the target view to be displayed has been given the name direct navigation link (DNL). One view is said to DNL to another. Our application only switches between views in the same application but the underlying DNL mechanism allows applications to switch between any views in the system. For example, the detail view in Contacts can navigate directly to the New message view in messaging.

Using DNLs appears to be the most attractive option; however, we must take into account that a view can be activated for a number of different reasons.

In the `MultiView1` example application, moving from list view to detail view is not the only way in which the details view can be activated. In the emulator, if you click the Applications button you jump directly to the application launcher. Similarly, on phones there are numerous ways in which you can jump out of the current application, for example using the task manager option. If you now come back to the `MultiView1` application from the application launcher (or task manager on a phone) the details view is activated. Not surprisingly, the application launcher knows nothing about your application-specific parameters, so these are not passed to your application's `View-ActivatedL()` method.

9.2.6 Sending Parameters to Detail Views

`MultiView2` Example

The `MultiView2` application demonstrates how you can package and send a parameter to the details view. We define a simple class, `TListDetailsDnlInfo`, to contain the parameters we wish to send. In the `MultiView2` example, a single integer is sent.

```
class TListDetailsDnlInfo
  {
public:
  TInt iItemIndex;
  };

const TUid KUidDisplaySpecificEntry = {0x00000001};

typedef TPckgBuf<TListDetailsDnlInfo>TListDetailsDnlInfoBuf;

void CAppSpecificListView::HandleListBoxEventL(
                         CQikListBox* aListBox,
                   TQikListBoxEvent aEventType,
                               TInt aItemIndex,
                                  TInt aSlotId)
  {
  switch (aEventType)
    {
    case EEventItemConfirmed:
    case EEventItemTapped:
      {
      TListDetailsDnlInfo entry;
      entry.iItemIndex=aItemIndex;
      TListDetailsDnlInfoBuf bb(entry);
      iQikAppUi.ActivateViewL(KViewIdDetailsView,
                   KUidDisplaySpecificEntry,bb);
      break;
      }

    default:
      break;
   }
  }
```

Pro tip

The definition and usage of the `TPckgBuf` is really a piece of syntax to enable the software to convert between a class and a byte stream in a type-safe way without needing to use casts or manually create the byte stream.

When the information is delivered to the details view, we need to convert the byte stream back into the original data structure and extract the parameters that have been packaged up. The following code demonstrates this happening:

```
void CAppSpecificDetailsView::ViewActivatedL(
               const TVwsViewId& aPrevViewId,
                 const TUid aCustomMessageId,
                const TDesC8& aCustomMessage)
  {
  TBuf<128>buf;
```

```
if (aCustomMessageId==KUidDisplaySpecificEntry)
  {
  TListDetailsDnlInfoBuf bb;
  bb.Copy(aCustomMessage);
  iEikonEnv->Format128(buf,R_STR_ACTIVATING,bb().iItemIndex);
  iEikonEnv->InfoMsg(buf);
  }
else
  {
  iEikonEnv->Format128(buf,R_STR_UNKNOWN_MSGID,
                       aCustomMessageId.iUid);
  iEikonEnv->InfoMsg(buf);
  }
}
```

At its simplest level, this code fragment shows how to implement the `ViewActivatedL()` method and accommodate both ways in which it can be called.

Pro tip

When called by the application launcher, the `aCustomMessageId` value is zero. For this reason, you should not define any application-specific `aCustomMessageId` value to be zero; in our example, we define it to be the value one.

In `MultiView2`, the example code does not perform any meaningful tasks with the passed parameter set: we do not set the details view to display anything other than the UI component's default values. In a real application, any values displayed should be synchronized with the internal state of the item defined by the passed parameters.

If a view is not actually passed any parameters, it needs another mechanism to determine which item is being referenced. Unfortunately, the solution to this problem varies depending on individual circumstances.

If a view is displaying read-only information, you may allow each of the UI components in your view to retain state value. When such a view is `ViewActivatedL()` and is passed some parameters, it sets up the UI components. When it is not passed any parameters, no action is required since the UI components retain the correct state value. This approach assumes that you do not destroy components on a `ViewDeactivated()` and attempt to re-create on the `ViewActivatedL()`. Unfortunately, this solution is of limited value since it is rare that a view contains just read-only information.

If your view displays modifiable content, you need a way of knowing to which item the details belong. Otherwise, your application cannot associate any changes with the correct item. If the `ViewActivatedL()` method is called with some parameters, that information is normally provided in those parameters. However, when no parameters are passed, that information is missing. Therefore the view or an external object needs to record which item is being displayed.

Again, there are numerous solutions for this problem. For example, when parameters are passed to the view they could be stored. When no parameters are passed, the view would use the previously stored parameters.

MultiView3 Example

We demonstrate using previously stored parameters in the `View-ActivatedL()` method of the `MultiView3` example:

```
void CAppSpecificDetailsView::ViewActivatedL(
                const TVwsViewId& aPrevViewId,
                  const TUid aCustomMessageId,
                 const TDesC8& aCustomMessage)
  {
  TListDetailsDnlInfoBuf bb;
  if (aCustomMessageId==KUidDisplaySpecificEntry)
    {
    bb.Copy(aCustomMessage);
    iSavedParams.Copy(aCustomMessage);
    }
  else
    {
    bb.Copy(iSavedParams);
    }
  TBuf<128>buf;
  iEikonEnv->Format128(buf,R_STR_ACTIVATING,bb().iItemIndex);
  iEikonEnv->InfoMsg(buf);
  }
```

This solves the problem but it is not a particularly elegant solution for the majority of applications.

Pro tip

If an application publishes the message IDs and parameters with the intention of allowing other applications to run the view, then it must have functionality similar to that described above. In contrast, if a view remains private to an application, then an alternative approach is possible.

MultiView4 Example

`MultiView4` demonstrates an alternative approach that you can use if the detail view is private to your application. In `MultiView4`, an external object, the engine, stores which item from the list is selected. The view switch code reduces to the following code fragment:

```
void CAppSpecificListView::HandleListBoxEventL(
                          CQikListBox* aListBox,
                    TQikListBoxEvent aEventType,
                               TInt aItemIndex,
                                  TInt aSlotId)
  {
  switch (aEventType)
    {
    case EEventItemConfirmed:
    case EEventItemTapped:
      iEngine->SetListViewIndex(aItemIndex);
      iQikAppUi.ActivateViewL(KViewIdDetailsView);
      break;

    default:
      break;
    }
  }
```

The equivalent `ViewActivatedL()` method of the details view would look something like this:

```
void CAppSpecificDetailsView::ViewActivatedL(
               const TVwsViewId& aPrevViewId,
                 const TUid aCustomMessageId,
                const TDesC8& aCustomMessage)
  {
  TBuf<128>buf;
  iEikonEnv->Format128(buf,R_STR_ACTIVATING,iEngine->ListViewIndex());
  iEikonEnv->InfoMsg(buf);
  }
```

This has considerably simplified the application; we have removed one object, `TListDetailsDnlInfo`, and have dispensed with the packing and unpacking of that object. There are a number of other advantages with this approach. For example, since the engine knows which item is being referenced, we do not have to pass an index value to get at the details of the current entry. In our example, this only saves a single parameter to a single function but when this pattern is repeated numerous times the cumulative effect can be quite substantial.

An additional benefit that may be utilized concerns the amount of memory required to maintain application state. For example, since we now store the selected item index within the engine, when the list view is `ViewDeactivated()` we can reset the list box and release all the memory used to store duplicate copies of the list entry names. When the list view is `ViewActivatedL()`, we can reconstruct the list box including setting the currently selected item, exactly as though the list box had remained untouched. The `MultiView4` application takes advantage of this benefit as the time taken to rebuild the list box when the list view is `ViewActivatedL()` is negligible.

Pro tip

The `MultiView4` application is typical of many applications. An engine supplies the content to be displayed and the UI displays the content. In addition to the actual data, it is reasonable for the engine to contain functionality or information that assists a particular UI component, in this case a value representing the currently selected list item.

Pushing functionality into the engine has numerous advantages. For example, it is much easier to write automatic test code to exercise the engine and, should you choose to port your application, the more information and processing that you can remove from the phone-specific UI, the easier the task will be.

9.2.7 Saving Modified Content

In views that allow a user to modify data, the user must be able to save or cancel those modifications. The view framework defines a virtual `SaveL()` method that should be implemented by views that wish to save any modifications.

The `SaveL()` method is called when a view is deactivated:

- by switching to the application launcher or task manager.
- by calling the `CQkViewBase::ActivatePreviousViewL (ESave)` method
- by calling the `CQkViewBase::SaveThenDnlToL()` method
- by a call to `CCoeAppUi::ActivateViewL()` method.

The `SaveL()` method is not called if the `CQikViewBase::ActivatePreviousViewL(ECancel)` method is called. While that method can be called directly, the framework processing of the `EQik-CmdGoBack` command calls this method on your behalf. If you have unsaved data, your application is responsible for prompting the user about unsaved changes.

The `MultiView4` example shows the `SaveL()` method in action:

```
void CAppSpecificDetailsView::SaveL()
  {
  TAppSpecificEntity entry(iEngine->CurrentListItem());
  CEikChoiceList* cl=LocateControlByUniqueHandle<
              CEikChoiceList>(EAppChoiceList1);
  entry.iState=(TAppEntityState)cl->CurrentItem();

  CQikNumberEditor* numEd = LocateControlByUniqueHandle<
                         CQikNumberEditor>(EAppEdwin1);
  entry.iPoints=numEd->Value();
  CQikSlider* sl = LocateControlByUniqueHandle<
                     CQikSlider>(EAppSlider1);
  entry.iOddsOfWinning=sl->CurrentValue();

  iEngine->UpdateCurrentListItem(entry);
  }
```

`MultiView4` has a typical `HandleCommandL()` method:

```
void CAppSpecificDetailsView::HandleCommandL(CQikCommand& aCommand)
  {
  switch (aCommand.Id())
    {
    case EAppCmdSave:
      ActivatePreviousViewL(ESave);
      break;

    case EQikCmdGoBack:
      if (DetailsHaveChanged())
        {
        CQikSaveChangesDialog::TExitOption ret =
              CQikSaveChangesDialog::RunDlgLD();
        if (ret==CQikSaveChangesDialog::EShut)
          break;
        if (ret==CQikSaveChangesDialog::ESave)
          {
          ActivatePreviousViewL(ESave);
          break;
          }
        }

    // FALL through to default:
    default:      // For example the back button...
      CQikViewBase::HandleCommandL(aCommand);
      break;
    }
  }
```

The framework code provides a standard dialog, `CQikSaveChangesDialog`, which you can use to prompt the user to save or cancel changes. This dialog can return one of three values:

- `CQikSaveChangesDialog::EShut`: the system closed the dialog, for example due to an incoming phone call. Your application should remain displaying the current view.
- `CQikSaveChangesDialog::ESave`: the user wants to save any changes and continue with the view change.
- `CQikSaveChangesDialog::ECancel`: the user does not want to save the changes; however, they do want to continue with the view change.

Pro tip

The example code deals only with the first two cases. The third case, `CQikSaveChangesDialog::ECancel`, is dealt with by the framework. We simply fall through the `switch` statement to the default case. This may be contrary to your preferred coding standards. It saves code but some argue that it is less maintainable. We use it often, provided that it is commented to clearly identify to which `case` label the fall-through should go. You can see this in the example code.

9.3 Dialogs in UIQ

In a classical desktop GUI application, you would typically display views of application data and present dialogs to perform actions on the data. On a desktop screen, dialog boxes can be large and contain many fields and controls.

Mobile phones have small screens, so we often replace the dialog with a separate full screen view, a detail view. You may think that there is little difference between a view and a full screen dialog; both present information to a user and both allow the data to be manipulated. To a large extent this is true but, at a detailed level, there are some differences between views and dialogs within UIQ 3.

9.3.1 The `CEikDialog` Class

Those familiar with other versions of Symbian OS, such as UIQ 2 have almost certainly used the `CEikDialog` class to present dialogs to the user. Although the `CEikDialog` class has been upgraded to support the

UIQ 3 command processing framework, it is deprecated. It should not be used because it may be removed from UIQ 3 in the future. We therefore recommend that, if possible, you phase out the usage of this class from your applications.

If you use any of the standard system-supplied dialogs that subclass `CEikDialog`, we recommend that you continue to use these dialogs as opposed to implementing your own versions of these dialogs.

9.3.2 The `CQikSimpleDialog` Class

Dialogs are classified as simple if they can display all of their controls on a single page and do not require tabs or scrolling.

As with regular views, simple dialogs can have different layouts and command sets for different screen modes. Your resource file defines the layouts and command sets.

> **Pro tip**
>
> There are a number of issues with `CQikSimpleDialog` in UIQ 3 which have been resolved for UIQ 3.1. In particular, attempting to run a `CQikSimpleDialog` asynchronously is not possible in UIQ 3 as there is no mechanism to cancel the asynchronous dialog. In UIQ 3.1, a new method has been added to the `CQikSimpleDialog` class called `CancelNonWaitingDialog()` and this allows it to be cancelled. While deprecated, the `CEikDialog` does fully support asynchronous or non-waiting dialogs.

9.3.3 The `CQikViewDialog` Class

Dialogs that do not fit into the simple classification, such as those that have a significant number of controls requiring the display to scroll or have a number of tabs-worth of controls, need to be implemented with the `CQikViewDialog` class as opposed to the `CQikSimpleDialog` class.

9.3.4 Converting `CEikDialog` to `CQikSimpleDialog`

Provided your `CEikDialog` is not too complex, it is simple to convert it to a `CQikSimpleDialog`. For example, a `CEikDialog`-derived About dialog may use the following resource:

```
RESOURCE QIK_COMMAND_LIST r_continue_command_list
  {
  items=
    {
```

```
    QIK_COMMAND
      {
      id = EAppCmdContinue;
      type = EQikCommandTypeDone;
      text = "Continue";
      }
    };
  }

RESOURCE DIALOG r_about_dialog
  {
  title="About Commands 1";
  flags=EEikDialogFlagWait;
  command_list=r_continue_command_list;
  items=
    {
    DLG_LINE
      {
      type=EEikCtLabel;
      id=1;
      control=LABEL
        {
        txt="";
        standard_font=EEikLabelFontAnnotation;
        };
      },
    DLG_LINE
      {
      type=EEikCtLabel;
      id=2;
      control=LABEL
        {
        txt="© ZingMagic Limited";
        standard_font=EEikLabelFontAnnotation;
        };
      },
    DLG_LINE
      {
      type=EEikCtLabel;
      id=3;
      control=LABEL
        {
        txt="www.zingmagic.com";
        standard_font=EEikLabelFontAnnotation;
        };
      }
    };
  }
```

The corresponding `CQikSimpleDialog` resource is:

```
RESOURCE QIK_DIALOG r_about_dialog
  {
  title = "About Commands 1";
  configurations =
    {
    QIK_DIALOG_CONFIGURATION
      {
```

```
      ui_config_mode = KQikPenStyleTouchPortrait;
      container = r_about_container;
      command_list = r_about_commands;
      },
    QIK_DIALOG_CONFIGURATION
      {
      ui_config_mode = KQikPenStyleTouchLandscape;
      container = r_about_container;
      command_list = r_about_commands;
      },
    QIK_DIALOG_CONFIGURATION
      {
      ui_config_mode = KQikSoftkeyStyleTouchPortrait;
      container = r_about_container;
      command_list = r_about_commands;
      },
    QIK_DIALOG_CONFIGURATION
      {
      ui_config_mode = KQikSoftkeyStylePortrait;
      container = r_about_container;
      command_list = r_about_commands;
      },
    QIK_DIALOG_CONFIGURATION
      {
      ui_config_mode = KQikSoftkeyStyleSmallPortrait;
      container = r_about_container;
      command_list = r_about_commands;
      }
    };
  }

RESOURCE QIK_COMMAND_LIST r_about_commands
  {
  items=
    {
    QIK_COMMAND
      {
      id = EAppCmdContinue;
      type = EQikCommandTypeDone;
      text = "Continue";
      }
    };
  }

RESOURCE QIK_SCROLLABLE_CONTAINER_SETTINGS
 r_about_container
  {
  controls =
    {

    QIK_CONTAINER_ITEM_CD_LI
      {
      type = EQikCtOnelineBuildingBlock;
      control = QIK_SYSTEM_BUILDING_BLOCK
        {
        content =
          {
          QIK_SLOT_CONTENT_DIRECT
```

```
        {
        slot_id = EQikItemSlot1;
        type = EEikCtLabel;
        unique_handle = EAppLabel1;
        control = LABEL
          {
          standard_font =
           EEikLabelFontAnnotation;
          horiz_align=EEikLabelAlignHCenter;
          txt = "";
          };
        }
      };
    };
  },

QIK_CONTAINER_ITEM_CD_LI
  {
  type = EQikCtOnelineBuildingBlock;
  control = QIK_SYSTEM_BUILDING_BLOCK
    {
    content =
      {
      QIK_SLOT_CONTENT_DIRECT
        {
        slot_id = EQikItemSlot1;
        type = EEikCtLabel;
        control = LABEL
          {
          standard_font =
           EEikLabelFontAnnotation;
          horiz_align=EEikLabelAlignHCenter;
          txt="© ZingMagic Limited";
          };
        }
      };
    };
  },

QIK_CONTAINER_ITEM_CD_LI
  {
  type = EQikCtOnelineBuildingBlock;
  control = QIK_SYSTEM_BUILDING_BLOCK
    {
    content =
      {
      QIK_SLOT_CONTENT_DIRECT
        {
        slot_id = EQikItemSlot1;
        type = EEikCtLabel;
        control = LABEL
          {
          standard_font =
           EEikLabelFontAnnotation;
          horiz_align=EEikLabelAlignHCenter;
          txt="www.zingmagic.com";
```

```
            };
          }
        };
      };
    }
  };
}
```

The `CEikDialog`-derived class used to present the About dialog is:

```
class CAboutDialog : public CEikDialog
  {
 protected:
  void PreLayoutDynInitL();
  };

const TInt KAppMajorVersion=1;
const TInt KAppMinorVersion=0;
const TInt KAppBuild=2;

void CAboutDialog::PreLayoutDynInitL(void)
  {
  TBuf<128>bb;
  iEikonEnv->Format128(bb,R_STR_VERSION,
             KAppMajorVersion,KAppMinorVersion,KAppBuild);
  static_cast<CEikLabel*>(Control(EAppLabel1))->SetTextL(bb);
  }
```

The `CQikSimpleDialog`-derived class used to present the About dialog is:

```
class CAboutDialog : public CQikSimpleDialog
  {
 protected:
  void PreLayoutDynInitL();
  };

const TInt KAppMajorVersion=1;
const TInt KAppMinorVersion=0;
const TInt KAppBuild=2;

void CAboutDialog::PreLayoutDynInitL()
  {
  TBuf<128>bb;
  iEikonEnv->Format128(bb,R_STR_VERSION,
             KAppMajorVersion,KAppMinorVersion,KAppBuild);
  CEikLabel* lbl = LocateControlByUniqueHandle<CEikLabel>(EAppLabel1);
  lbl->SetTextL(bb);
  }
```

There are a number of differences in the resource but the code is practically identical.

9.3.5 Views as Dialogs

One of the key differences between a modal dialog and a view is that the code execution is synchronous for a modal dialog and asynchronous for a view.

When a modal dialog is started (Figure 9.6a), the calling code stops, the modal dialog runs to completion and then returns to the requesting code. Data can be passed back directly to the original calling code.

In contrast, when a view change request is processed (Figure 9.6b), the code making the request continues to run. The requested view is displayed at some time in the future, meaning that code after the view change request will be run and there is no direct correspondence between the calling code and the new view. The new view has no direct means of returning data back to the original view. Such information has to be communicated through a component such as an engine. In essence, views act as modeless dialogs.

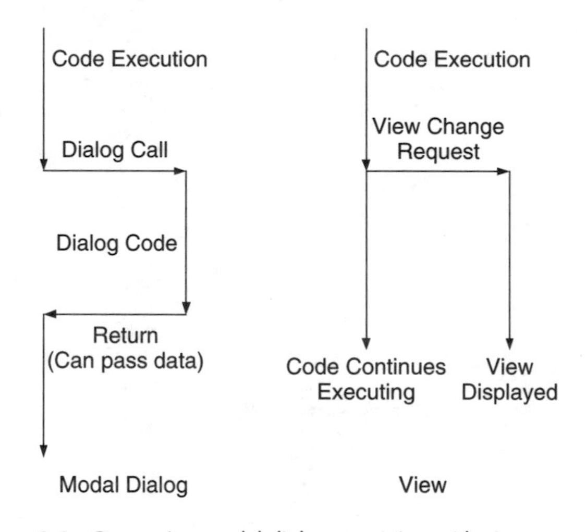

Figure 9.6 Comparing modal dialog execution with view execution

In pseudo-code, switching between views goes like this:

```
CActiveScheduler::RunL()
CCoeAppUi::HandleWsEventL()
ListView::CommandEntryPoint();
ActivateViewL(DetailsView);
ListView::CommandExitPoint();
...
CActiveScheduler::RunL()
CQikViewBase::HandleViewActivatedEvent()
DetailsView::ViewActivatedL()
```

In contrast, when we display a modal dialog, code execution goes along the lines of:

```
CActiveScheduler::RunL()
CCoeAppUi::HandleWsEventL()
ListView::CommandEntryPoint();
Dialog::ExecuteLD();
Dialog::PreLayoutDynInit();
Dialog::ExitDialog();
ListView::CommandExitPoint();
```

Recognizing that it is often much easier for application developers to handle the synchronous code flow allowed by the dialog approach, UIQ 3 supports a hybrid view and dialog. View dialogs combine the ability to support multiple configurations (from views) with the convenience of synchronous code flow (from dialogs).

`MultiView4` contains an example of such a view dialog:

```
class CDetailsAsAViewDialog : public CQikViewDialog
  {
protected:

  // From CQikViewBase.
  void HandleCommandL(CQikCommand& aCommand);
  void ViewActivatedL(const TVwsViewId& aPrevViewId,
                        const TUid aCustomMessageId,
                      const TDesC8& aCustomMessage);
  void ViewConstructL();
  void SaveL();

public:
  TInt RunDialogLD(CAppEngine* aEngine);

protected:
  CAppEngine* iEngine;
  };
```

The class derives from a `CQikViewDialog` which in turn derives from the `CQikMultiPageViewBase`, thus ultimately `CQikViewBase`. Most of the methods should be familiar to you; the only new one is:

```
TInt CDetailsAsAViewDialog::RunDialogLD(CAppEngine* aEngine)
  {
  iEngine=aEngine;
  return(ExecuteLD());
  }
```

The resources used to define the `CQikViewDialog` are identical to those we use when we display the details in a separate view.

The primary difference between regular views and CQikViewDialog-derived views relates to how they are displayed and removed.

A regular view is displayed and removed through a call to iQikAppUi.ActivateViewL(); with varying parameters. The CQikViewDialog is displayed with a call to CQikViewDialog::RunLD() or CQikViewDialog::ExecuteLD() and is removed though a call to CQikViewDialog::CloseDialog().

The following code fragment shows the list view running a CQikViewDialog:

```
void CAppSpecificListView::HandleCommandL(CQikCommand& aCommand)
  {
  switch (aCommand.Id())
    {
    case EAppCmdDetails:
      {
      CDetailsAsAViewDialog* q = new(ELeave)CDetailsAsAViewDialog;
      q->RunDialogLD(iEngine);
      break;
      }

    default:
      CQikViewBase::HandleCommandL(aCommand);
      break;
    }
  }
```

The following code performs a regular view switch:

```
void CAppSpecificListView::HandleListBoxEventL(
                            CQikListBox* aListBox,
                      TQikListBoxEvent aEventType,
                                  TInt aItemIndex,
                                     TInt aSlotId)
  {
  switch (aEventType)
    {
    case EEventItemConfirmed:
    case EEventItemTapped:
      iQikAppUi.ActivateViewL(KViewIdDetailsView);
      break;
    default:
      break;
    }
  }
```

If you run the MultiView4 application there are no discernible differences in the content displayed or interactions supported between a regular view and a CQikViewDialog.

10
Building an Application

In previous chapters, we have built some example applications to explore various aspects of the UIQ platform. In this chapter, we build on that knowledge with the primary objective of creating a GUI application, `SignedApp`, capable of passing the Symbian Signed Test Criteria which are presented in Chapter 14. As the application is developed, the impact of Symbian Signed is explained and the required metafiles are described.

The first stage, `SignedAppPhase1`, provides a generic application framework that can be adapted to any application that is intended to be Symbian Signed.

Next, we describe how to utilize the generic framework in building the `SignedAppPhase2` application. We add application-specific code including an application engine, category management, a list view and a detail view. We explain the use of INI files, externalization and internalization, with emphasis on the use of streams and stores. We then discuss preparations for submission to Symbian Signed. In Chapter 11, we extend the `SignedAppPhase2` application to `SignedAppPhase3` by adding some multimedia functionality.

10.1 Symbian Signed

Symbian Signed is a scheme run by Symbian and backed by the mobile phone industry. Symbian Signed applications follow industry-agreed quality guidelines and support Network Operator requirements for signed applications. Symbian Signed enables you to certify your application to be installed and run on UIQ 3 and S60 phones.

Symbian Signed is described in detail in Chapter 14, however, it is very important to consider your approach to Symbian Signed *before* you start coding. For most options in Symbian Signed, you will need a *Publisher ID*.

This is a digital certificate that is issued by a third-party trusted authority. It verifies your identity to the Test House and to end users installing your application.

10.1.1 Capabilities

Prior to coding, you usually write some form of functional specification which sets out the functions and features of your application. Generally your specification includes details of how an application should integrate with the mobile phone environment. It is important that you consider how easy it is to implement the required functions:

- Are the APIs available for the features you wish to include?
- Does your application perform functions that are considered to be sensitive and are protected by Symbian OS?

Capabilities define which protected APIs within Symbian OS v9 your application can successfully access. Without the required capability, an API usually returns an error (`KErrPermissionDenied = -46`) and does not perform the requested task. The SDK documentation indicates required capabilities for protected APIs.

Capabilities are divided into three groups:

- *User*, also known as *Basic*. The user can grant the capability to an unsigned application via an on-screen dialog box. A Symbian Signed application is automatically granted the required capabilities with no user interaction.
- *System*, previously known as *Extended*. The application must be Symbian Signed to get these capabilities.
- *Device Manufacturer*. You must have manufacturer support and Symbian Sign the application.

As we develop our `SignedApp` example, we add functions that require User capabilities. Symbian Signed is the best way to grant these capabilities, and is the approach that we have chosen.

You should review the functionality in your application and check for the capabilities that you need.

10.1.2 UIDs and Symbian Signed

The example applications in the previous chapters use UIDs in the range `0xE0000000` to `0xEFFFFFFF`. These values are reserved for internal development purposes only. Any applications that are to be distributed should not use these UID values. In particular, there is no formal mechanism to ensure that two developers do not use the same values from this particular range.

If you have chosen *not* to get your application Symbian Signed, you must obtain a UID from the Symbian OS v9 unprotected range `0xA0000000-0xAFFFFFFF` for your application.

If you *have* chosen to get your application Symbian Signed, your first requirement is to obtain a UID from the Symbian OS v9 protected range `0x20000000-0x2FFFFFFF`. If you change your mind and later decide you no longer want to go through the Symbian Signed tests, you have to change the UID since the application installer refuses to install applications with UID values in the protected range unless they are signed.

10.1.3 How to Get Symbian Signed

We completed our `SignedApp` project before the introduction of the new Open, Express and Certified Signed schemes in late 2007. Had those options been available, we would probably have chosen Express Signed, which grants access to the capabilities we need with the minimum overhead. In the event, we chose to follow an option equivalent to Certified Signed, in which we used our Publisher ID and an external Test House. The various ways you can get your application Symbian Signed are explained in Chapter 14.

Both Express and Certified Signed require a Publisher ID. Certified Signed also requires independent testing; a Test House performs the testing and facilitates the signing of our application using the Publisher ID. This process typically takes one week; we recommend that you allow at least two weeks for your first project. You will also have to accept legal agreements, which can take additional time if company lawyers become involved.

10.1.4 Symbian Signed Test Cases

All Symbian OS applications are expected to conform to basic criteria of soundness and stability, defined by Test Criteria. Symbian Signed provides a Test Criteria document detailing the tests to be performed and the expected results. Before embarking on a project you should read the Symbian Signed Test Criteria document, available from the Symbian Signed website, to ensure your functional specifications and the Test Criteria are achievable.

If you follow the Certified Signed option, these are the criteria the Test House will test against. If you follow the Express Signed option, you self-certify test results based on these criteria.

10.1.5 Important Note on Building the Applications

You can freely build the `SignedApp` example, with or without code changes, and run it in the UIQ emulator. Please note that we use the same application UID in each of the phases. This approach means that only one of the phases can be installed at any time.

`SignedAppPhase3` has been Symbian Signed and we provide the signed SIS file in the example download so that you can install it to a mobile phone.

To test `SignedAppPhase1`, `SignedAppPhase2`, any modification to or rebuild of `SignedAppPhase3` or your own application on a real mobile phone, you must use the Open Signed method. This uses developer certificates that are restricted to specific phones based on IMEI number.

Unfortunately, this also means that we are unable to include a suitable developer certificate in the example download, since Publisher IDs and therefore developer certificates are issued to named companies or organizations. To build and deploy the `SignedApp` examples, you will need to follow the Open Signed process. Section 10.4 explains how we prepare `SignedAppPhase2` for testing on real hardware.

10.2 Starting Our Project: `SignedAppPhase1`

Our first task is to generate the project framework for our application. IDEs provide wizards to automate some of this work, however, we generate the framework for `SignedAppPhase1` manually in order to help you to understand the components involved.

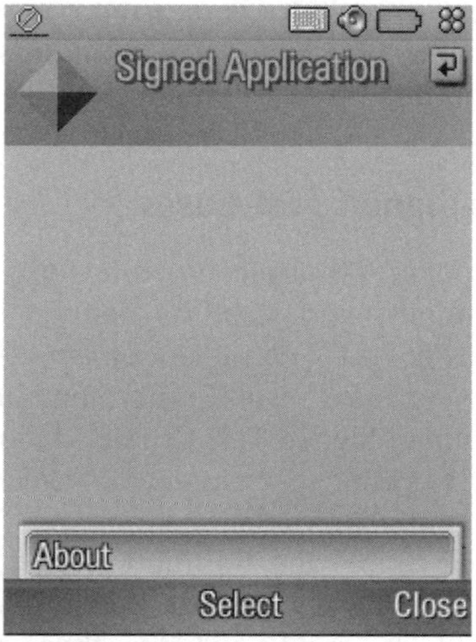

Figure 10.1 `SignedAppPhase1`

There is little substance to the application itself; the application space is blank with a single menu command (see Figure 10.1). The main thing is that it builds cleanly and runs on both the emulator and mobile phone. With replacement of the UID and renaming of the files, this framework can be applied to almost any application development.

10.2.1 Project Framework

Our previous examples used a very simple structure in which all files were held together in one folder, with just one subfolder for images. We now introduce the recommended folder structure which is commonly used for UIQ 3 application development. This structure is not mandatory, however it is widely used and you will benefit from adopting a well-defined environment.

Folder Structure

The folder structure for `SignedAppPhase1` is shown in Figure 10.2.

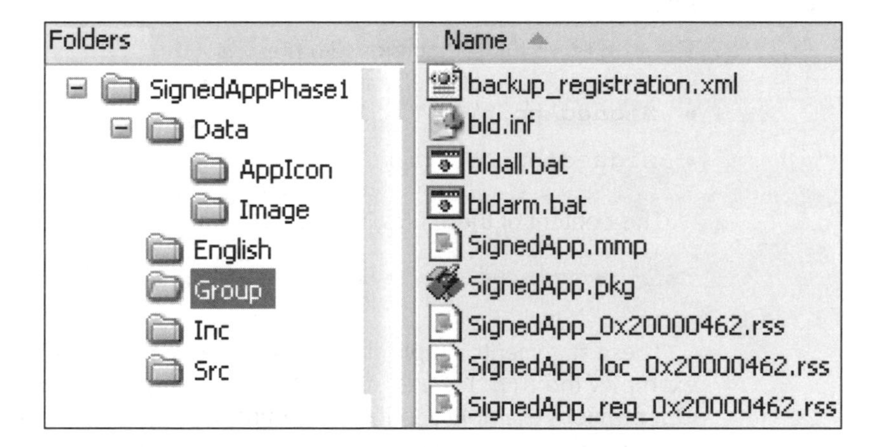

Figure 10.2 Application folder structure

The main change from our earlier example application is that the project maintenance files have been moved to the `Group` folder and the C++ code to the `Src` folder. The content of each folder is as follows:

- **`Data`**: This folder contains two sub-folders, one for the application icon and one for other graphics.
- **`English`**: Separate folders are used for each supported language. `English` contains files with the English language text for our application. In Chapter 13, we show how to add additional languages to your application.

- **`Group`**: This folder contains all the project maintenance metafiles.
- **`Inc`**: This folder contains the header files that need to be shared between the C++ code and resources.
- **`Src`**: This folder contains our application code. Currently our application is trivial and consists of a single source file and its associated header file. Later, we split this into various components.

> **Pro tip**
>
> It is good practice to make public only those entities that need to be public. Therefore:
>
> - public headers are kept in the `Inc` directory
> - private headers kept in the relevant source directory.

Language Folders

The `English` folder contains two files:

- `SignedApp.rls`
- `SignedApp_loc.rls`

The content of these files is a series of statements such as:

```
rls_string STR_R_CMD_ABOUT "About"
```

These statements simply define a set of tokens to have a value. In this example, the `STR_R_CMD_ABOUT` token is assigned the value '`About`'. Our previous example applications simply contained the text inline within the resource files. Using token–value pairs like this makes it easier to manage translations.

In Chapter 13, in our `localization` example, we add folders and text files for French and Simplified Chinese.

Group Folder

The `Group` folder contains the project maintenance and metafiles.

`backup_registration.xml`

The system backup mechanism looks for a file named `backup_registration.xml` within the private folder of your application. This file defines how your application interacts with the system backup.

Our backup registration file specifies that the application data is fully contained within its private data-caged folder and both it and the application should be backed up automatically by the system. This means that our application does not have to actively participate in backup.

The concepts of data caging and private folders are discussed in Chapter 4. The Symbian Signed test CON-04 in Test Criteria v2.11.0 (UNI-07 in Test Criteria v3.0.0) covers backup behavior and we discuss it in Chapter 14.

bld.inf

As with our previous examples, this informs the Symbian build environment for which targets it should create makefiles. It also specifies the MMP filename.

bldarm.bat

This is a simple command-line batch file that automates building the Arm version of our application using the GCCE compiler. An IDE would normally be able to perform this task at the click of a button.

bldall.bat

Building the executable program by using the `bldarm` batch file is a core task but it is not the only task required to construct an application suitable for deployment. In particular, you need to:

- package all your files into a single SIS file
- sign the SIS file so it can be deployed to a real mobile phone for test purposes
- sign the SIS file so it can be submitted to the test house.

This batch file encapsulates the entire build process for our application. We describe the content of this file later. In general, an IDE only performs part of this task, building the application. Packing and signing is not fully supported.

> **Pro tip**
>
> In Carbide 1.2, the project settings only allow you to specify a single certificate and key pair. Usually two certificate and key pairs are required, one for the SIS file you test and one for the SIS file you submit for signing. Therefore you need to change the settings and re-create the SIS file before submitting to a test house or Express signing. To eliminate any possible errors, we recommend using a batch file such as `bldall.bat` to perform these tasks.

`SignedApp.mmp`

This is the application master project file. We look at this file in more detail later on.

`SignedApp.pkg`

This is the application package file, specifying various attributes of your application including the list of files you wish to install to a target mobile phone.

`SignedApp_0x20000462.rss`

This file contains the application-specific resources. We have worked extensively with resource files in the preceding chapters.

`SignedApp_loc_0x20000462.rss`

This file contains the application registration localization information and is discussed later.

`SignedApp_reg_0x20000462.rss`

This contains the application registration information and is discussed later.

Naming Resource and Executable Files

It is not currently a Symbian Signed test requirement, but recommended best practice is that the binary files (such as EXEs and DLLs) of your application have a unique name.

In versions of Symbian OS prior to v9, most application executables were placed in their own folders but with the advent of Platform Security in Symbian OS v9, all executable files must be placed in the same folder, `\sys\bin`. If the executable is not in that folder, the OS will not load or run it. Name clashes are therefore much more likely than was the case with previous versions of Symbian OS.

To avoid name clashes, we recommend that any executable should have the application UID appended to the executable name. An end-user is highly unlikely to ever see the executable filename so there should not be any presentation issues.

There are two reasons why this affects resource filenames. Firstly, as the registration and standard resource files are also placed in common folders, any name clash that could occur with binary files can also occur with these files. If these files cannot be installed or we can't overwrite files where the name clashes, then problems will occur. Secondly, the application framework code derives your resource and multi-bitmap filename from the name of the binary file. Since the framework code opens these files on your behalf, if it cannot derive the filename, it will fail to open the files.

It is not mandatory to name your files like this, but as Symbian Signed evolves it may become mandatory. If you submit applications to a Symbian Signed test house and do not name binaries correctly, this is noted in the test report. We strongly recommend you adopt these naming policies.

Pro tip

Various files in a Symbian OS project are automatically generated, in particular MBG and RSG files. The full name of these files is generated from the incoming bitmap and resource filename respectively and is referenced by other source files in your project. If you change the names of your files mid-project, remember that changing all references is time consuming and prone to errors.

You may believe that you can rename the files to contain the application UIDs at the last stage, when you generate the PKG file. This is true for some of the files, for example the EXE. However, if you look at the content of `SignedApp_reg_0x20000462.rss`, you see it references `SignedApp_loc_0x20000462`. Simply renaming files in the PKG file will not cause the compiled content of the registration resource file to reference the correct file. Therefore we strongly recommend that you adopt the file naming convention from the beginning of the project.

Application Registration Files

The application registration file contains attributes of the application. The application launcher reads it to find out about the application.

Our application registration file contains the following:

```
// File type UID.
UID2 KUidAppRegistrationResourceFile

// Application UID.
UID3 0x20000462

RESOURCE APP_REGISTRATION_INFO
  {
  // Filename of application binary (minus extension).
  app_file = "SignedApp_0x20000462";

  // Location of the localisable icon/caption
  // definition file.
  localisable_resource_file =
              "\\Resource\\Apps\\SignedApp_loc_0x20000462";
  }
```

Developers familiar with S60 3rd Edition may notice that this file seems to differ from the S60 equivalent. For example, the `localisable_resource_id` field does not appear to be defined. As with many resources, the above code utilizes a number of default values. The `localisable_resource_id` default value is 1.

Depending on your organization's preferences you may wish to generate the `<SignedApp_loc_0x20000462.rsg>` header file, include it within this file and assign to the `localisable_resource_id` field the same name as assigned to the `LOCALISABLE_APP_INFO` resource, described later.

```
localisable_resource_id = R_APPLICATION_INFO;
```

Similarly, the application takes various other default properties such as:

- `KAppNotHidden`
- `KAppNotEmbeddable`
- `KAppDoesNotSupportNewFile`
- `KAppLaunchInForeground`.

Again, you may prefer to explicitly define these values, for example:

```
embeddability = KAppNotEmbeddable;
newfile = KAppDoesNotSupportNewFile;
```

You should refer to the UIQ 3 SDK documentation for the full range and meaning of these attributes.

Rather than relying on filename extensions, Symbian OS usually interprets files by their content. Most files native to Symbian OS contain a four-digit header comprising three UIDs and a checksum. This provides a fast and reliable way of identifying content.

Compiled application registration files contain this header, and the UID2 and UID3 statements provide the second and third digits. As previously explained, UID2 only has meaning in the context of UID1 and UID3 in the context of UID2.

Application registration file content is defined by the application launcher. It has defined UID2 to be the value `KUidAppRegistrationResourceFile (0x101F8021)` so it knows this file is an application registration file as opposed to any other file type. It has further defined UID3 of the file to take the UID of the application registration file with which it is associated.

The `app_file` statement specifies the name of the application executable. Since all executables must reside in the `\sys\bin` folder, further information about its path is not required.

Many applications present language-dependent or, more precisely, locale-dependent information to a user. Both icons and text can contain locale-specific information. The `localisable_resource_file` statement specifies where to find this information.

Application Registration Localization Files

The `SignedAppPhase1` application registration localization file contains the following content:

```
#define EViewIdPrimaryView 0x00000001

#ifdef LANGUAGE_01
#include "..\English\SignedApp_loc.rls"
#endif

// This file localises the applications icons and caption.
RESOURCE LOCALISABLE_APP_INFO
  {
  short_caption = STR_R_APP_SHORT_CAPTION;
  caption_and_icon =
    {
    CAPTION_AND_ICON_INFO
      {
      caption = STR_R_APP_LONG_CAPTION;
      number_of_icons = 3;
      icon_file =
      "\\Resource\\Apps\\SignedApp_icons_0x20000462.mbm";
      }
    };
  view_list =
    {
    VIEW_DATA
      {
      uid=EViewIdPrimaryView;
      screen_mode=0;
      caption_and_icon =
        {
        CAPTION_AND_ICON_INFO
          {
          }
        };
      },

    VIEW_DATA
      {
      uid=EViewIdPrimaryView;
      screen_mode=EQikScreenModeSmallPortrait;
      caption_and_icon =
        {
        CAPTION_AND_ICON_INFO
          {
          }
        };
      }
    };
  }
```

As with application registration files, developers familiar with S60 3rd Edition may notice some differences. For example, no name is assigned to the `LOCALISABLE_APP_INFO` resource. In S60 3rd edition, the name is used within the application registration file and so is required. In UIQ v3, we simply use the default value, 1. In contrast, S60 3rd Edition does not utilize the `view_list` fields. It uses the default values.

Application Captions At the top of the resource definition, you can see the `short_caption` and `caption_and_icon` fields. A caption is the application name presented to the user. The `caption_and_icon` structure defines the normal caption along with information about the application icons that can be used. In general, the short caption is used when the normal caption does not fit on the display.

Rather than containing quoted text, the caption names are represented by tokens. These tokens are converted into text when the file is compiled. The token value pairs are defined in the `#include '..\English\SignedApp_loc.rls'` file. As you may be able to guess, when alternate locale versions of the file are compiled, the tokens change to take on values specific to the alternate locale.

Pro tip

If you are only ever going to produce a single-locale version of your application, it is possible to replace the tokens with inline quoted text. However, we would still recommend using the token-based approach as it is negligible extra work at this stage and makes it easy for you to add additional locales later on.

Application Icons Our application needs three icons and transparency masks so that it can be displayed correctly in each application launcher state. For the application list view, we need an icon that is 18×18 pixels in size (see Figure 10.3).

A 40×40 pixel icon is used in the grid style launcher and in the application title bar; a 48×48 pixel icon is used when highlighted (see Figure 10.4).

The transparency mask specifies which part of the icon is visible (the white area in the mask) and which is transparent (the black area in the mask). You can also use an eight-bit grayscale mask and define areas that are semi-transparent. We define the compilation of these images into a single multi-bitmap (MBM) file in the MMP file.

You may wish to note that all masks in UIQ 3 are reversed compared to other Symbian platforms, including UIQ 2.

From UIQ 3.1, you can use SVG icons in the application launcher.

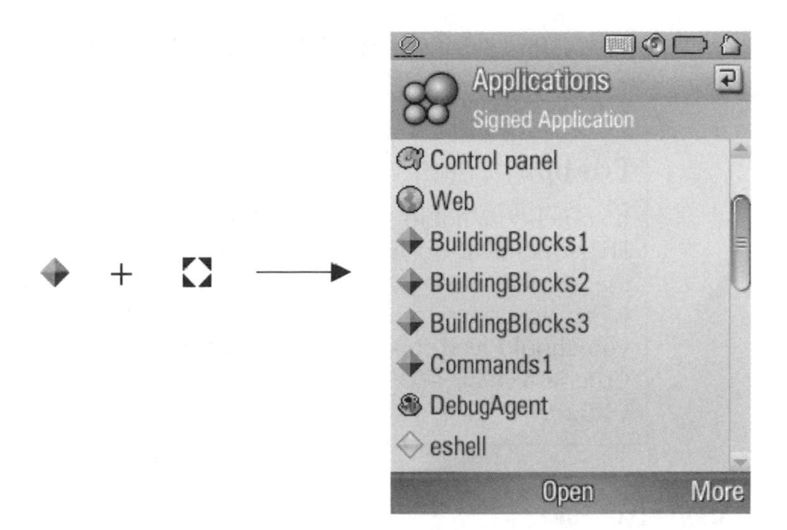

Figure 10.3 Application icon in Application Launcher list view

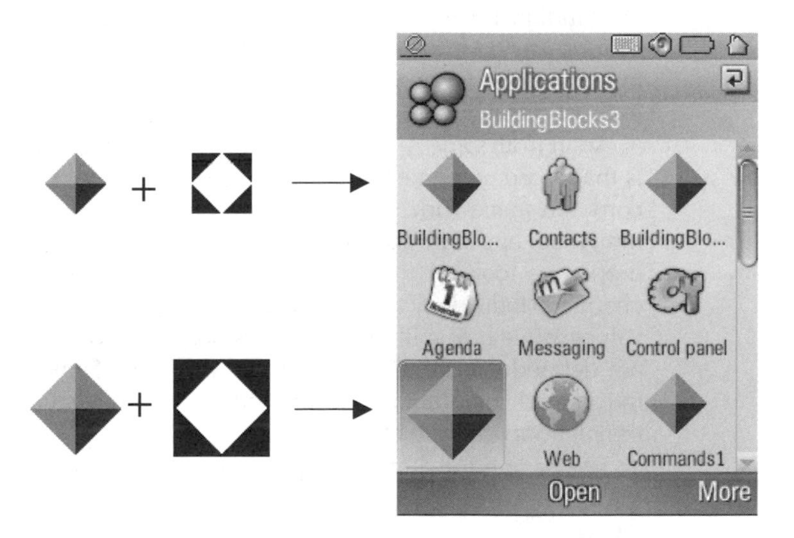

Figure 10.4 Application icon in Application Launcher grid view

The `view_list` Some UIQ 3 mobile phones are able to support multiple view configurations. For example, the Sony Ericsson P990i uses Pen Style configuration when the flip is open and Softkey Style Small when the flip is closed. To inform the Sony Ericsson P990i that your application supports Softkey Style Small, it is not enough to simply define a `KQikSoftkeyStyleSmallPortrait` view configuration in your primary resource file. You also have to tell the application launcher that it should display your application icon when the phone is in Softkey Style Small UI configuration. The `VIEW_DATA` structures inform the application launcher how your application supports the default and `EQikScreenModeSmallPortrait` screen modes.

> **Pro tip**
>
> If you choose not to support the Softkey Style Small mode of the Sony Ericsson P990i in your application, not only do you need to omit the second `VIEW_DATA` structure but you must also document this when you submit your application to Symbian Signed. In particular you should read, understand and comply with Symbian Signed Test Criteria test GEN-03 in Test Criteria v2.11.0 (UNI-10 in Test Criteria v3.0.0).

Application MMP File

A number of fundamental changes have been made with the introduction of Symbian OS v9. Perhaps the most significant of these is that applications are no longer polymorphic DLLs which are loaded as plug-ins to a common framework (the `AppRun` executable). They are now full executables in their own right.

Apart from some API changes, the most significant aspect of this change is that those properties of an application that were defined by the single common framework, such as stack and heap size, are now available for individual applications to define. While many of the common framework properties took on the system-wide default values, some values were changed. Individual applications may want to consider changing these values as well. As with previous versions of Symbian OS, these properties are defined through the application MMP file. Symbian OS v9 has also introduced some additional properties for an executable, most notably the platform security capabilities.

The `SignedAppPhase1` application contains the following within its MMP file:

```
TARGET              SignedApp_0x20000462.exe
TARGETTYPE          exe
UID                 0x0 0x20000462
EPOCSTACKSIZE       0x5000
CAPABILITY          NONE
```

As with UIQ 2, we define our primary target filename with the `TARGET` statement. Applications have changed to be of type EXE; the `TARGETTYPE` statement reflects this change.

Application UIDs As we previously discussed, Symbian OS has a strong tendency to interpret files by their content rather than extension. In particular, Symbian OS native files usually have a signature comprising three UIDs and a checksum. The `TARGETTYPE` statement is used to generate the value for UID1.

Prior to Symbian OS v9, applications were polymorphic DLLs. They were required to have UID2 set to `0x100039CE` to be recognized by the common application framework as a DLL that implemented the application-specific polymorphic DLL interface.

In Symbian OS v9, applications are executables in their own right. They no longer need to identify themselves as implementing a particular plug-in interface since they do not plug in to any other component. UID2 should now be set to 0.

UID3 takes the unique application ID.

`EPOCSTACKSIZE` The system-wide default value for the stack size is `0x2000` (8 KB). This value is defined within the Symbian OS tool chain that creates executables.

Prior to Symbian OS v9, the common application framework set the `EPOCSTACKSIZE` to a value of `0x5000` (20 KB). Your application therefore ran with a stack size of 20 KB on earlier versions of UIQ.

From Symbian OS v9, applications are executables in their own right and can therefore redefine their stack size if they need more memory.

Factors to Consider for Stack Size In general, default values should be used unless there is good reason to change them. Stack overflow issues are very hard to track down, however. For example, if you are porting known working applications from previous versions of UIQ or S60 they work with a stack size of 20 KB. Using the default of 8 KB is probably taking an unnecessary risk.

> **Pro tip**
>
> Stack overflow can sometimes manifest itself as a `KERN-EXEC: 3` panic, which is normally associated with `NULL` or invalid pointers.

If you are writing new applications, but using similar or shared functionality to previously known working applications, then setting the stack size to 20 KB is also reasonable.

Text on UIQ 3 mobile phones is stored as Unicode. The original 8 KB stack size was set in the tool chain when single-byte character sets was the standard build variant. A single filename buffer comprises 256 Unicode characters or 512 bytes. With an 8 KB stack, an absolute maximum of 16 filename buffers can be stored. In reality, it is less, as the stack is also used for various system runtime information as well as all the other automatics.

Perhaps 10 or 12 filename buffers would be the maximum, compared to the original design choice of around 20 buffers.

Adding code to reduce stack usage can needlessly complicate applications. For example, if all filename buffers had to be allocated from the heap, your application would need to manage those memory allocations and deal with potential allocation failures. This can easily add several kilobytes of code as well as still requiring the memory storage, thus the potential memory-saving benefits of a smaller stack are eroded rapidly.

Moving buffers previously allocated on the stack into object property can needlessly increase the size of object property. Unlike stack-based variables, the memory occupied by object property is not normally re-usable. In addition, simply saving a few kilobytes from the stack by moving it to object property does not change the overall memory resource usage of an application, it simply redistributes the memory amongst the different, logically separate regions of memory an application uses.

If written correctly, a large proportion of your UIQ v3 application comprises system-supplied API calls. It is difficult to know how little or how much stack they use, particularly if they manipulate text or filenames. In general, they are frugal; however, there are no explicit tools available to measure stack usage. To err on the size of caution is to be recommended.

Current UIQ 3 phones have more than 20 MB memory. Even at 20 MB, a 20 KB stack is just 0.1 % of available memory.

> **Pro tip**
>
> With practice, and with some care, it is relatively easy to write applications restricted to a 20 KB stack. In contrast, writing applications restricted to an 8 KB stack is somewhat harder. Therefore, if you feel that using the default 8 KB stack size is likely to be a bit too restrictive, we recommend that a 20 KB stack be used as opposed to any other value.

In exceptional circumstances you can define your stack size to be larger, however, you must exercise care. Whatever value you choose, the system reserves that amount of memory, making it unavailable to other applications. For example if you choose to have a 1 MB stack but only ever use 50 KB then the remaining 950 KB is unusable by the rest of the system. As a developer on any platform, but particularly on one with limited memory resources, you are responsible for efficiency and good behavior in your application to ensure the system and, ultimately, the user is not adversely affected.

`CAPABILITY` Statement The `CAPABILITY` keyword in the MMP file is used to list the set of platform security capabilities that our application needs in order to perform its functions. Our application does not require any capabilities yet since we have not added any code that accesses protected APIs.

It is strongly recommended that you start with:

```
CAPABILITY NONE
```

And then add capabilities as and when you determine they are required.

Pro tip

Some developers suggest that you start with a `CAPABILITY` statement such as:

```
CAPABILITY ALL -TCB
```

This says you want all capabilities except for the `Trusted Computing Base` capability. While this may solve any immediate problem attempting to use sensitive APIs, this approach should be avoided if at all possible.

As an application developer, you need to understand which APIs used by your application require which capabilities. In particular, when you come to submit an application via the Symbian Signed website you need to complete a declarative statement explaining which system capabilities your application uses and why.

In addition, `CAPABILITY ALL -TCB` also includes a number of system capabilities, such as `AllFiles`, `TCB` and `DRM`, that can only be granted by a sponsor such as a mobile phone manufacturer.

Application Icons The three application icons and masks that we saw in the previous section must be compiled into a single MBM file. The following code in the MMP file does this:

```
// Application icons
START BITMAP  SignedApp_icons_0x20000462.mbm
SOURCEPATH    \Data\AppIcon
TARGETPATH    \Resource\Apps
SOURCE        c16 AppIcon18x18.bmp
SOURCE        1 AppIcon18x18mask.bmp
SOURCE        c16 AppIcon40x40.bmp
SOURCE        1 AppIcon40x40mask.bmp
SOURCE        c16 AppIcon64x64.bmp
SOURCE        1 AppIcon64x64mask.bmp
END
```

You should always define masks as either 1 (black and white) or 8 (eight-bit grayscale) so that they work correctly. The icons are defined as c16, 16-bit color.

You can find out more about MMP files if you work through the examples presented earlier in this book.

10.2.2 Compiling Your Application

The source files that we have discussed are compiled into a PKG file, as shown in Figure 10.5. The MakeSIS tool creates a SIS file. Because we are using a protected range UID, you must first sign the SIS file before it can be installed on a mobile phone. Section 10.4 describes how to deploy for testing.

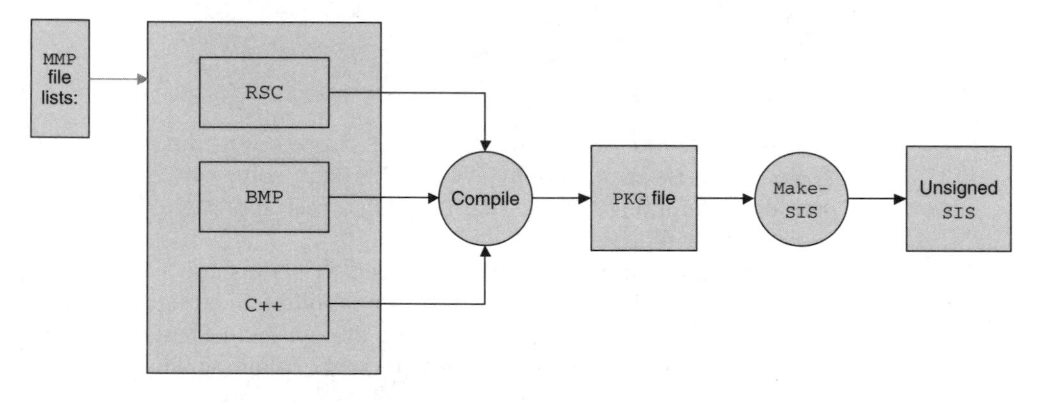

Figure 10.5 Compiling to a PKG file and creating a SIS file

10.3 SignedAppPhase2

SignedAppPhase2 takes the basic project framework that we created in SignedAppPhase1 and adds functionality to make a simple file manager application with the following features:

- display a list view of the files present in folder c:\Private\20000462
- sort the list by name, size, type, date
- filter the list by category
- display a detail view of filename, date and category and allow these to be updated
- provide a *Send as* facility to transfer files
- provide a registration feature using IMEI.

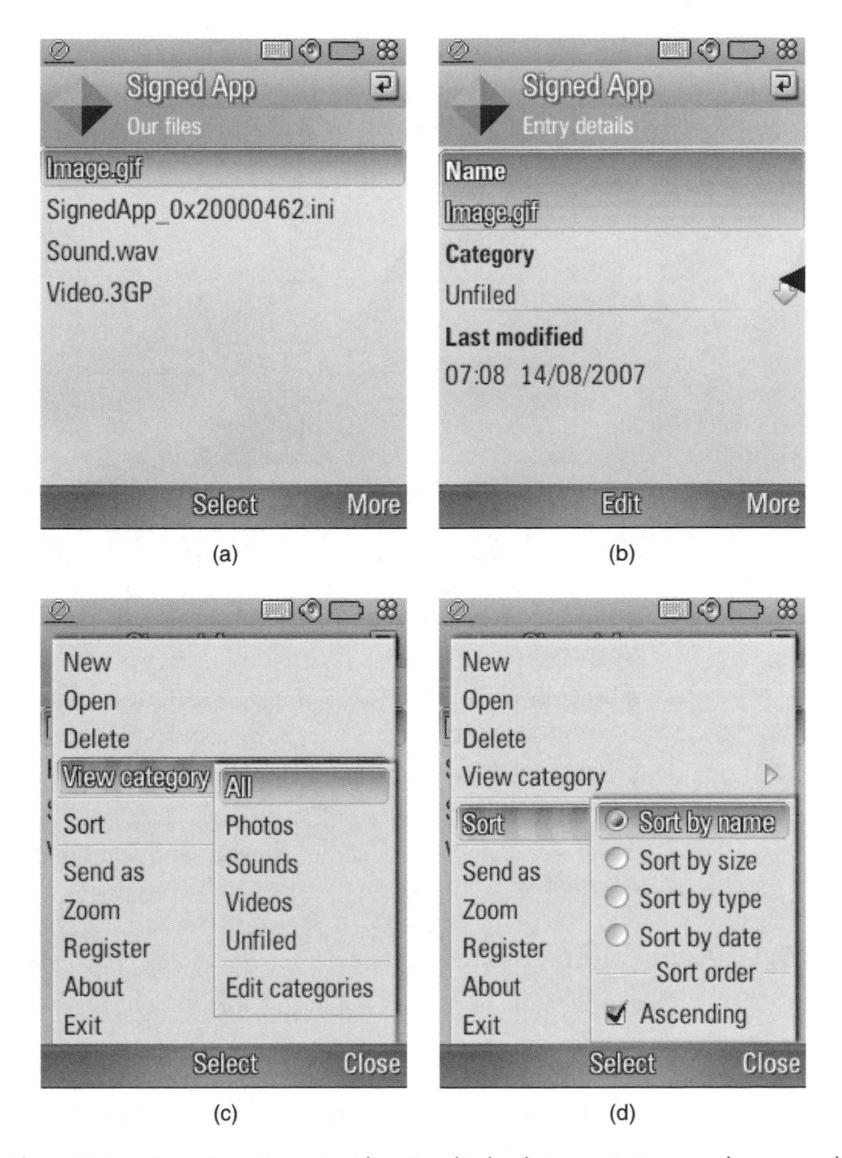

Figure 10.6 `SignedAppPhase2`: a) list view, b) detail view, c) category sub-menu and d) sort sub-menu

To help you understand the design and coding process, Figure 10.6 shows some screenshots of the final `SignedAppPhase2` application.

Prior to implementing the code for our application, we need to have an application design that implements the functional specification. You may have company procedures that tell you how to do this. For our application, we have chosen a minimalist design strategy, comprising a set of statements and a couple of diagrams.

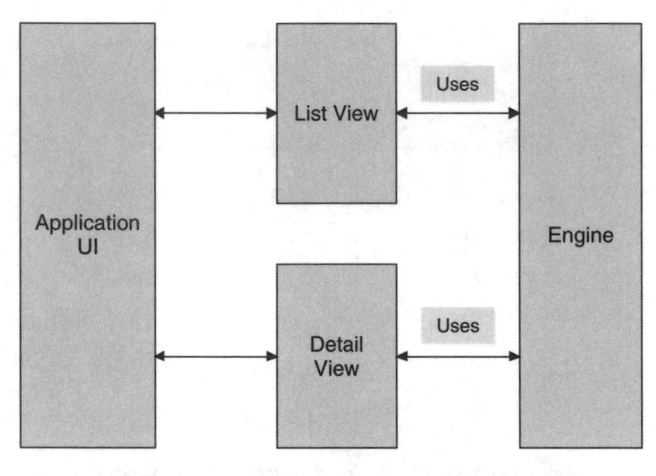

Figure 10.7 Application structure

The application (see Figure 10.7) comprises:

- an engine responsible for maintaining the data structures we wish to display, `Engine.cpp`
- a list view responsible for displaying a sorted and filtered list of entries, `ListView.cpp`
- a detail view responsible for displaying and editing the details of a specific entry, `DetailsView.cpp`
- a set of dialogs and some general application-framework code, `SignedApp.cpp`.

10.3.1 Application Engine

The application engine is responsible for:

- generating the list of files within a particular folder on the phone
- maintaining the set of categories defined within the application
- sorting and filtering entries depending on the user's chosen sort order and category.

Sorting and Filtering Entries

You may recall that the `Commands1` example application showed how we might use the list box to sort entries. We have chosen to move the sorting and filtering functionality into the engine for a number of reasons:

- List boxes do not support filtering by category very easily; you have to rebuild the list box from an underlying data structure. Entries that do not belong to the current category have to be stored outside the list box – we have to have some kind of engine.

- The sort comparison function in the `Commands1` example is inefficient. Every comparison requires a lookup from an externally stored variable, via thread local storage in the case of the `Commands1` application, to obtain a reference to the engine to know how to perform the sort. The alternative is to supply different sort comparison functions, one for every possible variation supported.
- Sorting by more than a single key is difficult to implement since there is no easy access to second-order key information.
- Sorting can leave. This requires you to implement some error handling and rollback, which is particularly unfavorable when other operations, such as rename operations, are considered.

In our implementation, we generate a set of all entries within a folder and store them within an `RArray`. Each entry is represented by a `TFolderEntry` object. We then generate an index referencing these entries.

To produce a filtered list, we simply build an index that references a subset of the underlying entries; the underlying entries remain static. To sort the underlying entries, we sort the index rather than the entries themselves (see Figure 10.8). To produce a sorted and filtered list, we simply apply the two operations to the index.

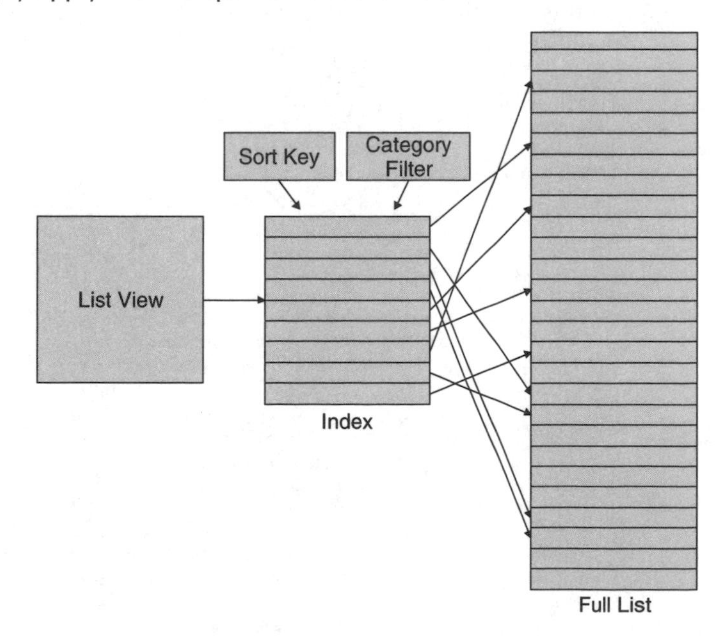

Figure 10.8 Using an index to reference a full list

The advantages of this approach are that:

- Filtering the list by category is quite simple. All that is required is to create an index that references only those entries belonging to a specific category. If the chosen category is All then all the entries exist within the index.
- The sort function is very efficient and properly encapsulated. By using an appropriate class, we can remove any external variable lookup and perform multi-key sorts very easily.
- Only a single comparison function is required for all possible sort variations.
- The swap function is very efficient; we only need to swap index entries and not the underlying data structures.
- The sort operation cannot leave.

The following method generates the new folder list, creates an index containing those entries within the currently selected category, sorts the index using the currently selected sort order and finally replaces the existing data structures with the new structures. This last operation is important to ensure we can roll back with the previous application state remaining intact. For example, if we simply replaced the `iFolderEntryList` with the new list before we successfully created a new index on this list, there are no guarantees that the old index is valid for this new list.

```
void CAppEngine::BuildDirectoryListL(const TDesC& aPath)

// Get a list of the files and folders
// within the indicated folder.

  {
  TParse parse;
  parse.Set(KWildCardName,&aPath,NULL);
  CDir *dirList;
  User::LeaveIfError(iFs.GetDir(parse.FullName(),
             KEntryAttNormal | KEntryAttReadOnly |
  KEntryAttHidden | KEntryAttSystem | KEntryAttDir,
                                 0,dirList));

  // Ensure some part of the application takes ownership of
  // the dirList before we can leave later on.
  CleanupStack::PushL(dirList);

  // Create a new container to store the entries.
  RArray<TFolderEntry>*folderEntryList=
                      new(ELeave)RArray<TFolderEntry>(8);

  // Not only do we need to ensure we delete
  // the actual allocated Rarray.
```

```
CleanupDeletePushL(folderEntryList);

// But we also have to ensure we call Close()
// before the delete to release any resources.
CleanupClosePushL(*folderEntryList);

// Add a set of entries to our new container.
TFolderEntry entry;
const TInt count=dirList->Count();
for (TInt i=0;i<count;i++)
  {
  // Construct a new entry.
  entry.Construct((*dirList)[i]);

  // Add to the new list of entries.
  folderEntryList->AppendL(entry);
  }

// We need an index onto the entries
// to perform efficient processing.
CArrayFixFlat<TInt>* index=
         BuildIndexL(folderEntryList,iCurrentCategory);

// Now sort the array of TInts dependant on
// the content they refer to, in the current
// mode + ascending/descending order.
TkeyEntryList key(folderEntryList,iSortType,iSortOrder);
index->Sort(key);

// Replace any existing list with the new list.
if (iFolderEntryList)
  {
  iFolderEntryList->Close();
  delete(iFolderEntryList);
  }
iFolderEntryList=folderEntryList;

// Firstly the Cleanup stack item that calls close().
CleanupStack::Pop(folderEntryList);

// Then the cleanup stack item that deletes
// the actual object.
CleanupStack::Pop(folderEntryList);

// Replace the index.
delete(iFolderEntryIndex);
iFolderEntryIndex=index;

// We have a new list, display from the top,
// if there is a top item.
iCurrentEntry=0;
LimitCurrentEntryVal();

// Update the path our entries exist within
// now all constructed.
iPath=aPath;

// While it may have been tempting to perform
```

```
// this immediately after the end of the loop above
// it is very important to match Push/Pops on
// the cleanup stack, so we simply defer to here.
CleanupStack::PopAndDestroy(dirList);

// Perform a self test to verify our internal
// structure is still self consistent.
__TEST_INVARIANT;
}
```

`RArray` Compared to `CArrayX` Classes

As we discussed in Chapter 4, there are two sets of collection classes in Symbian OS, the original `CArrayX` classes and the newer `RArray` classes. Both types of collection class are relatively easy to use so, while there are some minor performance differences that we outline in Chapter 13, our application demonstrates the use of both classes, enabling you to choose your preferred set.

Creating the Set of `TFolderEntry` Objects

The `BuildDirectoryListL()` method contains the following code to create and store the set of `TFolderEntry` objects:

```
// Create a new container to store the entries.
RArray<TFolderEntry>* folderEntryList=
                        new(ELeave)RArray<TFolderEntry>(8);

// Not only do we need to ensure
// we delete the actual allocated Rarray.
CleanupDeletePushL(folderEntryList);

// But we also have to ensure we call Close()
// before the delete to release any resources.
CleanupClosePushL(*folderEntryList);

// Add a set of entries to our new container.
TFolderEntry entry;
const TInt count=dirList->Count();
for (TInt i=0;i<count;i++)
  {
  // Construct a new entry
  entry.Construct((*dirList)[i]);

  // Add to the new list of entries.
  folderEntryList->AppendL(entry);
  }

//... remaining code removed for clarity.
```

The following points are worth noting:

- To cleanup the `RArray` fully, we have to call the `Close()` method of the object **and** delete the memory allocated to the actual object. The

`CleanupClosePushL()` templated function ensures the `Close()` method is called if a leave occurs. The `CleanupDeletePushL()` function ensures the object is destroyed. The order presented above is important to ensure the object is closed before being deleted.

- We use the `AppendL()` method. A common misuse of `RArray` classes is to use the `Append()` method and disregard the return value. The `CArrayX` classes don't allow you to do this since they don't support `Append()`. In all cases, the method fails to add entries if it runs out of memory. Your application needs to do something about this. By using `AppendL()`, the method leaves when this condition occurs.

Pro tip

You should remember that the `CleanupStack::PushL()` method has two overloads, one that takes a `TAny*` and the other a `CBase*`. Since an `RArray` does not derive from `CBase`, the `CleanupStack::PushL(TAny*)` overload would be applied to an `RArray` pointer. While using the `CleanupStack::PushL()` method actually works and performs the correct task, a common misconception is that the object destructor is called for all objects placed on the cleanup stack, even if they do not derive from `CBase`. For this reason, we recommend you use `CleanupDeletePushL()` to indicate a difference from regular `CleanupStack::PushL()` usage.

Building the Index

The index is created with the following code:

```
CArrayFixFlat<TInt>* CAppEngine::BuildIndexL(
  RArray<TFolderEntry>* aList,
  const TInt aCategory)
  {
  const TInt count=aList->Count();
  TInt granularity=count;
  if (!count)
    granularity=1;

  CArrayFixFlat<TInt>* index = new(ELeave)
         CArrayFixFlat<TInt>(granularity);
  CleanupStack::PushL(index);
  for (TInt i=0;i<count;i++)
    {
    if (aCategory==EAppCategoryAll ||
                    (*aList)[i].EntryCategory()==aCategory)
      index->AppendL(i);
    }
  index->Compress();
```

```
CleanupStack::Pop(index);
return(index);
}
```

This code fragment demonstrates various aspects of using collection classes that you should consider using in your applications.

We calculate a granularity for the array. By making the granularity equal to the number of entries that we put in the array, we ensure that a single memory allocation occurs. This means we are very efficient in terms of the number of memory allocations that occur, irrespective of the number of entries we are about to put into the array.

As an alternative to the above, you could choose to use the default granularity, then use the `SetReserveL()` method to ensure the array has space for the number of entries you are about to add. As with setting the granularity, using `SetReserveL()` ensures that only a single memory allocation occurs.

In our example, for the cases where only a subset of entries is added to the index due to the category filtering, we have some spare capacity within the array. Good application memory management requires we free up this spare capacity since we are not going to use it. We achieve this by using the `Compress()` method. Even if we used the system default value for the granularity we should still use the `Compress()` method to release any spare capacity. The `Compress()` method does not cause a re-allocation of the underlying memory; it simply returns unused memory to the heap allocator.

In our example code, the worst case scenario is that we have one memory allocation irrespective of the number of elements we put in our index. The typical default granularity is eight. This means that for every eight entries added to the index, a memory re-allocation and content copy occurs. Memory re-allocation is an expensive operation.

Sorting the Index

If we had chosen to use an `RArray` to store our index, we would need to use a `TLinearOrder` function to perform our sort. An example of this is presented in the `Commands1` application.

Since we are using a `CArrayX` class, we need to use a `TKey`-derived class to perform the sort. As with many Symbian OS objects, rather than have either blank or pure virtual functionality, some default functionality is implemented. In some cases this functionality may be sufficient, in others it may not. The `TKeyArrayFix` class is a classic example of this approach. The default implementation provides a significant set of functionality; however none is sufficient for our use. The standard C++ approach to extending functionality is to override virtual functions. In this case, we override the comparison function using the following object definition:

```
class TKeyEntryList : public TKeyArrayFix
  {
public:
  TKeyEntryList(RArray<TFolderEntry>*aFolderEntryList,
        const TInt aSortType, const TInt aDescending);
  TInt Compare(TInt aLeft,TInt aRight) const;
protected:
  RArray<TFolderEntry>* iFolderEntryList;
  TInt iSortOrder;
  };
```

The implementation of the `Compare()` function is:

```
TInt TKeyEntryList::Compare(TInt aLeft,TInt aRight) const
  {
  TFolderEntry& leftEntry=(*iFolderEntryList)
           [*static_cast<TInt*>(At(aLeft))];
  CONVERT_POINTER_TYPES conv;
  conv.thisIsAPointerToTheObject=At(aRight);
  TFolderEntry& rightEntry=(*iFolderEntryList)
        [(*conv.whichWeConvertToOneOfThese)];

  // Left item comes before right item.
  TInt ret=(-1);
  switch (iCmpType)
    {
    // Sort by file size, if they are
    // the same order by name.
    case EFolderEntrySortBySize:
      ret=leftEntry.EntrySize() - rightEntry.EntrySize();
      if (ret==0)
        ret=leftEntry.EntryName().CompareF(rightEntry.EntryName());
      break;

    // Sort by name.
    case EFolderEntrySortByName:
      ret=leftEntry.EntryName().CompareF(rightEntry.EntryName());
      break;

    // Sort by last modified date,
    // if they are the same order by name.
    case EFolderEntrySortByModified:
      if (leftEntry.EntryModified()>rightEntry.EntryModified())
        ret=1;
      else if (leftEntry.EntryModified()==rightEntry.EntryModified())
        ret=leftEntry.EntryName().CompareF(rightEntry.EntryName());
      break;
```

```
  default:
    break;
  }
  if (iSortOrder)
    ret=(-ret);          // Simply reverse the result
                         // to sort in descending order.
  return(ret);
  }
```

Since CArrayX classes can be used to store objects of practically any type and due to the original implementation, the At() method simply returns a TAny*, that is, a pointer to one of the elements in our CArrayX. Since our CArrayX is an array of integers, the TAny* needs to be converted into a TInt*. There are numerous ways in which we can convert pointer types. This application demonstrates two options, a static_cast to convert the left index and a union to convert the right index.

```
typedef union
  {
  TAny* thisIsAPointerToTheObject;
  TInt *whichWeConvertToOneOfThese;
  } CONVERT_POINTER_TYPES;
```

For our application, we read the index values as stored in the CArrayX and use those to get references to the underlying TFolderEntry objects. This particular task is difficult to achieve with a TLinearOrder as there is no obvious way to gain addressability to data stored outside the RArray being sorted.

Once we have determined which two entries are to be compared, we can perform the comparison function, returning a negative value if the left entry comes before the right entry, positive if the left entry comes after the right entry and zero if the entries are identical. Our application has chosen to implement some second-order key information. If we are sorting by size or modification date then matching entries are also sorted by name.

To cause the CArrayX to be sorted, we do this:

```
TKeyEntryList key(folderEntryList,iSortType,iSortOrder);
index->Sort(key);
```

10.3.2 List View

The list view displays the entries belonging to the index, in the order the index specifies. Rather than the UI code needing to know this or how to translate between the index and an actual entry, the engine presents a couple of functions for the UI to call:

```
TInt EntryCount() const;
const TFolderEntry& Entry(const TInt aIndex) const;
```

Constructing the list box reduces to:

```
CQikListBox* listbox = LocateControlByUniqueHandle<
                    CQikListBox>(EListViewListId);

// Get the listbox model.
MQikListBoxModel& model(listbox->Model());

// Remove any existing entries.
model.RemoveAllDataL();
model.ModelBeginUpdateLC();
const TInt count=iEngine->EntryCount();
for (TInt i=0;i<count;i++)
  {
  MQikListBoxData* lbData =
          model.NewDataL(MQikListBoxModel::EDataNormal);
  CleanupClosePushL(*lbData);
  lbData->AddTextL(iEngine->Entry(i).EntryName(),
                          EQikListBoxSlotText1);
  lbData->SetItemId(i);
  CleanupStack::PopAndDestroy(lbData);
  }

model.ModelEndUpdateL();
```

This should be familiar code from previous examples.

Sorting in the UI

As with the `Commands1` application, we have a cascade menu option presenting the sort options available to the user. We override the `DynInitOrDeleteCommandL()` method of the list view to ensure that the UI presentation of the sort type radio buttons and sort order check box match the internal values.

```
CQikCommand* CListView::DynInitOrDeleteCommandL(
  CQikCommand* aCommand,
  const CCoeControl& aControlAddingCommands)
  {
  TBool val=EFalse;
  switch (aCommand->Id())
    {
    // Determine which of the radio buttons
    // should be checked.
  case EAppCmdSortByName:
    if (iEngine->SortType()==EFolderEntrySortByName)
      val=ETrue;
    aCommand->SetChecked(val);
    break;
```

```
case EAppCmdSortBySize:
  if (iEngine->SortType()==EFolderEntrySortBySize)
    val=ETrue;
  aCommand->SetChecked(val);
  break;

case EAppCmdSortByType:
  if (iEngine->SortType()==EFolderEntrySortByType)
    val=ETrue;
  aCommand->SetChecked(val);
  break;

case EAppCmdSortByDate:
  if (iEngine->SortType()==EFolderEntrySortByModified)
    val=ETrue;
  aCommand->SetChecked(val);
  break;

  // Determine whether the single check box
  // type menu option should be checked.
case EAppCmdSortOrder:
  if (iEngine->SortOrder()==ESortOrderAscending)
    val=ETrue;
  aCommand->SetChecked(val);
  break;
default:
  break;
  }
return(aCommand);
}
```

The user request to sort the list causes a `CQikCommand` object to be delivered to the `HandleCommandL()` method in which we request the engine to sort the index:

```
void CListView::HandleCommandL(CQikCommand& aCommand)
  {
  switch (aCommand.Id())
    {
    case EAppCmdSortByName:
      SortListL(EFolderEntrySortByName,
                iEngine->SortOrder(),EAppCmdSortByName);
      break;
    case EAppCmdSortBySize:
      SortListL(EFolderEntrySortBySize,
                iEngine->SortOrder(),EAppCmdSortBySize);
      break;
    case EAppCmdSortByDate:
      SortListL(EFolderEntrySortByModified,
      iEngine->SortOrder(),EAppCmdSortByDate);
      break;
    case EAppCmdSortByType:
      SortListL(EFolderEntrySortByType,
                iEngine->SortOrder(),EAppCmdSortByType);
      break;
    case EAppCmdSortOrder:   // Swap between ascending and descending.
```

```
    SortListL(iEngine->SortType(),
            (TFolderEntrySortOrder)(ESortOrderAscending +
             ESortOrderDescending-iEngine->SortOrder()),
                                             EAppCmdSortOrder);
    break;
  default:                  // e.g. the back button...
    CQikViewBase::HandleCommandL(aCommand);
    break;
  }
}
```

Once sorted, we have to rebuild the list box. In our application, we have chosen to rebuild the list box from scratch as there is no easy way of re-ordering the list box elements. Since this operation can fail we need to implement some rollback to ensure that either the operation fully succeeds or the application returns to the state it was previously in.

```
void CListView::SortListL(
  const TFolderEntrySortType aType,
  const TFolderEntrySortOrder aOrder,
  const TInt aCmdId)
  {
  // Record current settings incase we need to roll back.
  TFolderEntrySortType oldType=iEngine->SortType();
  TFolderEntrySortOrder oldOrder=iEngine->SortOrder();

  // Update the list.
  SetCurrentEntry();
  iEngine->SortEntries(aType,aOrder);

  // Attempt to update the listbox display,
  // track the current entry...
  TRAPD(err,
    UpdateListBoxL();
    LocateControlByUniqueHandle<CQikListBox>
                   EListViewListId)->SetCurrentItemIndexL(
            iEngine->CurrentEntryIndex(),ETrue,EDrawNow);
  );

  // If the listbox display fails, rollback
  // to previous sort order.
  if (err!=KErrNone)
    {
    iEngine->SortEntries(oldType,oldOrder);
    CQikCommandManager& cm=CQikCommandManager::Static();

    // Restore the Radio button state within the UI.
    if (oldType!=aType)
      {
      TInt commandId=EAppCmdSortByName;
      if (oldType==EFolderEntrySortBySize)
        commandId=EAppCmdSortBySize;
      else if (oldType==EFolderEntrySortByModified)
        commandId=EAppCmdSortByDate;
      else if (oldType==EFolderEntrySortByType)
        commandId=EAppCmdSortByType;
```

```
    if (commandId!=aCmdId)
      { // Swap back the radio button state
      cm.SetChecked(*this,aCmdId,EFalse);
      cm.SetChecked(*this,commandId,ETrue);
      }
    }

  // Restore the Ascending check box within the UI.
  if (oldOrder!=aOrder)
    {
    TBool val=EFalse;
    if (oldOrder==ESortOrderAscending)
      val=ETrue;
    cm.SetChecked(*this,EAppCmdSortOrder,val);
    }
  User::Leave(err);
  }
}
```

The ability of the engine to sort entries without leaving is important here. To roll back fully, we need to put the application back to the state it was originally in before the sort operation started. The ability to sort the entries back to their previous state without leaving is part of the required rollback.

Tracking the Current Entry

In small-screen devices, such as mobile phones, the usability of an application is of paramount importance. When a list is sorted, the lines of content displayed in the list are changed. Rather than simply letting the focus remain on the same line (probably with different content), it is often preferable to either cause the focus to move with the content or perhaps to reset the focus to the top of the list. In most cases, causing the focus to move with the content provides a superior user experience since contextual information is maintained. The user can immediately see where the selected entry fits within the newly sorted list without having to search for it. After all, a primary reason to sort entries is to present contextual information about entries.

Our engine tracks the current entry in the sort method:

```
void CAppEngine::SortEntries(
  const TfolderEntrySortType aSortType,
  const TFolderEntrySortOrder aOrder)
  {
  TInt whichEntry=iFolderEntryIndex->At(iCurrentEntry);
  TKeyEntryList key(iFolderEntryList,aSortType,aOrder);
  iFolderEntryIndex->Sort(key);
  iSortType=aSortType;
  iSortOrder=aOrder;

  // Track the 'current' entry.
```

```
const TInt count=EntryCount();
for (TInt i=0;i<count;i++)
  {
  if (iFolderEntryIndex->At(i)==whichEntry)
    {
    iCurrentEntry=i;
    break;
    }
  }
}
```

10.3.3 Detail View

In a list and detail model, the detail view is used to view and edit individual entries. Our detail view has a set of controls which allow the user to change the name, assign a category and update the last modified date of an entry. Since we have chosen not to make the detail view public to external applications, we do not need to support DNL links to the detail view. We use the simplified approach of informing the engine which entry we wish to display before switching to the detail view:

```
void CListView::HandleListBoxEventL(
  CQikListBox* aListBox,
  TQikListBoxEvent aEventType,
  TInt aItemIndex,
  TInt aSlotId)
  {
  switch (aEventType)
    {
    case EEventItemConfirmed:
    case EEventItemTapped:
      iEngine->SetCurrentEntry(aItemIndex);
      iQikAppUi.ActivateViewL(KViewIdDetailsView);
      break;
    default:
      break;
    }
  }
```

The `ViewActivatedL()` method picks up the current entry and configures the component controls with data associated with that entry.

```
void CDetailsView::ViewActivatedL(
  const TVwsViewId& aPrevViewId,
  const TUid aCustomMessageId,
  const TDesC8& aCustomMessage)
  {
  const TFolderEntry& entry=iEngine->CurrentEntry();
  CEikEdwin* edwin = LocateControlByUniqueHandle<CEikEdwin>(EAppEdwin1);
  edwin->SetTextL(&entry.EntryName());
  CEikChoiceList* cl = LocateControlByUniqueHandle<
                CEikChoiceList>(EAppChoiceList1);
```

```
cl->SetArrayL(this);
cl->SetArrayExternalOwnership(ETrue);
TInt id=entry.EntryCategory();
TInt count=iEngine->CategoryListCount();
for (TInt i=1;i<count;i++)
  {
  if (iEngine->CategoryListAt(i).iCategoryId==id)
    {
    cl->SetCurrentItem(i-1);
    break;
    }
  }

CQikTimeAndDateEditor* dt = LocateControlByUniqueHandle<
                  CQikTimeAndDateEditor>(EAppDateTime1);
dt->SetTimeL(entry.EntryModified());

}
```

To support categories, we have to perform two tasks:

- provide a list of the categories the user can choose between
- set up the category to which the current entry belongs.

To provide the list of categories, the choice list needs an object to implement the `MDesCArray` interface. The `CDetailsView` object can provide that interface and does so with the following code:

```
TInt CDetailsView::MdcaCount() const
  {
  return(iEngine->CategoryListCount()-1);
  }

TPtrC CDetailsView::MdcaPoint(TInt aIndex) const
  {
  return(iEngine->CategoryListAt(aIndex+1).iCategoryName);
  }
```

You should have observed a small translation, by a factor of one, occurring in most of the code associated with categories. In general you have All, Unfiled and a set of application-specific categories. No entries should be set to belong to the All category, they should belong to either Unfiled or an application-specific category. The All category is simply a mechanism for the user to view items irrespective of which category they belong to. When we present the list of categories an entry can belong to, we should remove the All category. It is no accident that the All category is the first category entry created in the engine. With this arrangement, we can simply adjust counts and indices with the above code.

Updating Entries

One objective of the detail view is to enable a user to update an entry. As with the `MultiView4` application (see Section 9.2.6), if the user chooses to move back to the list view, we should check whether any of the details have been modified. Our application always prompts the user to save changes. You may prefer a Cancel and Done approach where Cancel prompts prior to abandoning any changes, while Done saves changes without any prompt.

Unlike the `MultiView4` application, changes to entries in this application have a more profound effect on the displayed list, for example:

- If you change the category to which the entry belongs and the list box is displaying a specific category (as opposed to the All category), the entry should no longer be listed since it no longer belongs to the category being displayed.
- If you change the entry name, it is unlikely the entry should remain in the same position within the sorted list box entries.

In the `MultiView4` application, we chose to `ActivatePreviousViewL(ESave)` to return to the previous view, calling the `CAppSpecificDetailsView::SaveL()` method when required.

`ActivatePreviousViewL(ESave)` calls the `SaveL()` method asynchronously. In particular, it calls `SaveL()` *after* it has called the `CListView::ViewActivatedL()` method. While appropriate for the `MultiView4` example, this is not satisfactory for our application. We need to update the list before we attempt to display it.

Since our application needs to update any changes before we switch back to the list view we should use the `SaveThenDnlToL(ParentView())` method. This calls the `SaveL()` method synchronously before the framework attempts to switch views:

```
void CDetailsView::HandleCommandL(CQikCommand& aCommand)

// Handle the commands coming in from the controls
// that can deliver cmds...
  {
  switch (aCommand.Id())
    {
    case EAppCmdAbout:
      iQikAppUi.HandleCommandL(EAppCmdAbout);
      break;

    case EQikCmdGoBack:
      if (DetailsHaveChanged())
        {
        CQikSaveChangesDialog::TexitOption
    ret=CQikSaveChangesDialog::RunDlgLD();
```

```
        if (ret==CQikSaveChangesDialog::EShut)
          break;
        if (ret==CQikSaveChangesDialog::ESave)
          {
          SaveThenDnlToL(ParentView());
          break;
          }
        }
    default: // e.g. the back button...
      CQikViewBase::HandleCommandL(aCommand);
      break;
    }
  }
```

Updating Entries in the Engine

The engine is responsible for actually updating a `TFolderEntry`. It has several tasks to perform:

- If the name has changed, the underlying filename also needs to be changed as well as the `TFolderEntry` content.
- If the last modified date has changed, the underlying file date should also be updated.
- If the category has changed, we may need to adjust the set of entries being displayed by the list box.
- If a change affects the current list box sort order, the entries should be sorted into their new order.

```
void CAppEngine::UpdateCurrentEntryL(
  const TFolderEntry& aUpdate)
  {
  TFolderEntry& curEntry=EntryListAt(
  iFolderEntryIndex->At(iCurrentEntry));
  TParse srcName;
  srcName.Set(curEntry.EntryName(),&iPath,NULL);
  TBool sortReqd=EFalse;
  if (curEntry.EntryName()!=aUpdate.EntryName())
    { // Names being changed.
    TParse destName;
    destName.Set(aUpdate.EntryName(),&iPath,NULL);
    User::LeaveIfError(iFs.Rename(srcName.FullName(),
                               destName.FullName()));
    curEntry.UpdateName(aUpdate.EntryName());
    sortReqd=ETrue;
    }

  if (curEntry.EntryModified()!=aUpdate.EntryModified())
    {
    User::LeaveIfError(iFs.SetModified(srcName.FullName(),
                                aUpdate.EntryModified()));
```

```
    curEntry.UpdateModified(aUpdate.EntryModified());
    if (iSortType==EFolderEntrySortByModified)
      sortReqd=ETrue;
    }

  if (curEntry.EntryCategory()!=aUpdate.EntryCategory())
    {
    TInt oldCategory=curEntry.EntryCategory();
    curEntry.UpdateCategory(aUpdate.EntryCategory());
    if (iCurrentCategory==oldCategory)
      {
      iFolderEntryIndex->Delete(iCurrentEntry);
      LimitCurrentEntryVal();
      }
    }

  if (sortReqd)
    SortEntries(iSortType,iSortOrder);
  }
```

Our sort operation tracks the current entry. This is useful because if the list needs to be sorted due to an entry being updated, we automatically handle any positional change of the current entry. When we return to the list view, the entry whose details we were manipulating remains the current entry, assuming it is not removed from the list due to a change in category. Maintaining this contextual information enhances usability.

Note that the engine code above makes an assumption regarding the UI. In particular, it assumes that it is not possible to move an entry into a category the UI is currently displaying, only out of the displayed category. In our application this is true since we only have a single list of items and no mechanism to change the details of items that are not displayed in the list. Other applications may have more than one list or an alternative mechanism to set entry categories. In this case, additional functionality may be required to ensure any change of category causes the entry to appear in all lists to which it belongs.

Deleting Entries

Our application supports the ability to delete entries. Currently we do not support any undo functionality, so it is important to ensure the user really wants to perform the delete operation. To delete an entry we need to:

- remove the entry from disk
- remove the entry from our `TFolderEntry` list
- update the index
- update the display.

The engine performs the first three tasks:

```
void CAppEngine::DeleteCurrentEntryL()
  {
  TFileName name;
  EntryFullName(name);
  User::LeaveIfError(iFs.Delete(name));
  TInt index=iFolderEntryIndex->At(iCurrentEntry);
  iFolderEntryList->Remove(index);
  iFolderEntryIndex->Delete(iCurrentEntry);
  const TInt count=EntryCount();
  for (TInt i=0;i<count;i++)
    {
    TInt val=iFolderEntryIndex->At(i);
    if (val>=index)
      (*iFolderEntryIndex)[i]=val-1;
    }
  LimitCurrentEntryVal();
  }
```

Rather than create a whole new index, filtered and sorted according to the user's current preferences, we manipulate the existing index. This has a couple of advantages: firstly, it is not possible to run out of memory attempting to build the new index and, secondly, it is considerably faster.

The UI performs the remaining tasks:

```
void CListView::DeleteCurrentEntryL()
  {
  SetCurrentEntry();
  const TFolderEntry& entry=iEngine->CurrentEntry();
  TBuf<256>bb;
  iEikonEnv->Format256(bb,R_STR_ABOUT_TO_DELETE,
                              &entry.EntryName());
  TBuf<64>bb2;
  iEikonEnv->ReadResourceL(bb2,R_STR_ATTENTION);
  if (iEikonEnv->QueryWinL(bb2,bb))
    {
    iEikonEnv->Format256(bb,R_STR_DELETED,&entry.EntryName());
    iEngine->DeleteCurrentEntryL();
    CQikListBox* listbox =
            LocateControlByUniqueHandle<CQikListBox>(
                                    EListViewListId);
    listbox->RemoveItemL(listbox->CurrentItemIndex());
    TInt i=iEngine->CurrentEntryIndex();
    if (i>=0)
      listbox->SetCurrentItemIndexL(i,ETrue,EDrawNow);
    iEikonEnv->InfoMsg(bb);
    UpdateCommandAvailability();
    }
  }
```

As a general rule, providing some positive feedback that an operation has completed is almost as important as requesting confirmation that an operation should be performed or presenting error information. Such positive feedback should be unobtrusive and preferably not require any user interaction. The use of `iEikonEnv->InfoMsg()` to display an informational message is ideal for this task as it automatically disappears after a couple of seconds.

10.3.4 Command Availability

Actions are usually performed on content; for example, you delete an entry. If no entries exist then many commands have little meaning and most applications reach a boundary condition.

For example, in our `CListView::DeleteCurrentEntryL()` method, the software assumes that it can always return a `TFolderEntry&`. If no entries exist, the current entry index is −1. Attempting to obtain the `TFolderEntry&` results in a `CBase-21` panic, indicating that the array index is out of range. Rather than relying on users not to attempt such operations, you need to be proactive and write code to ensure the application does not panic.

In general, there are two approaches:

- Add some code to verify that the operation can be performed without causing a panic; for example, make sure there is at least one entry that can be deleted.
- Remove the option from the UI, therefore preventing the user from choosing an option that cannot be performed.

We have chosen the second option for our application:

```
void CListView::UpdateCommandAvailability()
  {
  CQikCommandManager& cm=CQikCommandManager::Static();
  TBool avail=EFalse;
  if (iEngine->EntryCount()>0)
    avail=ETrue; // Some entries exist in the current filtered list.

  cm.SetAvailable(*this,EAppCmdOpen,avail);
  cm.SetAvailable(*this,EAppCmdDelete,avail);
  cm.SetAvailable(*this,EAppCmdDelete2,avail);
  cm.SetAvailable(*this,EAppCmdSendAs,avail);
  cm.SetAvailable(*this,EAppCmdSortCascade,avail);
  }
```

Don't forget that not all commands arrive by the user choosing a menu option. In our application, we have Delete in the menu and Delete assigned to the Cancel key through the command type `EQikCommandTypeDelete`. Both these delete commands need to be marked as unavailable.

Pro tip

If you define two identical commands to get one in the menu and one on a hardware key, make sure you don't get both commands in the More menu if the softkey algorithm changes. Make one command of type `EQikCpfFlagHardwarekeyOnly` so that it cannot go into the menu.

10.3.5 Category Management

Categories and entries are very closely linked. With the exception of changing the name of a category, any manipulation has an effect on either the list of entries displayed or the entries themselves.

Changing Categories

If a user chooses to display the set of entries belonging to a specific category, our application has to generate a new sorted index on those entries.

```
void CAppEngine::ChangeCategoryL(const TInt aHandle)
  {
  CArrayFixFlat<TInt>* index=BuildIndexL(iFolderEntryList,aHandle);
  TKeyEntryList key(iFolderEntryList,iSortType,iSortOrder);
  index->Sort(key);
  iCurrentCategory=aHandle;
  delete(iFolderEntryIndex);
  iFolderEntryIndex=index;
  iCurrentEntry=0;
  LimitCurrentEntryVal();
  }
```

The UI is responsible for updating the list box to display the set of entries that belong to the chosen category:

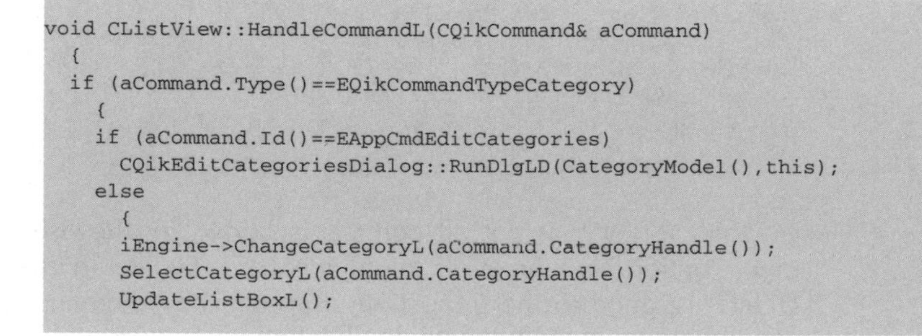

```
void CListView::HandleCommandL(CQikCommand& aCommand)
  {
  if (aCommand.Type()==EQikCommandTypeCategory)
    {
    if (aCommand.Id()==EAppCmdEditCategories)
      CQikEditCategoriesDialog::RunDlgLD(CategoryModel(),this);
    else
      {
      iEngine->ChangeCategoryL(aCommand.CategoryHandle());
      SelectCategoryL(aCommand.CategoryHandle());
      UpdateListBoxL();
```

```
      UpdateCommandAvailability();
      }
    }
  }
```

Note that, if the user selects a category containing zero entries, we need to ensure the user is not able to choose commands that could cause our application to panic. While subtly different from the case where no entries exist at all, by encapsulating entry count and referencing in the engine the UI does not need to know about this difference and the same logic can be used to determine command availability.

Deleting Categories

In our application, we allow the user to delete a category even if there are entries associated with that category. When this happens, we must use one of the following methods to tidy up afterwards:

- delete all the entries associated with the category
- accept that entries can belong to non-existent categories and therefore only appear within the All category until the user manually updates the entry.
- adjust any entries that belong to a category that is about to be deleted such that they belong to a different category. Our application implements this option by moving entries to the Unfiled category.

Our application also allows the user to delete the category that the list view is currently displaying. In this case, something reasonable has to be done to update the list view, such as displaying a different category of entries. Reverting to the All entries category is a safe option since this category cannot be deleted. We have implemented this option in our application:

```
TInt CAppEngine::DeleteCategoryL(const TInt aHandle)
  {
  TInt i;
  TInt ret=0;
  if (aHandle==iCurrentCategory)
    {
    CArrayFixFlat<TInt>* index = BuildIndexL(
           iFolderEntryList,EAppCategoryAll);
    TkeyEntryList key(iFolderEntryList,iSortType,iSortOrder);
    index->Sort(key);
    delete(iFolderEntryIndex);
    iFolderEntryIndex=index;
```

```
    iCurrentEntry=0;
    LimitCurrentEntryVal();
    iCurrentCategory=EAppCategoryAll;
    ret=1;
    }
  TInt count=EntryListCount();
  iFolderEntryIndex->SetReserveL(count);
  TKeyEntryList key(iFolderEntryList,iSortType,iSortOrder);
  for (i=0;i<count;i++)
    {
    TFolderEntry& entry=EntryListAt(i);
    if (entry.EntryCategory()==aHandle)
      {
      entry.UpdateCategory(EAppCategoryUnfiled);
      if (iCurrentCategory==EAppCategoryUnfiled)
        {
        if (iFolderEntryIndex->InsertIsqAllowDuplicatesL(
                                    i,key)<=iCurrentEntry)
          iCurrentEntry++;
        ret=(-1);
        }
      }
    }

  iFolderEntryIndex->Compress();
  count=CategoryListCount();
  for (i=0;i<count;i++)
    {
    TAppCategoryEntry& category=CategoryListAt(i);
    if (category.iCategoryId==aHandle)
      {
      iCategoryList->Remove(i);
      break;
      }
    }
  return(ret);
  }
```

As discussed, when we update entries that belong to the category being deleted, we move them to the Unfiled category. If the UI happens to be displaying the Unfiled category, these entries should become visible to the user. To make the entries visible, we have to expand the index. An expansion could fail due to running out of memory. In our application, we adopt an *all or nothing* approach by ensuring that the index is expanded so that it can contain all the required entries by using `iFolderEntryIndex->SetReserveL(count)`. This also provides a minor performance benefit because the index is only re-allocated once, in the `SetReserveL()` method, to contain any extra items.

The user interface presents entries to a user in the currently chosen sort order. The presentation order is dictated by the content of the index. When we add entries to the index, we should maintain the currently chosen sort order. This can be achieved either by appending entries to the end then sorting the full list or by inserting the entries in the correct place. Our application implements the latter and so uses the

`InsertIsqAllowDuplicatesL()` method. As with an `RArray`, the `CArrayX` reuses the entry comparison object to determine the insert position. The `InsertIsqAllowDuplicatesL()` method has a further useful property; it returns the position in the index at which the new item is inserted. We use this information to track the currently selected item. When a UI returns from the category management dialogs, the contextual information within the list box is preserved.

Pro tip

Due to the way views and controls work in UIQ 3, if you leave a choice list control in the detail view referencing a category other than the All or Unfiled categories and some categories are deleted, then even though the detail view is deactivated, your application panics with `USER 130` to indicate that the array index is out of bounds. A simple solution is to set the choice list's current item to reference the All category, as this cannot be deleted when the view is deactivated.

```
void CDetailsView::ViewDeactivated()
  {
  LocateControlByUniqueHandle<CEikChoiceList>(
          EAppChoiceList1)->SetCurrentItem(0);
  }
```

10.3.6 Send As

In some applications, including ours, the ability to transfer content from one device to another is a useful feature. The UIQ 3 framework contains a useful set of functionality called *Send as* to perform this task.

The Send As framework code has a number of properties:

- It presents a standard interface to the user.
- It is easy to use.
- It supports a range of transports.
- No Symbian OS platform security capabilities are required to successfully use the Send as framework code.

This code is all you need to support the sending of files via the transports which accept attachments:

```
void CListView::SendFileAsL()

// Send the file, via a user chosen transport.
```

```
{
CQikSendAsLogic* sendAs=CQikSendAsLogic::NewL();
CleanupStack::PushL(sendAs);
TFileName fullName;
iEngine->EntryFullName(fullName);
sendAs->AddAttachmentL(fullName);
CleanupStack::Pop(sendAs);
CQikSendAsDialog::RunDlgLD(sendAs,KUidMtmQuerySupportAttachments);
}
```

On a real mobile phone, this typically means infrared, Bluetooth and MMS transports.

You should note that ownership of the `SendAs` object is taken immediately by the `RunDlgLD()` method and, in particular, before that method can leave. Therefore it is important to remove the `SendAs` object from the cleanup stack before calling `RunDlgLD()`. This is not typical Symbian OS behavior.

This is a minimal implementation of sending files. The UIQ 3 SDK documentation has further information about the `CQikSendAsLogic` and `CQikSendAsDialog` objects which you can review if the above is not sufficient for your requirements. For example, you may wish to add some Subject or Body text, particularly if the file is being transferred via MMS.

10.3.7 Registration Using a Phone IMEI

A common distribution mechanism is to provide free downloads of an application in which the functionality is restricted. The user can purchase a full license, at which time an unlock key or registration code is supplied. The user can enter this code into the downloaded version, converting it into a full version of the application. It is not necessary to uninstall the free version or install a separate full version.

Many distribution sites support the 'dynamic registration' model, where the registration code is based on some unique piece of information associated with a mobile phone. In the case of GSM and UMTS phones, this is usually the International Mobile Equipment Identity number (IMEI).

Obtaining the IMEI of the phone has to be performed by calling an asynchronous function supplied by the `CTelephony` class. An active object class must be defined to submit the asynchronous request and handle the completion event, but this is relatively simple, and the entire object definition and functionality is:

```
class CGetPhoneIMEI : public CActive
  {
protected:

  void RunL();
  void DoCancel();
```

```
public:
  ~CGetPhoneIMEI();
  CGetPhoneIMEI(CAppSpecificUi* aAppUi);

  void ConstructL();
protected:
  CAppSpecificUi* iAppUi;
  CTelephony::TPhoneIdV1Pckg iBuf;
  CTelephony::TPhoneIdV1 iData;
  CTelephony* iTelephony;
  };

CGetPhoneIMEI::~CGetPhoneIMEI()
  {
  Cancel();
  delete(iTelephony);
  }

CGetPhoneIMEI::CGetPhoneIMEI(CAppSpecificUi* aAppUi) :
                    CActive(CActive::EPriorityHigh),
                         iAppUi(aAppUi),iBuf(iData)
  {
  CActiveScheduler::Add(this);
  }

void CGetPhoneIMEI::ConstructL()
// Start the request to obtain the IMEI.
  {
  iTelephony=CTelephony::NewL();
  iTelephony->GetPhoneId(iStatus,iBuf);
  SetActive();
  }

void CGetPhoneIMEI::RunL()
  {
  iAppUi->SetIMEI(iData.iSerialNumber);
  delete(this);
  }

void CGetPhoneIMEI::DoCancel()
  {
  iTelephony->CancelAsync(CTelephony::EGetPhoneIdCancel);
  }
```

For more information about Symbian OS active objects, please consult Chapter 4.

The code above demonstrates several points:

- We add the active object at a priority of `CActive::EPriority-High`. This does not mean that the request takes precedence over any other system request; it means that as soon as the information is delivered one of the first active objects to be checked is this one. Obtaining the phone IMEI is a fast operation that is only asynchronous by virtue of communicating with a server, therefore this active object gets to run before most of our other active objects.

- The `RunL()` method executes a `delete(this);` once we have obtained the IMEI there is no further use for the active object. Deleting itself within the `RunL()` is perfectly safe since no further object property access is required and the active scheduler is especially designed to ensure it can handle objects removing themselves from the active object queue within the `RunL()`.

Using the `CGetPhoneImei` object is just as easy:

```
#ifdef __WINS__
  iPhoneImei=_L("98-765432-123456-7");
#else
  CGetPhoneIMEI* imei=new(ELeave)CGetPhoneIMEI(this);
  CleanupStack::PushL(imei);
  imei->ConstructL();
  CleanupStack::Pop(imei);
#endif
```

Note that the IMEI is hard-coded for the WINS build because the telephony components are not fully supported by `CTelephony` for the Symbian OS Windows emulator.

Once the `ConstructL()` method, or more particularly, the code `iTelephony = CTelephony::NewL()` within the `ConstructL()` method, completes successfully, the active object is set up and its `RunL()` method is called on completion of the asynchronous event. The `CGetPhoneImei` takes ownership of itself – it deletes itself when `RunL()` gets called. We do not need to take ownership of `CGetPhoneImei` anywhere else within the system.

Once the IMEI has been retrieved, the `CGetPhoneImei` object calls the `iAppUi->SetIMEI()` method which simply stores the IMEI in the `iPhoneImei` property. While there is still a small gap between the application starting up and the IMEI being available, in most situations an application can be constructed such that this gap is irrelevant.

Prompting for a Registration Code

Our application contains a Register menu option which presents a dialog to the user (see Figure 10.9). Within the dialog we display the currently obtained IMEI and request the user to enter the registration code.

Pro tip

Displaying a mobile phone's IMEI is very useful for support purposes. A common support issue with dynamic registration is that it relies on the customer to correctly enter their IMEI on a website at the time of purchase. Naturally, a percentage of customers fail to do this.

> Having an easy way for customers to obtain their IMEI – by reading it from your Register dialog rather than entering `*#06#` into the phone keypad – makes it easier.

Figure 10.9 Registration dialog box

Obtaining the IMEI and CAPABILTIES

For some early UIQ 3 phones, it is necessary to have two capabilities in order to obtain the IMEI. This is because these products used a version of Symbian OS v9 that had an internal dependency within the telephony software which required applications to have the `ReadDeviceData` and `NetworkServices` capabilities. This dependency has now been removed and IMEI can be freely retrieved on most UIQ 3 phones.

If you have a requirement to support the earlier phones, you need to add the `ReadDeviceData` and `NetworkServices` capabilities to your application. If, like ours, your application is to be Symbian Signed, then adding these capabilities has little effect on the testing process. Since `ReadDeviceData` is a System capability, you must obtain the capability via Symbian Signed if you need to retrieve IMEI on the very early UIQ 3 phones.

10.3.8 Loading and Saving

In general applications have two types of data. The first type is the actual information being presented to the user. The second type is how the information is presented. For example, in a Contacts or Agenda application, the first type of data would be the contact records or agenda entries. The second type of data would be the zoom level, category or sort order that has been chosen to present the data.

When applications close, most users expect both types of data to be preserved. When going back to the application, they expect it to appear exactly as it last was, almost as though the application had not been closed.

The UIQ 3 view framework provides substantial support for record-based applications through the `SaveL()` method of the `CQikViewBase` class. As users move away from a view, the application can choose how to handle any updates that may have occurred. In our application, the `CDetailsView` updates the information stored on disk (such as renaming a file) as well as any internal representation information. Database-type applications would normally update the database record at this time.

While the underlying data is normally updated when a view change occurs, it is unusual for the presentation information to be updated as well. In general, presentation data is only stored when applications are closed.

Closing Applications

The current UIQ 3 style guidelines recommend that applications do not present a Close or Exit menu option to the user. On some UIQ 3 mobile phones, a task manager is available which enables users to close applications. On all UIQ 3 mobile phones, if memory resources become low then the system requests applications to shut down, releasing resources.

In both cases, your application is sent an `EEikCmdExit` command. To facilitate testing of any functionality associated with this command, it is common practice to add an `EQikCmdFlagDebugOnly` type command to the application:

```
QIK_COMMAND
  {
  id = EEikCmdExit;
  type = EQikCommandTypeScreen;
  stateFlags = EQikCmdFlagDebugOnly;
  groupId = EAppCmdMiscGroup;
  priority = EAppCmdLastPriority;
  text = "Exit";
  }
```

Pro tip

You are free to vary from the UIQ 3 recommended guidelines if you feel this is appropriate to your application. Games, in particular, often have a Close menu option of some description, especially those that choose to run full screen. It is harder to navigate to the task manager from a full-screen application; therefore a Close option can be useful. Your application will not fail Symbian Signed if you include a Close menu option.

Where to Store Presentation Information

If your application does not store presentation information along with the primary data, an alternative location for presentation information is required.

Symbian OS supports a wide range of file services. You can choose from basic file services, where your application reads and writes information directly to the file in a format of your choice, through to abstract database record services.

One set of file services provided is called *stream stores*. Your application serializes its internal object network to an external stream to save information and de-serializes that external stream into the internal object network to load information.

A detailed discussion about stream stores and their variations is beyond the scope of this book. The UIQ v3 SDK documentation contains a significant amount of information on this topic. Additional information regarding the structure and usage of streams and stores can be found in *Symbian OS C++ for Mobile Phones*.

For the purposes of our application, a stream store is quite simply a file that contains a binary stream of data which can be loaded and saved conveniently.

`ExternalizeL()`

In an object-oriented environment, encapsulation is a common metaphor. When applied to saving content to a stream store, an object encapsulates its save functionality within an externalization method. The implementation of the `ExternalizeL()` method for the `TAppCategoryEntry` object is:

```
void TAppCategoryEntry::ExternalizeL(
  RWriteStream& aStream) const
  {
  aStream.WriteUint8L(KAppCategoryEntryStreamVersion);

  // iCategoryId.
  aStream.WriteInt32L(iCategoryId);

  // iCategoryName.
  aStream.WriteUint8L(iCategoryName.Length());
  aStream.WriteL((TText*)(iCategoryName.Ptr()),
                      iCategoryName.Length());
  }
```

Here we are storing a single byte for some version information, four bytes for the category ID, a single byte for the length of a text string and a number of two-byte entities, one for each character in the text string.

It is up to you to ensure that the number of bytes being stored is sufficient to represent the range of values an item can contain. For example, the version number value is extremely unlikely to ever exceed 10. The value 10 can be represented in a single byte; therefore we only store a single byte.

Pro tip

The number of bytes stored to represent a value is a compromise between minimizing the required disk storage compared to ease of use and maintainability. We could quite easily store four bytes for the version number. Doing so means we store three bytes which are always 0. For every category we have, we would store three bytes of redundant information. In our application, we have a very small set of categories so this is not significant. Examine the needs of your application and store the number of bytes that you need and no more.

Unlike traditional raw-file-based loading and saving, by using stream stores we don't need to worry about ensuring any underlying buffering is sufficient, whether entities cross buffer boundaries, the endian-ness in which data is stored, exactly where data is being placed or indeed if the underlying data is being compressed, encrypted or both. The stream store takes care of all those details. Similarly when we come to load the data all these issues are handled by the framework code. This makes the application code significantly easier to write and maintain.

In our application, we have a set of categories owned by the engine object and stored within an `RArray`. A single category entry is saved using the above code. To save the entire set of category entries, the engine needs to call the `ExternalizeL()` method on each of the `TAppCategoryEntry` objects.

Symbian OS contains a template for the << operator. This template expects to see a method name of `ExternalizeL()` taking a single `RWriteStream&` parameter belonging to the object to which the << operator is applied. If we choose to call our object-saving function `ExternalizeL(RWriteStream&)` then we can choose whether to use the << operator or call the method directly. In our application we have named our function so that we can use the << operator. The engine can save the set of categories with the following code:

```
TInt count=iCategoryList->Count();
aStream.WriteInt16L(count);
for (i=0;i<count;i++)
  aStream<<((*iCategoryList)[i]);
```

The identical functionality could have been performed with:

```
TInt count=iCategoryList->Count();
aStream.WriteInt16L(count);
for (i=0;i<count;i++)
  (*iCategoryList)[i].ExternalizeL(aStream);
```

To a large extent it is personal preference as to which of the above you use within an application. One point of detail that may influence your decision is the fact that the code performing the store can leave. When using the C++ << operator, it is not particularly obvious that << can leave.

InternalizeL()

The `InternalizeL()` method of our `TAppCategoryEntry` is:

```
void TAppCategoryEntry::InternalizeL(RReadStream& aStream)
  {
  TInt version=aStream.ReadUint8L();
  if (version<=KAppCategoryEntryStreamVersion)
    {
    // iCategoryId.
    iCategoryId=aStream.ReadInt32L();

    // iCategoryName.
    iCategoryName.SetLength(aStream.ReadUint8L());
    aStream.ReadL((TText*)(iCategoryName.Ptr()),
                          iCategoryName.Length());
    }
  else
    User::Leave(KErrNotSupported);
  }
```

When compared with the `ExternalizeL()` method you should observe that items are internalized in the same order they were externalized, that is version information, category ID, then category name. Your application must ensure that the amount of content it reads for a particular field matches the amount of data it stored. For example, since we stored a single byte for the version information, we need to read a single byte so the stream remains aligned.

As with `ExternalizeL()`, Symbian OS contains a template for the >> operator. This template expects to see a method name of `InternalizeL()` taking a single `RReadStream&` parameter belonging to the object to which it is applied. If we choose to call our object-loading function `InternalizeL(RReadStream&)` then we can choose whether to use the >> operator or call the method directly. We have named our function such that we can use the >> operator. The engine can load the set of categories with the following code:

```
TInt count=aStream.ReadInt16L();
for (i=0;i<count;i++)
  {
  TAppCategoryEntry catEntry;
  aStream>>catEntry;
  catList->AppendL(catEntry);
  }
```

Again, whether you use >> or `InternalizeL()` is down to personal preference. As with the << operator you should be aware that the >> operator can leave.

> **Pro tip**
>
> It is possible to call your load and save methods anything you like, or indeed to have methods that take a different set of parameters should your application require it. The potential problem with this approach is that you cannot use the << or >> operators and programmers familiar with many of the other Symbian naming conventions find it more difficult to understand the functionality.

Streams and Version Information

It is up to your application what information it chooses to save to a stream and what order that information is stored in.

One piece of information we recommend storing at various points is some class or structure version information. Should you ever need to update a particular object, a version number within the stream data enables you to remain backwards compatible. Since it tends to be individual objects that change, it is common to store a version number with an object as opposed to storing a single higher-level version number. This is another example of encapsulation and is demonstrated in our example application.

> **Pro tip**
>
> There is a trade-off with storing version numbers. If you have a large collection of a single type of object and you store a version number for every instance of that object, you need one more byte for each instance you save to disk. Conversely, if you have a single version number stored outside the collection you have a bit more work to do when loading the content. In particular, you have to pass the version information to the `InternalizeL()` method in some fashion. Doing

so may mean you change the `InternalizeL()` method so that you can no longer use the >> operator.

How to Approach Writing the `InternalizeL()` and `ExternalizeL()` Code

The `ExternalizeL()` code is quite straightforward. In most cases you simply need to run though your object network and save the property you deem is required such that the application can be re-started as though it had never exited.

On the other hand, the `InternalizeL()` method tends to be much more complicated. In general, as well as loading the information, it has to create the object network. Because of this we strongly recommend that you write the `InternalizeL()` method first. This allows you to define the most convenient order in which content is saved, so that you can reconstruct your object network.

As previously discussed, it is often not sufficient to simply leave when an error occurs. Apart from cleaning up any resources, your application generally needs to roll back to a stable state. The `InternalizeL()` method is no different. In our application, we need to create the category list, folder entry list and index; only when all three are successfully constructed can we replace any existing versions of those structures with the new ones.

Pro tip

`InternalizeL()` rollback functionality is particularly important if your application supports multiple data sources of different types, for example word-processor documents and spreadsheets. Rollback means that the user remains able to view and manipulate the current document in the case that loading a new document fails. For our application, it is somewhat less critical since we do not support multiple documents. However, in an object-oriented environment, objects, particularly if they want to be reused, should try to avoid any assumptions about the environment in which they are used.

Even with our example application, you can imagine a future version that allows a user to move around the folder hierarchy. It will therefore need to display previously visited folder content with folder-specific sort and category information.

When you come to consider some of the Symbian Signed Test Criteria, you should observe that rollback of the `InternalizeL()` method gains additional significance. In particular, our application becomes quite robust and can survive a number of the error conditions that it is expected to handle.

Purpose of Internalizing

A modern user expectation is that an application resumes from the point at which it was left. However, at a detail level it may be more important to save some states than others. For example, if your application supports a sort function it would be quite reasonable to expect the application to retain that information and present its data in the last sort order chosen by the user. However, saving exactly which entry had the highlight when the application exited may be a step too far.

For our example application we have:

- a list of user-defined categories
- a list of folder entries with some associated data
- some presentation information, such as currently chosen category and sort order.

We have chosen to internalize all those details we consider important but have chosen not to internalize everything, such as which entry within the list view contains the highlight or whether the list or detail view was the active view. We consider this acceptable for our application. Your application may have different requirements.

`CAppEngine::InternalizeL()` Functionality

Our example application is slightly atypical since it is viewing the content of the file system. Other file management applications may be able to view and manipulate the same content, particularly if we are viewing publicly accessible folders. Consequently, we need to be somewhat defensive in our programming approach to internalizing our application state.

To recap, our key requirements are to internalize the set of categories, internalize our folder information and present the list of content previously chosen by the user.

When we come to internalize our folder content, we need to be aware of several issues:

- additional folder entries may have been created
- some of the folder entries we originally listed may have been renamed or removed.

Our application handles these conditions by storing the category information along with a filename in our stream store. When we internalize the folder entry list, rather than simply taking the list of entries we externalized, we have to create a list of what is currently on the disk, locate those entries in our internalized data and assign the category information.

```
TParse parse;
parse.Set(KWildCardName,&path,NULL);
CDir *dirList;
User::LeaveIfError(iFs.GetDir(parse.FullName(),
                          KEntryAttNormal|KEntryAttReadOnly|
              KEntryAttHidden|KEntryAttSystem|KEntryAttDir,
                                                  0, dirList));
CleanupStack::PushL(dirList);

// Add a set of entries to the new container.
TFolderEntry entry;
count=dirList->Count();
for (i=0;i<count;i++)
  {
  // Construct a new entry.
  entry.Construct((*dirList)[i]);

  // Add to the new list of entries.
  folderEntryList->AppendL(entry);
  }

// Now we have finished with the dirList and
// its top of cleanup stack we can delete it.
CleanupStack::PopAndDestroy(dirList);

// Load the list of entries we knew about before exit.
count=aStream.ReadInt32L();
RArray<TFolderEntry>* savedEntryList =
                        new(ELeave)RArray<TFolderEntry>(8);
CleanupClosePushL(*savedEntryList);
for (i=0;i<count;i++)
  {
  aStream>>entry;
  savedEntryList->AppendL(entry);
  }

// Update our new entry list with
// saved category entry information.
TInt entryCount=folderEntryList->Count();
for (i=0;i<entryCount;i++)
  {
  TFolderEntry& entry=(*folderEntryList)[i];
  const TDesC& name=entry.EntryName();
  for (TInt j=0;j<count;j++)
    {
    TFolderEntry& entry2=(*savedEntryList)[j];
    if (name==entry2.EntryName())
      { // Found the entry we saved the category info for.
      entry.UpdateCategory(entry2.EntryCategory());
      break;
      }
    }
  }
CleanupStack::PopAndDestroy(savedEntryList);
```

In terms of defensive programming, we don't assume that the list of files that exists on the disk matches the list of files we last knew about. Files that have been deleted are not added to the list presented to the user. Newly created or renamed files appear but are placed in the Unfiled category.

You may want to go further with this type of activity in a generic file management type application. For example, if a user tasks away to another application then tasks back, the other application may have created a file within the folder being viewed. Indeed it is possible for other applications to create files in the background or for new files to be sent to a mobile phone while your application is in the foreground. Should you wish to investigate supporting this type of functionality within an application, we suggest you study the `RFs::NotifyChange()` and `CEikAppUi::HandleForegroundEventL()` methods.

Stream Stores and INI Files

INI files are simply one of the possible variants of a stream store, in this case a Dictionary store. In such a store, numerous streams can be saved independently of each other. To gain access to a particular stream, a UID is associated with the stream. A table of UID and stream pairs is saved within an index (the dictionary), which in turn is saved within the store.

Saving content to an INI file can be achieved with the following code:

```
void CAppSpecificUi::SaveIniFileL(const TInt aVersion)
  {
  CEikApplication* app=(CEikApplication*)Document()->Application();
  CDictionaryStore* iniFile=app->OpenIniFileLC(iEikonEnv->FsSession());
  RDictionaryWriteStream writeStream;
  writeStream.AssignLC(*iniFile,KUidIniFilePrefs);

  // Top level version number
  writeStream.WriteUint8L(aVersion);

  // now all the content
  SaveIniFilePreferencesL(writeStream);

  writeStream.CommitL();
  CleanupStack::PopAndDestroy(); // writeStream
  iniFile->CommitL();
  ...
  }
```

The framework code supports the `OpenIniFileLC()` method. This method automatically determines in which folder to place the INI file and what the filename should be. For our application, the INI file will be called `SignedApp_0x20000462.ini` and it is placed in our application's private folder "`\private\20000462`" on the internal disk (C drive). While the application does not need to know this, you should be aware of what is going on as this affects Symbian Signed testing.

In our application, we only have a single stream of data identified by the UID `KUidIniFilePrefs`. This UID is private to our application and to the INI file; as such it is common to use the value 1 although this is presented symbolically to the reader.

As we discussed previously, version numbers are very useful to allow for backwards compatibility in future versions of the application. We store a top-level version number in our stream.

In our application, we have chosen to store all configuration information except for the user-entered registration code in a single place, the engine. We separated out the registration code to show that it is not necessary to store all configuration information within an engine, if that is not appropriate for your application.

You can distribute application state in many places if you want, for example, you can have the list view sort order information stored in the list view, the zoom size stored in the details view, and so on. In this case, you have to locate all those objects and ask them to save their configuration data. On application start-up, you have to locate all the new versions of those objects, (which may not actually have been created yet, so beware) and tell them to read their configuration information.

Pro tip

- While storing all configuration information within the engine gives rise to an increased API to the engine, this simplifies many application programming tasks. All loading and saving can be achieved by the engine, all configuration parameters can be transferred simply by making the engine available to those objects that need them, and the software can be automatically tested more easily.
- A current known issue in the Windows emulator causes an application to terminate with a `CBase 65` panic after you have saved content to the INI file. The framework code in the emulator is inadvertently popping too many items from the cleanup stack. This panic does not occur on real mobile phones. The `CBase 65` panic occurs before the framework code attempts to clean up all application resources and check there are no resources unaccounted for. We strongly recommend that you periodically disable the saving code to allow the framework to perform this check. This exposes unaccounted resource errors that are hard to find and typically result in `CONE 8` or `ALLOC` panics earlier in your development cycle.

Loading our INI File

The following code loads the stream we stored to our INI file:

```
void CAppSpecificUi::LoadIniFile()
  {
  TRAPD(err,
      CEikApplication* app=(CEikApplication*)Document()->Application();
      CDictionaryStore* iniFile=app->OpenIniFileLC(
                            iEikonEnv->FsSession());
      if (iniFile->IsPresentL(KUidIniFilePrefs))
        {
        RDictionaryReadStream readStream;
        readStream.OpenLC(*iniFile,KUidIniFilePrefs);

        // load all the content
        TInt version=readStream.ReadUint8L();

     LoadIniFilePreferencesL(version,readStream);

        CleanupStack::PopAndDestroy(); // readStream
        }
      CleanupStack::PopAndDestroy(iniFile);
      );
  }
```

As with saving content to an INI file, the framework provides the functionality to locate and open our application-specific INI file. If the INI file does not exist, the `OpenIniFileLC()` method will leave. Our code fragment then checks to see if the stream with the UID value `KUidIniFilePrefs` exists. If it does, it proceeds to load the stream content.

In common with most other applications, this application should be capable of continuing if no INI file exists or if the expected stream is not present. In general, this can be achieved by simply setting a `TRAP` for the functionality and ignoring any error, as presented above.

When to Load an INI File

The first time an application is started on a mobile phone, it should set up any default parameters and user preferences. If the application is started subsequently, it should override those default preferences with any stored ones.

In practice, INI files tend to be small, so loading them is quite fast. Rather than attempting to code for all possible conditions individually, a perfectly reasonable approach is to set up all the default values each time an application is started, usually within object construction, then attempt to override them with the values stored in the INI file. Our application construction code therefore looks like this:

```
void CAppSpecificUi::ConstructL()
  {
  CQikAppUi::ConstructL();
  iEngine=new(ELeave)CAppEngine(iEikonEnv->FsSession(),
                                    EQikCmdZoomLevel2);
  iEngine->ConstructL();

  // Setup the default category list.
  TBuf<KCategoryNameMaxLength> bb;
  for (TInt i=0;i<EAppCategoryLastItem;i++)
    {
    iEikonEnv->ReadResource(bb,R_STR_CATEGORY_NAME_1+i);
    iEngine->AddCategoryL(i,bb);
    }

  // Locate our private path -
  // where our example files have been installed to.
  TFileName installedDrive(Application()->BitmapStoreName());
  TFileName path;
  iEikonEnv->FsSession().PrivatePath(path);
  TParse parse;
  parse.Set(path,&installedDrive,NULL);
  iEngine->SetDefaultPathL(parse.DriveAndPath());

  // Override any default settings from the INI file.
  LoadIniFile();
  ...
  }
```

It is very important that you understand the importance of rollback in the `CAppEngine::InternalizeL(RReadStream& aStream)` method. If there is a failure to load the INI file fully and the engine rolls back properly, our application can at least be started, even if some user configuration information is lost. If the engine does not roll back properly it is likely to be left in an inconsistent state and is likely to panic. Not being able to start the application is even worse than losing some information!

Saving Streams to your INI File

Depending on your application structure, you may have varying requirements as to when information is externalized. For example, you may have configured your application to partition user preferences across views and save these individual preferences within different streams in a single INI file. When the application leaves a view, you would update the stream associated with the view. In our application we have chosen to store all user preferences within volatile RAM – in the engine – until such time as the application is told to exit. This makes the internalizing and externalizing processes easier but the user preferences are lost if the application crashes. The methodology you choose should be appropriate to the requirements of your application. For the vast majority of applications, it is adequate only to save preferences when the application exits.

Even if your application does not have an Exit or Close menu option, the UIQ framework may request your application to exit, for example, when memory gets low or when a user requests a task manager application to close an application.

When requested to exit, the method `CAppSpecificUi::HandleCommandL()` sends an `EEikCmdExit` command:

```
void CAppSpecificUi::HandleCommandL(TInt aCommand)
  {
  switch (aCommand)
    {
    case EEikCmdExit:
      TRAPD(junk,SaveIniFileL(KIniFileStreamVersion));
      Exit();
      break;
    ...
    default:
      break;
    }
  }
```

In general, if requested by the system to exit you should do so. For our application, the worst-case scenario is that we are unable to externalize our user preferences. If we simply allowed the `SaveIniFileL()` method to leave, a system error message would be displayed. However that error message is only visible if the application is in the foreground. When the system is requesting an application to shut down, it is unlikely that the application is in the foreground. If the user makes the exit request from within the task manager application, the application is in the background. Therefore the user is unlikely to see the error message or be able to respond to it. In practice, the task manager's Close function appears to fail with no visual clue as to what has occurred.

In addition to this, Symbian Signed test CON-01 in Test Criteria v2.11.0 (UNI-05 in Test Criteria v3.0.0) requires that applications can be closed through the task list. If choosing the task manager's Close function does not cause the application to shut down, irrespective of any error condition, your application will fail against the Symbian Signed Test Criteria.

In practice, there are only two ways your application can fail to save the INI file. It may run out of memory attempting to create objects representing the open streams within the dictionary store or the internal disk may become full. While both scenarios have a probability of occurring, they are very rare in the real world. To eliminate any risk of not exiting, we choose to `TRAP` any leave and ignore any error condition prior to exiting. If this is truly problematic for your application, you must adopt a different strategy for saving user preferences.

10.3.9 Running `SignedAppPhase2`

`SignedAppPhase2` appears in the application launcher as `SignedApp`. When you first launch it, the list is empty. This is because we have not yet populated the folder at which `SignedAppPhase2` looks. We add some data and make it possible to open and create new files later on, in `SignedAppPhase3`.

You can find some suitable files in `SignedAppPhase3\Example Files` and copy them to the application folder in the emulator `<path>\epoc32\winscw\c\Private\20000462` where `<path>` is the path to where the emulator is installed on your PC.

The New and Open commands in the menu are not yet implemented in `SignedAppPhase2`.

10.4 Building your Application for Deployment

Now that we have an application framework with some meaningful functionality, we can deploy it to a real mobile phone to ensure that it performs adequately.

If you are going to Symbian Sign your application, you need to test it first on real mobile phones. Symbian provides a low overhead self-signing option known as Open Signed, which is based on 'developer certificates'. You cannot use Open Signed to sign an application for final release, because developer certificates are restricted to specific phones based on IMEI number. Open Signed is available in online mode for developers who have no Publisher ID and a more advanced offline mode for those with a Publisher ID. Open Signed is described in detail in Chapter 14.

10.4.1 Open Signed Without Publisher ID

To test on a single mobile phone. you can follow the Open Signed online procedure and upload your application to the Symbian Signed portal for signing. The portal uses your email address to check that the application's UID belongs to you. Since `SignedAppPhase2` uses our UID, you must either change the UID to one in the test range `0xE0000000` to `0xEFFFFFFF` or obtain your own UID in the protected range `0x20000000` to `0x2FFFFFFF` and apply it to `SignedAppPhase2`.

To sign the application, you simply upload it to the Symbian Signed portal together with some information including the IMEI of your test phone. You may only specify a single IMEI. You can download the signed application when you receive the email that tells you it is ready. The signed application is valid for 36 months.

10.4.2 Open Signed with Publisher ID

To test on multiple mobile phones, you can use the Open Signed offline procedure and sign with a developer certificate. Section 14.4 tells you how to obtain a developer certificate, `DevCert.cer`. You must have a Publisher ID to obtain a developer certificate. We use the `signsis` tool to sign applications.

If you recall, we have chosen to send our application to a test house when it is ready for final testing and signing. The copy that we send to the test house must be signed with our Publisher ID. Because we completed `SignedAppPhase3` prior to the introduction of Express Signed and Certified Signed, we are using an ACS Publisher Certificate which was issued by VeriSign.

> **Pro tip**
>
> For the time being, certificates issued by VeriSign remain valid for Symbian Signed. New certificates should be obtained from the new certificate authority, TC TrustCenter.

The `bldall.bat` file in our group folder is responsible for creating the application SIS file. It contains the following statements:

```
rem Build the releasable SIS file

call makesis SignedApp.pkg SignedApp_uiq3_001.sis

call signsis -s SignedApp_uiq3_001.sis
        SignedApp_DevCert_uiq_3_0_v_1_00_01.sis
                                    DevCert.cer
                    ACSPublisherCertificate.key

call signsis -s SignedApp_uiq3_001.sis
        SignedApp_ACSCert_uiq_3_0_v_1_00_01.sis
                    ACSPublisherCertificate.cer
                    ACSPublisherCertificate.key
```

You may notice that the above batch file creates two SIS files, one signed with the developer certificate and one signed with the ACS Publisher Certificate. Currently, there is no mechanism for testing the SIS file signed with the ACS Publisher Certificate until it has been signed by a test house, because the SIS cannot be installed on a mobile phone. The nearest you can get is to test a version signed with a developer certificate (see Figure 10.10). If we generate both SIS files at the same time, we minimize the risk of there being any differences between the developer certificate and ACS Publisher Certificate versions of the SIS file.

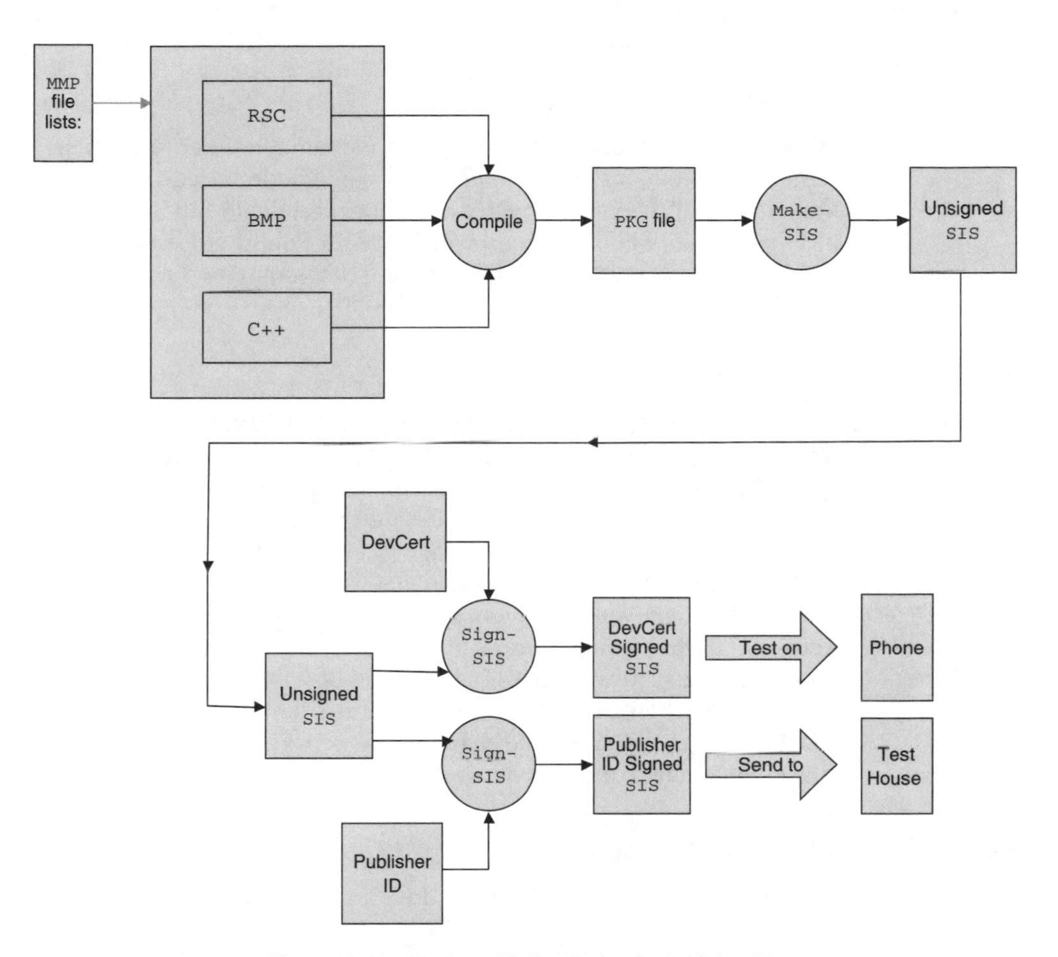

Figure 10.10 Signing with DevCert and a Publisher ID

10.4.3 SIS Filename

You may have noticed that we appear to have a rather odd-looking filename for the SIS file. When you come to submit the application to a test house via the Symbian Signed portal, you are required to name the SIS appropriately. In particular, you need to specify the target platform, in our case UIQ 3.0 and above, and provide the version number of the application. The version number **must** match the version number in any About screen and other submitted content. The filenames used conform to the current Symbian Signed web portal requirements. While you are free to use any name you like at this stage, you must ensure that the SIS files are correctly named prior to submission.

Pro tip

We recommend that you update the SIS filename in order to track the signatures it contains. In our application we include 'DevCert' in the filename for SIS files signed with a developer certificate and 'ACSCert' in filenames for SIS files signed with our ACS Publisher Certificate. In this way, you can distinguish between the different files.

10.4.4 Symbian Signed Test Criteria

A core objective of our example application is to ensure we can pass the Symbian Signed Test Criteria. At this stage of our application development, we have a functional application that can be deployed to a target mobile phone. While later chapters add further functionality, it is appropriate to run through the test criteria at this stage to check that we comply and to plan the correction of any non-compliance issues.

The test criteria are presented in full in Chapter 14 together with a discussion of their impact upon our `SignedAppPhase2` application. Test Criteria v2.11.0 was current at the time of writing and was used by the test house when `SignedAppPhase3` was submitted for testing.

At this stage, we failed the Symbian Signed Test Criteria and must make changes in `SignedAppPhase3` to pass the following tests.

PKG-03 (UNI-08): File creation location

Files should only be created on the drive to which the application was installed.

Our application has chosen to store configuration information within an INI file. The application framework creates that file on our behalf, currently always on the internal drive C irrespective of the disk on which the application is installed.

An exception is available, EX-006, which covers the case where the system code creates files on our behalf. To ensure we do not fail this test it must be documented within a `readme.txt` file. This file must be submitted along with the application so that the exception is clear to the tester.

This criterion is UNI-08 in Test Criteria v3.0.0. The exception case may no longer apply.

CON-02: Privacy statement dialog

A Symbian OS v9 application must display a dialog informing a user about the usage of various OS functions if such functionality exists in an application.

Since our application supports Send as, which can send content via MMS, Bluetooth or infrared, our application requires a privacy statement dialog. Without such a dialog, our application will fail this test. We need to add such a dialog in a later phase of our application development.

This criterion is expected to be removed in Test Criteria v3.0.0.

CON-03: Billable events

Our application supports Send as. One of the transports that a user can choose on a real mobile phone is MMS. Even though our application passes control to the UIQ Messaging application to send the MMS, our application must provide a billable-event warning dialog. To be sure of passing this test, we therefore adjust our application to present a billable-event dialog prior to using the Send as functionality.

This criterion is expected to be removed in Test Criteria v3.0.0.

11

Multimedia

Mobile applications frequently include multimedia content. We use images, animations, audio and video to enrich the user experience.

This chapter gives you an introduction to using multimedia and covers some of the common application-level multimedia requirements:

- load images
- store images
- perform some basic image manipulation
- use the camera to take a photograph
- load and play back audio clips
- record and save audio clips
- load and playback a video clip
- use the FM radio via the Tuner API.

11.1 `SignedAppPhase3`

In the previous chapter, we developed a simple file management application, `SignedAppPhase2`. We use this as the starting point for our multimedia examples and add multimedia functions to create `SignedAppPhase3` (see Figure 11.1).

The New command (Figure 11.2) enables us to record a sound clip or take a photograph. The Open command allows us to open the sample media. The folder `SignedAppPhase3\ExampleFiles` contains sample media files. Copy them to the application folder in the emulator

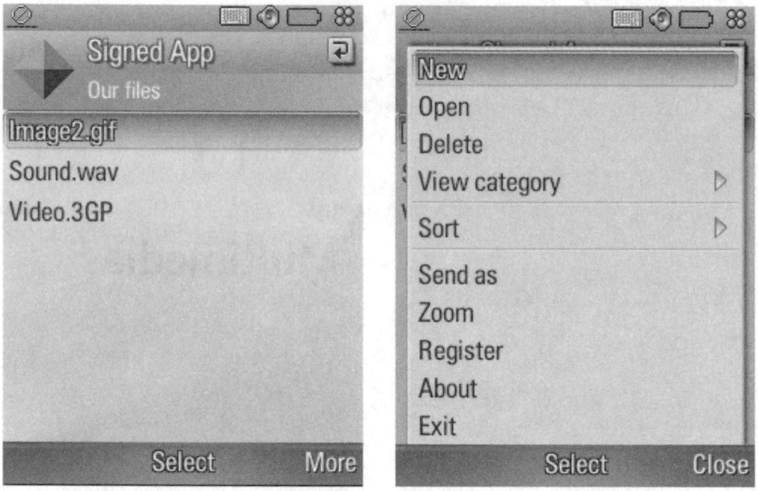

(a) (b)

Figure 11.1 `SignedAppPhase3` a) list view and b) menu

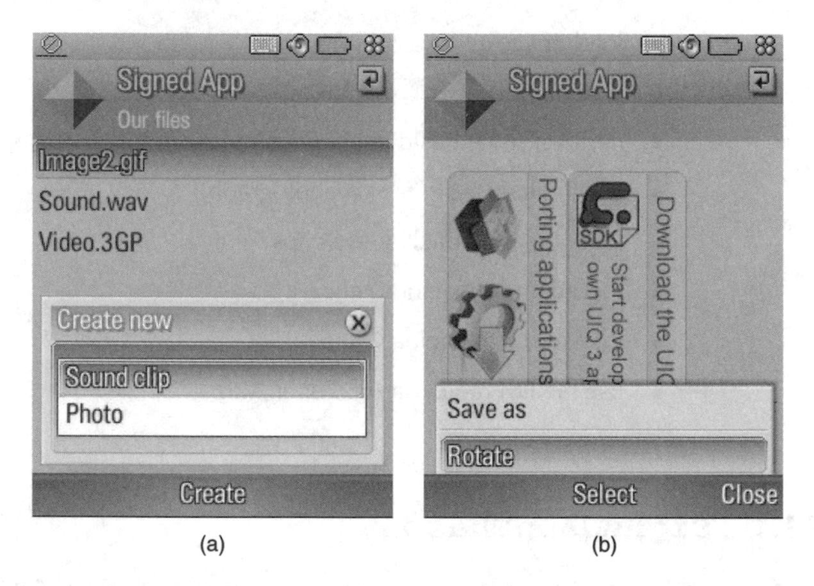

Figure 11.2 New command and image display with rotate function

`<path>\epoc32\winscw\c\Private\20000462` where `<path>` is the path to where the emulator is installed on your PC.

After adding the multimedia features to `SignedAppPhase3`, we submitted it to the test house and obtained Symbian Signed. The signed application is included in the download.

11.2 Symbian Signed Requirements

11.2.1 Capabilities

Prior to proceeding with any development work, we should consider any implications that Symbian OS platform security may impose. Capabilities are described in Chapter 14. Multimedia APIs may require your application to have one or more of these three capabilities:

- `UserEnvironment` (User)
- `SurroundingsDD` (System)
- `MultimediaDD` (System).

Of these, `UserEnvironment` can be granted by a user; `SurroundingsDD` and `MultimediaDD` can be granted by passing Symbian Signed. `MultimediaDD` used to be a Device Manufacturer capability. You must use Open Signed with a Publisher ID to test an application that uses `MultimediaDD`.

To work out which capabilities you need, consult the SDK documentation for the classes that your application uses. Occasionally, you may need to use trial and error where the SDK is unclear.

The trailing `DD` in both the `SurroundingsDD` and the `MultimediaDD` capabilities, along with their description, is an indication that these capabilities are only required if we need to gain direct access to the internals of device drivers. The vast majority of applications simply want to use the functionality provided by the device drivers and do not need such low-level access.

The `MultimediaDD` capability is only required if applications wish to manipulate the priority or preferences of the audio record or playback channel it has opened. Few third-party applications should manipulate these features, as they may affect the incoming phone call ringing or other alarm type sounds. Symbian Signed test Criterion GEN-01 requires that applications do not adversely affect the use of other system features. Blocking an incoming phone call ring is likely to cause your application to fail this test.

No capabilities are required for the image manipulation, audio and video playback example applications in this chapter. `UserEnvironment` is required for sound and video recording and using the camera.

11.2.2 Symbian Signed Test Criteria

When we assessed `SignedAppPhase2` against the Symbian Signed Test Criteria in Section 10.4.4, we identified that further work was required to pass all the test cases. We have added the privacy statement for CON-02

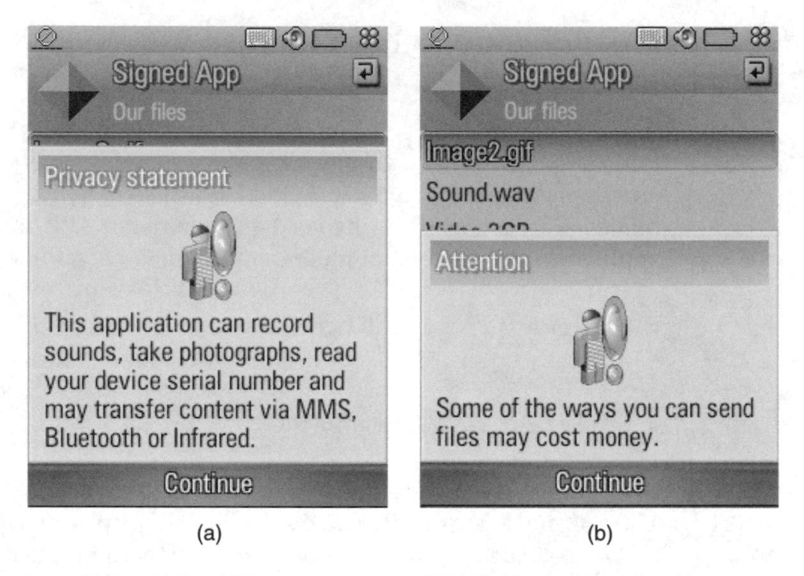

Figure 11.3 a) Privacy statement and b) billable event warning dialog

and billable-event dialog for CON-03 (see Figure 11.3). Although these tests from Test Criteria v2.11.0 are likely to be removed in v3.0.0, it will always be important that you check your application against the current test criteria and make any necessary changes.

Once we have added the multimedia functionality described in this chapter, our application is ready for final testing and submission via the Symbian Signed website.

11.3 Images

Symbian OS provides an Image Conversion Library (ICL) which offers support for image file encoding and decoding, plus scaling and rotation of bitmaps. The ICL provides encoding and decoding support for the image types listed in Table 11.1.

Since the ICL uses a plug-in architecture to perform decoding tasks (see Figure 11.4), the set of decoders on your mobile phone may vary from the set within the SDK. The Sony Ericsson P1i supports BMP, GIF, JPEG, MBM, PNG and WBMP formats. We recommend that you check the support and test your application on the mobile phones that you wish to support.

11.3.1 Image Decoding

Image decoding is the process of loading an image and converting it into the internal `CFbsBitmap` format used by Symbian OS applications to display it on screen. Input formats include BMP, JPG and GIF.

Table 11.1 Image types supported by ICL

Format	Encode support	Decode support
BMP (Bitmap)	Yes	Yes
EXIF (Exchangeable Image File format)	Yes	Yes
GIF (Graphics Interchange Format)	Single frame, no transparency	Single- and multi-frame, bitmap mask support
JPEG (Joint Photographic Experts Group)	Yes	Yes
MBM (Multi-Bitmap)	Single frame	Single- and multi-frame
MNG (Multiple image Network Graphic)	Yes	Yes
PNG (Portable Network Graphics)	No transparency	Bitmap mask support
SMS OTA (Over the Air)	No	Yes
TIFF (Tagged Image File Format)	No	LittleEndian and BigEndian sub-type support
WBMP (Wireless Bitmap)	No	Yes
ICO (Icon)	No	Single- and multi-frame
WMF (Windows Meta File)	No	STD, APM and CLP sub-type support

The image decoding classes use asynchronous methods to perform the actual processing. Instead of using threads, they use active objects. This has several consequences:

- Our application must have an active scheduler in the thread performing the image decoding. A standard UIQ application primary thread always has a scheduler as it is provided by the application framework.
- Our application needs to wrap the image-processing functions in an active object.
- Our application can process other events between the time an image-processing event is started and the time it completes.

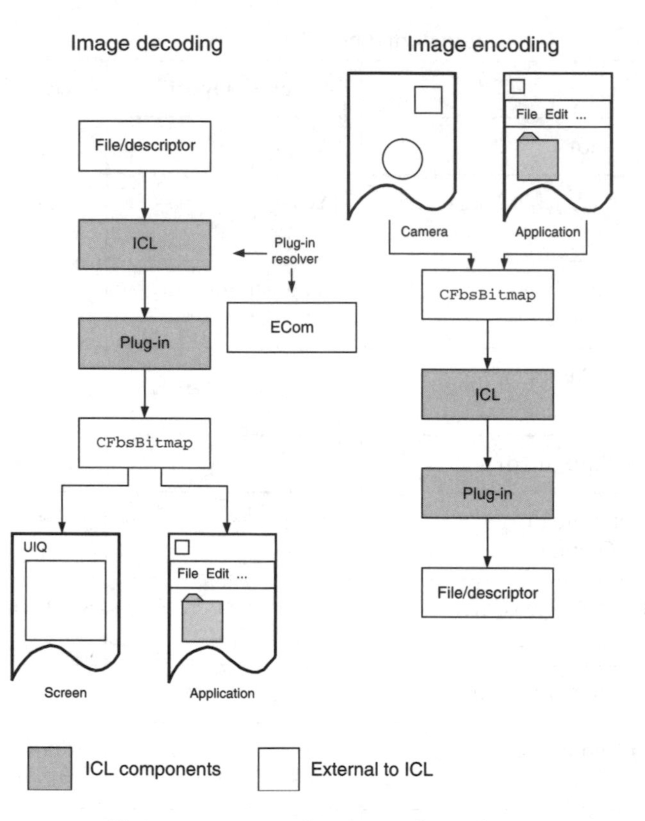

Figure 11.4 Image decoding and encoding

The `CImageDecoder` class provides the basic image decoding functionality. Since the ICL comprises a set of plug-ins, typically one plug-in per image format, a mechanism of associating the correct plug-in with the content is required. Variants of the `CImageDecoder` class allow applications to specify different attributes. For our application, we simply want to be able to load an image file, irrespective of its underlying format. This can be achieved by requesting the `CImageDecoder` to use the underlying plug-ins to determine the file type on our behalf.

Image Size and Color Depth

Once the `CImageDecoder` object has been created, we should request information about the underlying image. We are particularly interested in the image size and color depth attributes.

Prior to loading the image, we need to create space, in the form of a `CFbsBitmap` object, to store the image data. The `CFbsBitmap` needs to be the same width and height as the original image.

Color depth refers to the number of different colors that each pixel can represent. GIF images use eight bits per pixel, so each pixel can represent one color from a palette of 256, but BMP and JPG formats are capable of representing millions of colors. Similarly, the screen has a color depth and this will vary between different models of mobile phone. The first UIQ phone, the Sony Ericsson P800, had a 4096-color screen. The MOTO Z8 has a 16-million-color screen.

When loading, the values representing colors in the image need to be translated to the equivalent value for those colors used by the screen, particularly if an image format uses a palette to store color information. Optimal display speed is achieved when the color depth of an image matches the display mode of the window within which the image is displayed. This is because no translation of the individual pixels is required between the `CFbsBitmap` data and the data displayed within a window. Since windows are normally created using a default display mode, it is preferable to create the `CFbsBitmap` with the same display mode and only translate the image once, when it is loaded.

Translating the image when it is loaded is only possible if the underlying plug-in supports color translations. This is indicated by the `TFrameInfo::ECanDither` flag returned within the `TFrameInfo` object. We use this in our example application. If it is important for your application to always load images into the current display mode but the plug-in does not support `TFrameInfo::ECanDither`, simply create a `CFbsBitmap` of the required color depth then copy the image into a second `CFbsBitmap` after it has been loaded. The copy operation in the form of the `DrawBitmap()` method automatically performs the color-depth translation.

Loading the Image

The code for our image loading is:

```
CImageLoader::~CImageLoader()
  {
  Cancel();
  }

CImageLoader::CImageLoader(CImageDisplayControl& aDisplay):
     CActive(CActive::EPriorityStandard),iDisplay(aDisplay)
  {
  CActiveScheduler::Add(this);
  }

void CImageLoader::StartLoadImageL(
  const TFileName& aFileName)
  {
  iLoader=CImageDecoder::FileNewL(CEikonEnv::Static()
                          ->FsSession(),aFileName);

  // Create a bitmap within which to store
  // the loaded image.
```

```
TFrameInfo frameInfo(iLoader->FrameInfo(0));
  TInt ret=KErrNoMemory;
  iBitmap=new CFbsBitmap();
  if (iBitmap)
    {
    if (frameInfo.iFlags&TFrameInfo::ECanDither)
      ret=iBitmap->Create(frameInfo.iOverallSizeInPixels,
     CEikonEnv::Static()->ScreenDevice()->DisplayMode());
    else
      ret=iBitmap->Create(frameInfo.iOverallSizeInPixels,
                              frameInfo.iFrameDisplayMode);
    }

  if (ret!=KErrNone)
    {
    delete(iBitmap);
    iBitmap=NULL;
    delete(iLoader);
    iLoader=NULL;
    User::Leave(ret);
    }

  // Start loading the image into our bitmap,
  // RunL() called when complete.
  iLoader->Convert(&iStatus,*iBitmap,0);
  SetActive();
  }

void CImageLoader::RunL()
  {
  iDisplay.LoadComplete(iStatus.Int(),iBitmap);
  iBitmap=NULL;                    // We no longer own it.
  delete(iLoader);
  iLoader=NULL;
  }

void CImageLoader::DoCancel()
  {
  iLoader->Cancel();
  delete(iBitmap);
  iBitmap=NULL;
  delete(iLoader);
  iLoader=NULL;
  }
```

As you can see, this is a relatively simple active object. The `CImageDisplayControl` object referenced by the constructor would create and own this object. When an image is required to be loaded, the `StartLoadImageL()` method is called, passing the fully defined file name. At some point in the future, the `RunL()` method is called by the active scheduler. The image has been fully loaded by the time this method is called. In our application, we call back to our owning object, informing it of the success or otherwise of the load request and pass any created bitmap object. The CImageDisplay-Control::LoadComplete(iStatus.Int(),iBitmap) method is called back.

StartLoadImageL() demonstrates that simply leaving if we fail to create objects is usually not sufficient. In this example, if we called iBitmap = new(ELeave) CFbsBitmap() and the leave occurred, we would have a CImageDecoder object created and stored in the iLoader property. It is likely that if the StartLoadImageL() did leave, we would not tidy up or delete the CImageLoader object; instead, we would probably inform the user of a problem and allow them to try again.

Trying again means calling StartLoadImageL() again. This would create a second CImageDecoder object and store the handle in iLoader overwriting any previous object handle, which leads to a memory leak of the original iLoader. An alternative implementation of StartLoadImageL() using the cleanup stack is:

```
void CImageLoader::StartLoadImageL(
  const TFileName& aFileName)
  {
  CImageDecoder*decoder=CImageDecoder::FileNewL(
            CEikonEnv::Static()->FsSession(),aFileName);
  CleanupStack::PushL(decoder);

  // Create a bitmap within which
  // to store the loaded image.
  TFrameInfo frameInfo(decoder->FrameInfo(0));
  CFbsBitmap* bmp=new(ELeave)CFbsBitmap();
  CleanupStack::PushL(bmp);
  if (frameInfo.iFlags&TFrameInfo::ECanDither)
    User::LeaveIfError(
              bmp->Create(
                         frameInfo.iOverallSizeInPixels,
                     CEikonEnv::Static()->ScreenDevice()
                                  ->DisplayMode()));
  else
    User::LeaveIfError(
              bmp->Create(
                         frameInfo.iOverallSizeInPixels,
                          frameInfo.iFrameDisplayMode));

  // Only now we can't leave can we take ownership
  // of the objects in object property.
  iBitmap=bmp;
  iLoader=decoder;
  CleanupStack::Pop(2);

  // Start loading the image into our bitmap,
  // RunL() called when complete.
  iLoader->Convert(&iStatus,*iBitmap,0);
  SetActive();
  }
```

Both implementations perform the same task. In general, the second version is shorter, faster and arguably easier to follow since there are fewer branches to understand.

> **Pro tip**
>
> A common error when creating bitmaps is failing to take any notice of the return value of the `Create()` method. Unlike many objects which leave, if there are errors creating internal resources, the `CFbsBitmap::Create()` method returns an error code. Failure to take notice of the error code usually causes your application to panic if it attempts to use the `CFbsBitmap`, for example, to display it on screen.

11.3.2 Image Encoding

Image encoding is the process of taking a bitmap used internally by an application and converting it to a specified output format. Internal bitmaps are represented by the `CFbsBitmap` object. Output formats include BMP, JPG and GIF. Since the ICL uses a plug-in architecture to perform the encoding tasks, the set of encoders on each model of mobile phone may vary from the set within the SDK.

Image formats typically comprise a primary type along with a set of sub-types. These types and sub-types are properties of the set of plug-ins that support image saving. In our application, we enumerate this information, presenting it to the user to make the choice of saved format (see Figure 11.5). The information is presented within a dialog comprising a text editor to accept the target file name and two choice lists, one for the primary type and one for the sub-types associated with that primary type.

The set of primary file types presented to the user is generated with:

```
void CImageSaverDialog::PreLayoutDynInitL()
  {
  CImageEncoder::GetImageTypesL(iImageTypes);
  CEikChoiceList* cl=static_cast<CEikChoiceList*>(Control(2));
  CDesCArrayFlat* items=new(ELeave)CDesC16ArrayFlat(4);
  CleanupStack::PushL(items);
  TInt count=iImageTypes.Count();
  for (TInt i=0;i<count;i++)
    items->AppendL(iImageTypes[i]->Description());
  cl->SetArrayL(items);
  CleanupStack::Pop(items);
  cl->SetCurrentItem(0);
  }
```

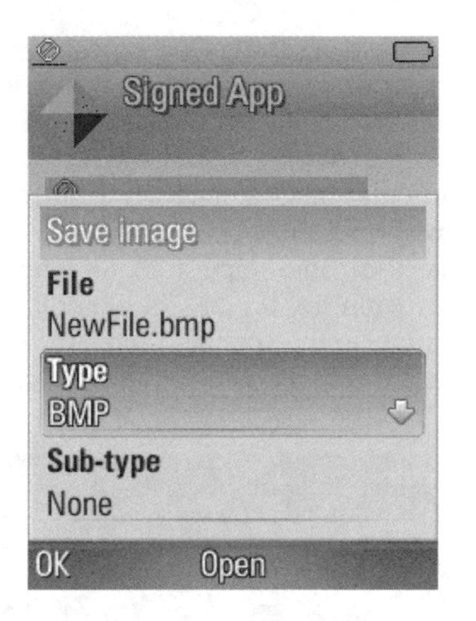

Figure 11.5 Save image dialog in `SignedAppPhase3`

And the set of image sub-types is generated with the following code:

```
void CImageSaverDialog::SetImageSubTypesL()
  {
  CEikChoiceList* cl=static_cast<CEikChoiceList*>(Control(2));
  TUid type=iImageTypes[cl->CurrentItem()]->ImageType();

  // Create the sub-types list based on primary type.
  cl=static_cast<EikChoiceList*>(Control(3));
  TRAPD(err, CImageEncoder::GetImageSubTypesL(type,iImageSubTypes));

  if (err==KErrNotFound)
    { // No sub-types are reported via an error!
    cl->SetArrayL(R_IMAGE_SAVE_NO_SUB_TYPES_ARRAY);
    cl->SetCurrentItem(0);
    }
  else
    {
    User::LeaveIfError(err);

    // Create list of sub-types to display to the user.
    CDesCArrayFlat* items=new(ELeave)CDesCArrayFlat(4);
    CleanupStack::PushL(items);
    TInt count=iImageSubTypes.Count();
    for (TInt i=0;i<count;i++)
      items->AppendL(iImageSubTypes[i]->Description());
    cl->SetArrayL(items);          // Takes ownership.
    CleanupStack::Pop(items);
    cl->SetCurrentItem(0);
    }
  }
```

In the case of sub-types it is possible that no sub-types for a particular primary type are supported. In this case, the `GetImageSubTypesL()` method leaves with the error code `KErrNotFound`. In our application, we choose to display a choice list containing a single entry entitled None. If a particular plug-in supports a range of sub-types, we can enumerate them as shown above.

Once the name, type and sub-type information has been collected, we can begin the process of encoding the image, storing the resultant information in the specified file. As with decoding, the encoding operation is asynchronous and our application wraps this within an active object:

```
CImageSaver::~CImageSaver()
  {
  Cancel();
  }

CImageSaver::CImageSaver(CImageDisplayControl& aDisplay) :
    CActive(CActive::EPriorityStandard),iDisplay(aDisplay)
  {
  CActiveScheduler::Add(this);
  }

void CImageSaver::StartSaveImageL(
  const TFileName& aFileName,
  const TUid& aImageType,
  const TUid& aImageSubType,
  CFbsBitmap* aBitmap)
  {
  iSaver=CImageEncoder::FileNewL(
                          CEikonEnv::Static()->FsSession(),
                      aFileName,CImageEncoder::EOptionNone,
                                aImageType, aImageSubType);
  iSaver->Convert(&iStatus,*aBitmap);
  SetActive();
  }

void CImageSaver::RunL()
  {
  iDisplay.SaveComplete(iStatus.Int());
  delete(iSaver);
  iSaver=NULL;
  }

void CImageSaver::DoCancel()
  {
  iSaver->Cancel();
  delete(iSaver);
  iSaver=NULL;
  }
```

As with image decoding, the active object is not too difficult to understand. The `CImageDisplayControl` object referenced by the constructor would create and own this object. When an image is required to be saved, the `StartSaveImageL()` method is called, passing the fully defined file name, type, subtype and image data. In due course, the `RunL()` method is called by the active scheduler. The image has been saved by the time this method is called. In our application, we call back to its owning object, informing it of the success or failure of the save request. The `CImageDisplayControl::SaveComplete(iStatus.Int())` method is called back.

11.3.3 Image Processing

The multimedia framework provides a number of additional image-processing functions such as rotation, scaling and mirroring. In our example, we demonstrate the usage of the `CBitmapRotator` class to perform bitmap rotation. Mirroring uses the same class but different parameters. Scaling is performed by the `CBitmapScaler` class which follows an identical pattern to the usage of the `CBitmapRotator` class.

Our application supports rotating an image. As with encoding and decoding, the image rotation is performed asynchronously. We wrap this within an active object as shown:

```
CImageRotator::~CImageRotator()
  {
  Cancel();
  }

CImageRotator::CImageRotator(CImageDisplayControl& aDisplay) :
        CActive(CActive::EPriorityStandard),iDisplay(aDisplay)
  {
  CActiveScheduler::Add(this);
  }

void CImageRotator::StartRotateImageL(
  CFbsBitmap* aBitmap)
  {
  iRotator=CBitmapRotator::NewL();

  // Take ownership now we cant leave.
  iBitmap=aBitmap;

  // Start rotating the image into our bitmap.
  // RunL() called when complete.
  iRotator->Rotate(&iStatus,*iBitmap,
  CBitmapRotator::ERotation90DegreesClockwise);
  SetActive();
  }

void CImageRotator::RunL()
  {
  iDisplay.RotateComplete(iStatus.Int(),iBitmap);
  iBitmap=NULL;                  // We no longer own it.
  delete(iRotator);
  iRotator=NULL;
  }
```

```
void CImageRotator::DoCancel()
  {
  iRotator->Cancel();
  delete(iBitmap);
  iBitmap=NULL;
  delete(iRotator);
  iRotator=NULL;
  }
```

In our example, we have chosen to store the rotated image in the same `CFbsBitmap` as it was originally stored. This reduces the amount of memory required to store the image since a second target `CFbsBitmap` is not required but, because the operation is asynchronous, it is entirely possible that a redraw event occurs while in the middle of the rotation. Attempting to draw from a partially rotated bitmap will, at best, produce an apparently corrupt image. In our application, we avoid this by transferring ownership to the `CImageRotator` object and passing it back when the image rotation completes. In this way, we only ever draw from the `CFbsBitmap` when no rotation is in progress.

As previously mentioned, both image rotation and image saving are asynchronous operations. The main reason for this is to ensure your application remains responsive to other events. This includes the possibility of a user choosing either the Rotate or Save menu options before the first request has completed. Attempting to rotate an image while in the middle of saving may result in a panic and is not recommended. A `CBase-42` panic occurs if you attempt to start a rotate (or save) operation while the current rotate (or save) operation is in progress since an active object cannot represent two simultaneous outstanding events.

Robust applications need to handle this situation. One Symbian Signed test, GEN-02, is for the test house to try and break your application. A panic usually results in failure.

In the case of rotating or saving images, the time between starting the operation and it completing is relatively small. The likelihood of a user managing to request a second operation while the first is running is very low. In such cases, rather than presenting a Cancel type dialog it is usually acceptable to simply ignore the second request. Almost no-one in the real world will ever manage to hit the condition and notice that their request was ignored, yet we remove the possibility of our application crashing via a panic.

To support this requirement, rather than run a state machine to track our internal status, we can simply use the active object state. Therefore our `RotateImageL()` code that wraps around the `CImageRotator` class is:

```
void CImageDisplayControl::RotateImageL()
  {
```

```
if (iBitmap && !iImageRotator->IsActive() &&
                  !iImageSaver->IsActive())
  {
  iImageRotator->StartRotateImageL(iBitmap);
  iBitmap=NULL;
  DrawNow();             // Bitmap removed from display,
                         // while being manipulated.
  }
}
```

11.3.4 Summary

In this section, we have discovered how to add some basic image-processing support to our application, apart from the ability to load Symbian native MBM files.

As you have seen, the image-processing functionality is largely asynchronous. Robust applications need to track the state of the image-processing subsystems to ensure the application remains in a stable state irrespective of user actions. This type of application programming is typical of real-time systems.

11.4 Alternative Image Support: `CQikContent`

As an alternative to using the low-level image conversion library to load and display images, UIQ provides a powerful UI-level object, `CQikContent`, designed to simplify programming by hiding a number of the complexities that we encountered in Section 11.3. We have already seen `CQikContent` in action within the `ListView2` example application. We present further examples of its use in the `QContent` example which we discuss in this section.

At a high level, details such as the graphics file format are often a distraction from the task in hand. For example, if we wish to display a graphic icon to represent a particular piece of information, we are only interested in what information we are trying to present and not which file format the graphic icon happens to be stored in.

Similarly, details about how to actually display the image, for example, whether there is a bitmap and icon pair or simply a bitmap and whether to scale or crop the image are implementation details. Objects that want to display images strongly prefer such details to be dealt with by the image class itself leaving them to focus on which images to display.

A `CQikContent` object encapsulates these concepts to aid the higher level programming tasks. The `QContent` application demonstrates numerous ways in which we can construct a `CQikContent` object.

11.4.1 Construction from an MBM File

Probably the most common way to construct a CQikContent object is by referencing a bitmap and mask pair stored within the application MBM file.

```
icon=CQikContent::NewL(NULL, KMbmFile,EMbmQcontentIcon2,
                                    EMbmQcontentIcon2mask);
```

As a shortcut to referencing the application-specific MBM file, the KMbmFile descriptor can be set to the value "*". The underlying system code explicitly checks for this value and works out your full MBM file name from the application name and the file system path from which the application was loaded. The remaining enums, EMbmQcontent-Icon2 and EMbmQcontentIcon2mask are automatically generated in an application-specific MBG file when you compile your image file list defined in the MMP file

11.4.2 Construction from a Resource File

Rather than explicitly define which images are required within your C++ code, it is possible to define the images in your application resource file and use that information to construct a CQikContent. The QContent example demonstrates the usage of both the QIK_CONTENT and QIK_CONTENT_MBM resource structures:

```
RESOURCE QIK_CONTENT r_qik_content_icon
  {
  uri = "c:\\Private\\EDEAD023\\image.gif";
  }

RESOURCE QIK_CONTENT_MBM r_command_icon
  {
  bmpid=EMbmQcontentIcon0;
  bmpmask=EMbmQcontentIcon0mask;
  }
```

These resources are read by the following code fragments:

```
// From a QIK_CONTENT.
{
TResourceReader rr;
HBufC8* q=iEikonEnv>AllocReadResourceAsDes8LC(
                          R_QIK_CONTENT_ICON);
rr.SetBuffer(q);
CQikContent* icon=CQikContent::NewL(NULL,rr);
CleanupStack::PopAndDestroy(q);
}
//  ... other code removed for clarity.
```

```
// From a QIK_CONTENT_MBM.
{
TResourceReader rr;
HBufC8* q=iEikonEnv->AllocReadResourceAsDes8LC(
                                    R_COMMAND_ICON);
rr.SetBuffer(q);
CQikContent* icon=CQikContent::NewL(NULL,rr);
CleanupStack::PopAndDestroy(q);
}
```

The first constructs a `CQikContent` containing an image loaded from a GIF format file named `image.gif` which is stored at the root of the C drive. The second loads an image and bitmap mask from the application-specific MBM file.

11.4.3 Construction from a `CGulIcon`

Your application may already be using icons in the form of `CGulIcon` objects. Rather than fully construct a `CQikContent` using the original source images of a `CGulIcon`, you can clone the `CGulIcon` into a `CQikContent`. Cloning is considerably faster than constructing the objects from scratch as little underlying bitmap manipulation is required.

The following code fragment constructs a `CGulIcon` and clones it into a `CQikContent`.

```
// Create GulIcon for example purposes only.
CGulIcon* gi=iEikonEnv->CreateIconL(KMbmFile,EMbmQcontentIcon3,
                                        EMbmQcontentIcon3mask);
CleanupStack::PushL(gi);

// Create a CQikConent from a CGulIcon.
CQikContent* icon=new(ELeave)CQikContent;
CleanupStack::PushL(icon);
icon->CloneL(gi);
CleanupStack::Pop(icon);

// For this example application purposes only.
CleanupStack::PopAndDestroy(gi);
```

11.4.4 Construction from a `CQikContent`

Your application may need to display the same graphic image in several different places. Rather than loading the graphic image each time a new `CQikContent` object is required, you can load the images once then clone the originals to generate the other `CQikContent` objects.

The following code demonstrates how you might create a `masterIcon` then clone it.

```
CQikContent* masterIcon=CQikContent::NewL(NULL,  KMbmFile,
                 EMbmQcontentIcon2, EMbmQcontentIcon2mask);
CQikContent* icon=masterIcon->CloneL();
```

11.4.5 Construction from an External File

A core feature of the `CQikContent` class is the ability to load images from image files that are not native to Symbian. We have already seen how this can be achieved using a `QIK_CONTENT` resource definition. The following code fragment shows how you might load images directly in C++:

```
icon1=CQikContent::NewL(NULL,_L("c:\\Image.bmp"));
icon2=CQikContent::NewL(NULL,_L("c:\\Image.gif"));
icon3=CQikContent::NewL(NULL,_L("c:\\Image.jpg"));
icon4=CQikContent::NewL(NULL,_L("c:\\Image.png"));
icon5=CQikContent::NewL(NULL,_L("c:\\Image.wbmp"));
```

In this code fragment we load five images, each from a different file format.

As we stated at the beginning of the chapter, the list of image formats supported by a particular phone is dependent on the set of image decoders available on that phone.

If you observe the `QContent` application closely, you should notice that those icons we load from MBM files are displayed immediately whereas those loaded from file formats that are not native to Symbian OS are displayed after a slight delay. This tiny delay is due to the `CQikContent` code having to perform some reasonably computationally intensive operations to load and decode the external image file formats. This same delay is observed in our `SignedAppPhase3` application when we use the `CImageDecoder` classes.

11.5 Camera

Symbian OS provides an abstract onboard camera API called ECam, which is implemented by handset manufacturers according to the capabilities of each particular mobile phone. ECam provides a hardware-agnostic interface for applications to communicate with and control any onboard camera hardware. This API allows applications to:

- control camera settings, for example setting contrast, brightness and zoom settings
- capture images
- capture video clips.

The precise settings available are dependent on the hardware within a phone. The API accommodates variations in hardware through various informational services.

Since cameras capture information about the user and the current user environment, applications that wish to use the camera API are required to have the `UserEnvironment` capability.

11.5.1 Symbian Signed Requirements

Symbian Signed test GEN-01 requires that applications do not affect the use of system features or other applications. In the case of the camera resource, this means that our application should relinquish control of the camera if the user chooses to task away from our application. When returning to our application it would be quite reasonable to take back control of the camera as if we were using it. This ensures that we do not affect any other application that requires usage of the camera, such as the Camera application itself.

Not all phones contain cameras. Attempting to use the camera APIs on such phones will, at best, not work and, at worst, panic our application. Panics cause applications to fail Symbian Signed testing. Our application does not depend on the camera and contains functionality that still makes sense on mobile phones that have no camera. We should therefore support phones that do not contain a camera. This may not be possible if your application must use the camera, in which case you should disable application installation onto phones except those that contain cameras.

In our application, we report that the functionality is not available should the user attempt to use the camera on a mobile phone that does not have one:

```
TInt count=CCamera::CamerasAvailable();
if (count<1)
  User::Leave(KErrNotSupported);
```

11.5.2 Using the Camera

To use the camera to take a photograph, our application needs to:

- create a `CCamera` object
- reserve usage of the camera, ensuring our application has exclusive use of it
- switch the camera on
- optionally configure the camera settings
- optionally display images within a viewfinder

- capture an image
- switch power off
- release the camera, so other applications can use it.

As with the image-processing classes, the camera class uses asynchronous methods to perform the actual processing. Using active objects has the following consequences:

- Our application must have an active scheduler in the thread performing the camera operations. A standard UIQ application primary thread always has a scheduler as it is provided by the application framework.
- Event completion is indicated through call backs. We have to implement the callback interface.
- Our application can process other events between the time the camera object is created and the time a preview image is available to capture. We must ensure our application handles such events appropriately.

Camera Callback Interface

The camera API supports two callback interfaces, `MCameraObserver` and `MCameraObserver2`. Neither of these interfaces has been deprecated. The main difference is in the information delivered as part of the captured image. Our application demonstrates the usage of both these interfaces but the interface chosen depends on a compile time `#define` within the `CameraView.cpp` file. The final application uses the `MCameraObserver2` interface, which we recommend, although you may wish to check with the mobile phone manufacturer.

> **Pro tip**
>
> We recommend using the `MCameraObserver2` interface as this is the more recent API.

Capturing an Image

When the camera view receives `ViewActivatedL()`, it starts the camera. When `ViewDeactivated()` is called, we shut down usage of the camera. This ensures we comply with the Symbian Signed GEN-01 test.

Firstly, we ensure there is a least one camera on the phone, and leave if none exists. This ensures that our application does not panic if users attempt to take photos on phones that do not contain cameras.

```
void CCameraDisplayControl::StartViewFinderL()
  {
  // Check to see if we have a camera on the current phone.
  TInt count=CCamera::CamerasAvailable();
  if (count<1)
    User::Leave(KErrNotSupported);

#ifdef USE_CAMERA_OBSERVER_INTERFACE
  iCamera=CCamera::NewL(*this,0);
#else
  iCamera=CCamera::NewL(*this,0,EPriorityNormal);
#endif

  iCamera->Reserve();
  }
```

Secondly, we create an instance of the `CCamera` object; the `CCameraDisplayControl` class implements the `MCameraObserver` or `MCameraObserver2` interface depending on the compile time `#ifdef`.

Using the `MCameraObserver` Interface

The following section explains the processes and callbacks that occur if an application chooses to use the `MCameraObserver` interface.

Our first requirement for using the camera is to `Reserve()` usage. This is an asynchronous function which calls back the `ReserveComplete()` method.

```
void CCameraDisplayControl::ReserveComplete(TInt aError)
  {
  if (aError==KErrNone)
    {
    iCamera->PowerOn(); // Calls back PowerOnComplete().
    }
  else
    {
    TBuf<128>bb;
    iEikonEnv->Format128(bb,R_STR_CAMERA_RESERVE_ERROR,
                                                    aError);
    iEikonEnv->InfoMsg(bb);
    CloseCamera();
    }
  }
```

If we are able to reserve usage, we then need to power on the camera by calling the `PowerOn()` method. This method is asynchronous and eventually calls back the `PowerOnComplete()` method.

```
void CCameraDisplayControl::PowerOnComplete(TInt aError)
  {
  TCameraInfo info;
  info.iImageFormatsSupported=0;
```

```
if (aError==KErrNone)
  {
  iState=ECameraStatePoweredOn;
  iCamera->CameraInfo(info);
  TRAP(aError,
  // Tell view finder draw directly to our scn space
      TRect rect(Rect());
      rect.Move(PositionRelativeToScreen());
      iCamera->StartViewFinderDirectL(
              iEikonEnv->WsSession(),
           *iEikonEnv->ScreenDevice(),
                     Window(),rect);

     iCamera->PrepareImageCaptureL(CCamera::EFormatJpeg,
                      info.iNumImageSizesSupported-1);
     );
  }
if (aError!=KErrNone)
  {
  TBuf<128>bb;
  iEikonEnv->Format128(bb,R_STR_CAMERA_POWER_ERROR,
              aError,info.iImageFormatsSupported);
  iEikonEnv->InfoMsg(bb);
  CloseCamera();
  }
}
```

Assuming that the power-on request completed without error, we now request the camera to present content directly to the screen to act as a viewfinder. This is achieved using the `StartViewFinder-DirectL()` method. Since this method uses direct screen access, we need to adjust the rectangle our window occupies relative to the screen, such that the output from the camera is positioned correctly for our application.

Finally, we tell the camera to `PrepareImageCaptureL()`. By calling this method now, we minimize any latency between attempting to take a photo and the image actually being captured.

We must provide some sort of *shutter* button for the user to take a photo. In our application, this is achieved by tapping on the screen, depressing the Action key, entering the numeric value *5* or pressing Enter. Do not forget that some UIQ 3 mobile phones do not have touch-sensitive screens and may present a standard phone keypad. Therefore, you should consider presenting a number of ways (commands) to achieve the same task.

Whatever input mechanism is used, the `TakePictureL()` method is called:

```
void CCameraDisplayControl::TakePictureL()
 {
 if (iState==ECameraStateTakingPicture)
   User::Leave(KErrInUse);
```

```
if (iState!=ECameraStatePoweredOn)
  User::Leave(KErrNotSupported);

iCamera->CaptureImage(); // Calls back ImageBufferReady()
iState=ECameraStateTakingPicture;
}
```

Since we are in a real-time system, we need to ensure the application is in the correct state to capture images. We achieve this by recording the states our application progresses through and validating that we are able to take a photo before calling the `CaptureImage()` method. This eventually calls back the `ImageReady()` method.

```
void CCameraDisplayControl::ImageReady(
  CFbsBitmap* aBitmap,
  HBufC8* aData,
  TInt aError)
  {
  if (aError==KErrNone)
    {
    RFile file;
    _LIT(KPhotoJpg,"Photo.jpg");
    aError=file.Replace(iEikonEnv->FsSession(),
                                         KPhotoJpg,
                EFileShareExclusive| EFileWrite);
    if (aError==KErrNone)
      {
      aError=file.Write(*aData);
      file.Close();
      if (aError==KErrNone)
        {
        TRAP(aError, iEngine->AddEntryL(KPhotoJpg,
                               EAppCategoryImages);
            );
        iEikonEnv->InfoMsg(R_STR_PHOTO_SAVED);
        }
      }

    if (aError!=KErrNone)
      iEikonEnv->FsSession().Delete(KPhotoJpg);
    }

  if (aError!=KErrNone)
    {
    TBuf<128>bb;
    iEikonEnv->Format128(bb,R_STR_CAMERA_IMAGE_ERROR,
                                              aError);
    iEikonEnv->InfoMsg(bb);
    }

  CloseCamera();
  Static_cast<CQikAppUi*>(iEikonEnv->AppUi())->
              ActivateViewL(KViewIdListView);
  }
```

In this code extract, we save the data to file, tell the engine there is a new entry, close down the camera and switch back to our list view.

> **Pro tip**
>
> The *5* key on a numeric keypad is often used as a *Fire* or *Enter* button. If you have used Java applications on a mobile phone, pressing the *5* usually results in an action.

Using the `MCameraObserver2` Interface

The following section explains the processes and callbacks that occur using the `MCameraObserver2` interface. The actions to perform are the same as when using the `MCameraObserver` interface but it is how they are performed that is slightly different.

Our first task is to reserve usage of the camera. Calling the `Reserve()` function results in a callback to the `HandleEvent()` method as opposed to the `ReserveComplete()` method:

```
void CCameraDisplayControl::HandleEvent(
  const TECAMEvent& aEvent)
  {
  TBuf<128>bb;
  if (aEvent.iEventType==KUidECamEventReserveComplete)
    {        // Called back after a Reserve() completes.
    if (aEvent.iErrorCode==KErrNone)
      {
      iCamera->PowerOn();
      // Calls back HandleEvent() with
      // KUidECamEventPowerOnComplete.
      }
    else
      {
      iEikonEnv->Format128(bb,R_STR_CAMERA_RESERVE_ERROR,
                                     aEvent.iErrorCode);
      iEikonEnv->InfoMsg(bb);
      CloseCamera();
      }
    }
  else
  // Omitted for clarity, see below ...
  }
```

As before, we simply progress to calling the `PowerOn()` method to switch the camera on. Completion of the `PowerOn()` request calls back the `HandleEvent()` method a second time. To work out which event is generating the callback, we check the `aEvent.iEvent` field of the passed `TECAMEvent` object.

```
void CCameraDisplayControl::HandleEvent(
  const TECAMEvent& aEvent)
  {
  TBuf<128>bb;
  if (aEvent.iEventType==KUidECamEventReserveComplete)
    {              // Omitted for clarity, see above ...
    }
  else if (aEvent.iEventType ==
                              KUidECamEventPowerOnComplete)
    { // Called back after a PowerOn() completes.
    TInt err=aEvent.iErrorCode;
    TCameraInfo info;
    info.iImageFormatsSupported=0;
    if (err==KErrNone)
      {
      iState=ECameraStatePoweredOn;
      iCamera->CameraInfo(info);
      TRAP(err,
         TRect rect(Rect());
         rect.Move(PositionRelativeToScreen());
         iCamera->StartViewFinderDirectL(
                  iEikonEnv->WsSession(),
               *iEikonEnv->ScreenDevice(),
                           Window(),rect);
         iCamera->PrepareImageCaptureL(
                                      Camera::EFormatJpeg,
                         info.iNumImageSizesSupported-1);
         );
      }
    if (err!=KErrNone)
      {
      iEikonEnv>Format128(bb,R_STR_CAMERA_POWER_ERROR,
                   err,info.iImageFormatsSupported);
      iEikonEnv->InfoMsg(bb);
      CloseCamera();
      }
    }
  }
```

As before, we start the viewfinder and tell the camera to prepare for image capture. When the user selects one of the input mechanisms supported to take a photo, the `TakePictureL()` method is called. Since we are implementing the `MCameraObserver2` interface, the `ImageBufferReady()` method is called back when the photo has been taken.

```
void CCameraDisplayControl::ImageBufferReady(
  MCameraBuffer& aCameraBuffer,
  TInt aError)
  {
  if (aError==KErrNone)
    {
    RFile file;
    _LIT(KPhotoJpg2,"Photo.jpg");
    aError=file.Replace(iEikonEnv->FsSession(),KPhotoJpg2,
                           EFileShareExclusive|EFileWrite);
```

```
    if (aError==KErrNone)
      {
      aError=file.Write(*aCameraBuffer.DataL(
             aCameraBuffer.iIndexOfFirstFrameInBuffer));
      file.Close();
      if (aError==KErrNone)
        {
        TRAP(aError,
          iEngine->AddEntryL(KPhotoJpg2,EAppCategoryImages);
          );
        iEikonEnv->InfoMsg(R_STR_PHOTO_SAVED);
        }
      }
    if (aError!=KErrNone)
      iEikonEnv->FsSession().Delete(KPhotoJpg2);
    }

  if (aError!=KErrNone)
    {
    TBuf<128>bb;
    iEikonEnv->Format128(bb,R_STR_CAMERA_IMAGE_ERROR,aError);
    iEikonEnv->InfoMsg(bb);
    }

  aCameraBuffer.Release();
  CloseCamera();
  Static_cast<CQikAppUi*>(iEikonEnv->AppUi())->
              ActivateViewL(KViewIdListView);
  }
```

In this code fragment, we save the data to file, tell the engine there is a new entry, close down the camera and switch back to our list view.

You should note that in both the `ImageRead()` and `ImageBuffer-Ready()` method code fragments included above, we have removed the code that generates the fully formed drive, path and file name for the sake of clarity. Our application always creates a file called `Photo.jpg` in our private folder.

Image Formats and Sizes

In our application, we have chosen not to enumerate the supported image types and sizes or allow a user to choose between them. The Sony Ericsson P990i supports capturing JPG images, so we have chosen to hard code this option. We use image size 640 × 480 pixels.

To enumerate the supported image formats and sizes associated with them you would perform an action similar to:

```
TInt i;
TInt mask=0x01;
for (i=0;i<17;i++,mask<<=1)
  {
  if (info.iImageFormatsSupported&mask)
```

```
    { // This format is supported.
    CCamera::TFormat format=(CCamera::TFormat)mask;
    TSize size;
    for (TInt j=0;j<info.iNumImageSizesSupported;i++)
      {
      iCamera->EnumerateCaptureSizes(size,j,format);
      if (size.iWidth>0 && size.iHeight>0)
        { // This format can generate this size image...
        }
      }
    }
  }
```

It should be noted that the `iNumImageSizesSupported` field actually means the maximum number of supported image sizes irrespective of format, therefore we need to check that the returned size is not zero, should we want to use the size information.

11.5.3 `CQikCameraCaptureDlg` Class

As an alternative to the callback interface, you may find the `CQikCameraCaptureDlg` is adequate for your application. To use it, simply call:

```
CFbsBitmap* image=NULL;
CQikCameraCaptureDlg::RunDlgLD(image);
```

This presents a dialog enabling a user to take a photo (see Figure 11.6). For the application to process as required, a copy of the image is returned as a `CFbsBitmap`.

Figure 11.6 Camera dialog

Since cameras capture information about the user and the current user environment, applications that wish to use the `CQikCameraCaptureDlg` are required to have the `UserEnvironment` capability. If you choose to use this class you should also ensure you have an appropriate privacy statement dialog displayed when your application is run for the first time.

11.5.4 Summary

We have seen how to add some support for taking photographs within our application. Good application behavior requires that we relinquish control of the camera should the user task to another application. As we have seen, this task is not especially difficult, particularly if the application is structured appropriately.

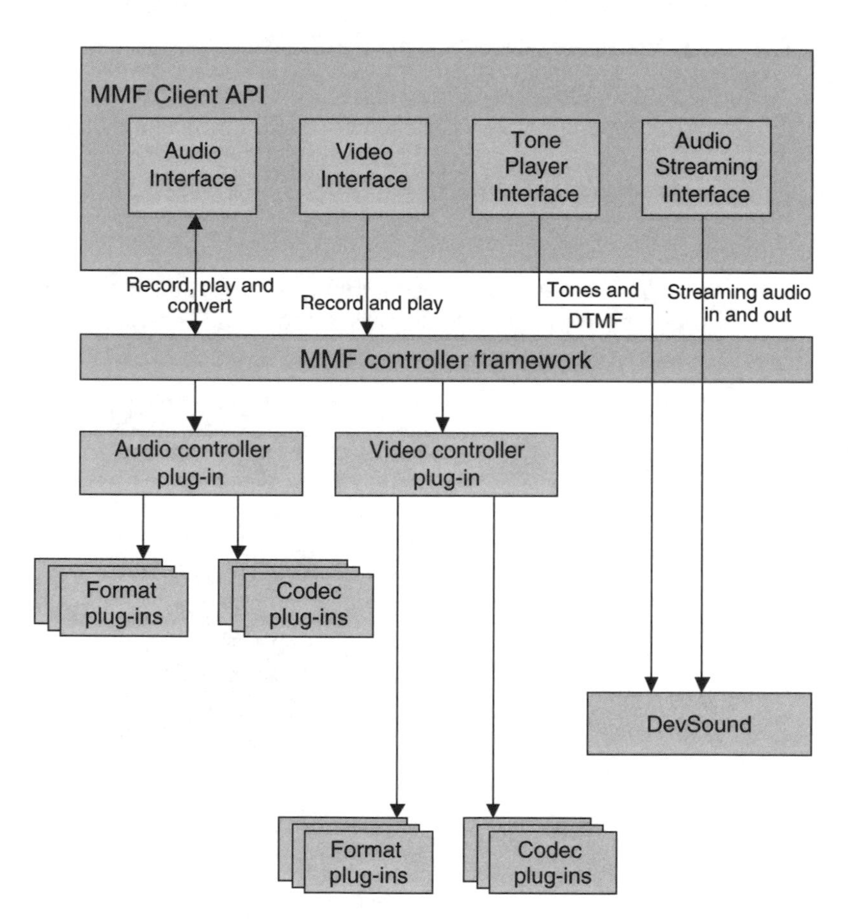

Figure 11.7 Multimedia framework

11.6 Multimedia Framework (MMF)

A large number of audio and video file formats have been developed, many of which are widely used on mobile phones today. The multimedia framework is designed to accommodate these different formats by use of a plug-in architecture (see Figure 11.7). Each plug-in is responsible for supporting a specific data type and presenting a common interface to the framework. Writing plug-ins is a specialist task and is beyond the scope of this book.

The audio file formats supported as standard by the multimedia framework are AU, WAV and raw audio. Each of these file formats has variants such as mono/stereo, sample rate and bits-per-sample. Further details are available in the SDK documentation. MMF supports video: however, the phone manufacturer must add the desired plug-ins.

For example, the following audio types are supported on both the Sony Ericsson P1i and MOTO Z8: AAC (various types), AMR-NB, iMelody, MP3 and MIDI. We recommend that you check and test that the mobile phones you are targeting work correctly with the multimedia formats that your application uses.

The MMF makes it possible for your application to work with different format files. Apart from setting any necessary parameters, you need know nothing about the encoding and decoding of the formats.

The tone player and audio streaming APIs are part of the MMF but plug-ins are not used.

If your application contains audio clips, you should check that the audio format you are using is available on the mobile phones you wish to support.

The audio and video playback classes use asynchronous methods to perform the actual processing; in particular, they use active objects. This has several consequences:

- Our application must have an active scheduler in the thread performing the playback. A standard UIQ application primary thread always has a scheduler as it is provided by the application framework.
- Event completion is indicated through callbacks so we have to implement a callback interface. In general, we also have to run a state machine to manage the playback states.
- Our application can process other events between the time a load, record or playback request is started and the time it completes.

11.7 Audio

In this section, we extend our example application code to play and record audio clips.

11.7.1 Playing Audio Clips

Symbian OS supports a number of mechanisms to play back audio. The easiest of these to use is encapsulated by the `CMdaAudioPlayerUtility` object.

Playing back sound involves two tasks:

- loading the sound data to play
- supplying that data to the underlying hardware to generate the sound.

Loading Audio Data

Loading audio content from file is usually more difficult than simply opening a file and reading its content. In general, audio files contain information indicating how the audio data has been stored and compressed. To play back a sound successfully, this information needs to be interpreted and acted upon. The multimedia framework automatically interprets audio file information by determining which of the plug-ins is prepared to perform this task on its behalf. Once the correct plug-in has been identified, the audio content can be prepared for playback.

While this approach removes the need for individual applications to understand audio file formats, the disadvantage is that loading audio files can take a long time. The multimedia framework takes this into account within the API presented by the `CMdaAudioPlayerUtility` object. However, this makes the object harder to use since some application-level co-ordination is required, particularly between loading and playing a particular audio clip.

Playing Back Audio Data

Once an audio clip has been loaded, it can be played back. Short audio clips might be used to confirm user actions or to alert the user to events. Audio playback may last a long time, however. In the case of a music player, where a user may use the mobile phone to perform other tasks, such as sending messages, at the same time, then the user will task away to another application. Applications must therefore be able to play audio clips in the background. In common with loading audio clips, the multimedia framework takes this into account within the API presented by the `CMdaAudioPlayerUtility` object.

Usage of the `CMdaAudioPlayerUtility` object requires that we implement the `MMdaAudioPlayerCallback` interface. The interface comprises two methods:

```
void MapcInitComplete(TInt aError,
            const TTimeIntervalMicroSeconds& aDuration);
void MapcPlayComplete(TInt aError);
```

To simply load and play back an audio clip, the following code is sufficient:

```
void CListView::TidyUpSoundPlayer()
  // Tidy up sound player resources.
  {
  if (iSoundPlayer)
    {
    iSoundPlayer->Close();
    delete(iSoundPlayer);
    iSoundPlayer=NULL;
    }
  }

void CListView::MapcInitComplete(
  TInt aError,
  const TTimeIntervalMicroSeconds& aDuration)
  {
  if (aError)
    TidyUpSoundPlayer();
  else
    iSoundPlayer->Play();
  }

void CListView::MapcPlayComplete(TInt aError)
  // The sound we were playing has finished,
  // successfully or otherwise.
  {
  TidyUpSoundPlayer();
  }

void CListView::PlayAudioEntryL()
  // The entry has been classified as an audio entry.
  {
  TFileName name;
  iEngine->EntryFullName(name);
  iSoundPlayer=CMdaAudioPlayerUtility::NewFilePlayerL(
                                          name,*this);
  }
```

This code will successfully load and play back audio clips but it does not handle all of the events that could take place. In common with other real-time components, one of the reasons for audio playback to occur in the background, or asynchronously, is to allow other events to occur. The user may choose to select a second audio clip to play before the first has completed. In this case, the `PlayAudioEntryL()` method is called before the first audio clip has finished playing. Apart from overwriting the `iSoundPlayer` property with a new object handle, numerous undesirable downstream consequences are possible when the audio clip playback completes. At best, the application will survive until application exit when a memory leak will be detected. At worst, your application will panic.

A simple solution to this problem, and one that is adequate for our application, is to change the `PlayAudioEntryL()` method to cancel any currently playing audio clip before attempting to load and play another audio clip. We also need to adjust the `TidyUpSoundPlayer()` method to take into account an additional real-time state: the audio clip is being played back at the time the method is called:

```
void CListView::TidyUpSoundPlayer()
  // Tidy up sound player resources.
  {
  if (iSoundPlayer)
    {
    // Cancel any playback, harmless if not playing.
    // Will panic if we have not completed the initialization yet.
    if (iSoundInitialized)
      iSoundPlayer->Stop();
    iSoundPlayer->Close();
    delete(iSoundPlayer);
    iSoundPlayer=NULL;
    }
  }

void CListView::MapcInitComplete(
  TInt aError,
  const TTimeIntervalMicroSeconds& aDuration)
  // The sound has been loaded.
  {
  if (aError)
    {
    TidyUpSoundPlayer();
    return;
    }
  iSoundInitialized=ETrue;
  iSoundPlayer->Play();// Calls back MapcPlayComplete().
  }

void CListView::MapcPlayComplete(TInt aError)
  // The sound we were playing has finished, successfully or otherwise.
  {
  TidyUpSoundPlayer();
  }

void CListView::PlayAudioEntryL()
  // The entry has been classified as an audio entry.
  // Play it back.
  {
  // If any sound is currently playing we need to ensure
  // we stop it in this application since we re-use the
  // iSoundPlayer property for a different sound.
  TidyUpSoundPlayer();

  // Create a sound player to play back the
  // currently selected file.
  TFileName name;
  iEngine->EntryFullName(name);
  iLevel=ESoundLevelMedium;
  iSoundInitialized=EFalse;
```

```
  iSoundPlayer=CMdaAudioPlayerUtility::NewFilePlayerL(
                                          name,*this);
  }
```

Another event that can occur is that the application is asked to exit while in the middle of playing back an audio clip. To accommodate such an event, we add a `CListView` destructor:

```
CListView::~CListView()
  {
  TidyUpSoundPlayer();
  }
```

Synchronizing Audio Clips with Events

A common requirement is for audio clips to be synchronized with an event occurring within the system. If a significant amount of time is required to load a particular audio clip before it can be played back, the event may have passed. In some cases, playing a sound a half or one second after an event has occurred may be acceptable, in others it may not. Additionally, if a particular audio clip is played relatively frequently the amount of processing required to continually load the audio clip results in slower overall system performance and reduced battery life.

To reduce the delay between requesting a sound to be played and the sound being heard, audio clips can be pre-loaded, using one `CMdaAudioPlayerUtility` object per audio clip. The primary disadvantage to this is approach is that a significant amount of RAM is required to hold the audio clip data. UIQ 3 provides a specific class, `CMMFDevSound`, for low-latency audio playback, although it also uses RAM to hold the clip.

Playing Back Tones

Audio tones, such as sine waves or DTMF telephony signals can be played back using the `CMdaAudioToneUtility` class. This class follows a very similar pattern to the `CMdaAudioPlayerUtility` class. The object is instantiated and data is prepared asynchronously. Once the data has been prepared, the `MatoPrepareComplete()` method is called back. This is equivalent to the `MapcInitComplete()` callback. When asynchronous playback completes, the `MatoPlayComplete()` method is called. This is equivalent to the `MapcPlayComplete()` callback.

Further details of this class are available in the UIQ 3 SDK documentation.

Streaming Audio

Audio streaming enables applications to play audio without the entire audio clip necessarily being present. The audio data is held in buffers and playback takes place from them. During playback, incoming data is added to the buffers. To provide uninterrupted sound output, it is important the buffers are, on average, filled faster than the data rate of the sound playback. Large buffers enable playback to continue for longer if the data is interrupted. The `CMdaAudioOutputStream` class is used to stream audio content. As with the `CMdaAudioToneUtility`, this class follows a very similar pattern to the `CMdaAudioPlayerUtility` class.

Further details of this class are available in the UIQ 3 SDK documentation.

11.7.2 Audio Recording

Recording of audio clips is achieved using the `CMdaAudioRecorderUtility` object. In theory, you can use this class to record audio in any of the audio formats for which a controller plug-in exists, however, we were restricted to recording to WAV on the emulator and AMR on a phone. You should carefully check the audio recording support that is available on the phones that you are targeting.

Our application presents a list of the available plug-ins along with their configuration parameters (see Figure 11.8).

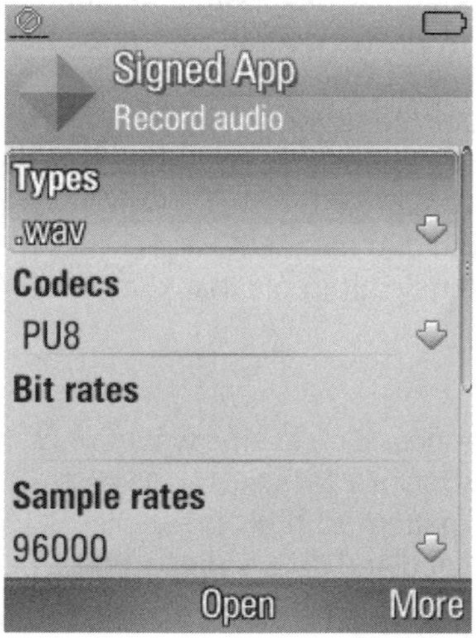

Figure 11.8 `SignedAppPhase3` record sound clip function

The list of plug-ins is presented as file name extensions and is generated by the following code:

```
CDesCArrayFlat* array=new(ELeave)CDesCArrayFlat(4);
CleanupStack::PushL(array);

RMMFControllerImplInfoArray controllers;
```

```
CleanupResetAndDestroyPushL(controllers);

RArray<TUid> mediaIds;
mediaIds.Append(KUidMediaTypeAudio);
CleanupClosePushL(mediaIds);

CMMFControllerPluginSelectionParameters* select =
       CMMFControllerPluginSelectionParameters::NewLC();
select->SetMediaIdsL(mediaIds,
        CMMFPluginSelectionParameters::EAllowOnlySuppliedMediaIds);

CMMFFormatSelectionParameters* selectRead =
                 CMMFFormatSelectionParameters::NewLC();
select->SetRequiredPlayFormatSupportL(*selectRead);

CMMFFormatSelectionParameters* selectWrite =
                 CMMFFormatSelectionParameters::NewLC();
select->SetRequiredRecordFormatSupportL(*selectWrite);

select->ListImplementationsL(controllers);

TBuf<KMaxFileName> bb;
TInt count=controllers.Count();
for (TInt i=0;i<count;i++)
  {
  const RMMFFormatImplInfoArray& recordFormats =
                      controllers[i]->RecordFormats();
  TInt rCount=recordFormats.Count();
  for (TInt j=0;j<rCount;j++)
    {
    const CDesC8Array& extensions = recordFormats[j]
                         ->SupportedFileExtensions();
    TInt eCount=extensions.Count();
    for (TInt k=0;k<eCount;k++)
      {
      bb.Copy(extensions[k]);
      array->AppendL(bb);
      }
    }
  }
CleanupStack::PopAndDestroy(5);
```

The code fragment asks the system to generate a list of all the controllers. For all controllers, we request a list of recording formats. From each of the formats, we request the file name extension most commonly associated with the format. In our code, we need to convert the extension from ASCII to Unicode so we can present it to the user. The code that performs the conversion is as follows:

```
bb.Copy(extensions[k]);
```

It assumes all characters in the filename extension have a one-to-one mapping between ASCII and Unicode. This is always true for the filename extensions we are manipulating.

Prior to being able to add any data to the four remaining choice lists, we have to create a `CMdaAudioRecorderUtility` object and open a file with the required file name extension:

```
iRecorder=CMdaAudioRecorderUtility::NewL(*this);
iRecorder->OpenFileL(iFileName);
```

The 'open file' request is asynchronous. When complete, it calls back the `MoscoStateChangeEvent()` method which is defined in the `MMdaObjectStateChangeObserver` interface definition.

Once we receive the callback informing us the file has been successfully created, we can request the data type, bit rate, sample rate and channel information required to seed the remainder of our choice lists. It should be noted that if we request some information but none is available, the data supply methods leave with a `KErrNotSupported` error.

```
void CAudioRecordView::CreateFormatChoicesL(void)

  // Create the data sets from which the choice
  // list content is generated.
  {
  TInt err;
  TRAP(err,iRecorder->GetSupportedDestinationDataTypesL(iDataTypes));
  if (err!=KErrNotSupported)
    User::LeaveIfError(err);
  TRAP(err,iRecorder->GetSupportedBitRatesL(iBitRates));

  if (err!=KErrNotSupported)
    User::LeaveIfError(err);
  TRAP(err,iRecorder->GetSupportedSampleRatesL(iSampleRates));
  if (err!=KErrNotSupported)
    User::LeaveIfError(err);
  TRAP(err,iRecorder->GetSupportedNumberOfChannelsL(iChannels));
  if (err!=KErrNotSupported)
    User::LeaveIfError(err);
  }
```

Since we are presenting the information within a UI we are able to handle blank choice lists. Therefore we need to `TRAP()` any leave and continue collecting information as shown in the code fragment above.

Once we have collected the information we can inform the choice lists to display the first entry:

```
  CEikChoiceList* cl;
  cl=LocateControlByUniqueHandle<CEikChoiceList>(EAppChoiceList2);
  if (iDataTypes.Count()>0)
    cl->SetCurrentItem(0);
  cl->DrawNow();
```

```
cl=LocateControlByUniqueHandle<CEikChoiceList>(EAppChoiceList3);
if (iBitRates.Count()>0)
  cl->SetCurrentItem(0);
cl->DrawNow();

cl=LocateControlByUniqueHandle<CEikChoiceList>(EAppChoiceList4);
if (iSampleRates.Count()>0)
  cl->SetCurrentItem(0);
cl->DrawNow();

cl=LocateControlByUniqueHandle<CEikChoiceList>(EAppChoiceList5);
if (iChannels.Count()>0)
  cl->SetCurrentItem(0);
cl->DrawNow();
```

Our application allows the user to choose between any of the parameters for which the underlying plug-ins report is available.

If the user chooses a different file extension we have to close down the current file and `CMdaAudioRecorderUtility` object, create a new one and replace the recording parameters with those associated with this new format. The following code detects the change in the Types list:

```
void CAudioRecordView::HandleControlEventL(
 CCoeControl* aControl,
  TCoeEvent aEventType)
  {
  if (aEventType==EEventStateChanged)
    {
    CEikChoiceList* cl=LocateControlByUniqueHandle<
                  CEikChoiceList>(EAppChoiceList1);
    if (cl==aControl)
      {           // Audio types changed content...
      OpenRecordFileL();
      }
    }
  }
```

When the user is ready to record, we need to set up the recording parameters and start recording:

```
void CAudioRecordView::StartRecordingL()

  // Set up the recorder with the user's chosen parameters
  // and begin recording.
  {
#ifdef __WINS__
  iRecorder->SetDestinationDataTypeL(KMMFFourCCCodePCMU8);

#else
  iRecorder->SetDestinationDataTypeL(KMMFFourCCCodeAMR);
  iRecorder->SetDestinationBitRateL(12200);
```

```
  iRecorder->SetDestinationSampleRateL(8000);
  iRecorder->SetDestinationNumberOfChannelsL(
                 TMdaAudioDataSettings::EChannelsMono);
  iRecorder->SetGain(iRecorder->MaxGain());

#endif

  iRecorder->SetPosition(TTimeIntervalMicroSeconds(0));
  TRAPD(err,iRecorder->RecordL());
  }
```

Recording is stopped by calling:

```
iRecorder->Stop();
```

11.7.3 Summary

We have seen how to add audio playback and recording to applications. Your application must track the current state of the audio subsystems so you do not attempt to perform an action while a previously requested action is still running.

11.8 Video

Symbian OS supports some basic video services, including the ability to record and play back video clips. As with the audio and image components of the multimedia framework, support for video is based on plug-ins. Unlike the audio and image components Symbian OS does not provide any standard plug-ins; this task is delegated to the mobile phone manufacturer. In practice, most mobile phones support the 3GPP file format standard since this has been specially developed for UMTS mobile phones. The Sony Ericsson P1i and MOTO Z8 phones both support 3GPP and MPEG-4.

11.8.1 Loading Video Clips

Prior to playing back a video clip, we have to load it and prepare it for playback. Since loading a video clip can take a long time, this task is performed asynchronously by the multimedia framework code.

Video clips can be considered as a large number of bitmaps that are drawn on the screen in rapid succession. To present smooth playback, these images need to be displayed at a constant rate and remain synchronized with any audio component. Rather than requiring each application to perform such tasks, applications should use the `CVideo-PlayerUtility` class.

Since the `CVideoPlayerUtility` class performs playback for us, it needs to know where on the screen the playback will occur. In our application, we want playback to occur within the rectangle belonging to the window where we are displaying any other output. The code to create a `CVideoPlayerUtility` and start loading a video clip is:

```
void CVideoDisplayControl::NewVideoSelectedL()
  // Open the entry currently selected in the list view.
  {
  TRect rect(Rect());
  rect.Move(PositionRelativeToScreen());

  iVideoPlayer=CVideoPlayerUtility::NewL(*this,
                              EMdaPriorityNormal,
                     EMdaPriorityPreferenceNone,
                         iEikonEnv->WsSession(),
                     *iEikonEnv->ScreenDevice(),
                              Window(),rect,rect);

  TFileName name;
  iEngine->EntryFullName(name);
  TRAPD(err,
         iVideoPlayer->OpenFileL(name);
    iEikonEnv->BusyMsgL(R_STR_LOADING);
       );
  if (err!=KErrNone)
    {
    CloseVideo();
    User::Leave(err);
    }
  }
```

Once the `iVideoPlayer->OpenFileL(name)` request completes, the `MvpuoOpenComplete()` method is called back.

```
void CVideoDisplayControl::MvpuoOpenComplete(
  TInt aError)
  {
  if (aError==KErrNone)
    iVideoPlayer->Prepare();
  else
    {
    iEikonEnv->BusyMsgCancel();
    TBuf<128>bb;
    iEikonEnv->Format128(bb,R_STR_OPEN_ERROR,aError);
    iEikonEnv->InfoMsg(bb);
    }
  }
```

Prior to playing back the video clip or querying any of its properties, we have to call the `CVideoPlayerUtility:: Prepare()` method. This in turn calls back the `MvpuoPrepareComplete()` method.

```
void CVideoDisplayControl::MvpuoPrepareComplete(
  TInt aError)
  {
  iEikonEnv->BusyMsgCancel();
  if (aError==KErrNone)
    {
    TRAP(aError,
           iVideoPlayer->SetPositionL(0);
      iVideoPlayer->GetFrameL(EColor64K);
            iState=EVidDispStateGetFrame;
        );
    }
  if (aError!=KErrNone)
    {
    TBuf<128>bb;
    iEikonEnv->Format128(bb,R_STR_PREPARE_ERROR,aError);
    iEikonEnv->InfoMsg(bb);
    }
  }
```

In our application, we have chosen to load the first frame of the video clip to display on screen while the video is not being played back. We achieve this with the `SetPositionL()` and `GetFrameL()` methods. The `GetFrameL()` method is asynchronous, eventually calling back the `MvpuoFrameReady()` method.

```
void CVideoDisplayControl::MvpuoFrameReady(
  CFbsBitmap& aFrame,
  TInt aError)
  {
  iState=EVidDispStateIdle;
  if (aError==KErrNone)
    {
    iBitmap=(&aFrame);
    DrawNow();
    }
  if (aError!=KErrNone)
    {
    TBuf<128>bb;
    bb.Format(_L("FrameErr %d"),aError);
    iEikonEnv->InfoMsg(bb);
    }
  }
```

As you can see, this method presents a `CFbsBitmap` handle that we can use to display the first frame as our primary content.

You may have noticed that we track the various `CVideoPlayerUtility` states through the `iState` property. In our application, we have defined the following states:

- `EVidDispStateClosed`: the `CVideoPlayerUtility` class is closed and we are not performing any video-related tasks.

- `EVidDispStateGetFrame`: the `CVideoPlayerUtility` has successfully been opened and a call to obtain the first frame to display as our primary content has been made.
- `EVidDispStateIdle`: our application is currently idle, displaying the first frame. We must be in this state to begin playback of a video clip.
- `EVidDispStatePlaying`: our application is currently playing back the video clip.

We use the `iState` information in two places: firstly to know when to draw the bitmap frame and, secondly, to know when we are able to actually play the video clip.

11.8.2 Playing Video Clips

Once successfully loaded, we can play the video clip with the following code:

```
void CVideoDisplayControl::PlayL()
  // Start the video clip playback.
  {
  if (iState!=EVidDispStateIdle)
    User::Leave(KErrNotSupported);

  iVideoPlayer->SetPositionL(0);     // From the start.
  iVideoPlayer->Play();

  iState=EVidDispStatePlaying;
  DrawNow();
  }
```

When playback completes, the `MvpuoPlayComplete()` method is called:

```
void CVideoDisplayControl::MvpuoPlayComplete(
   TInt aError)
  // Called back by the framework when
  // iVideoPlayer->Play() completes.
  {

  // We want to be able to replay
  // so don't close the image down here.
  iState=EVidDispStateIdle;
  DrawNow();
  }
```

In our application, we remove the single static frame displayed while the video clip is not playing, and replace it with a red square so that you can more easily see when playback commences. This helps demonstrate how you might construct a state machine to handle various states that occur in systems such as video playback.

As with the asynchronous functions in audio and image processing, we need to add some functionality to handle events such as a second playback request while the first is running. When we handled this condition for image rotation or saving, we chose to simply ignore the request. Since video clips are often minutes or even hours long, it is very easy for a user to choose the Replay menu option while playback is already in progress. As a minimum, we should present some kind of feedback to the user. In more consumer-friendly applications, we might choose to interpret the Replay event to mean *'cancel playing the current clip and restart from the beginning'*. We have chosen to simply present some feedback to the user. An alternative option could be to disable the Replay menu option while the video clip is playing and restore it at a later date. Whichever option you choose needs to fit with your application requirements. Not handling the condition is likely to cause problems.

11.8.3 Recording Video Clips

The `CVideoRecorderUtility` class facilitates video recording. As with video playback, the recording class requires you to:

- create a `CVideoRecorderUtility` object providing it with a class implementing the `MVideoRecorderUtilityObserver` interface
- open a file into which the video content is recorded
- prepare the recorder for recording
- record the content.

We have not implemented this in our application, but the required process follows an almost identical pattern to video playback. Selection of the format in which to save the video is very similar to the selection of a format in which to save images. In particular, the `CVideoRecorderUtility::GetSupportedVideoTypesL()` method scans the set of available plug-ins to generate the set of supported video types.

Pro tip

Since recording video clips captures information about the user and the current user environment, applications that wish to record video content are required to have the `UserEnvironment` capability. Recording video necessarily requires that you have a camera to capture the image stream. Applications should not panic if they are run on phones that do not have a camera, even if the user inadvertently selects such a feature.

11.9 Tuner API

The Tuner API allows your application to control a radio tuner and play and record audio from it. The Tuner API is available in current Sony Ericsson phones, such as the P1i, that have an FM tuner. It allows you to:

- find out about the tuner
- take control, for example, by selecting a frequency
- play the audio
- record the audio
- use any Radio Data System (RDS) data.

11.9.1 Tuning

The `CMMTunerUtility` provides the basis for the Tuner API. This class allows you to perform tuning and acts as a factory for the other classes that form the Tuner API. To use it, you must derive from `MMMTunerObserver`.

You should first check that a tuner is present. The `CMMTuner-Utility::TunersAvailable()` method indicates how many tuners are present on the phone. If it returns a value greater than 0, you can call `MMTunerUtility::NewL()`.

The `CMMTunerUtility::GetCapabilities()` method can be used to retrieve the capabilities of the tuner.

Some phones require an external antenna (usually the handsfree cable) to be attached before the tuner can be used. Also, you should never attempt to use the tuner if the phone is in 'flight mode' (the PDA and media player functions operate but radio transmitters and receivers are disabled).

You can determine the requirements of the tuner using the `Addi-tionalFunctions` member of class `TTunerCapabilities`; this is a bit field of values from `TTunerCapabilities::TTunerFunctions`. You should check the bits `ETunerFunctionRequiresAntenna` and `ETunerFunctionAvailableInFlightMode`.

```
TTunerCapabilities caps;
User::LeaveIfError(CMMTunerUtility::GetCapabilities(0, caps));
if (caps.iAdditionalFunctions & TTunerCapabilities::
                ETunerFunctionAvailableInFlightMode)
  {
  // We can use the tuner regardless of the flight mode setting.
  ...
  }
else
  {
  TBool flightMode;
```

```
User::LeaveIfError(iTuner->GetFlightMode(flightMode));
if (flightMode)
  {
  // We cannot use the tuner until flight mode is disabled.
  ...
  }
else
  {
  // Flight mode is disabled; we can use the tuner.
  ...
  }
}
```

You can check the state of the external antenna using the `IsAntennaAttached()` method. If you try to use a tuner when either the flight mode or the antenna state does not allow, it results in an error of `KErrNotReady`.

Unlike the ECAM API discussed in Section 11.5, the tuner does not have explicit control over the power state of the tuner hardware; there is no `PowerOn()` method. This is because the Tuner API allows multiple clients to use the same tuner hardware concurrently. The hardware remains powered on so long as at least one client requires it.

If the flight mode and antenna states allow the tuner to be used, you can start using it by issuing a `Tune()` request:

```
iTuner->Tune(TFrequency(98500000),CMMTunerUtility::ETunerBandFm);
```

The class `TFrequency` is used to represent the required radio frequency. The unit of the frequency is Hertz; so the example above corresponds to 98.5 MHz in the FM band.

You should check the tuner capabilities to determine which bands are supported, and whether you should use frequencies or channels; you can check the valid range of frequencies or channels for a given band using the `GetFrequencyBandRange()` and `GetChannelRange()` methods.

After making any tune or search call, you receive a callback indicating success or otherwise to the `MToTuneComplete()` method. You must not make any further tuner calls until you have received this callback.

You can search for frequencies or channels that carry stations using an appropriate overload of the `StationSeek()` method.

Tuner Notifications

There are a number of events for which you can request notification when using the tuner. Each notification category has an associated callback class and request method in `CMMTunerUtility`.

Tuner-state change notifications can be requested using the Notify-Change() method; you must derive from class MMMTunerChange-Observer to receive these notifications. This provides the following notifications:

- changes of tuner state – that is, if the tuner hardware is powered up, if it is playing and if someone is recording from it
- changes to the currently tuned frequency or channel
- antenna attachment and detachment notifications
- flight mode change notifications.

If you derive from MMMSignalStrengthObserver, you can call its NotifySignalStrength() method to request notifications when the signal strength changes. Similarly, you can request notifications when the audio reception changes between stereo and mono by deriving from MMTunerStereoObserver and calling its NotifyStereoChange() method.

Control of the Tuner

When you instantiate the tuner utility you can specify a tuner access priority. This is used to arbitrate between multiple applications attempting to use the tuner at the same time. Only one application is granted control of the tuner at any given time; only that application can retune it or otherwise change its state.

When you call any method in the tuner utility that requires control of the tuner, a request for control is made for you. When control of the tuner is granted, you receive a callback to MToTunerEvent():

```
void RadioTuneEngine::MToTunerEvent(
                    MMMTunerObserver::TEventType aType,
                TInt aError, TAny* /*aAdditionalInfo*/)
  {
  if (aType == MMMTunerObserver::EControlEvent)
    {
    if (aError == KErrNone)
      {
      // We have been granted control of the tuner.
      iHaveControl = ETrue;
      }
    else
      {
      // Some other tuner client has pre-empted us.
      iHaveControl = EFalse;
      }
    }
  else if (aError != KErrNone)
    {
```

```
    // Handle the error.
    }
  }
```

An event type of `MMMTunerObserver::EControlEvent` indicates that you have been granted or denied control of the tuner.

Once granted control of the tuner, you keep it until either you release it explicitly, by calling `ReleaseTunerControl()`, or you are pre-empted by another application that has a higher priority for access to the tuner.

Note that having control of the tuner does not prevent other applications playing or recording the tuner audio. Permission to do this is handled by the system-wide audio policy, in exactly the same way as for any other audio playback or recording.

Closing the Tuner

When you have finished using the tuner utility, you can call the `Close()` method to release any tuner resources that have been allocated to you. Any ongoing tuner audio playback or recording is terminated.

Before deleting your `CMMTunerUtility` instance, you must ensure that any tuner utilities that you have created using the `GetXxx-Utility()` methods have been deleted first. Failure to do so leads to undefined behavior – most likely a crash.

11.9.2 Tuner Audio Playback

You use class `CMMTunerAudioPlayerUtility` to play audio from the tuner; it is quite straightforward. This class is instantiated by `CMM-TunerUtility`.

```
iPlayer = iTuner->GetTunerAudioPlayerUtilityL(*this);
```

You must derive from the `MMMTunerAudioPlayerObserver` class in order to use it. The `CMMTunerUtility` object used to instantiate the tuner player utility (`iTuner`) must persist until after the player utility instance is deleted.

First, you must initialize the utility using the `InitializeL()` method. After initializing the utility, you receive a callback to `MTapoInitial-izeComplete()`.

```
iPlayer->InitializeL();
...
void CRadioTuneEngine::MTapoInitializeComplete(TInt aError)
  {
  if (aError == KErrNone)
    {
```

```
    iPlayer->Play();
    }
  else
    {
    // Handle the error.
    }
  }
```

Playback continues indefinitely until it is stopped or some error occurs. If an error occurs during playback, you receive a callback to the `MTapoPlayEvent()` method. The type of event is determined by the `aEvent` parameter, which takes values from enumeration `MMMTunerAudioPlayerObserver::TEventType`. This is either `ETunerEvent` which indicates some problem with the tuner, such as the antenna being removed, or `EAudioEvent`, which may mean that the audio policy has revoked your permission to play the audio.

The class `CMMTunerAudioPlayerUtility` also provides a number of methods to change the volume and balance of the tuner audio and some other methods common with the standard audio-clip-player utility.

11.9.3 Tuner Audio Recording

The `CMMTunerAudioRecorderUtility` class can be used to record audio from a tuner. It is very similar to the `CMdaAudioRecorderUtility` class described in Section 11.7.2. The main difference is that you must derive from a different class, `MMMTunerAudioRecorderObserver`, to use it.

The tuner audio recorder utility is instantiated using the `CMMTunerUtility::GetTunerAudioRecorderUtilityL()` method.

The tuner player and recorder utilities can be used simultaneously. However, before you attempt to play and record audio from the radio at the same time, you should use the `CMMTunerUtility::GetCapabilities()` method to check that the `TunerFunctionSimultaneousPlayAndRecord` capability is supported.

As with the tuner audio player utility, the `CMMTunerUtility` instance used to instantiate a `CMMTunerAudioRecorderUtility` must not be deleted before the recorder utility itself is deleted.

12

Communications

Mobile phones, by their very nature, incorporate a wide range of communications technologies. In this chapter, we present an overview of these technologies and their characteristics. We then provide example code demonstrating these types of communication:

- sockets
- Bluetooth sockets, together with more information on using sockets in general
- HTTP
- messaging including Send As
- telephony.

12.1 Communications Technologies

12.1.1 Wide Area Networking

Current UIQ 3 mobile phones support 2G (GSM) and 3G (UMTS) communications. We show how to perform data communication using sockets and give an introduction to managing voice calls.

Circuit-Switched Network

The most basic way to connect to the Internet and company networks is over a modem connected to a landline. The modem establishes a telephone call to the remote server which is *circuit-switched,* meaning that a telephone circuit is maintained for the duration of the connection, irrespective of whether data flows or not. The connection has a fixed bandwidth and is normally charged on the basis of how long the connection is held.

In mobile networks, this service is known as circuit-switched data (CSD) and runs at a basic speed of 9.6 kbps, though 14.4 kbps is frequently supported. High-speed, circuit-switched data (HSCSD) is supported by a limited number of mobile operators and provides faster connections. The Sony Ericsson P1i supports HSCSD at 28.8 kbps downlink and 14.4 kbps uplink.

CSD and HSCSD are rarely used in modern solutions although UIQ 3 still allows CSD and HSCSD Internet Access Points (IAP) to be configured and used.

Packet-Switched Network

Packet-switched methods provide faster connections (see Table 12.1) and much more efficient utilization of network resources. Data is transmitted in small packets, as needed, much like IP on the Internet. Capacity is only used when data is being sent or received, which means that it is possible to be *constantly* connected so that applications have immediate access to networked servers. The service is typically charged by the amount of data transferred, although *unlimited* subscriptions (subject to a *fair use policy*) are increasingly common.

GPRS (General Packet Radio Service) is the original implementation in GSM networks. EDGE (Enhanced Data Rates for Global Evolution) introduced greater throughput using more advanced modulation techniques.

UMTS (Universal Mobile Telecommunications System) networks typically launched with a 384 kbps connection speed. High Speed Packet Access (HSPA) implements a high speed shared channel in UMTS. Initially available in the downlink direction only, HSDPA technology increases the downlink speed to 3.6 Mbps, 7.2 Mbps and up to 14.4 Mbps in the future. Complementary HSUPA technology increases the uplink speed from the basic 384 kbps up to 5.76 Mbps.

Table 12.1 Data transfer speeds

Technology	Downlink	Uplink
GPRS (CS-2)	53.6 kbps	26.8 kbps
EDGE	247.4 kbps	123.7 kbps
UMTS	384 kbps	384 kbps
HSDPA	Up to 14.4 Mbps	384 kbps
HSUPA	Up to 14.4 Mbps	Up to 5.76 Mbps

At the time of writing, most HSDPA phones and networks offer 3.6 Mbps, while HSUPA is just being introduced at 2 Mbps. All HSUPA phones are expected to include HSDPA, since downlink speed is almost always more important to the end user than uplink. The MOTO Z8 supports HSDPA at 3.6 Mbps.

A mobile phone normally operates at the best available data rate, which may be constrained by the subscription, coverage and network traffic. Your application simply selects a packet-switched IAP.

Mobile networks are typically configured to transfer a data connection from UMTS/HSPA to GPRS/EDGE if the mobile phone moves out of UMTS coverage and in to GSM-only coverage. The connection returns to UMTS/HSPA if the network and the mobile phone are configured to support the transition back again. The allocated IP number and Domain Name Resolution (DNS) servers remain the same irrespective of which network is serving the mobile phone.

Latency Latency may be measured as the *Round Trip Time* (RTT) for a data packet from the mobile phone to be sent to a server on the network and a response packet arriving back. The Ping utility performs such a test. For example, pinging ***www.sonyericsson.com*** via a GPRS connection has an RTT of around 750 ms, depending on network traffic.

EDGE, UMTS and HSPA each introduce additional network features to improve RTT. The same ping test using UMTS has an RTT of around 200 ms, while HSDPA reduces it to around 120 ms or less.

Access Point Names The required service is selected using an Access Point Name (APN). APNs are defined by the mobile network operator and are frequently pre-configured on the mobile phone. Many mobile operators configure separate APNs for Internet, MMS services and WAP browsing. It is possible for multiple connections to be in progress at the same time; such devices are said to be multi-homing.

Many applications require access to the Internet and can select the appropriate connection using the `RConnection` API.

Internet Access Mobile network operators frequently place limits on Internet access:

- limited access to domains, pages within those domains or content within pages in so-called *walled gardens*
- content filtering to prevent children accessing unsuitable material
- restriction of certain protocols; for example SMTP to external servers may be barred in order to protect against spammers.

IP Addresses In a mobile network, the IP address of a device is usually dynamically allocated. This is very similar to IP address allocation within a private company LAN via a DHCP server. Therefore, each time you connect, you are likely to be allocated a different IP address.

Network Address Translation (NAT) Most networks still use IPv4 and must therefore manage the limited number of available IP addresses. They typically employ network address translation (NAT) to reduce the need for globally unique IP addresses. A typical network uses addresses from the private IP address ranges, such as 192.168.x.x or 10.x.x.x. When a device on the local network communicates with the global Internet, the NAT translates the local IP address to a global IP address.

To perform the translation, the NAT needs to maintain mapping tables between the local and global addresses, otherwise it would not be able to route any returning content to the request originator. NAT mappings are automatically created and destroyed without the user knowing. Common inactivity timeouts for NAT mappings are 40–180 seconds for UDP mappings and 30–60 minutes for TCP mappings.

The NAT mapping tables are created on *uplink* packets from devices, that is, the mapping is created when a phone within the mobile network attempts to communicate with the outside world. Therefore, a server outside the mobile network is unable to initiate connections to mobile phones within the mobile network, even if it knows the local IP address that has been allocated to a phone, for example through a phone sending the server an SMS containing its IP address. In practice, communication must be initiated by the mobile phone unless your application's server component resides inside the mobile operator's network.

You should also note that some network protocols have specific requirements if it is expected that NAT will be involved. The most significant of such protocols is an IPSec-based virtual private network (VPN) which does not work unless the network layers are enclosed in UDP packets.

Always-on connections If you are building an application that presents an *always-on* metaphor to the user, such as a push email solution, various aspects of a mobile network and phone usage should be considered:

- Applications should not assume the connection will always remain open. Inactivity timeouts or loss of radio signal may occur in the network and connections may be interrupted when the mobile phone has no network coverage.
- Some phones may have *flight* modes where no radio activity is possible even though the phone is switched on and otherwise operating normally.
- Applications must be designed to be battery friendly, allowing a system to be in a power-save mode for as much of the time as

possible. Background activity should be carefully considered and minimized.

- If an application uses polling techniques to *keep alive* a connection, you should consider the effect that various poll rates have on data volumes and battery life compared to the usefulness of keeping a connection alive.
- Other communications activity on the mobile phone may suspend data connectivity. When on GSM, a voice call will cause any GPRS connection to be suspended.

Roaming When roaming on another mobile network, there is generally no need to change any settings. The APN of the home network still connects to the same services. The connection is established from the roaming network back through the home network. The IP number and DNS are allocated from the home network.

12.1.2 Personal and Local Area Networking

Infrared and Bluetooth are short-range communication technologies that are typically used to connect two mobile phones. Some UIQ 3 phones include Wireless LAN compliant with the popular IEEE 802.11 standard.

Infrared

Infrared provides communications over a short distance, about one meter maximum, with a clear line of sight between the two devices. Speeds up to 115 kbps are possible on UIQ 3 mobile phones. The protocol stack used is IrDA, standardized by the Infrared Data Association.

Infrared is most commonly used to transfer, or *beam*, contact details and appointments from one mobile phone to another. The communication is easily set up and does not cost anything.

OBEX, serial and socket communication are supported.

Bluetooth Technology

Bluetooth is a wireless technology operating in the 2.4 GHz license-free radio band, often called the industrial, scientific and medical (ISM) band. Most phones have a range of up to 10 meters, though 100 meters is also possible. Typical speed is 768 kbps, but can be increased up to 2.1 Mbps using EDR (Enhanced Data Rate).

Bluetooth technology defines a way of discovering and connecting to other devices. Bluetooth *profiles* indicate the services that are supported. Dial-Up Networking profile enables a Bluetooth-equipped laptop to connect to the Internet using the mobile phone as a modem. Headset profile enables the mobile phone to work with a Bluetooth headset. Communication over Bluetooth technology is free.

OBEX, serial and socket communication are supported. Section 12.4 provides an example where we open sockets over Bluetooth technology and send text strings.

Wireless LAN

Wireless LAN (WLAN) is a common way of connecting to corporate networks and to the Internet. The Sony Ericsson P990i and P1i phones include WLAN to the IEEE 802.11b standard, part of the WiFi family. IEEE 802.11b provides connectivity up to 11 Mbps and operates in the same 2.4 GHz ISM band as Bluetooth technology. The connection speed drops to 5.5, 2 and 1 Mbps as the mobile phone moves away from the access point. Maximum range is approximately 100 meters indoors and 400 meters outdoors, depending on obstacles and interference. Additional protocols are used for authentication and encryption.

WLAN is commonly used to make a home ADSL (broadband Internet) connection available to multiple computers. WEP (Wireless Equivalent Privacy) or WPA-Personal (Wireless Protected Access, Personal) is generally enabled to provide authentication of users connecting to the WLAN and encryption of the data packets. To connect, you must know the SSID (Service Set ID) and the network password. Since the cost of the broadband connection is typically fixed and does not increase as more devices are added via WLAN, many users regard such access from a mobile phone to be free, not least because GPRS-type packet data charges are avoided.

Internet hotspots in cafes, hotels and airports generally do not employ any authentication at WLAN level. Simply discovering and connecting to the SSID is sufficient. However, without further authentication, connectivity may be strictly limited to, say the service provider's or hotel's website. Full Internet access is gained after entering a username or password, or buying a fixed period of access from the service provider. A default web page is provided for the user to buy access or log on. Since there is typically no encryption, any sensitive transaction must be protected at a higher level, such as secure web pages.

When deployed for corporate LAN access, WPA is used with a defined authentication technique. Configuration information is required from the network administrator.

When connecting to the Internet from home or a hotspot, NAT is used to manage the IP addresses, so you meet the same issues as we described for GPRS connections, including those for *always-on* type applications.

WLAN is IP-based, therefore, the `Sockets` examples in this chapter apply.

12.1.3 Multi-homing

Multi-homing is the ability of a mobile phone to manage multiple connections at once. The need arises as mobile operators typically segregate access to their MMS, WAP and Internet services. The mobile phone must connect to each service separately.

In older GPRS phones, where only one connection could be in progress at once, the connections were prioritized so that, say, an MMS connection would suspend any ongoing Internet connection.

UMTS phones are typically capable of establishing more than one connection at a time. The addition of Bluetooth PAN and WLAN gives rise to the possibility of additional concurrent connections. Applications need to select which connection they want to use.

In UIQ 3, connections are defined as IAPs. When you open a socket, you must select the required IAP. You can do this in a number of ways:

- take the user-configured default in UIQ 3
- let UIQ 3 present a list of IAPs to the user at connect time
- select a particular connection, for example, defined in the user preferences for your application.

As a mobile phone moves around, the available networks (especially PAN and WLAN) change. Your application may wish to manage this situation and:

- support specific change requests from a user
- propose changes to a user if a change is detected
- automatically change behind the scenes, for example, via some kind of configurable change policy.

Detecting changes is more complicated but may offer your users a superior experience. You can use the `RConnection` API to either determine the current IAP availability or receive notifications about changes in bearer availability. Your application must take into account that the IP number and DNS servers will change when switching between Bluetooth PAN, WLAN and the mobile network.

12.1.4 Messaging-based Communication

Messaging can be used as a transport to convey communication, including specifically coded messages that are sent and received by applications rather than by humans. UIQ 3 provides a Send As interface, which we used in `SignedAppPhase2` (Section 10.3) to send files. In this chapter, we provide a further Send As example and also show you how to scan the inbox for messages and send messages directly.

SMS

A single SMS message can convey 160 seven-bit characters, 140 eight-bit characters or 70 16-bit characters. SMS messages are typically charged per message, though the popularity of the medium means that many users buy a set number of messages per month at a reduced rate. Longer messages are possible using concatenated SMS, where the message is broken into a number of underlying SMS blocks, which are reassembled to recreate the full message at the far end. The sender is charged for each underlying block.

MMS

MMS enables the user to send text, audio, images and video from the mobile phone. The service is charged per message and the typical maximum size of the content that is conveyed is in the range 100–300 KB.

Email

POP3, IMAP and SMTP Internet email accounts can be configured and used on UIQ 3 mobile phones. Both body text and attachments can be conveyed using email messages.

12.1.5 Voice Communication

Applications can set up and manage voice calls. The ETel ISV API enables your application to monitor phone status and manage calls.

12.2 Symbian OS Communications Architecture

12.2.1 ESOCK

ESOCK (see Figure 12.1) provides the framework for accessing various lower-level protocols in Symbian OS. It provides the sockets API for making connections and the connection management API for selecting, monitoring and configuring connections.

IP network access is possible over several bearers, including Bluetooth, infrared, WLAN, GPRS and CSD connections. It is also possible for multiple connections to be in progress at the same time, for example the MMS messaging server can connect over UMTS to the network operator's MMS server at the same time that an Internet connection is in progress for web browsing.

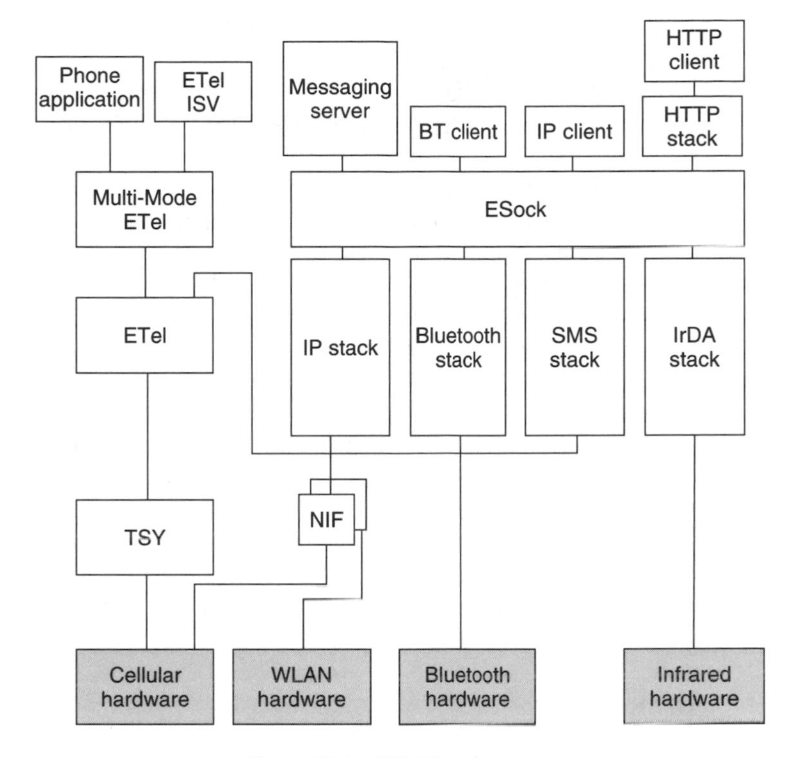

Figure 12.1 ESOCK architecture

UIQ 3 provides the GUI for setting up the required parameters for each connection. IAPs can also be preconfigured or set up remotely. The IAP parameters are stored in the communications database, `CommsDat`. An application called `Connection Manager` allows advanced users to monitor and control connections.

12.2.2 Symbian Signed Requirements

If your application incorporates a communications technology within its feature set, it is very likely to require some capabilities. For example, applications that wish to transfer data using infrared or Bluetooth technology require the `LocalServices` capability. Applications wishing to communicate using mobile network services, such as GPRS or SMS, require `NetworkServices`. If you want to use the messaging architecture, you are likely to need `ReadUserData` and `WriteUserData` capabilities.

These capabilities can be granted by a user upon application installation but we recommend that commercial applications are Symbian Signed so the user does not have to worry about granting capabilities. As we discuss in Chapter 10, this has numerous implications with respect to application UIDs, file naming and Test Criteria conformance.

Although unlikely, it is possible that your application may require `NetworkControl` capabilities. For example, you may need to configure the network interface. At the time of writing, this is a Device Manufacturer capability and so you should determine whether you will be granted the required platform approval before proceeding with your application development. In the new Open Signed and Certified Signed schemes, you can get `NetworkControl` without manufacturer approval. Symbian Signed is explained in detail in Chapter 14.

We have not designed the example applications in this chapter to be Symbian Signed. In general, they are not complete applications, but rather focus on demonstrating each topic.

You should be aware that debugging communications applications is difficult. This is partly due to their real-time nature and partly due to the differences between a development PC and a mobile phone. A common technique is to log information to a file or the UI. Our example applications simply log information to a standard UIQ list box, which is then displayed as the primary UI element. For some types of problem solving, you may find it useful to run your application in the emulator and use PC-based analysis tools such as Wireshark.

12.3 Sockets

Sockets provide a high-level interface to various communications protocols, allowing the transfer of information between endpoints. While originally developed to support the TCP/IP protocol, the concepts can also be applied to other communications protocols. In Symbian OS, sockets are used for TCP, UDP, Bluetooth and infrared communications.

A socket represents one end of a connection between applications. Usually, the applications reside on different devices. In that case, data written to a socket on one device can be read from a socket on the remote device.

12.3.1 Connected and Connectionless Sockets

A socket is classified as either connected (or connection-oriented) or connectionless. For connected sockets, a path between the two endpoints must be established. Once a connection is established, data can flow between the endpoints without every transaction having to know the address of the remote endpoint. Data flow is usually ordered and delivery is deterministic. On the other hand, a connectionless socket forms the path between the endpoints for every item of data transferred; data flow is not normally ordered and often little knowledge about the successful delivery, or otherwise, is known. TCP implements connected sockets; UDP uses connectionless sockets.

12.3.2 ESOCK API

Socket-based communications in Symbian OS utilize the following classes.

RSocketServ

This class represents the connection between the client application and the server process within Symbian OS. The server process actually implements the communication protocols and provides:

- general enquiry functions about the available communications protocols
- the mechanism by which communication sub-sessions are instantiated.

RHostResolver

This class is responsible for name resolution services. If a protocol supports a DNS service, you use an RHostResolver class to gain access to those services.

RSocket

This class represents the communications protocol endpoint.

RConnection

This class is responsible for managing the network interface – in effect, it is the transport layer.

RNetDatabase

Some protocols have a database associated with them, for example, LM-IAS with the infrared protocol. This class enables access to such databases.

RSocketServ Class

Before any socket-based communications can take place you need to establish an IPC connection to the Symbian OS socket server process. RSocketServ is the client-side object encapsulating your IPC connection to that process. Once the IPC connection is established, you can make requests, for example, to enumerate the communication protocols available or create an endpoint.

The Sockets1 example application enumerates the available communications protocols and displays them as a list box:

```
void CAppSpecificListView::ViewConstructL()
  {
  ViewConstructFromResourceL(R_APP_VIEW_CONFIGURATIONS);
  User::LeaveIfError(iSocketServ.Connect());
```

```
CQikListBox* listbox=LocateControlByUniqueHandle<
        CQikListBox>(EAppSpecificListViewListId);
MQikListBoxModel& model(listbox->Model());
model.ModelBeginUpdateLC();
TBuf<128>bb;
TUint count;
User::LeaveIfError(iSocketServ.NumProtocols(count));
TProtocolDesc info;
for (TUint i=0;i<count;i++)
  {
  User::LeaveIfError(iSocketServ.GetProtocolInfo(
                                    i+1,info));
  MQikListBoxData* lbData=model.NewDataL(
          MQikListBoxModel::EDataNormal);
  CleanupClosePushL(*lbData);
  lbData->AddTextL(info.iName,EQikListBoxSlotText1);
  _LIT(KTypeProtAddr,"Type:%u, Prot:%u, Addr:%u");
  bb.Format(KTypeProtAddr,info.iSockType,
                    info.iProtocol,info.iAddrFamily);
  lbData->AddTextL(bb,EQikListBoxSlotText2);
  CleanupStack::PopAndDestroy(lbData);
  }
model.ModelEndUpdateL();

iEikonEnv->ReadResourceL(bb,R_STR_APP_TITLE);
ViewContext()->AddTextL(1,bb);
}
```

The emulator typically displays the content shown in Figure 12.2.

Figure 12.2 shows that a protocol named TCP exists, has a type of `KSockStream` `(Type:1)`, implements the `KProtocolInetTcp` protocol `(Prot:6)`, and supports the `KAfInet` address family (`Addr:2048`).

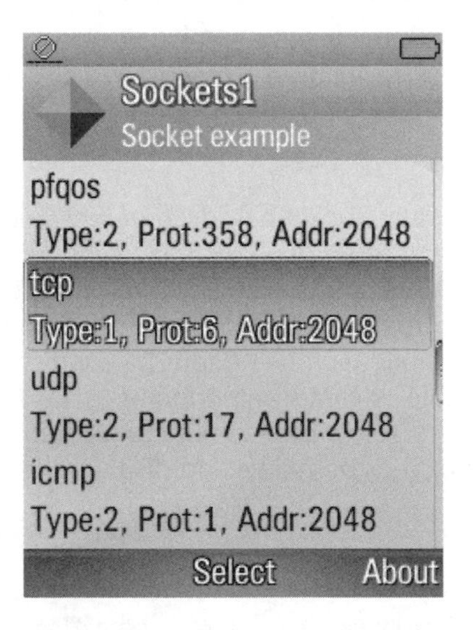

Figure 12.2 `Sockets1` example application

Similarly, a protocol named udp exists, is of type KSockDatagram, implements the KProtocolInetUdp protocol and supports the KAfInet address family.

RHostResolver Class

In simple terms, we obtain a connection to a remote endpoint by specifying the address of that endpoint. In practice, this is often harder to achieve as the address of the remote endpoint may not be easily known. In TCP/IP, the address is often fixed; it is the IP address. For Bluetooth and infrared connections, there is no permanent network and connections are typically created as and when they are required.

In practice, Bluetooth and infrared connections are preceded by a discovery phase, where one device searches for other devices within range that offer the required services. With TCP and UDP, servers are often known by a textual name which needs to be converted to an underlying IP address via the Domain Name Service (DNS). The RHostResolver object encapsulates the functionality required to determine the remote endpoint address.

Obtaining endpoint addresses can take time. Therefore, while both synchronous and asynchronous variants of the required functions exist, we recommend that you use the asynchronous variants. Doing so requires you to use active objects, as shown in the Sockets1 example application:

```
void CSocketEngine::Connect()

  // Attempt to connect to the server
  // defined by iServerName.
  {
  TInetAddr addr;
  if (addr.Input(iServerName)==KErrNone)
    Connect(addr.Address());       // Server name is already
                                   // a valid IP address.
  else
    {  // Need to look up IP addr using DNS.
    TInt err=iHostResolver.Open(iSocketServer,KAfInet,
                                     KProtocolInetUdp);
    if (err==KErrNone)
      {
      iHostResolver.GetByName(iServerName,
                    iNameEntry, iStatus);
      }
    else
      { // Deal with all errors in the RunL().
      TRequestStatus* q=(&iStatus);
      User::RequestComplete(q,err);
      }
    iState=ESeLookingUp;        // Now started request.
    SetActive();
    }
  }
```

Our objective is to establish a socket-based connection to the remote server. If the `iServerName` field already contains an IP address, such as 192.168.0.1, we do not need to resolve the name: it is already resolved! We can use the address immediately. Conversely if the `iServerName` field contains some text, such as ***www.uiq.com***, we need to convert the text into an IP address. We need to use DNS to translate the textual address into an IP address.

Our application requests the `RHostResolver` to convert the textual name into a resolved `TNameEntry` object asynchronously by calling the `iHostResolver.GetByName()` method. When the asynchronous request completes, our `RunL()` method is called.

```
void CSocketEngine::RunL()
  {
  switch (iState)
    {
    case ESeLookingUp:        // Name look up has completed.
      iHostResolver.Close();
      iState=ESeNotConnected;
      if (iStatus==KErrNone)
        {
        // DNS look up successful -  attempt
        // to connect to that IP address.
        Connect(TInetAddr::Cast(iNameEntry().iAddr).Address());
        }
      else
        {
        // Error handling code.
        }
      break;

    // Other cases, removed for clarity.
    }
  }
```

We extract the IP address from the `TNameEntry` object with the `TInetAddr::Cast(iNameEntry().iAddr).Address()` code, passing the resultant address to our socket connect method:

```
void CSocketEngine::Connect(const TUint32 aAddr)
  {
  TInt err=iSocket.Open(iSocketServer,
                 KAfInet,KSockStream, KProtocolInetTcp);
  if (err==KErrNone)
    {
    iAddress.SetPort(iPort);
    iAddress.SetAddress(aAddr);
// Initiate socket connection (async)
    iSocket.Connect(iAddress,iStatus);
    }
  else
    {
    // Deal with errors in RunL().
```

```
    TRequestStatus* q=(&iStatus);
    User::RequestComplete(q,err);
    }
  iState=ESeConnecting;
  SetActive();
  }
```

The socket connection request is asynchronous and so calls `RunL()` when it completes:

```
void CSocketEngine::RunL()
  {
  switch (iState)
    {
    case ESeConnecting:
      if (iStatus==KErrNone)
        { // Sucessfully connected..
        iState=ESeConnected;
        }
      else
        { // Connect attempt failed.
        iSocket.Close();
        iState=ESeNotConnected;
        }
      break;

      // Other cases, removed for clarity.
    }
  }
```

If we successfully connect, we can perform read and write operations as defined by the service we have connected to. In the case of our example application, we have chosen to connect to port 80 at ***www.uiq.com***. This is the standard port for a server to deliver web pages via a higher-level protocol called HTTP (see Section 12.5).

This example application does not perform any `RSocket` read or write operations. For an example of how you may use those methods, you may wish to read the Bluetooth example application in Section 12.4.

Figure 12.3 shows the log output when we run the `Sockets1` example in the emulator and connect to the ***www.uiq.com*** server. You should be able to see that we originally had the textual name ***www.uiq.com***. This has been converted by the `RHostResolver` into the IP address 195.84.17.233.

Capabilities

In order to successfully use the `RHostResolver`, our application needs `NetworkServices` capabilities. Without that capability, you will get the rather unexpected error code `5120` (no response from the DNS server) when you call `GetByName()`.

Figure 12.3 `Sockets1` log output

`RConnection` Class

As you can see in the `Sockets1` example, there is no specific requirement to define which connection preferences or access point should be used to establish a physical connection. In the emulator, there is often only a single access point which is used automatically.

As long as there is at least one access point defined, UIQ 3 has a preferred access point. The user can change the preferred access point in Internet account settings (see Figure 12.4).

The user can also tell UIQ 3 to display the connection dialog for manual selection. This setting is also stored in the communications

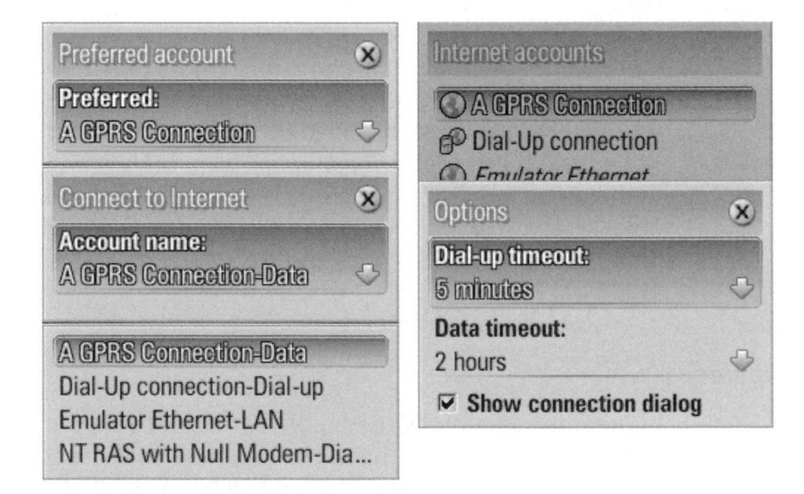

Figure 12.4 Internet account settings

database CommsDat. Sony Ericsson phones allow the user to create groups of access points, where a group consists of a list of IAPs.

In our Sockets1 example, we used the RHostResolver::Open (RSocketServ&, TUint, TUint) variant of the open method. If we wanted more control over which access point should be used we can use the RHostResolver::Open(RSocketServ&, TUint, TUint, RConnection&) variant.

Similarly our application uses the RSocket::Open(RSocket-Serv&, TUint, TUint, TUint) overload of the open method. An alternative is to use the RSocket::Open(RSocketServ&, TUint, TUint, TUint, RConnection&) variant.

A detailed explanation and examples of using RConnection and its interaction with CommsDat is beyond the scope of this book. The Connection Management section of the UIQ 3 SDK documentation has further information about using RConnection, including some usage guidelines.

12.3.3 Secure Sockets

Secure sockets enable applications to transfer encrypted content over a public network with both endpoints having the opportunity to authenticate each other.

In Symbian OS, secure socket support is provided by a set of plug-ins to the socket server. The default plug-ins support Transport Layer Security v1.0 (TLS1.0) and the Secure Socket Layer v3.0 (SSL3.0). A full specification of these protocols is contained within RFC 2246.

A secure socket implementation secures an already connected socket. In the Sockets1 example, once we successfully established a socket-level connection, we could transfer content. Prior to transferring content over a secure socket, a further step is required:

```
void CSecureSocketEngine::RunL()

  // One of our requests has completed.
  {
  switch (iState)
    {
    case ESeConnecting:
      if (iStatus==KErrNone)
        {
        TRAPD(err,
          _LIT(KTLS1,"TLS1.0");
          iSecureSocket = CSecureSocket::NewL(iSocket,KTLS1);
        );

      if (err!=KErrNone)
        {
        iSocket.Close();
```

```
      iState=ESeNotConnected;
      }
    else
      { // Perform a client handshake with the
        // server to establish security.

      iSecureSocket->StartClientHandshake(iStatus);
      iState=ESeClientHandshake;
      SetActive();
      }
    }
  else
    { // Connect attempt failed, allow
      // retry or perhaps a different address.

    iSocket.Close();
    iState=ESeNotConnected;
    }

    break;

  // Other cases removed for clarity.
  }
}
```

We create a `CSecureSocket` object, passing the connected socket handle and an indication as to which protocol is to be used. Assuming that we successfully create the `CSecureSocket` object, we call the asynchronous `StartClientHandshake()` method.

A Security Information dialog is displayed in the emulator (see Figure 12.5). Selecting Yes enables the `StartClientHandshake()` functionality to continue.

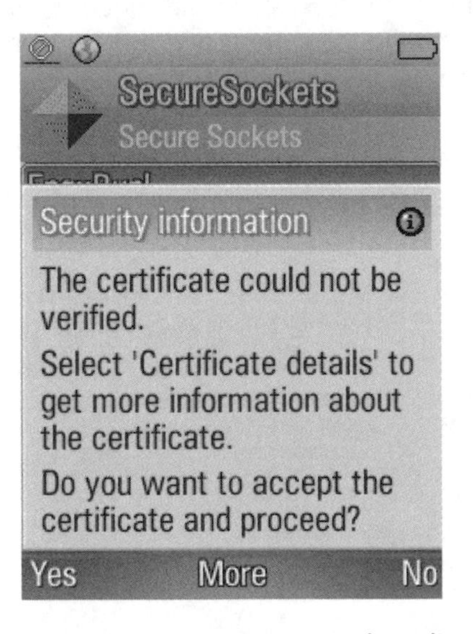

Figure 12.5 `SecureSockets` example application

Eventually, the `StartClientHandshake()` function completes its task and our `RunL()` is called:

```
void CSecureSocketEngine::RunL()

  // One of our requests has completed.
  {
  switch (iState)
    {  // Establishing secure connection.

    case ESeClientHandshake:
      if (iStatus==KErrNone)
        {
        iSecureSocket->RecvOneOrMore(
                              iRecvBuffer,iStatus,iXfrLen);
        iState=ESeReading;
        SetActive();
        }
      else
        { // Failed to get secure connection.
        iSecureSocket->Close();
        delete(iSecureSocket);
        iSecureSocket=NULL;
        iSocket.Close();
        iState=ESeNotConnected;
        }
      break;

    // Other statements removed for clarity.
    }
  }
```

If the request completes successfully, we can start reading and writing content over the secure socket. Rather than use the `RSocket` methods, you need to use the equivalent `CSecureSocket` methods as shown in the code fragment.

Code using secure sockets works only on a phone, not in the emulator, so you need to test this part of your code on a phone.

12.4 Bluetooth Technology

Generally, Bluetooth connections are established in an ad hoc fashion, for example, when a mobile phone and a Bluetooth headset come in to range.

As with most communications systems, one device acts as a server, listening out for incoming connection requests, while the other initiates the connection. In a Bluetooth connection, either device can act as the server and the other acts as the connection initiator. In our example application, we let the user decide, via a menu option, which device should listen for incoming connection requests. In a commercial-grade application, it is possible for both devices to be set up to listen for connection requests. Whenever an action occurs on one device that requires a connection to be made, that device can change from listening for incoming connections to initiating the connection.

12.4.1 Capabilities

Our example application requires the `LocalServices` capability. This is necessary to use both the `CSdpAgent` and `RSocket` classes.

12.4.2 Use of Active Objects

The Bluetooth example requires extensive use of active objects. Using active objects has several consequences:

- Our application must have an active scheduler in the thread performing the Bluetooth communications. A standard UIQ application primary thread always has a scheduler.
- Our application can process other events between the time a Bluetooth request is started and the time it completes.

A general design philosophy with active objects is that one active object represents one event source. The events that we need to manage are:

- asynchronous writing (transmission) of data
- asynchronous reading (reception) of data
- asynchronous connection establishment.

This suggests that we might need three active objects. If we look more closely, we discover that while both asynchronous read and write operations can be outstanding simultaneously, neither the read nor write can be outstanding at the same time as the connection request – you cannot read or write if there is no remote endpoint to read from or write to! In our application, we have chosen to merge the connection and read events and handle them with a single active object, demonstrating that this is reasonable. A potential disadvantage to this approach is that we need to run a slightly more complex state machine to track the current status.

12.4.3 Accepting Incoming Connection Requests

To accept an incoming Bluetooth connection request, the application needs to perform a number of tasks:

- determine a free port on which the connection can be made
- decide the security policy for the application

- publish the availability and service type being offered by the application.

The class CBtConnectionListener is used to accept incoming connection requests. The ConstructL() method of this class is:

```
void CBtConnectionListener::ConstructL()
  { // Any write requests are handled
    // by this active object.
  iWriteChannel=new(ELeave)CBtWriter(iSocket,iObserver);
  User::LeaveIfError(iListener.Open(
                          iSocketServer,KRFCOMMDesC));
  TInt port;
  User::LeaveIfError(iListener.GetOpt(
                       KRFCOMMGetAvailableServerChannel,
                                    KSolBtRFCOMM,port));
  TBTSockAddr addr;
  addr.SetPort(port);

  // Set up the security settings for a point to point BT connection.
  TBTServiceSecurity serviceSecurity;
  serviceSecurity.SetUid(KUidServiceSDP);
  serviceSecurity.SetAuthentication(EFalse);
  serviceSecurity.SetEncryption(EFalse);
  serviceSecurity.SetAuthorisation(ETrue);
  serviceSecurity.SetDenied(EFalse);
  addr.SetSecurity(serviceSecurity);

  // Now request we start listening on the indicated
  // port for connection requests.
  User::LeaveIfError(iListener.Bind(addr));
  User::LeaveIfError(iListener.Listen(1));

  // Connect to the service discovery subsystem.
  User::LeaveIfError(iSdp.Connect());
  User::LeaveIfError(iSdpDatabase.Open(iSdp));

  // We want a serial port service.
  iSdpDatabase.CreateServiceRecordL(
                 KSerialPortServiceClass,iRecordHandle);

  // With the following protocol description this needs to match the
  // lSerialPortAttributeList[]structure - since that's what we're going
  // to parse against.
  CSdpAttrValueDES* sdpAttr=
                        CSdpAttrValueDES::NewDESL(NULL);
  CleanupStack::PushL(sdpAttr);
  TBuf8<1> bb8;
  bb8.Append(port);
  sdpAttr
    ->StartListL()
    ->BuildDESL()
    ->StartListL()
```

```
    ->BuildUUIDL(KL2CAP)
    ->EndListL()

    ->BuildDESL()
    ->StartListL()
    ->BuildUUIDL(KRFCOMM)
    ->BuildUintL(bb8)
    ->EndListL()
    ->EndListL();

// And add that to the SDP database record.
iSdpDatabase.UpdateAttributeL(iRecordHandle,
            KSdpAttrIdProtocolDescriptorList,*sdpAttr);
CleanupStack::PopAndDestroy(sdpAttr);

// Add short human readable name for the service.
 _LIT(KUiqBtExample,"UiqBtExample");
iSdpDatabase.UpdateAttributeL(iRecordHandle,
               KSdpAttrIdBasePrimaryLanguage+
                 KSdpAttrIdOffsetServiceName,
                                KUiqBtExample);

// Add human readable description for the service.
_LIT(KUIQBookExampleBlueToothservice,"UIQ,
        Book Example, BlueTooth service");
iSdpDatabase.UpdateAttributeL(iRecordHandle,
               KSdpAttrIdBasePrimaryLanguage+
          KSdpAttrIdOffsetServiceDescription,
            KUIQBookExampleBlueToothservice);
}
```

Firstly, we open a listening `RSocket`, specifying the `RFCOMM` protocol. We then request a free port on which we can listen. It is important to note that the port number can vary between requests, depending on the state of the rest of the phone. After defining the security policy that we wish our application to implement, we bind the listening socket to the port and security policy, via a `TBTSockAddr` class. We then start listening for incoming connection requests by calling `iListener.Listen(1)`. The parameter `1` specifies a queue of one slot for outstanding connection requests. This is sufficient for our example since we only support a single point-to-point connection.

For a remote device to successfully establish a socket-based connection, it needs to know the port on which the server is listening. In our application, this port is dynamically allocated, so a mechanism is required to communicate this information prior to the socket connection attempt.

The Bluetooth specification defines a Service Discovery Database which facilitates the discovery of services supplied by a device. A server typically publishes the services it supports within the Service Discovery Database. Remote devices read this database to determine if there is an appropriate service to which they can connect.

In our application, we publish a record which includes the port information. The remote device locates this database record and extracts the port information, thereby discovering the port to which it must connect.

Access to the Service Discovery Protocol Database is via a Symbian OS process. We therefore have to connect to this process before we can open a handle to the actual database:

```
User::LeaveIfError(iSdp.Connect());
User::LeaveIfError(iSdpDatabase.Open(iSdp));
```

Once we have an open handle to the database, we can create a record within the database:

```
iSdpDatabase.CreateServiceRecordL(
                KSerialPortServiceClass,iRecordHandle);
```

`CreateServiceRecordL()` returns a handle to the record within the `iRecordHandle` property. In our application, we create a record with service type attribute `KSerialPortServiceClass`. A full description of Bluetooth service classes is beyond the scope of this book, but a service class provides an indication of the capabilities of the service and, more particularly, defines the attributes which must appear along with the service class, such that the remote device can parse the service record and understand the content.

Service classes are identified by unique numbers. The Bluetooth specification pre-defines a set of service class and attribute values. In our application, we have chosen to use the pre-defined Serial Port Service class. This in turn defines the attributes and values required within the record data.

The code below creates the record content containing the correct attributes and values for a serial port service database record:

```
CSdpAttrValueDES* sdpAttr=CSdpAttrValueDES::NewDESL(NULL);
CleanupStack::PushL(sdpAttr);
TBuf8<1> bb8;
bb8.Append(port);        // The port to connect to.
sdpAttr
  ->StartListL()
  ->BuildDESL()
  ->StartListL()
  ->BuildUUIDL(KL2CAP)
  ->EndListL()

  ->BuildDESL()
  ->StartListL()
  ->BuildUUIDL(KRFCOMM)
  ->BuildUintL(bb8)
  ->EndListL()
  ->EndListL();

// And add that to the SDP database record.
```

```
iSdpDatabase.UpdateAttributeL(iRecordHandle,
          KSdpAttrIdProtocolDescriptorList,
                                   *sdpAttr);
CleanupStack::PopAndDestroy(sdpAttr);
```

Your application decides the security settings. We have chosen `SetAuthorisation(ETrue)` meaning that a user is prompted to accept the connection attempts, however we have `SetAuthentication(EFalse)` meaning that no pin number exchange is required. Your choice of security settings should reflect the importance of the information being transferred. Since our application only transfers text strings and displays them on screen, this low level of security is appropriate. However, if your application transfers information that would otherwise remain private to a user, such as contact details or address book entries, the level of security should be higher, almost certainly requiring authentication and possibly requiring encryption.

12.4.4 Listening for Incoming Connections

Once our service has been published, we can start listening for incoming socket-level connection requests:

```
void CBtConnectionListener::StartListening()

  // Start the acceptance of a connection from
  // a remote Bluetooth device.
  {
  iSocket.Open(iSocketServer);
  SetAvailability(0xFF);
  iListener.Accept(iSocket,iStatus);
  SetActive();
  iState=EListenerStateConnecting;
  }
```

Firstly, we open a blank `RSocket` into which any incoming connection request is transferred. If you recall, the size of our listening queue is 1. The first pending connection from that queue is assigned to this newly opened blank socket. After successful completion of the `Accept()` request:

- The `iListener` socket can be used to make further connections by issuing further `Accept()` requests on blank sockets. Our application only supports a single connection at a time so we have not implemented such functionality.
- The `iSocket` object represents a connection between devices and can be used to transfer content using the read and write methods.

In our state machine, we set `iState` to `EListenerState-Connecting`. This enables us to know how to handle any cancel requests or event completions. When the `Accept()` request completes, our `RunL()` method is called:

```
void CBtConnectionListener::RunL()
  // A socket event has occurred.
  {
  switch (iState)
    {
    case EListenerStateConnecting:
      SetAvailability(0x00);      // No more connections at this time.
      if (iStatus.Int()==KErrNone)
        {
        iState=EListenerStateConnected;
        Read();
        iObserver->StatusInfo(EBtsConnectionEstablished,
                                                 KErrNone);
        }
      else
        { // Shutdown and start again.
        _LIT(KListenError,"Listen Error");
        iObserver->LogInfo(KListenError);
        iState=EListenerStateIdle;
        iSocket.Close();
        StartListening();       // Re-try connection.
        }
      break;

      // Other events removed for clarity.
    }
  }
```

If we connected successfully, we then issue a read request to accept any data the remote device may wish to transfer. On failure, we close down the blank socket and retry for a connection.

Processing Incoming Data

The format and content of data transferred between devices is specific to the applications involved. A socket, whether it is a Bluetooth, infrared or TCP socket, simply transfers a stream of bytes between endpoints. Our application transfers discrete packets where each packet contains one unit of information: a textual message. In general, the receiving phone needs to perform some processing of the data to produce the desired output. Our application is required to display one line of text for each message. It therefore needs to process any received data into the individual text messages for display.

Some applications may choose not to transfer data within discrete packets. They simply send a byte stream containing commands and content. The commands and content may span one or more buffers of data. Conversely a buffer may contain more than one command and associated content. For example, an application that transfers photographs from one device to another is unlikely to send the entire photograph in a single buffer in a RAM-limited environment such as a mobile phone.

The content being transferred dictates how your application chooses to implement the read and write operations. Our example application has been defined to transfer text messages, which are no longer than 128 bytes. We could read whatever data is available from the socket using the `RecvOneOrMore()` method and then assemble it back into text messages, before passing it to the display code. Instead, we adopt a slightly different approach.

Given the constraints of our application, we can send a small piece of information at the start of every packet of data to be transferred. By doing this, we can easily perform a quick validation check and know how much more data to read from the socket in order to obtain the remainder of the packet. In effect, we use the socket to assemble all the data belonging to the packet before it is delivered to our application.

The read code comprises two methods:

```
void CBtConnection::Read()

  // Start the read of a packet header.
  {
  iSocket.Recv(iFrameHdr,0,iStatus);
  SetActive();
  iReadingState=EReadingHeader;
  }

void CBtConnection::ReadContent(void)

  // Start a read of data content.
  {
  iDataPtr.Set((TUint8*)iData.Ptr(),0,iFrameHdr[5]);
  iSocket.Recv(iDataPtr,0,iStatus);
  SetActive();
  iReadingState=EReadingContent;
  }
```

The `Read()` method requests exactly six bytes of information, that is, the size of the `iFrameHdr` descriptor. Once the header arrives, our `RunL()` method is called:

```
void CBtConnectionListener::RunL()

  // A socket event has occurred.
  {
  switch (iState)
    {
    case EListenerStateConnected:
```

```
      // Data arrived for us to process.
      if (iReadingState==EReadingHeader)
        {
        TInt error=iStatus.Int();
        if (error==KErrNone)
          { // We have got the header.
          TPtr8 q((TUint8*)iFrameHdr.Ptr(),5,5);
          if (q==KBluetoothHeaderText)
            {
            ReadContent();          // Get the frame data.
            break;
            }

          // Header is broken shut down and re-start.
          error=KErrCorrupt;
          }
        iObserver->StatusInfo(EBtsDisconnected,error);
        iState=EListenerStateIdle;
        iSocket.Close();
        StartListening();          // Re-try connection.
        }
      else
        // .....removed for clarity.
      break;

     // Other cases removed for clarity.
   }
 }
```

If we successfully receive the header and it passes the validation check, we can use the length information transferred as part of this header to read the rest of the packet by calling the `ReadContent()` method.

Efficiencies and Complexities of this Approach

As with most software tasks, there are trade-offs between various approaches. Using the above approach to read content means that, for every packet transferred, we perform exactly two `Read()` requests: we are delivered two events and thus execute `RunL()` twice. Had we chosen to structure our application to use `RSocket.RecvOneOrMore()`, the chances are that in the vast majority of cases only a single read would be required to deliver the whole packet, however this is not guaranteed. We would therefore need some logic to process the incoming data into text messages, caching partial messages, before they can be delivered to the UI component. Both approaches have advantages and disadvantages, and you should consider which is best for your application.

> **Pro tip**
>
> There is nothing of particular significance in the size or content of the header information that we have chosen to transfer at the start

of every packet. We are simply using the five characters defined by `KBluetoothHeaderText` as a defensive programming technique to give us much higher confidence in the interpretation of the sixth byte we receive; the length indicator. The header text information can be any length you choose, including zero where you simply receive a length byte. Longer information increases the likelihood of ensuring that the content is what you are expecting to receive.

12.4.5 Creating a Connection

To create a connection with a remote device we need to:

- discover what devices are currently available within the local vicinity
- connect to and read the selected device's Service Discovery Database
- read and process any records associated with the service class to which our application has been defined to connect
- obtain the port number on which the remote device is listening for socket-level connection requests
- request a socket-level connection.

12.4.6 Discovering Devices

The following code is used to discover what devices are available within the local vicinity:

```
void CBtConnectionCreator::StartDiscoveryL()
  {
  iRemotePort=(-1);
  CBTDeviceArray* btDeviceArray = new(ELeave)
                          CBTDeviceArray(2);
  CleanupStack::PushL(btDeviceArray);
  TRAPD(err,
    CQBTUISelectDialog* dlg = CQBTUISelectDialog::
                             NewL(btDeviceArray);
    if (dlg->RunDlgLD(KQBTUISelectDlgFlagNone))
      {
      if (btDeviceArray->Count()>0)
        {
        CBTDevice* dev=btDeviceArray->At(0);
        iRemoteDevAddr=dev->BDAddr();
        iRemoteDevClass=dev->DeviceClass();
        TBTDeviceName8 name(dev->DeviceName());
        iRemoteDevName = BTDeviceNameConverter::
                              ToUnicodeL(name);
        iSdpAgent=CSdpAgent::NewL(*this,iRemoteDevAddr);
        iSdpSearchPattern=CSdpSearchPattern::NewL();
        iSdpSearchPattern->AddL(
```

```
                                    KSerialPortServiceClass);
        iSdpAgent->SetRecordFilterL(*iSdpSearchPattern);

        // Calls back NextRecordRequestComplete().
        iSdpAgent->NextRecordRequestL();
        iState=ECreatorStateFindService;
        iStatus=KRequestPending;
        SetActive();
        }
      }
    );

  if (err!=KErrNone)
    {
    delete(iSdpAgent);
    iSdpAgent=NULL;
    delete(iSdpSearchPattern);
    iSdpSearchPattern=NULL;
    iObserver->StatusInfo(EBtsConnectionEstablished,err);
    iState=ECreatorStateIdle;
    }
  CleanupStack::PopAndDestroy(btDeviceArray);
  User::LeaveIfError(err);
  }
```

We launch the standard UIQ 3 Bluetooth device selection dialog using `CQBTUISelectDialog`. This performs the local search and presents a list of devices from which the user can choose. Once a device has been chosen, we can use the low-level Bluetooth device address to connect to the Service Discovery Database on the remote device. Since our connection listener is defined as providing a `KSerialPortServiceClass`, we should filter the database records to contain only those associated with this service class.

As shown in the previous code fragment, the search for and delivery of database records is asynchronous. Rather than present any direct active object interface, the `CSdpAgent` performs its processing behind the scenes, only calling us back when required. From an application-level perspective, we have a single active event running. Our application chooses to represent this by setting up a pseudo-event source and recording the current state with the following code:

```
iState=ECreatorStateFindService;
iStatus=KRequestPending;
SetActive();
```

In this way, canceling the active search can be managed and it does not matter which of the various callbacks fail or succeed. We can handle the completion of the search in a single place, our `RunL()`.

The `CSdpAgent` requires that we implement the `MSdpAgentNotifier` interface. As each individual record from the remote Service

Discovery Database is retrieved so the `NextRecordRequest-Complete()` method is called:

```
void CBtConnectionCreator::NextRecordRequestComplete(
                                              TInt aError,
                          TSdpServRecordHandle aHandle,
                               TInt aTotalRecordsCount)
  {
  if (aError==KErrEof)  // Scanned all records - see if it found
                        // the data we're  looking for.
    {
    TBuf<32>bb;
    _LIT(KErrEof,"KErrEof %d");
    bb.Format(KErrEof,iRemotePort);
    iObserver->LogInfo(bb);
    RequestComplete(iRemotePort==(-1) ? KErrNotFound : KErrNone);
    }
  else if (aTotalRecordsCount==0)
    RequestComplete(KErrNotFound);
  else if (aError!=KErrNone)
    RequestComplete(aError);
  else
    {
    // We want attributes of the record.
    TRAPD(err,iSdpAgent->AttributeRequestL(aHandle,
                KSdpAttrIdProtocolDescriptorList));
    if (err!=KErrNone)
      RequestComplete(err);
    }
  }
```

If there are no more records, `aError` is set to `KErrEof`. By this time, we anticipate that we will have located the `KSerialPortService-Class` record describing the connection on which the remote copy of our application is listening. Our application validates that we have obtained a remote port value. If so, the search has been successful.

When we are delivered a record, we need to fetch the attributes associated with the record. As a reminder, the attributes contain the port information we are searching for. Once fetched, the `AttributeRequest-Result()` method is called back:

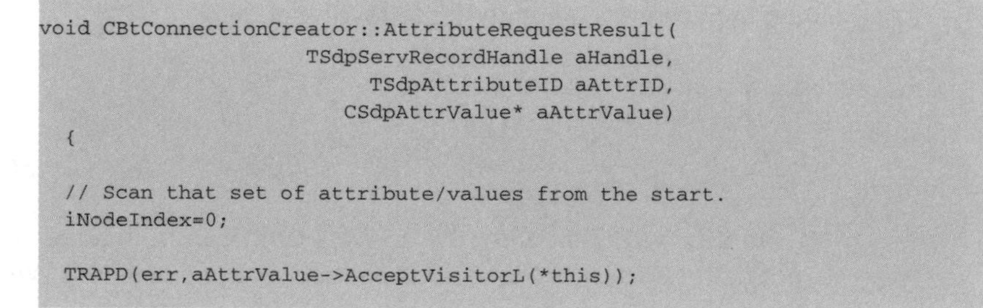

```
void CBtConnectionCreator::AttributeRequestResult(
                     TSdpServRecordHandle aHandle,
                          TSdpAttributeID aAttrID,
                        CSdpAttrValue* aAttrValue)
  {

  // Scan that set of attribute/values from the start.
  iNodeIndex=0;

  TRAPD(err,aAttrValue->AcceptVisitorL(*this));
```

```
if (err!=KErrNone)
  RequestComplete(err);  // Can't run AcceptVisitorL().

else if (lSerialPortAttributeList[iNodeIndex].iAction==
                                          EAttrItemFinished)
  {
  iRemotePort=iExtractedValue;

 // Reached end of struct and all as expected -
 // so we have located the port.
  TBuf<32>bb;
  _LIT(KPortIndex,"Port %d, Index %d");
  bb.Format(KPortIndex,iRemotePort,iNodeIndex);
  iObserver->LogInfo(bb);
  }
}
```

To parse the attributes we call the `AcceptVisitorL()` method. We are expected to implement the `MSdpAttributeValueVisitor` interface to handle callbacks generated during the parsing. In summary, for each node in the attributes, the `VisitAttributeValueL()` method is called back.

```
void CBtConnectionCreator::VisitAttributeValueL(
    CSdpAttrValue &aValue,
    TSdpElementType aType)

  {
  TBuf<32>bb;
  bb.Format(_L("VisitAttributeValueL %d"),iNodeIndex);
  iObserver->LogInfo(bb);

  switch (lSerialPortAttributeList[iNodeIndex].iAction)
    {
    case EAttrItemCheckType:
      if (lSerialPortAttributeList[iNodeIndex].iElementType!=aType)
        User::Leave(KErrGeneral);
      break;

    case EAttrItemCheckTypeAndValue:
      // Check type and values are as expected.
      if (lSerialPortAttributeList[iNodeIndex].iElementType!=aType)
        User::Leave(KErrGeneral);

      // Check value is as expected.
      switch (aValue.Type())
        {
        case ETypeUint:
          if (aValue.Uint() != lSerialPortAttributeList[
                                        iNodeIndex].iValue)
            User::Leave(KErrArgument);
          break;
```

```
      case ETypeInt:
      if (aValue.Int()!=lSerialPortAttributeList[
                              iNodeIndex].iValue)
        User::Leave(KErrArgument);
      break;

    case ETypeBoolean:
      if (aValue.Bool()!=lSerialPortAttributeList[
                              iNodeIndex].iValue)
        User::Leave(KErrArgument);
      break;

    case ETypeUUID:
      if (aValue.UUID()!=TUUID(lSerialPortAttributeList[
                                iNodeIndex].iValue))
        User::Leave(KErrArgument);
      break;

     default:
       break;
    }
    break;
  case EAttrItemCheckEnd:
  case EAttrItemFinished:   // Too many items in list.
    User::Leave(KErrGeneral);
    break;
  case EAttrItemReadValue:
  // Check type is as expected and read remote value.
    if (lSerialPortAttributeList[iNodeIndex].iElementType!=aType)
      User::Leave(KErrGeneral);
    // Possibly found the remote port info.
    iExtractedValue=aValue.Uint();

   _LIT(KEAttrItemReadValue,"EAttrItemReadValue: %d");
    bb.Format(KEAttrItemReadValue,iExtractedValue);
    iObserver->LogInfo(bb);
    break;
  default:
    break;
  }
 NextNodeL();
 }
```

Since we are expecting a specific structure, the one defined by the connection listener, we can run a small state machine to validate each of the nodes delivered. The more validation that we perform, the more likely it is that our application will correctly identify the required record.

If all goes well and the attribute structure matches the one we are expecting, we have extracted the remote port published by the connection listener. Upon completion of the search, we signal our pseudo-event source. This eventually calls our `RunL()`:

```
void CBtConnectionCreator::RunL()
  {
```

```
switch (iState)
  {
  case ECreatorStateFindService:
    delete(iSdpAgent);
    iSdpAgent=NULL;
    delete(iSdpSearchPattern);
    iSdpSearchPattern=NULL;
    if (iStatus.Int()==KErrNone)
      { // Sucessfully discovered the remote port.
      Connect();
      }
    else
      { // Report we failed to connect.
      iObserver->StatusInfo(EBtsConnectionEstablished,iStatus.Int());
      iState=ECreatorStateIdle;
      }
    break;

  // Removed for clarity....
  }
}
```

If the search completed successfully, we can proceed to create a socket connection:

```
void CDtConnectionCreator::Connect()

  // Issue the connect request now we have
  // the remote device address and port.
  {
  iSocket.Open(iSocketServer,KRFCOMMDesC);
  TBTServiceSecurity serviceSecurity;
  serviceSecurity.SetAuthentication(EFalse);
  serviceSecurity.SetEncryption(EFalse);
  serviceSecurity.SetAuthorisation(ETrue);
  serviceSecurity.SetDenied(EFalse);

  TBTSockAddr addr;
  addr.SetBTAddr(iRemoteDevAddr);
  addr.SetPort(iRemotePort);  // As fetched from remote.

  addr.SetSecurity(serviceSecurity);
  iSocket.Connect(addr,iStatus);
  SetActive();
  iState=ECreatorStateConnecting;
  }
```

As with the connection listener, we open an `RFCOMM` socket and set up the required security. Since we are establishing a Bluetooth connection, we need to specify the Bluetooth device address, obtained from the characteristics of the device chosen by the user, and the port obtained from the Service Discovery Database search.

Once the connection attempt is successful, we can issue a `Read()` request awaiting data from the remote device.

12.4.7 Reading and Writing with Sockets

Sockets are simply the endpoints of a logical connection. Whatever your application writes to a socket will be available to read at the other end. A socket, whether it be a Bluetooth, infrared or TCP-type socket, does not place any restrictions on or interpret content.

Sockets in Symbian OS, in common with other systems, expect to be presented with a stream of bytes to transfer. That stream of bytes is delivered to the remote end. Symbian OS uses Unicode to represent text. In particular, it uses the default UTF-16 encoding scheme which uses two bytes to represent a single character. Due to this mismatch in element size, we need to convert the text to a stream of bytes.

For simplicity, our application has assumed that any text that needs to be transferred only contains characters within the ASCII character set. By making this assumption we can convert between 8- and 16-bit descriptors simply by using the `Copy()` methods.

This approach is unlikely to be acceptable in a commercial-quality application since information loss is likely. One solution is to send each character as two bytes. In such a scheme, both ends need to agree in which order the low and high bytes are transferred. One disadvantage of this approach is the amount of redundancy in the data if the majority of characters belong to the ASCII character set.

An alternative approach is to use the UTF-8 encoding scheme. In such a solution, the UTF-16 encoded text is converted to UTF-8. The UTF-8 data is now in a form suitable for direct transfer via sockets. The receiving end would convert the UTF-8 encoded text back into UTF-16 resulting in lossless transfer of Unicode text between phones. Symbian OS contains support for translating between character sets in the form of the `CnvUtfConverter` class. This class is documented in the UIQ 3 SDK so we do not discuss it further here.

12.5 HyperText Transfer Protocol

The HyperText Transfer protocol (HTTP) is an industry-standard communications protocol for the transfer of content between devices. The full protocol specification is defined in RFC 2616 and is available at ***www.ietf.org***.

HTTP is said to be an application-level protocol. It places few restrictions on the mechanism by which content is transferred between devices other than assuming the transport is *reliable*. In practice, the TCP/IP protocol is normally used to transfer HTTP data between devices.

HTTP is usually associated with transferring web pages between a server and a browser, however that is only one of many possible uses of the protocol. Applications are free to define their own uses.

The HTTP specification defines a number of high-level commands which can be transferred between devices. The two most common are called `GET` and `POST`. A `GET` command is usually sent by a client device when it wants to retrieve information from the remote device. The `POST` request is typically used to send data to a remote device. Since both commands can send information with the request and accept data as part of the command response, it is possible to send data with a `GET` request and retrieve information with a `POST`.

12.5.1 Universal Resource Identifier

In order to fetch or store content, it needs to be identified. A text string called a Universal Resource Identifier (URI) is used to identify items. Strictly speaking a URI is different from a Universal Resource Location (URL), however many servers provide defaults so the subtle differences between the two terms have become eroded. Today the term URL is frequently used to cover URI, Universal Resource Name (URN) and URL.

In general, a URL comprises a number of constituent parts including:

- the scheme, such as HTTP, FTP, FILE, MAILTO
- a remote computer name
- a port on the remote computer
- a path on the remote computer
- a query string
- fragment information.

The rules describing which components are optional, which are mandatory and what default values are supplied are complex and beyond the scope of this book. Suffice to say, a correctly formed URL is required to successfully transfer content between devices.

12.5.2 HTTP Headers

The HTTP protocol is designed to be a flexible and extensible application-level protocol. In such systems, care is required to ensure both endpoints understand the limitations of the other, particularly if general-purpose applications such as web servers and browsers are communicating with each other. For example, if a server can deliver images, the client needs to understand in which format they are delivered and, probably, restrict the server to only supply formats that the client can process. The mechanism by which the devices can communicate preferences, limitations, status and other control information is through the use of headers.

12.5.3 HTTP Transactions

To actually perform a GET or POST type request you need to:

- specify a URL
- define the set of headers
- define which operation you are performing (GET or POST)
- supply any data associated with the operation
- perform the operation
- receive a response from the remote device.

This set of discrete actions is combined into a single *transaction*. An application is said to be performing a GET or POST transaction.

12.5.4 String Pools

Despite only ever being processed by computers, many communications protocols are defined using English language structure and text. For example, the HTTP protocol sends the textual characters 'G', 'E', 'T' to inform the remote computer a GET command has been issued, similarly the characters 'P', 'O', 'S', 'T' are used to indicate a POST command. A remote computer has to convert between this variable-length, human-readable text (often called tokens) and an internal action. Since such tokens are static and do not vary between spoken languages, Symbian OS facilitates compile-time generation of sets of tokens that can be processed efficiently. In Symbian OS, these are called *string pools*. The HTTP protocol contains a significant number of these tokens. Rather than individual applications having to manage a string pool, they can use the HTTP-specific string pool supplied as part of Symbian OS.

12.5.5 Capabilities

Our example application uses the default transport, TCP, to transfer HTTP data between phones. The HTTP framework uses the sockets interface to supply the TCP functionality. In order to open a TCP-style socket, an application is required to have `NetworkServices` capability. Therefore our application needs the `NetworkServices` capability to perform HTTP transfers.

12.5.6 HTTP and Mobile Phones

You may recall the earlier discussion around mobile networks and, in particular, Network Address Translation (NAT). HTTP works well in the mobile network environment since the mobile phone initiates the data transfer regardless of whether it is sending or receiving information. It always sends the *uplink* packet required by NAT to form the address translation table entry.

12.5.7 HTTP GET

Bringing all this together to perform a simple HTTP GET request results in the following code taken from the `Http1` example application:

```
void CHttpEngine::SetHeaderL(
      RHTTPHeaders aHeaders,
             TInt aHdrField,
    const TDesC8& aHdrValue)

  {
  RStringF str=iSession.StringPool().OpenFStringL(aHdrValue);
  CleanupClosePushL(str);
  THTTPHdrVal val(str);
  aHeaders.SetFieldL(iSession.StringPool().StringF(
          aHdrField,RHTTPSession::GetTable()),val);
  CleanupStack::PopAndDestroy(&str);
  }

void CHttpEngine::GetContentL(const TDesC8& aUrl)
  {
  TUriParser8 uri;
  User::LeaveIfError(uri.Parse(aUrl));
  RStringF method = iSession.StringPool().StringF(
            HTTP::EGET,RHTTPSession::GetTable());
  iTransaction = iSession.OpenTransactionL(uri,*this,method);
  RHTTPHeaders hdrs=iTransaction.Request().GetHeaderCollection();

  _LIT8(KUserAgent,"UIQ Http Example (1.0)");
  SetHeaderL(hdrs,HTTP::EUserAgent,KUserAgent);
  _LIT8(KAccept,"text/*");
  SetHeaderL(hdrs,HTTP::EAccept,KAccept);

  iTransaction.SubmitL();
  }
```

As you can see, we generate a fully parsed URI in the `uri` object. We also obtain the string token identified by the `HTTP::EGET` enumeration from the HTTP-specific string pool. We use these objects to create the transaction within which the HTTP GET request is performed. The HTTP infrastructure automatically adds those headers defined by RFC 2616 as being mandatory for an HTTP 1.1 client, so the application does not need to add these headers itself. In general, applications only need to focus on the headers associated with acceptable content types. In our example application, we have specified that we can only accept text content. We have not, however, defined what character set is acceptable.

Explaining the use of all HTTP headers is beyond the scope of this book, however the example application shows how you can add application-specific headers to a transaction.

The HTTP transaction processing classes use asynchronous methods to perform the actual processing. Rather than using threads, they use active objects. This has several consequences:

- Our application must have an active scheduler in the thread performing the HTTP transaction. A standard UIQ application primary thread always has a scheduler.
- Our application can process other events between the time an HTTP transaction is started and the time it completes.

HTTP transaction processing requires us to implement the `MHTTPTransactionCallback` interface since that is the mechanism used to deliver events and content from the HTTP framework to individual applications:

```
void CHttpEngine::MHFRunL(RHTTPTransaction aTransaction,
                              const THTTPEvent& aEvent)

  // Note: this is not allowed to leave on some
  // events... despite it being an L function.
  {
  switch (aEvent.iStatus)
    {
    case THTTPEvent::EGotResponseHeaders:
      {// HTTP response headers have been received.
      _LIT(KEGotResponseHeaders,"EGotResponseHeaders");
      iObserver->LogInfo(KEGotResponseHeaders);
      RHTTPResponse resp=aTransaction.Response();
      iStatusCode=resp.StatusCode();
      iStatusText.Copy(resp.StatusText().DesC());
      break;
      }
    case THTTPEvent::EGotResponseBodyData:
      {// Some (more) body data has been received.
      _LIT(KEGotResponseBodyData,"EGotRespBodyData");
      iObserver->LogInfo(KEGotResponseBodyData);
      MHTTPDataSupplier* body=aTransaction.Response().Body();
      TPtrC8 bodyData;
      body->GetNextDataPart(bodyData);
      iObserver->ReportBody(bodyData);
      body->ReleaseData();
      break;
      }

    case THTTPEvent::EResponseComplete:
      // Transaction complete (got all content etc).
      _LIT(KEResponseComplete,"EResponseComplete");
      iObserver->LogInfo(KEResponseComplete);
      break;
```

```
    case THTTPEvent::ESucceeded:
      // Transaction finished - all ok.
      aTransaction.Close();
      iObserver->ReportEvent(KErrNone,iStatusCode,iStatusText);
      iIsBusy=EFalse;
      break;

    case THTTPEvent::EFailed:
      // Transaction finished - something fell over.
      aTransaction.Close();
      iObserver->ReportEvent(KErrGeneral,iStatusCode,iStatusText);
      iIsBusy=EFalse;
      break;

      case THTTPEvent::ERedirectedPermanently:
        _LIT(KERedirectedPermanently,"ERedirectedPermanently");
        iObserver->LogInfo(KERedirectedPermanently);
        break;

      case THTTPEvent::ERedirectedTemporarily:
        _LIT(KERedirectedTemporarily,"ERedirectedTemporarily");
        iObserver->LogInfo(KERedirectedTemporarily);
        break;

      default:
        { // -46 means no NetworkServices CAPABILITY !
        TBuf<64>bb;
        _LIT(KUnhandled,"Unhandled: %d");
        bb.Format(KUnhandled,aEvent.iStatus);
        iObserver->LogInfo(bb);
        break;
        }
    }
  }

TInt CHttpEngine::MHFRunError(TInt aError,
           RHTTPTransaction aTransaction,
                const THTTPEvent& aEvent)

  // Called by internal active object if error occurs.
  {
  TBuf<64>bb;
  _LIT(KMHFRunError,"MHFRunError: %d,%d");
  bb.Format(KMHFRunError,aError,aEvent.iStatus);
  iObserver->LogInfo(bb);

  return(KErrNone);
  }
```

In our application, we have chosen to capture and display events as and when they occur. We place the latest event at the top of the list. Figure 12.6 shows the result of performing an HTTP GET on the URL ***www.uiq.com***.

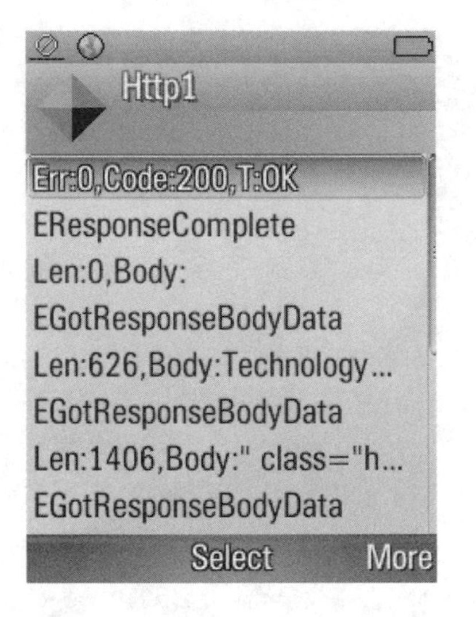

Figure 12.6 Http1 example application

Figure 12.6 shows that we received several `EGotResponseBodyData` events. We received multiple `EGotResponseBodyData` events containing 1406 characters followed by a single `EGotResponseBodyData` containing 626 characters. Finally, we received an `EGotResponseBodyData` event with zero-length data.

Once all the `EGotResponseBodyData` events have been delivered, an `EResponseComplete` followed by an `ESucceeded` event are delivered by the framework. Our application has extracted the HTTP status code, 200, and status text message, `OK`, from the transaction response headers. If you are familiar with the HTTP specification, you will recognize that the code and text values also indicate that the GET transaction has completed successfully.

If your `MHFRunL()` method leaves for any reason, the HTTP framework calls the `MHFRunError()` method to enable your application to handle the error condition. We simply choose to add a new log message.

Our application has only implemented a sample of the full range of possible events defined by the `THTTPEvent` class. A commercial-grade application needs to understand how each of the events can occur and deal with them appropriately. The full list of events can be found in the `thttpevent.h` file.

12.5.8 HTTP POST

To perform any truly meaningful tasks with HTTP POST you need to have a remote computer that can understand the body sent as part of the post. In general, this means you need to have some kind of system dedicated to the task in hand. Since this book is generic in nature, our example application does not perform any specific task using HTTP POST. Rather, it uses the fact that most web servers at least give you a response if you send them a POST request. This enables us to demonstrate how you might construct an application that uses HTTP POST requests without having to supply a specific HTTP server.

We have chosen to separate the functionality associated with HTTP GET and POST into two separate *engines*. This is done to aid in the understanding of what is required to perform these tasks. It is possible to combine the two engines should your specific application need to perform HTTP GET and POST transactions.

As with HTTP GET, HTTP POST needs to bring together the concepts of URIs, headers and transactions as demonstrated by the following code:

```
void CHttpPostEngine::PostContentL(
  const TDesC8& aUrl,
  HBufC8* aData)
  {
  iData=aData;

  TUriParser8 uri;
  User::LeaveIfError(uri.Parse(aUrl));

  RStringF method=iSession.StringPool().StringF(
          HTTP::EPOST, RHTTPSession::GetTable());
  iTransaction = iSession.OpenTransactionL(uri,
                                   *this,method);

  RHTTPHeaders hdrs=iTransaction.Request().
                     GetHeaderCollection();

  _LIT8(KUserAgent,"UIQ Http Example (1.0)");
  SetHeaderL(hdrs,HTTP::EUserAgent,KUserAgent);

  _LIT8(KContentType,"text/plain");
  SetHeaderL(hdrs,HTTP::EContentType,KContentType);

  iTransaction.Request().SetBody(*this);
  iTransaction.SubmitL();
  }
```

The main differences between a GET and POST transaction are:

- the method name supplied to the transaction
- the headers used
- the fact we have some content to transfer.

We inform the POST transaction that we have some content to transfer using the `iTransaction.Request().SetBody(*this)` method, which requires us to implement the `MHTTPDataSupplier` interface. Some applications may have large quantities of data to transfer and prefer to supply this data over several blocks. In our application, we have chosen to supply all the content in a single block.

The `OverallDataSize()` should report the total amount of data that needs to be transferred; `GetNextDataPart()` supplies one data block at a time. When all blocks have been supplied, `GetNextDataPart()` should return `ETrue`. Until that time, it should report `EFalse`, indicating further content is available.

The implementation of the `MHTTPDataSupplier` interface in our application is:

```
TBool CHttpPostEngine::GetNextDataPart(
  TPtrC8& aDataPart)
  {
  aDataPart.Set(*iData);
  return(ETrue);
  }

void CHttpPostEngine::ReleaseData(void)
  {
  delete(iData);
  iData=NULL;
  }

TInt CHttpPostEngine::OverallDataSize(void)
  {
  if (!iData)
    return(0);
  return((*iData).Length());
  }

TInt CHttpPostEngine::Reset(void)
  {
  return(KErrNotSupported);
  }
```

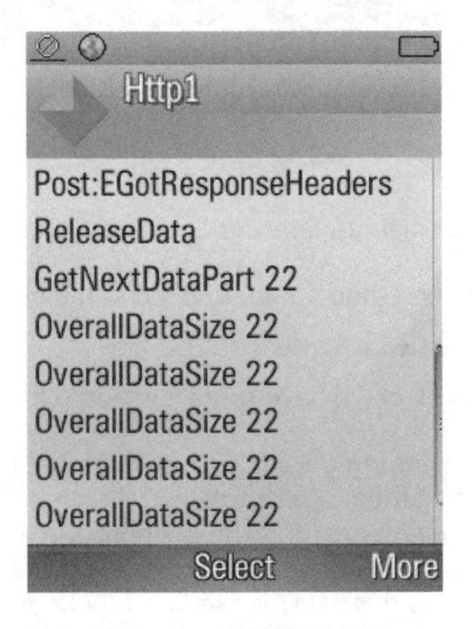

Figure 12.7 HTTP POST log from `Http1`

To monitor and handle events generated by the HTTP framework, we also need to implement the `MHTTPTransactionCallback` interface. Apart from a few points of detail, the implementation for a POST transaction is the same as described previously for the GET transaction.

The result of performing an HTTP POST request on the URL ***www.uiq.com*** (simply to demonstrate the code functioning) generates a log along the lines of that shown in Figure 12.7.

Note that since we sent the POST request to ***www.uiq.com***, the application gets response headers and a response body since that is what a typical web server sends.

12.6 Messaging Architecture

Symbian OS contains an extensive system component called the messaging architecture or message store (see Figure 12.8). This component facilitates the creation, sending, receiving and storing of messages, irrespective of which mechanism is used to transfer the messages.

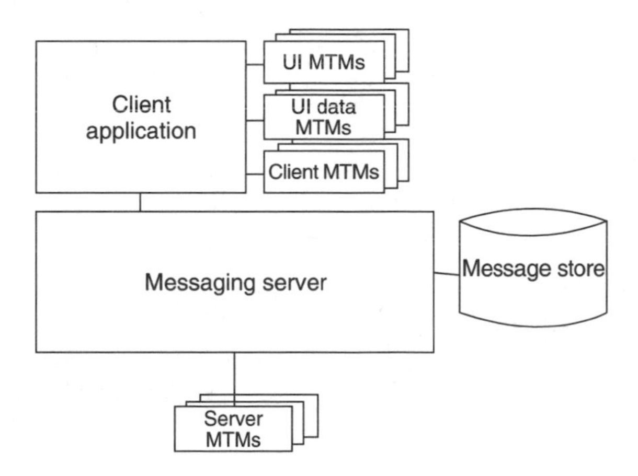

Figure 12.8 Symbian OS messaging architecture

12.6.1 Message Server

Symbian OS support for messaging is based around a client–server process model. The server manages the actual messages through a set of plug-ins, often called Message Type Modules (MTMs). Client access to the message server is through the `CMsvSession` object which encapsulates the IPC connection with the server.

A primary task of the message server is to provide thread-safe shared access to all entries within the overall message store.

12.6.2 Message Entries

The message server manages a set of entries. An individual entry can be one of the following types:

- folder
- message
- attachment
- service.

Entries may have child entries and they are, therefore, managed within a tree-like structure (see Figure 12.9) similar to files and folders within a file system. All entries are referenced by a unique ID, encapsulated by the `TMsvId` object.

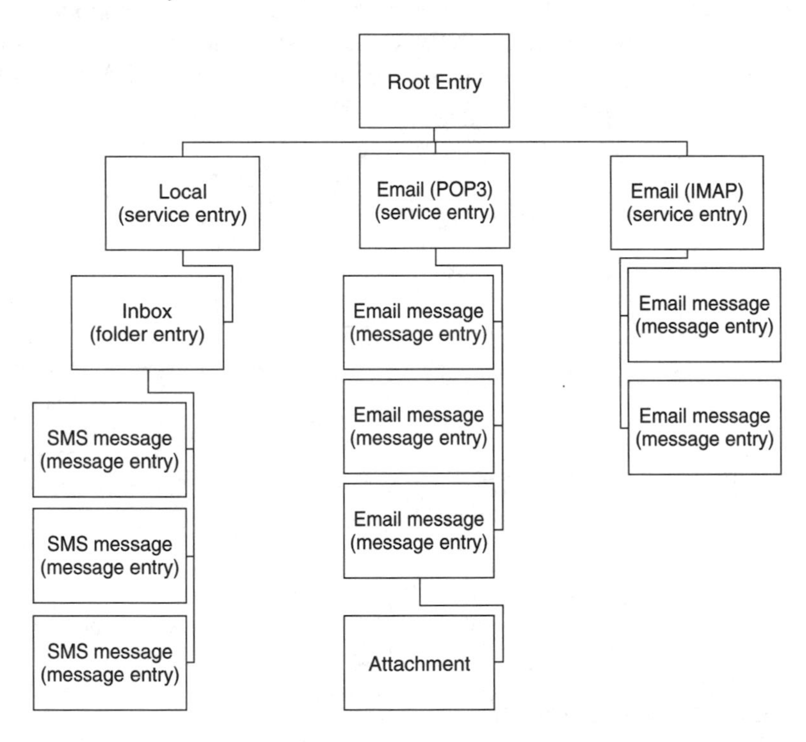

Figure 12.9 Messaging tree structure

The message server pre-defines a number of `TMsvId` IDs (entries), including the following:

- `KMsvLocalServiceIndexEntryId`
- `KMsvGlobalInBoxIndexEntryId`

- `KMsvGlobalOutBoxIndexEntryId`
- `KMsvDraftEntryId`
- `KMsvSentEntryId`
- `KMsvDeletedEntryFolderEntryId`.

At the root of the tree is the `KMsvRootIndexEntryId`, which leads to the `KMsvLocalServiceIndexEntryId` entry. At a minimum this contains the Inbox, Outbox, Drafts and Sent items folders. You can see this in Figure 12.9.

The Symbian OS messaging architecture is abstract in design, allowing many functions to be performed irrespective of entry type. These include navigating between parent and child entries, moving or copying entries between folders and deleting entries.

This abstract design facilitates the concepts of global Inbox and global Outbox. All entries, irrespective of specific entry details can be stored within the same folder. For example, you may have observed SMS messages, Bluetooth and infrared messages all stored together in your phone's Inbox.

The SMS MTM stores *incoming* messages in the global Inbox. This allows you to use the synchronous methods in `CMsvEntry` such as `ChangeL()` and `CreateL()`. For the POP, IMAP and SMTP MTMs, however, their *incoming* messages may be stored under the service entry (or potentially in a sub-folder in the service entry, for IMAP-subscribed folders). Because they are under the service entry, you have to use the asynchronous methods of `CMsvEntry`, as the MTM's server implementation has to deal with requests such as `DeleteL()` and `ChangeL()`. However, all outgoing messages are generally stored in the Drafts, Outbox and Sent folders regardless of the MTM.

12.6.3 Message Entry Storage

In general, the content of a single message entry is distributed over three storage locations (see Figure 12.10):

- a message server index entry represented by a `TMsvEntry`
- some persistent file storage represented by a `CMsvStore`
- a folder within the file system.

TMsvEntry

All entries, regardless of type, have an associated `TMsvEntry`. From an application programming perspective, each folder within the message store comprises an array or index of `TMsvEntry` objects.

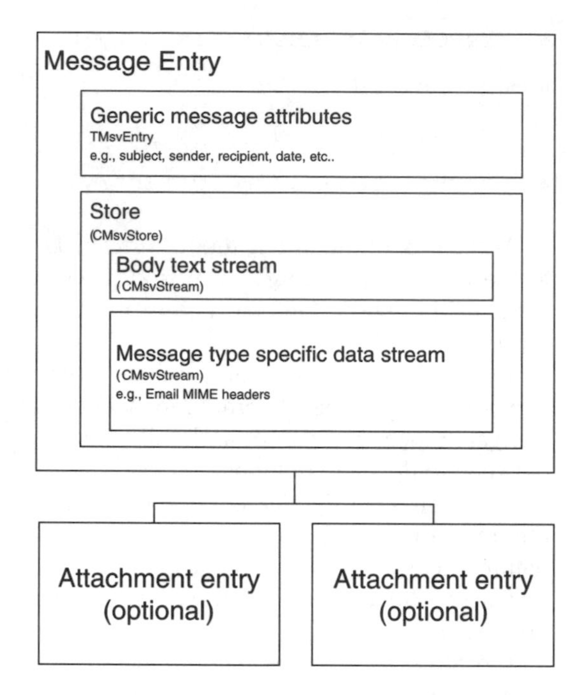

Figure 12.10 Message entry structure

If you look at the properties of a `TMsvEntry` (from `msvstd.h`) it shows a number of useful fields:

```
class TMsvEntry
  {
  private:
    TMsvId iId;
    TMsvId iParentId;
    TInt32 iData;
    TInt32 iPcSyncCount;
    TInt32 iReserved;  // Reserved for future proofing.

  public:
    TMsvId iServiceId;
    TMsvId iRelatedId;
    TUid   iType;
    TUid   iMtm;
    TTime  iDate;
    TInt32 iSize;
    TInt32 iError;
    TInt32 iBioType;
    TInt32 iMtmData1;
    TInt32 iMtmData2;
    TInt32 iMtmData3;
    TPtrC iDescription;
    TPtrC iDetails;
  };
```

Not surprisingly, the `iId` field contains the unique ID for a specific `TMsvEntry`. The `iParentId` contains the unique ID for the parent, allowing us to traverse the tree hierarchy.

The `iType` field can take one of the following values allowing applications to distinguish between different types of entry:

- `KUidMsvRootEntry`
- `KUidMsvServiceEntry`
- `KUidMsvFolderEntry`
- `KUidMsvMessageEntry`
- `KUidMsvAttachmentEntry`.

The `iMtm` field allows applications to determine the MTM with which the entry is associated. This is a key piece of information that enables applications to process entries fully.

A `TMsvEntry` also contains some generic fields such as `iDate`, `iSize`, `iDescription` and `iDetails`. These are sufficient for applications to display summary information, for example, in a list view of a particular folder, without having to fetch any other data associated with an entry. `iDetails` typically stores the address, while `iDescription` usually stores the subject for an email. The contents of these fields are MTM-specific.

CMsvStore

While some information is stored within the `TMsvEntry`, the remainder is saved within a file store entry. The format depends on what type of entry is being represented and is typically used to store headers and body text should such items be present in a particular message type.

File System Folder All entries are associated with a specific folder within the file system where they may choose to store further information. As with the `CMsvStore`, the usage of this type of storage is optional and is MTM-specific. A typical usage may be to store the actual attachment files associated with attachment entries.

CMsvEntry

A `CMsvEntry` encapsulates the functionality to obtain all the component parts of an individual entry. In particular, a `CMsvEntry` does not represent a specific entry, only the ability to access the component parts of an entry.

A `CMsvEntry` can often be considered as an `Iterator`. Each folder effectively comprises a set of `TMsvIds`. A `CMsvEntry` can be set to each of these IDs in turn to be able to access information about each entry.

Pro tip

Since a CMsvEntry object is expensive in terms of the resources it uses, we generally recommend that an application minimizes the number of CMsvEntry objects it creates and destroys. A CMsvEntry is re-used simply by assigning a different TMsvId to it. It is entirely possible not to use CMsvEntry objects at all. If you find you are having performance problems, we suggest that you investigate removing the use of CMsvEntry, and utilize other methods of the CMsvSession class.

12.6.4 Messaging Example Application

Our messaging application brings together many of the above objects to demonstrate how you may go about scanning the Inbox to look for a particular set of entries and creating and sending an entry. We use SMS as an example because it is a common requirement and relatively simple to implement.

Our example application requires the NetworkServices, ReadUserData and WriteUserData capabilities to be able to read the user Inbox and transmit SMS messages.

A first pass at scanning the Inbox to list all SMS entries might result in the following code:

```
void CMessageEngine::ConstructL()
  {

  // Get IPC connection to the message server process.
  iMessageServer=CMsvSession::OpenSyncL(*this);

  // We need a client-side CClientMtmRegistry  to obtain MTM objects.
  iMtmRegistry = CClientMtmRegistry::NewL(*iMessageServer);

  // Required to send SMS entries.
  iSendSelection=new(ELeave)CMsvEntrySelection();

  // The one CMsvEntry we should endeavor to reuse.
  iMsvEntry=CMsvEntry::NewL(*iMessageServer,
                KMsvGlobalInBoxIndexEntryId,
                   TMsvSelectionOrdering());

  TBuf<256>bb;

  // Scan the inbox looking for SMS messages.
  CMsvEntrySelection* inboxEntries=iMsvEntry->
             ChildrenWithMtmL(KUidMsgTypeSMS);
  CleanupStack::PushL(inboxEntries);
  TInt count=inboxEntries->Count();
```

```
// Display number of SMS entries we have been informed exist.
_LIT(KSMSEntries,"%d SMS entries");
bb.Format(KSMSEntries,count);
iObserver->LogInfo(bb);

// Display who each SMS is from and the content as
// an example of accessing these items.
for (TInt i=0;i<count;i++)
  {
  CMsvEntry* qq=iMessageServer->GetEntryL(inboxEntries->At(i));
  TMsvEntry msvEntry=qq->Entry();
  delete(qq);

  // Demonstrate it's an SMS msg that's arrived.
  if (msvEntry.iMtm!=KUidMsgTypeSMS)
    continue;

  // Obtain some info about the SMS in the Inbox.
  CSmsClientMtm* smsMtm=static_cast<CSmsClientMtm*>(
           iMtmRegistry->NewMtmL(KUidMsgTypeSMS));
  CleanupStack::PushL(smsMtm);
  smsMtm->SwitchCurrentEntryL(inboxEntries->At(i));
  smsMtm->LoadMessageL();

// Look at the SMS header to see who it's from.
  CSmsHeader& smsHdr=smsMtm->SmsHeader();
  TPtrC msgFrom(smsHdr.FromAddress());
  bb.Format(_L("SMS from: %S"),&msgFrom);
  iObserver->LogInfo(bb);

  // We can look at the body to see the content.
  TPtrC msgBody(smsMtm->Body().Read(0));
  bb.Format(_L("SMS body: %S"),&msgBody);
  iObserver->LogInfo(bb);

  CleanupStack::PopAndDestroy(smsMtm);
  }
CleanupStack::PopAndDestroy(inboxEntries);
}
```

Our first task is to obtain an IPC connection to the message server process. This is closely followed by the requirement to connect to the MTM registry. The `CClientMtmRegistry` facilitates the creation of the client MTM objects specific to the message type. These are required to be able to obtain information specific to the message type.

Our application creates a `CMsvEntry` referencing the `KMsv-GlobalInBoxIndexEntryId`. You should recall that a number of folders within the message store have pre-defined IDs. Our application uses this information to create a starting point.

We now request the set of entries belonging to the `CMsvEntry`, where the `iMtm` field of the child entries is set to `KUidMsgTypeSMS`. Since the `CMsvEntry` is referencing the Inbox and only SMS messages belong to the `KUidMsgTypeSMS`, this extracts all SMS messages within the Inbox.

A `CMsvEntry` has a number of methods capable of generating a selection:

- `CMsvEntrySelection* ChildrenL() const;`
- `CMsvEntrySelection* ChildrenWithServiceL(TMsvId aServiceId) const;`
- `CMsvEntrySelection* ChildrenWithMtmL (TUid aMtm) const;`
- `CMsvEntrySelection* ChildrenWithTypeL (TUid aType) const;`

As you can see, all the variants return a `CMsvEntrySelection` object. This is simply an array of `TMsvIds`.

From the `CMsvEntrySelection` array, we can simply iterate though each of the child entries that matched our selection criteria. In this first attempt, we create a `CMsvEntry` to access the generic properties of the entry described by a `TMsvEntry`. We also obtain some SMS-specific properties, such as the sender and body of a message.

Some Optimizations

As previously described, creating `CMsvEntry` objects is a resource-intensive operation. Similarly, creating and deleting client MTM objects is inefficient. Rather than creating and deleting such objects each time around the loop, they are designed to be re-used by updating the entry that they represent. In the case of a `CMsvEntry`, we use the `SetEntryL()` method: in the case of the `CSmsClientMtm`, we use the `SwitchCurrent-`
`EntryL()` method.

The revised code is:

```
void CMessageEngine::ConstructL()
  {

  // Get IPC connection to the message server process.
  iMessageServer=CMsvSession::OpenSyncL(*this);

  // We need a client side CClientMtmRegistry
  // to obtain MTM objects.
  iMtmRegistry = CClientMtmRegistry::NewL(*iMessageServer);

  // Required to send SMS entries.
  iSendSelection=new(ELeave)CMsvEntrySelection();
```

```
// The one CMsvEntry we should endeavor to reuse.
iMsvEntry=CMsvEntry::NewL(*iMessageServer,
             KMsvGlobalInBoxIndexEntryId,
                TMsvSelectionOrdering());

TBuf<256>bb;

// Scan the inbox looking for SMS messages.
CMsvEntrySelection* inboxEntries=iMsvEntry->
           ChildrenWithMtmL(KUidMsgTypeSMS);
CleanupStack::PushL(inboxEntries);
TInt count=inboxEntries->Count();

// Display number of SMS entries we have
// been informed exist.
_LIT(KSMSEntries,"%d SMS entries");
bb.Format(KSMSEntries,count);
iObserver->LogInfo(bb);

// Demonstrates how you should go about
// reusing the CMsvEntry.
CSmsClientMtm* smsMtm=static_cast<CSmsClientMtm*>(
           iMtmRegistry->NewMtmL(KUidMsgTypeSMS));
CleanupStack::PushL(smsMtm);
for (TInt i=0;i<count;i++)
  {
  iMsvEntry->SetEntryL(inboxEntries->At(i));
  const TMsvEntry& tEntry=iMsvEntry->Entry();

  // Demonstrate its an SMS msg thats arrived.
  if (tEntry.iMtm!=KUidMsgTypeSMS)
    continue;

  // Type should be KUidMsvMessageEntry,
  // iServiceId will be KMsvLocalServiceIndexEntryId.
  _LIT(KTypeServiceId,"Type: %d, ServiceId %d");
  bb.Format(KTypeServiceId,tEntry.iType.iUid,tEntry.iServiceId);
  iObserver->LogInfo(bb);

  // Obtain some info about the SMS in the Inbox.
  smsMtm->SwitchCurrentEntryL(inboxEntries->At(i));
  smsMtm->LoadMessageL();

  // Look at the SMS header to see who its from.
  CSmsHeader& smsHdr=smsMtm->SmsHeader();
  TPtrC msgFrom(smsHdr.FromAddress());
  _LIT(KSMSfrom,"SMS from: %S");
  bb.Format(KSMSfrom,&msgFrom);
  iObserver->LogInfo(bb);

  // We can look at the body to see the content.
  TPtrC msgBody(smsMtm->Body().Read(0));
  _LIT(KSMSbody,"SMS body: %S");
  bb.Format(KSMSbody,&msgBody);
  iObserver->LogInfo(bb);
  }
CleanupStack::PopAndDestroy(smsMtm);
```

```
CleanupStack::PopAndDestroy(inboxEntries);
}
```

The above code demonstrates the recommended approach to using `CMsvEntry` objects. In particular, it shows how to reuse the object to reference different entries.

Sending Messages

While our example application demonstrates sending an SMS, the general principles involved can be applied to any other message type.

To send a message we need to:

- create an entry in the message store to represent the message
- add any required body text and set up the MTM-specific information
- move the entry to the Outbox and schedule the time for it to be sent.

Since a number of the operations involved can take an extended time to complete, most of the functions used to send a message are asynchronous. In a more general interface, some functions need to be able to report progress information, such as reporting which is the current message being downloaded. The message server encapsulates the progress reporting within a `CMsvOperation` object returned by a number of asynchronous methods.

Creating a New Message

The following code starts the construction of a new message:

```
TBool CMessageEngine::StartSendText(
          const TDesC& aMessageBody,
           const TDesC& aSmsAddress)
  {
  if (IsActive())
    return(EFalse);  // We can't send the info just yet.

  // Reset list of Ids of messages we wish
  // to send - as we are creating a new one.
  DeleteComponents();
  iSendSelection->Reset();
  iMessageBody=aMessageBody;
  iDestinationAddress=aSmsAddress;

  // Set up a blank SMS message in the message server.
  // The created SMS has no body or destination
  // telephone number at this stage.
  TMsvEntry newEntry;
```

```
newEntry.iServiceId=KMsvLocalServiceIndexEntryId;
newEntry.iRelatedId=0;
// set up the type of the entry: a message.
newEntry.iType=KUidMsvMessageEntry;
newEntry.iMtm=KUidMsgTypeSMS;           // It's an SMS.
newEntry.iDate.UniversalTime();
newEntry.iSize=0;
newEntry.iError=0;
newEntry.iBioType=0;
newEntry.iMtmData1=0;
newEntry.iMtmData2=0;
newEntry.iMtmData3=0;
newEntry.SetInPreparation(ETrue);

TRAPD(err,
  // Represents an entry we are manipulating.
  iMsvEntry=CMsvEntry::NewL(*iMessageServer,
                      KMsvDraftEntryIdValue,
                   TMsvSelectionOrdering());
  iOperation=iMsvEntry->CreateL(newEntry,iStatus);
  );

iState=ESmsSendCreateMessage;
SetActive();

if (err!=KErrNone)
  { // Simulate request completion on error.
  TRequestStatus* q=(&iStatus);
  User::RequestComplete(q,err);
  }
return(ETrue);
}
```

The `CMsvEntry::CreateL()` method is both asynchronous and returns a `CMsvOperation`. The `CMsvOperation` can be used to monitor the progress of the message creation. In practice, this operation is quite fast so no specific user feedback needs to be provided. When the message creation completes, our `RunL()` is called:

```
void CMessageEngine::RunL()
  {
  TBuf<128>bb;
  TMsvLocalOperationProgress progress;
  TInt err=iStatus.Int();
  switch (iState)
    {
    case ESmsSendCreateMessage:
      if (err==KErrNone)
        {
        progress = McliUtils::GetLocalProgressL(*iOperation);
        if (progress.iError!=KErrNone)
          { // Some messaging error being reported.
          err=progress.iError;
          }
        else
          {
```

```
        TRAP(err,
          // In general we must not assume that iMsvEntry has not
          // changed, especially  in an async system, so we set it here.
          iMsvEntry->SetEntryL(progress.iId);
          DeleteOperation();
          MsgCreationCompleteL(progress.iId);
          );
        }
      }
    if (err!=KErrNone)
      {
      DeleteOperation();
      _LIT(KCreateErr,"Create err %d");
      bb.Format(KCreateErr,err);
      iObserver->LogInfo(bb);
      iState=ESmsSendIdle;
      }
    break;

    // Remainder removed for clarity.
  }
}
```

If we successfully create the message, we ensure that our single global `iMsvEntry` is referencing the correct entry and proceed to complete the message. At this time the message entry is marked as being *in preparation*.

```
void CMessageEngine::MsgCreationCompleteL(const TMsvId aId)
  {
  if (iMtm==NULL || iMsvEntry->Entry().iMtm != iMtm->
                                Entry().Entry().iMtm)
    { // We don't have an MTM or the MTM for this entry
      // is different to one we currently have.
    delete(iMtm);
    iMtm=NULL;
    iMtm=iMtmRegistry->NewMtmL(iMsvEntry->Entry().iMtm);
    }

  // Set indicated entry as current one.
  iMtm->SetCurrentEntryL(iMsvEntry);

  TMsvEntry tEntry=iMtm->Entry().Entry();

  // Set message body from our msg text.
  CRichText& mtmBody=iMtm->Body();
  mtmBody.Reset();
  mtmBody.InsertL(0,iMessageBody);

  // Set the destination address.
  tEntry.iDetails.Set(iDestinationAddress);

  tEntry.SetInPreparation(EFalse);
```

```
// We are no longer preparing msg.
// We are now waiting to send.
tEntry.SetSendingState(KMsvSendStateWaiting);

tEntry.iDate.UniversalTime();

CSmsClientMtm* smsMtm = static_cast<CSmsClientMtm*>(iMtm);
smsMtm->RestoreServiceAndSettingsL();

// CSmsHeader encapsulates data specific for sms
// messages, like service center number and
// options for sending.
CSmsSettings* sendOptions=CSmsSettings::NewL();
CleanupStack::PushL(sendOptions);
sendOptions->SetStatusReportHandling(
             CSmsSettings::EMoveReportToInboxVisible);
sendOptions->CopyL(smsMtm->ServiceSettings());
sendOptions->SetDelivery(ESmsDeliveryImmediately);

CSmsHeader& header=smsMtm->SmsHeader();
header.SetSmsSettingsL(*sendOptions);
CleanupStack::PopAndDestroy(); // sendOptions

// If no SMC address, attempt to use a
// default SMC address - if none give up.
if (!header.Message().ServiceCenterAddress().Length())
  {
  CSmsSettings& serviceSettings=smsMtm->ServiceSettings();
  // If no SMC - give up here.
  if (!serviceSettings.ServiceCenterCount())
    User::Leave(KErrCouldNotConnect);

  header.Message().SetServiceCenterAddressL(
          serviceSettings.GetServiceCenter(
       serviceSettings.DefaultServiceCenter()).Address());
  if (!header.Message().ServiceCenterAddress().Length())
    User::Leave(KErrCouldNotConnect);
  }

// Add our recipient to the list, takes in two TDesCs,
// first is real address and the second is an alias -
// works also without the alias parameter.
smsMtm->AddAddresseeL(iDestinationAddress,tEntry.iDetails);

// Save message to server.
smsMtm->SaveMessageL();

// Move the message from the drafts folder
// to the outbox folder.
iMsvEntry->SetEntryL(tEntry.Parent());
iOperation=iMsvEntry->MoveL(tEntry.Id(),
                 KMsvGlobalOutBoxIndexEntryId,iStatus);
iState=ESmsSendMoveMessageToOutBox;
SetActive();
}
```

At this stage, we set up the body of the SMS to be the content we will transfer. The message is changed from *in preparation* and the sending state is set to *waiting to be sent*. Since the message is an SMS, we need to obtain a `CSmsClientMtm` to be able to set SMS-specific information such as the SMSC address.

Finally, we need to ensure the entry being manipulated is updated on the server side. Until the call to `SaveMessageL()`, the message information is on the client side. The `SaveMessageL()` method flushes the updated information server side.

At this point, we have a fully constructed message which we need to move from the Drafts folder to the Outbox folder. The move request is asynchronous and calls our `RunL()` when it completes.

```
void CMessageEngine::RunL()
  {
  TBuf<128>bb;
  TMsvLocalOperationProgress progress;
  TInt err=iStatus.Int();
  switch (iState)
    {
  case ESmsSendMoveMessageToOutBox:
    if (err==KErrNone)
      {
      progress = McliUtils::GetLocalProgressL(*iOperation);
      if (progress.iError!=KErrNone)
        { // Some messaging error being reported.
        err=progress.iError;
        }
      else
        { // Moving to outbox has completed successfully.
          // Start sending the message.
        TRAP(err,
          // Add our message to the selection.
          iSendSelection->AppendL(progress.iId);
          DeleteOperation();
          TBuf8<4> junk;
          iOperation=iMtm->InvokeAsyncFunctionL(
                     ESmsMtmCommandScheduleCopy,
                  *iSendSelection,junk,iStatus);
          iState=ESmsSendTransmittingMessage;
          SetActive();
          );
        }
      }
    if (err!=KErrNone)
      {
      DeleteOperation();
      _LIT(KMoveErr,"Move err %d");
      bb.Format(KMoveErr,err);
      iObserver->LogInfo(bb);
      iState=ESmsSendIdle;
      }
    break;
```

```
    // Other cases removed for clarity.

    }
  }
```

Once the SMS has been successfully moved to the Outbox, we need to schedule its transmission. This is achieved by calling the `InvokeAsyncFunctionL()` method of the `CSmsClientMtm`. Since this method presents a generic interface, it takes a set of `TMsvIds` on which to operate. While we only have a single entry to send, we still have to pass it encapsulated within a `CMsvEntrySelection`. As the name implies, the `InvokeAsyncFunctionL()` is asynchronous and therefore calls our `RunL()` when it completes.

```
void CMessageEngine::RunL()
  {
  TBuf<128>bb;
  TMsvLocalOperationProgress progress;
  TInt err=iStatus.Int();
  switch (iState)
    {
    // Finished ESmsMtmCommandScheduleCopy.
    case ESmsSendTransmittingMessage:

      DeleteOperation();
      _LIT(KScheduledOK,"Scheduled OK");
      _LIT(KScheduledErr,"Schedule err %d");
      if (err==KErrNone)
        bb=KScheduledOK;
      else
        bb.Format(KScheduledErr,err);
      iObserver->LogInfo(bb);
      iState=ESmsSendIdle;
      break;

    // Other cases removed for clarity.
    }
  }
```

Note that this means the scheduling operation has completed and not that the SMS has actually been sent yet. The actual transfer depends on the scheduling information chosen. In our example, we have requested the SMS to be sent immediately since we set the `CSmsSettings` delivery options to be `ESmsDeliveryImmediately` and the `iDate` field to the current time.

When our SMS is actually sent, we receive status information from the message server via the `HandleSessionEventL()` method:

```
void CMessageEngine::HandleSessionEventL(
                TMsvSessionEvent aEvent,
                        TAny* aArg1,
```

```
                              TAny* aArg2,
                              TAny* aArg3)
{
if (aEvent==EMsvEntriesMoved)
  {

  // arg2 is the TMsvId of the new parent.

  TMsvId* entryId=static_cast<TMsvId*>(aArg2);
  if (*entryId==KMsvSentEntryId)
    {
      // Items have been moved to the Sent items
      // folder. aArg1 is a CMsvEntrySelection
      // (a list of entries that have been moved).
    CMsvEntrySelection* selection=static_cast<
                  CMsvEntrySelection*>(aArg1);

    TBuf<128>bb;
    TInt count=selection->Count();
    _LIT(KMovedToSentFolder,"%d moved to sent folder");
    bb.Format(KMovedToSentFolder,count);
    iObserver->LogInfo(bb);

    for (TInt i=0;i<count;i++)
      {
      _LIT(KMovedId,"Moved Id %d");
      bb.Format(KMovedId,selection->At(i));
      iObserver->LogInfo(bb);
      }
    }
  }
}
```

When we send entries, we are sent an `EMsvEntriesMoved` event since the message is moved from the Outbox to the Sent items folder. The example code shows us processing an `EMsvEntriesMoved` event. In general, we would record the `TMsvId`s of the entries we requested to be sent. When we are informed that entries have moved, we can compare the saved `TMsvId`s with those that have been moved to know whether our entry has been transmitted.

In applications that automatically create and send SMS messages to perform communications activity, you may wish to automatically delete messages after they have been transferred. You need to be quite sure that your messages are uniquely identifiable, and that your application does not delete any other SMS messages. The `CMsvEntry::DeleteL()` method can be used to delete entries.

Monitoring the Inbox

Monitoring the Inbox requires us to handle `EMsvEntriesCreated` delivered to the `HandleSessionEventL()` method, ensuring that the entry is created within the Inbox.

The following code fragment outlines the functionality you may incorporate to monitor the Inbox:

```
void CMessageEngine::HandleSessionEventL(
                TMsvSessionEvent aEvent,
                            TAny* aArg1,
                            TAny* aArg2,
                            TAny* aArg3)
  {
  if (aEvent==EMsvEntriesCreated)
    { // New entry been created in the message server.

    // Obtain the id from the session event.
    TMsvId* msvId=static_cast<TMsvId*>(aArg2);

    if (*msvId==KMsvGlobalInBoxIndexEntryId)
      { // New entries created in Inbox folder.
      CMsvEntrySelection* entries=static_cast<
                CMsvEntrySelection*>(aArg1));
      TInt count=entries->Count();
      for (TInt i=0;i<count;i++)
        { // Look at each individual entry.
        }
      }
    }
 }
```

You should note that this code is checking the global Inbox. In general, email messages are not stored in the global Inbox but rather within a folder associated with the email account.

Many applications need to monitor for message server events all the time; however, you can improve performance if there are periods when you do not need to monitor. You can turn off event call backs by calling:

```
CMsvSession::SetReceiveEntryEvents(EFalse);
```

12.7 The Send As Interface

While interfacing directly with the messaging architecture allows full control over all aspects of messaging, it is a large and complex environment. Many application requirements are met by the much simpler interface provided by the Send As interface.

In Chapter 10, we used Send As to add a file transfer function to our application. In that example, we restricted the transports to those that supported attachments, because we had a file to transfer. SMS does not support attachments.

The code below shows how you can use Send As to send SMS messages from your application:

```
void CAppSpecificListView::SendTextAsL()
```

```
{
CQikSendAsLogic* sendAs=CQikSendAsLogic::NewL();
CleanupStack::PushL(sendAs);

// Set the body text to be content of the currently
// selected list box entry.
CQikListBox* listbox=LocateControlByUniqueHandle<
        CQikListBox>(EAppSpecificListViewListId);
MQikListBoxData* lbData=listbox->Model().
           RetrieveDataL(listbox->CurrentItemIndex());
CleanupClosePushL(*lbData);
sendAs->SetBodyTextL(lbData->Text(EQikListBoxSlotText1));
CleanupStack::PopAndDestroy(lbData);

CQikSendAsDialog::RunDlgLD(sendAs);
CleanupStack::Pop(sendAs);
}
```

As you can see, this is considerably simpler than interfacing with the messaging architecture.

Table 12.2 summarizes the type of content that can be transported by each bearer.

Table 12.2 Content transported

Bearer	Supports body text	Supports attachment
Bluetooth	×	✓
Infrared	×	✓
SMS	✓	×
MMS	✓	✓
Email	✓	✓

12.8 Telephony

Access to the telephone functions is provided by `ETel ISV`. An abstract telephony class, `CTelephony`, is provided to present a simplified and uniform API for third-party developers.

We have already encountered the `CTelephony` class in Chapter 10. In that example, we used the `GetPhoneId()` method to obtain the device IMEI for registration purposes, however `CTelephony` provides a wide range of other services which can be split into two groups:

- phone and network status and information provision
- ability to dial and answer voice calls.

12.8.1 Phone Information

The CTelephony class provides a wealth of information about the mobile phone. Table 12.3 shows some of the more common items that you may need.

Table 12.3 CTelephony class methods: system information

Method	Required Capabilities	Description
GetPhoneId()	None	Reads the IMEI of the mobile phone.
GetFlightMode()	None	Retrieves the current flight mode status.
GetIndicator()	None	Retrieves the battery-charging, network availability and call-in-progress indicators.
GetBatteryInfo()	None	Retrieves the battery status and charge level.
GetLineStatus()	None	Finds out if the line is free (no voice call), has a call in progress or has a call in progress and one holding.
GetOperatorName()	None	Retrieves the name of the operator to which the phone is currently registered.
GetCallForwardingStatus()	NetworkServices, ReadDeviceData	Retrieves the Call Forwarding supplementary services status.
GetCallWaitingStatus()	NetworkServices, ReadDeviceData	Finds out if a call is waiting.

12.8.2 Managing Voice Calls

You can use the CTelephony class to dial and answer one or more simultaneous voice calls. The CTelephony class can also provide information on the current call status, call duration and start time (see Table 12.4).

Table 12.4 CTelephony class methods: call status

Method	Required Capabilities	Description
DialNewCall()	NetworkServices	Makes a new outgoing call.
Hold()	NetworkServices	Places a call on hold (must be a call initiated via CTelephony).
Resume()	NetworkServices	Takes a call off hold (must be a call initiated via CTelephony).
Swap()	NetworkServices	Swaps held and active calls (both calls must be initiated via CTelephony).
Hangup()	NetworkServices	Ends call (must be a call initiated via CTelephony).
AnswerIncomingCall()	NetworkServices	Answers incoming (ringing) voice call.

Further information is available within the UIQ 3 SDK documentation, which also includes numerous example code fragments demonstrating how to use each of the methods provided.

13
Refining Your Application

This chapter shows you how to make your application suitable for global deployment. It then presents some hints and tips to help you build applications that are more resilient, reliable and efficient.

Firstly, we develop the `SignedAppPhase2` application by demonstrating how to support languages based both on Latin and non-Latin characters within a commercial-grade application. This example exists in the `Localization` folder.

Next, we consider aspects of internationalization beyond the translation of application text. We must correctly format data to match the conventions used in various countries.

We then take a look at application performance. With our `Performance` example application, we generate some figures that demonstrate the relative speed of some Symbian OS features. From these figures and an understanding of the functionality being performed, application developers can make informed decisions about certain aspects of application development within the Symbian OS environment.

Finally we look at some other considerations, such as backup and restore.

13.1 Localization of Application Languages

This section shows you how to translate your application into multiple languages. We take `SignedAppPhase2` as the starting point for this example.

The `Localization` project contains the application changes, along with translated text files. This project is not intended to be complete but it does show you how to go about localization.

Mobile phones are used worldwide, creating the need for applications in many different languages. Symbian OS has considerable built-in support for multi-lingual applications; in particular, it is possible to build a single distributable program that supports multiple languages. By adding extra languages, you can expand the market for your application and increase user satisfaction.

A specific language comprises a number of unique characters. The English alphabet has 26 characters. Swedish adds ä, å and ö to make a 29-character alphabet. These types of alphabet, including uppercase and lowercase, are easily defined within an eight-bit (0–255) character set. However, other languages, such as Chinese, require thousands of characters and a multi-byte approach is needed.

This first type is given the generic name of single-byte character sets (SBCS), the best known of which is ASCII. Strictly speaking, very few systems use the original ASCII set today. Commonly used single-byte character sets include ISO-8859-1 and ISO-8859-15, which are supersets of ASCII and incorporate most western European characters.

To represent character sets with more than 256 characters, double-byte character sets (DBCS) and multi-byte character sets (MBCS) were developed. For example, Shift-JIS is a common MBCS used for Japanese. Over time many other character encodings were developed for different purposes on different computing devices. It was eventually recognized that a unified character encoding standard was required to be able to transfer data around without loss of information.

Unicode was developed to provide a single character set that covered all known characters. It is designed to inter-operate well with the ISO-8859-1 character set, as both character sets use the same numerical value to represent the same character.

Unicode defines a relatively small number of simple encoding schemes:

- UTF-32 stores each character as 32 bits.
- UTF-16 stores each character as one or two 16-bit values.
- UTF-8 stores each character using between one and four 8-bit values.

UTF-16 is the default encoding for Unicode and is designed such that practically all characters can be represented with a single 16-bit value. Symbian OS uses this encoding and a `TText` is defined as an `unsigned short int`.

While UTF-8 is really a replacement for older MBCS encodings, its design enables compression of data, particularly if most characters belong to the original ASCII set (values below hex 128).

13.1.1 Translating Application Text

When translating our application text we need to understand which character sets are used to display our text. It is also important to specify the formats required back from a translator.

If our required languages are supported by the ISO-8859-1 character set, then apart from actually translating the text, very little has to be done. In general, we can use our favourite text editor; the RLS files within the language folders remain standard eight-bit text files and the default options of the resource compiler tools perform the correct task.

However, let us assume we wish to support Simplified Chinese. The characters in that language no longer belong to the ISO-8859-1 character set; most standard text editors cannot handle such characters and we cannot represent them within a standard eight-bit text file.

To support a language such as Simplified Chinese we need to:

- create an RLS file which contains the tokens and actual text as UTF-8 encoded files
- tell the resource compiler that the RLS files should be interpreted as UTF-8 encoded files.

Numerous editors such as Notepad++, Scite and Windows Notepad are capable of creating UTF-8 files. The Symbian resource compiler is, however, very sensitive to the format of such files. In particular, when an editor such as Notepad generates a UTF-8 encoded file, it adds a three-byte file type header to the text file which the Symbian resource compiler cannot manage.

Two solutions are available to you:

- Use an editor that does not add the three-byte file type header to the UTF-8 encoded file.
- Use a tool that removes the three-byte header prior to passing the UTF-8 file to the resource compiler.

Since this is a common problem, we wrote a very simple command-line tool to remove the three-byte header. The entire tool as source code is:

```
#include "stdafx.h"
#include "stdio.h"

int main(int argc, char* argv[])
  {
  puts("Utf8 conv");
  puts(argv[1]);
  puts(" to ");
  puts(argv[2]);
  FILE* src=fopen(argv[1],"r");
  fseek(src,3,0);    // Remove the first 3 chars - UTF8 hdr.
  FILE* dest=fopen(argv[2],"w");
  unsigned char bb[1024];
  while(1)
    {
    size_t ret=fread(&bb[0],1,1024,src);
```

```
    if (ret>0)
      fwrite(&bb[0],1,ret,dest);
    if (ret<1024)
      break;
    }
  fclose(src);
  fclose(dest);
  return 0;
  }
```

The source code is available within the `Utf8ConvSrc` folder of the `Localization` folder.

13.1.2 Example of a Translated Application

The `Localization` project demonstrates how you may go about producing a multi-lingual application. In this example we show how to support French and Simplified Chinese as well as English.

You should note that:

- the translations are not complete
- while the text is valid, it may not be correctly presented in all contexts.

The `Localization` project is intended to demonstrate how you might go about adding languages to a project. Producing a fully localized Symbian Signed application is beyond the scope of this book.

Adding Language Folders

In our previous example applications, we had a folder called `English`, containing two RLS files, `SignedApp.rls` and `SignedApp_loc.rls`. You should recall these files contain token–value pair statements such as:

```
rls_string STR_R_CMD_NEW        "New"
```

To produce a language variant of an application we need to write an equivalent file, but replace the "`value`" component with its translated equivalent.

For example, the French may translate to:

```
rls_string STR_R_CMD_NEW        "Créer nouveau"
```

The Simplified Chinese may translate to:

```
rls_string STR_R_CMD_NEW            "新游"
```

> **Pro tip**
>
> We recommend splitting the language-specific components into separate folders; in our example we have `English`, `French` and `Simplified Chinese` folders. This significantly eases the production and maintenance of different languages.

Use of RLS Files

The examples in the previous chapters placed all text directly within the application RSS file. While valid, this creates work for translators as they have to be able to read and understand such files to locate and translate the text elements. The examples in this chapter split the text into separate token–value RLS files. If your application is only going to support a single language this creates some additional work for little benefit. However as soon as you need to translate your application, the power of RLS files should become evident.

> **Pro tip**
>
> It is not a requirement to split language content into RLS files. An equally valid approach is to have multiple RSS files containing inline text where the inline text is translated. Since the RSS files contain structure as well as language, the maintenance task of multiple RSS files is greater than using RSS with RLS files. For this reason, it is recommended that you split the actual language text out of your RSS files.
>
> Using the RLS filename extension is not mandatory; no tools fail if your file does not have this extension. Even within the various Symbian development environments there is a variety of filename extensions in use.

RLS Filenames

For our example application, we have chosen to use Notepad to create the UTF-8 encoded RLS files containing our application text. To be able to differentiate our Notepad-format UTF-8 source files we have chosen to prefix our filenames with `utf8_`. These UTF-8 encoded files are converted to UTF-8 encoded files compatible with the Symbian resource compiler using our command line tool `utf8_conv.exe`, included in the `Localization` folder. The process is wrapped in a batch file `utf8conv.bat`.

If we used a different UTF-8 text editor, we may be able to create Symbian resource compiler compatible files directly. It is unlikely that such files would need a visual indication of their format; that is, no leading "`utf8_`" would be required.

Our `utf8conv.bat` file creates files without the leading "`utf8_`". In this way, regardless of which mechanism is used to generate the RLS files, the resource compiler input filenames are the same.

13.1.3 Which Files to Encode as UTF-8?

UTF-8 is a multi-byte encoding scheme. Different values are encoded with a variable number of bytes. When decoding, a UTF-8 file reader has to know how to distinguish between values encoded with one, two, three or four bytes. UTF-8 defines the top bit of an eight-bit byte as a flag to indicate that a value needs a subsequent byte to decode correctly. The hex value `0x61` (the letter 'a') does not have the top bit set so a UTF-8 decoder uses the remaining bottom seven bits as the value. In contrast, the hex value `0xE0` (the letter 'à') has the top bit set. If a decoder sees such a value, it uses the bottom seven bits and reads the next byte as part of a single decoded value. The letter à cannot simply be stored as its hex value `0xE0;` it needs to be encoded over two bytes.

The significance of this is that if we tell the resource compiler that it has to deal with UTF-8 encoded RLS files to handle Simplified Chinese, we also have to convert any RLS files that contain text outside the `0x00` to `0x7F` hex value range.

> **Pro tip**
>
> In our example application we have chosen to convert the English RLS file to UTF-8. In practice, all resource string files, including languages, such as English, which use only basic characters, should be UTF-8 encoded. This greatly simplifies the process of building and packaging your application for distribution.

13.1.4 Informing the Resource Compiler about UTF-8 Encoded Files

Now we can write UTF-8 encoded files and convert them to be compatible with the Symbian resource compiler, we need to inform the resource compiler that the input files are UTF-8 encoded. This is achieved using the `CHARACTER_SET UTF8` statement inside the RSS files. Our application registration localization file now looks like this:

```
#include <AppInfo.rh>
#include <Qikon.hrh>
```

```
CHARACTER_SET UTF8

#define EViewIdPrimaryView 0x00000001

#ifdef LANGUAGE_01
#include "..\English\SignedApp_loc.rls"
#endif
#ifdef LANGUAGE_02
#include "..\French\SignedApp_loc.rls"
#endif
#ifdef LANGUAGE_31
#include "..\SimplifiedChinese\SignedApp_loc.rls"
#endif

// This file localizes the application's icons and caption.
RESOURCE LOCALISABLE_APP_INFO r_application_info
  {
  short_caption = STR_R_APP_SHORT_CAPTION;
  caption_and_icon =
    {
    CAPTION_AND_ICON_INFO
      {
      caption = STR_R_APP_LONG_CAPTION;
      number_of_icons = 3;
      icon_file ="\\Resource\\Apps\\SignedApp_icons_0x20000462.mbm";
      }
    };
  view_list =
    {
    VIEW_DATA
      {
      uid=EViewIdPrimaryView;
      screen_mode=0;
      caption_and_icon =
        {
        CAPTION_AND_ICON_INFO
          {
          }
        };
      },

    // By adding this in we tell P990s
    // we can run in flip-closed mode.
    VIEW_DATA
      {
      uid=EViewIdPrimaryView;
      screen_mode=EQikScreenModeSmallPortrait;
      caption_and_icon =
        {
        CAPTION_AND_ICON_INFO
          {
          }
        };
      }
    };
  }
```

13.1.5 Building a Multi-Lingual Application

To actually build the multi-lingual application, we need to update our MMP file.

Currently we have statements associated with each resource file such as:

```
SOURCEPATH      .
START RESOURCE SignedApp_loc_0x20000462.rss
HEADER
LANG            01
TARGETPATH      \Resource\Apps
END
```

This informs the resource compiler to compile the RSS file once, define the macro `LANGUAGE_01`, and generate an output file that has the R01 filename extension. In Symbian OS, the English language is assigned the value 01.

To add French and Simplified Chinese we need to update these directives. French has the language ID of 02; Simplified Chinese is 31. These values are defined by the `TLanguage enum`.

Our resource statements now look like this:

```
SOURCEPATH      .
START RESOURCE SignedApp_loc_0x20000462.rss
HEADER
LANG            01 02 31
TARGETPATH      \Resource\Apps
END
```

This informs the resource compiler to compile the RSS three times, once for each of the three languages we have defined. The macros `LANGUAGE_01`, `LANGUAGE_02` and `LANGUAGE_31` are defined as appropriate. The result is three files, with extensions R01, R02 and R31.

Looking back at the original RLS file statements, you see:

```
#ifdef LANGUAGE_01
#include "..\English\SignedApp_loc.rls"
#endif

#ifdef LANGUAGE_02
#include "..\French\SignedApp_loc.rls"
#endif

#ifdef LANGUAGE_31
#include "..\SimplifiedChinese\SignedApp_loc.rls"
#endif
```

We include a different set of token–value pairs from a language-specific RLS file, depending on which of the three times the file is compiled.

> **Pro tip**
>
> If you build a multi-lingual application using the CodeWarrior IDE, it continually overwrites the `R01` file instead of generating the R01, R02, and R31 files. To correctly build the language files, you need to use a command-line build or a different IDE such as Carbide.c++.
>
> Testing Simplified Chinese on the emulator is quite difficult since the Chinese fonts are missing. Your application displays either blank lines or lines containing a series of square boxes.

13.1.6 Deploying a Multi-Lingual Application

Applications are deployed as SIS files. Such a file usually contains all the components for your application to operate correctly. SIS files are created by a tool called `MakeSIS`. They take as input a PKG file. One of the fields in the PKG file is the textual name of your application presented to the user when they install the application. If you have a multi-lingual application the name may vary between languages. As discussed above, regular eight-bit text files cannot contain characters such as Simplified Chinese. Therefore a PKG file can either be an eight-bit text file or, unlike RLS files, a 16-bit Unicode file.

The Unicode files created by Notepad and those handled by the `MakeSIS` tool are compatible so you can create Unicode PKG files with Notepad without any other tool being required.

Our `Localization` PKG file contains the following content:

```
; Specify the languages we support; items must appear in
; this order subsequently.
&EN,FR,ZH

; List of localised vendor names - one per language.
; in order EN,FR,ZH...
%{"MyCompany","MyCompany","MyCompany"}

; The non-localized, globally unique vendor name
;(mandatory).
:"ZingMagic Limited"

; What our app name is displayed as, in order EN,FR,ZH
; NOTE this is only intended to be example text, "Signed App" does not
; translate to the text presented here (Chess).
#{"Signed App","Échecs","国际象棋"},(0x20000462),1,00,01,TYPE=SA

; ProductID for UIQ 3.0
; Product/platform version UID, Major, Minor, Build, Product ID
[0x101F6300], 3, 0, 0,
  {"UIQ30ProductID","UIQ30ProductID"."UIQ30ProductID"}
```

```
; The language-independent files we install.
"..\..\..\epoc32\release\gcce\urel\SignedApp_0x20000462.exe"-
    "!:\sys\bin\SignedApp_0x20000462.exe"
"..\..\..\epoc32\data\z\resource\apps\SignedApp_icons_0x20000462.mbm"-
    "!:\resource\apps\SignedApp_icons_0x20000462.mbm"
"..\..\..\epoc32\data\z\resource\apps\SignedApp_0x20000462.mbm"-
    "!:\resource\apps\SignedApp_0x20000462.mbm"
"..\..\..\epoc32\data\z\private\10003a3f\apps\SignedApp_reg_0x20000462.rsc"
    -"!:\private\10003a3f\import\apps\SignedApp_reg_0x20000462.rsc"

; Backup registration.
"backup_registration.xml"-"!:\Private\20000462\backup_registration.xml"

; The language resource files, in order EN,FR,ZH
; The registration file localised component.
{
"..\..\..\epoc32\data\z\resource\apps\SignedApp_loc_0x20000462.r01"
"..\..\..\epoc32\data\z\resource\apps\SignedApp_loc_0x20000462.r02"
"..\..\..\epoc32\data\z\resource\apps\SignedApp_loc_0x20000462.r31"
}-"!:\resource\apps\SignedApp_loc_0x20000462.rsc"

; The primary app resource file.
{
"..\..\..\epoc32\data\z\resource\apps\SignedApp_0x20000462.r01"
"..\..\..\epoc32\data\z\resource\apps\SignedApp_0x20000462.r02"
"..\..\..\epoc32\data\z\resource\apps\SignedApp_0x20000462.r31"
}-"!:\resource\apps\SignedApp_0x20000462.rsc"
```

13.1.7 Two-Character Language Codes

The PKG file defines what languages our application installation supports using two-character language codes, for example, EN for English, FR for French and ZH for Simplified Chinese. The full set of language codes is defined in the *Language code table* section of the Package File Format chapter of the SDK documentation.

A customer is prompted to choose which language variant they wish to use on installation. Since our application supports three languages the customer effectively chooses between index values 0, 1 or 2. This index choice drives which components are extracted; that is, if the user chooses Simplified Chinese, since that option is index value 2, the text at index 2 from the application name statement is used:

```
#{"Signed App","Échecs","国际象棋"},(0x20000462),1,00,01,TYPE=SA
```

Similarly the R31 files are chosen as they are at index 2 of the list of language files available to the installer.

> **Pro tip**
>
> If you choose Simplified Chinese when an application is installed, and your phone does not support Simplified Chinese, your menu options

contain a number of blank lines or square boxes where the text is supposed to be. This is perfectly normal because the Simplified Chinese fonts are missing. We recommend that you test on the appropriate Chinese versions of your target mobile phones.

13.2 Internationalization

Internationalization is more than simply translating the text strings in your application into different languages; it also involves changes in the way information is presented. For example, the date 03/04/07 is interpreted as 3rd April 2007 in the UK, 4th of March 2007 in the USA and 7th April 2003 in Sweden, China and Japan. Such ambiguities need to be resolved if an application is to present the intended meaning when used on different phones.

Strictly speaking, internationalization is the process of enabling your application to support locales. A locale is composed of a language plus standards and format data. The `Localization` example showed you how to support different language text. This section focuses on how to support standards and formats.

The class `TLocale` encapsulates most of the standards and formats information. As a simple example, we might want to improve on the display of a number by adding separators.

The following code would produce a string `"10200"`.

```
TInt val=10200;
buf.Num(val);
```

An improvement to produce the string “`10,200`” might be:

```
TInt thousands=10;
TInt rest=200;
buf.Format(_L("%d,%d"),thousands,rest);
```

However, this code assumes that the `','` character is the correct thousands separator and that the concept of thousands separators is correct for all locales.

Looking at the `TLocale` class, you can see the method `ThousandsSeparator()` which reports the character to use. Therefore an improvement might be:

```
TLocale locale;
TInt thousands=10;
TInt rest=200;
buf.Format(_L("%d%c%d"),thousands,locale.ThousandsSeparator(),rest);
```

The TLocale class shows that the following data items are dependent on the current locale:

- date and time formats
- clock display
- number display and formats
- currency display and formats
- workdays and start of week.

UIQ 3 phones are typically set to the correct locale as part of customization. The user can view and change locale settings in the Control Panel, Device, Number Formats and Time & Date settings.

If you are displaying these data items to users and you want them to be correctly understood, then you must apply locale information correctly.

Often a standard UIQ 3 control can be used to display your content for you. Such controls usually take the compact form of your information and format it correctly on your behalf.

13.2.1 Date and Time Formatting

Symbian OS provides extensive facilities to format date and time information in a locale-independent fashion.

```
TTime time;
time.HomeTime();
time.FormatL(bb,_L("%*D%*N%*Y%1 %2 %3, %-B%:0%J%:1%T%:3%+B"));
```

The above code fragment would produce "6 Jun 07, 7:30 pm" if run with a UK locale, "Jun 6 07, 7:30 pm" if run with a USA locale, and "07 Jun 6, 7:30 pm" if run with a Chinese locale. The application code does not have to change for this to happen. A full description and examples of usage of the format string is available in the UIQ 3 SDK documentation.

13.2.2 Number and Currency Display

TLocale also defines what character a particular locale is expecting an application to use as a thousands separator. Other characteristics such as how negative currency amounts should be displayed, where currency symbols should be placed and whether space characters are required between currency symbols and values can all be queried from the TLocale information. Unlike date and time formatting, Symbian OS does not provide any specific functionality to help with number and currency formatting. For example, TLocale tells you that negative currency amounts should be displayed as (£200) as opposed to, say, -£200, but your application must do the formatting itself.

13.2.3 Resource Files

In general, we recommend that no more than a single string formatting entity is placed in a resource string. For example, different locales have different rules as to the placement of nouns and verbs within a sentence. Placing more than a single formatting entity within a resource implies that the order of those two items is fixed across all locales. This is not universally true.

> **Pro tip**
>
> If you are familiar with the S60 environment you may have used the `StringLoader` class to handle multiple formatting entities in a locale-independent fashion. The `StringLoader` class is specific to S60 and, therefore, does not exist on the UIQ platform.

13.3 Application Performance

13.3.1 Computational Capability of Modern Mobile Phones

Mobile phones, like personal computers, get ever faster CPUs, more memory, better screens, longer battery life, and so on. In turn, software applications become more sophisticated. One example of this is the high-quality video playback on the MOTO Z8 phone. In this section, we run some test cases to get an idea of the capability of UIQ 3 mobile phones.

The `Performance` example application contains the source code used to generate the performance information that we present in this chapter. Numerous factors dictate the exact numbers you obtain, such as which compiler you use, which hardware revision you have and what other activities are occurring on the phone under test, for example, battery charging; after all, it's a multi-tasking operating system. We used the standard GCCE compiler, with no additional optimizations enabled. Our results were obtained by running the application numerous times on a standard Sony Ericsson W950i and averaging the values obtained.

The objective of the exercise is to give a sense of scale to the various activities an application can perform. It is not intended to be a definitive set of data and will not accurately predict how a real-world application will perform. Consequently, the numbers presented are rounded up or down to aid in the comprehension of the results.

Table 13.1 relates to how many function calls of varying types a mobile phone performs per second. We also include values for integer and floating-point calculations to give a general feel for the performance.

Table 13.1 Performance

Test	**Result**
Integer operations: Add and subtract	23,000,000 operations/second
Floating-point operations: Multiply and square root	360,000 operations/second
Calling a standard C-like function	5,500,000 calls/second
Calling a non-virtual method belonging to a class	5,200,000 calls/second
Calling a pure virtual method	4,500,000 calls/second
Calling a non-virtual method under a `TRAP` harness	380,000 calls/second
`RunL()`	70,000/second

Rather than placing too much emphasis on the exact values, most of the useful information is obtained by comparing the relative values. For example, the above figures inform us that you can perform approximately 70 method calls, 330 integer operations or five floating-point operations in the same time as it takes to perform a single `RunL()`.

Of particular significance from the above information is that if you partition a task using active objects but only perform a few method calls or calculations per partition, the vast majority of the time taken to perform a task is taken in scheduling the `RunL()` itself.

For example, in a spreadsheet-type application, if a single cell were evaluated in each `RunL()`, and each cell contained an average of five floating-point calculations, then 50 % of the elapsed time would be spent calculating the cell content and the other 50 % spent in the overhead of calling the `RunL()`. Apart from considerably slowing down the calculation, by executing twice as much code, you reduce the battery life. Both effects are undesirable. In this example, we can easily reduce the `RunL()` overhead by calculating multiple cells in each `RunL()`. If we calculated 100 cells per `RunL()`, instead of one, then approximately 99 % of the time is used calculating the result and only 1 % in the `RunL()` overhead. The spreadsheet would be calculated at almost optimal speed.

The performance table suggests we can perform 500 floating-point operations, or calculate 100 cells, in approximately 0.001 seconds. In other words we can check for other events, such as key presses (by virtue of partitioning the operation with active objects), approximately every thousandth of a second in such an environment.

The rate at which `RunL()` is called effectively defines the responsiveness of your application. As we discussed earlier, being able to respond to user-generated events within 0.1 seconds, when an otherwise computationally intensive task is running, is perfectly acceptable to the user. In our example spreadsheet application, we can calculate 10,000 cells in 0.1 seconds.

The above implementation analysis depends on your application-specific event sources. For example, if an application has a high-frequency timer delivering events, such as an animation timer ticking every 0.04 seconds, you need to ensure that other operations within your application can complete in that timescale for the timer events to be processed at a uniform rate.

You should recall that the `RunL()` of one active object cannot be interrupted by the `RunL()` of another active object belonging to the same active scheduler. If you had a spreadsheet recalculation whose `RunL()` method took 0.1 seconds to complete, regardless of the rate at which you set up any timer, the `RunL()` of the timer active object is not called until your application returns control to the active scheduler. While a multi-threaded environment can help to avoid such blocking issues, if the task of the timer is to interact with the underlying data being recalculated, different synchronization issues arise. Active objects resolve many synchronization issues but rely on the programmer to partition long-running tasks correctly.

Pro tip

The entire point of partitioning tasks is to allow other events to be processed between the partitions. A core objective is to be able to respond to user-generated events within an acceptable timeframe. To get optimal performance from an application, some thought and care needs to be exercised when partitioning tasks.

You may wish to note that the Pro and OEM editions of Carbide.c++ IDE contain profiling tools, which you may prefer to use in order to obtain better performance information about your application.

13.3.2 Performance Issues in Coding

Most of us are familiar with general programming practices and how to structure some code to increase performance at one level or another.

For example, most programmers would recognize that repeatedly calling functions within loops to obtain the same value is inefficient.

For example, the first piece of code shown below should be replaced with the second:

```
for (TInt i=0;i<aObj->Count();i++)
  {
  // Perform task.
  }

const TInt count= aObj->Count();
for (TInt i=0;i<count;i++)
  {
  // Perform task.
  }
```

We consider some of these general programming practices but this section is not exhaustive since these types of performance issue are well documented in many general-purpose programming books. We focus more on aspects specific to Symbian OS.

As a general rule, many performance issues can be solved by reducing the amount of code being executed to achieve the same end result. This in turn means you need to understand the code you are writing and the consequences of the lines of code in your application. As with programming on any platform, it is unlikely that there is a single right or wrong way to achieve a given task. The points below are issues you should consider when writing your application.

For all of the points below there are situations where it is perfectly reasonable and very sensible not to use any suggested *optimal* route. As a programmer, you need to examine your specific situation and apply any rules within the context of that situation.

Constructors

Even innocuous-looking declarations such as the following cause code to be executed:

```
TRect rect;
```

In this case, the constructor is called to set each of the four integers within the class to 0. If you happen to declare the `TRect` inside a loop then, for each iteration of the loop, the constructor is called. In some cases, this may be exactly what is required. However, in most cases an assignment of values to the four member variables quickly follows the object declaration. By simply declaring the object outside the loop, you get a 50 % increase in performance since only half as many assignments occur.

For example, replacing the first piece of code with the second reduces the number of memory assignments within the loop by half:

```
for (TInt i=0;i<count;i++)
  {
  TRect rect; // Performs 4 memory assignments.

  rect.iTl.iX=val1;
  rect.iBr.iX=val1+width;
  rect.iTl.iY=val2;
  rect.iBr.iY=val2+height;
  }

TRect rect;
for (TInt i=0;i<count;i++)
  {
  rect.iTl.iX=val1;
  rect.iBr.iX=val1+width;
  rect.iTl.iY=val2;
  rect.iBr.iY=val2+height;
  }
```

This example may be trivial but as a general rule you should have a clear understanding of exactly what tasks an object constructor performs, particularly objects that are instantiated through the Symbian two-phase construction idiom.

Pass by Reference Compared with Pass by Value

Compare the following functions:

```
void CExample::PassByReference(TRect& aRect)
  {
  if (aRect.Width()>aRect.Height())
    {                           // Perform functionality.
    }
  }

void CExample::PassByValue(TRect aRect)
  {
  if (aRect.Width()>aRect.Height())
    {                           // Perform functionality.
    }
  }
```

Superficially the functions look identical and perform the same task; however, the `PassByReference()` method call is more efficient to set up. Only a single integer value has to be calculated and passed, compared to four for the `PassByValue()` method. In practice, there are few cases where `PassByValue()` is preferable to `PassByReference()`, especially since we can define `PassByReference()` as follows:

```
void CExample::PassByReference(const TRect& aRect);
```

This eliminates the chance that the implementation of `PassByReference()` can accidentally modify the contents of the passed parameters.

Return by Reference Compared with Return by Value

Compare the following functions:

```
_LIT(KExampleName,"c:\\ExampleFile.txt");
void CExample::ReturnByReference(TFileName& aName)
  {
  aName=KExampleName;
  }

TFileName CExample::ReturnByValue()
  {
  return(KExampleName);
  }
```

Again, both perform the same task, however the `ReturnByReference()` is usually more efficient. In both the above cases, a significant amount of content is being moved from one place to another. In the case of the `ReturnByValue()` method, an entire object is being constructed and *returned* to the caller. A third, significantly more efficient variant would be:

```
const TFileName& CExample::ReturnByReference();
```

In this case, only a single four-byte entity is returned to the calling code. No copy of the content itself is occurring. While not applicable in every situation, this third case is often more than sufficient.

Use of `const` Keyword

By declaring variables, methods and parameters as `const`, you are informing the reader and the compiler about some attributes of the entity. Different compilers are able to use this type of information in different ways. Rather than simply assuming a compiler will process the code correctly, we recommend that you give as many hints as possible through the use of language keywords.

In the return by reference section, one option was to use:

```
const TFileName& CExample::ReturnByReference();
```

Here the `const` keyword is used to ensure that the caller cannot easily change the content of the variable being returned.

In contrast, you should be careful about simply adding the `const` keyword to existing methods, particularly virtual ones, since doing so changes the method signature. Unless all overloads of the virtual method are changed at the same time any overloaded methods will not match the signature.

Kernel Executive Calls

A kernel executive call is one where user-side code is allowed to enter the processor privileged mode in order to access kernel-side resources in a controlled fashion. A significant amount of code is executed to change between modes. One of the most common cases of this in Symbian OS applications is through the use of the `TRAP()` and `TRAPD()` macros.

As we saw in the function call table presented earlier, calling a method under a `TRAPD` results in a 93 % decrease in the number of method calls that can be made per second. Given its overhead, the `TRAP` macro should be used sparingly and, as with object constructors, you should declare them outside of loops wherever possible. Conversely, you can execute approximately the same number of `TRAPD`s as floating-point operations per second. In that context, it is perhaps not the slowest operation you can perform, and so adding complexity to the natural flow of an application simply to move a `TRAPD` outside a loop may not generate much performance benefit.

Context Switches

As in most operating systems, a context switch between threads requires processor time. An operating system has to move the calling thread from the *currently running* list to either the *ready to run* or *suspended* lists. It then has to move the thread being called to the *currently running* list. A number of other housekeeping tasks, such as memory mapping, are also needed before the called thread can resume execution. This means that the number of context switches should be kept to a reasonable level. As with many aspects of programming this is often a trade-off between efficiency and resource usage. A good example of this trade-off is reading the contents of a file. Reading a single byte at a time causes a large number of context switches; in contrast, reading the entire contents in one read operation may require a lot of memory.

Our `Performance` application generates the results shown in Table 13.2.

There is a significant overhead in performing context switches to the file server in both read and write operations. For example, reading 100 bytes at a time results in a throughput of only around 5 % of that achieved compared to reading 10,000 bytes at a time.

On the other hand, you should note that reading files in very large blocks may adversely affect the performance of the rest of the phone, since no other application can access the disk while your request is being serviced.

Table 13.2 Input–output performance

Test	Result
Writing content to file, 100 bytes per write request	72 KB/second
Writing content to file, 10,000 bytes per write request	235 KB/second
Reading content from file, 100 bytes per request	270 KB/second
Reading content from file, 1000 bytes per read request	1.5 MB/second
Reading content from file, 10,000 bytes per read request	5.2 MB/second

Pro tip

Unless there are other conditions your application needs to consider, or it is convenient to structure your application in this way, there is reduced value in using the asynchronous file read and write operations in most normal cases on modern UIQ phones, especially when low volumes of data are to be processed. Even with a modest 1 KB buffer, applications can read around 1.5 MB per second from disk.

Connections to Servers

The construction of an open connection between your application and a server within the system requires considerable resources. Assuming that the server is running then context switches are required to form the connection. Kernel resources are required to monitor and maintain the connection and the server needs to set up a sub-session to represent the new client connection. You should therefore avoid the frequent opening and closing of connections. As always, there is a trade-off between having a permanent open connection to a server, which uses resources, compared to only having a connection as and when required.

For example, most applications have a permanent open connection to the file server and should reuse that single connection. This reuse is demonstrated within our `Performance` example where we obtain the file server handle via the `Eikon` environment. In contrast, a connection to the communications server is normally created and destroyed each time communications are required. This is demonstrated by the example applications presented in Chapter 12.

Memory Management

The Sony Ericsson P990i has 64 MB RAM and the Sony Ericsson P1i has 128 MB. Although this may seem large, a mobile phone needs to run many applications that users expect to be graphically and functionally rich. Memory is therefore one of the resources that your application must use very carefully.

Applications are still responsible for managing their own memory requirements and should aim to use as little memory as possible to perform their tasks. Good memory management not only leads to less memory being used but also superior application performance.

For example, you may have a word processor application that can handle large documents. If you store the content in a single memory cell and then insert a character at the beginning of the document, it is likely that you would have to adjust every other character position by copying the memory. In contrast, if you store every character within its own memory cell you could eliminate the memory copying; however, a heap containing a large number of allocated cells brings its own performance problems. For this particular example, a memory management scheme half way between these two extremes is preferable. Indeed, Symbian OS has support for such a scheme, called segmented memory management, which is available through the `CArrayX` classes.

> **Pro tip**
>
> One of the biggest factors influencing the performance of an application is the algorithm design. A highly optimal implementation of a poor algorithm is no substitute for a superior algorithm. This is also true in memory management. If the lower-level data structures are correct, the application tends to look after itself at a higher level.

Heap Usage and Fragmentation While minimizing the amount of heap being used is one consideration, another is heap fragmentation. When a heap becomes very fragmented, application performance can suffer. In Symbian OS, a heap is made up of two lists, one containing the list of allocated cells and the other containing the list of free cells. When a heap becomes fragmented, these lists become long. Traversing the list of free cells to find a cell capable of containing a particular amount takes longer. In addition, each heap cell has an overhead associated with it that is required to maintain the lists. An allocated heap cell has a four-byte header overhead, therefore a heap cell big enough to contain 16 bytes has 25 % overhead, whereas a heap cell big enough to contain

256 bytes only has a 1.5 % overhead. Having a large number of small heap cells results in considerable memory wastage as well as performance issues.

Heap Usage and File Names A file name buffer, `TFileName`, is a good example of a data structure that should be considered carefully. On the one hand it needs to be defined to be large enough to store the longest possible file name, 256 characters. On the other hand few, if any filenames, including the drive and path information, exceed more than around 64 characters on a real phone. In this case, perhaps 192 characters' worth of heap memory (384 bytes) are left unused for every `TFileName` defined. If you have a data structure that needs to contain a filename, then you should consider how many of these data structures you might have in your application.

For example, in our `SignedAppPhase3` application we store an array of folder entries, and each entry contains a `TFileName`. Assuming an average full filename of 64 characters and around 32 bytes of storage for the remainder of the structure, over 60 % of the heap memory required to store a single entry is not used. It is actually worse than that in the `SignedAppPhase3` example since we don't store the drive and path name within the individual entry `TFileName`. Therefore maybe only 16 characters of the full `TFileName` are used. We could be wasting up to 90 % of the heap memory allocated to store each entry!

Wasting up to 90 % of the memory required to store a data structure could be acceptable if the amounts of memory being considered are small. For the `SignedAppPhase3` example we may be wasting 350 bytes per entry. Since we only have a small number of entries, around 10, we would be wasting around 3.5 KB of heap memory. The advantage here is reduced application complexity.

In contrast, if you are writing a proper file manager application where a folder may contain hundreds or perhaps even thousands of files, you can easily waste hundreds of kilobytes of heap memory. In our current scheme, 1000 entries would waste 359 KB! To avoid this type of issue you should consider whether to use a `TFileName` at all. A preferable solution would be to store each of the filenames in their own appropriately sized memory cell.

Heap Usage and Variable Arrays Variable arrays, either in the form of the `RArray` or `CArrayX` classes, are commonly used within Symbian OS to store content. Both sets of classes are designed to be able to store an unlimited number of entries of arbitrary size. In both cases, incorrect class usage can cause performance problems or excessive heap usage.

Performance problems can arise if the granularity is not appropriate for the usage of the class. For example, if we set the granularity to one, meaning that we only extend the array to accommodate one more element each time it has to expand, and then add 100 entries, this will cause the underlying heap cell containing the content to be resized 100 times. It is highly likely that each resize will require a memory reallocation, hence a copy of the data from one heap cell to another.

In contrast, excessive heap usage can occur if we set the granularity to 100 but only ever add a single entry. 99 % of the allocated memory is left unused.

In the case of variable arrays, careful use of array granularity, `SetReserveL()` and `Compress()` methods should be used to optimize performance and memory usage.

Table 13.3 summarizes a typical set of test results obtained from our `Performance` application.

Table 13.3 Array performance

Test	**Time taken**
Insert 1,000,000 entries into a `CArrayX` with a granularity of 1	4.85 seconds
Insert 1,000,000 entries into a `CArrayX` with a granularity of 100,000	1.74 seconds
Insert 1,000,000 entries into an `RArray` with a granularity of 1	5.42 seconds
Insert 1,000,000 entries into an `RArray` with a granularity of 100,000	0.57 seconds

Again, rather than looking at the absolute results, you should observe that in the case of `CArrayX` objects, around 65 % of the time is spent dealing with the memory reallocation. In the case of `RArray` objects, it is much worse with about 90 % of the time spent handling the memory reallocation.

Heap Usage and Flat Arrays When using flat arrays, that is, arrays where the content is stored within a single memory cell, some consideration should be given to the size a single memory cell may reach. In our `SignedAppPhase3` application, if we had 1000 entries we would require a single memory cell in excess of 500 KB. Attempting to add an entry means that we must find a free cell in excess of 500 KB to expand into, making a total of more than 1 MB. The default `EPOCHEAPSIZE` is 1 MB, so you will get an `Out of Memory` situation.

In this scenario, you should try to ensure you do not allow a single memory cell to grow too large. In our particular example, each of our entries is about 550 bytes. Breaking an entry up by making the

TFileName field of the entry a separate memory cell would solve several of the issues we have discussed here.

> **Pro tip**
>
> Very few applications should have a requirement to change the default heap size settings. If you find your application is running out of heap, it is just as likely that a change of implementation will solve your memory problem as increasing the heap size. The current choice of data structure in the SignedAppPhase3 example could be responsible for up to 90% of the allocated memory being wasted. Changing the implementation is preferable to adjusting the maximum heap size.

RArray Compared with CArray

One of the reasons cited for moving over to using RArray objects is the superior performance they offer. Until now there has been very little hard evidence available on exactly the performance differences. Table 13.4 shows that RArray objects are approximately three times faster than CArrayX classes, both at appending entries and accessing them.

Table 13.4 Array performance

Test	Time taken
10,000,000 array entry accesses on a CArrayX class	10.58 seconds
10,000,000 array entry accesses on an RArray class	3.50 seconds
10,000,000 array entry accesses on a standard C array based on object properties	1.22 seconds
Insert 1,000,000 entries into a CArrayX with a granularity of 100,000	1.74 seconds
Insert 1,000,000 entries into an RArray object with a granularity of 100,000	0.57 seconds

> **Pro tip**
>
> While these figures demonstrate that RArray objects offer superior performance, array access is fast in both cases. The slowest array access still offers approximately 1,000,000 array accesses per second on a phone. In the rather unlikely case that array access is the performance bottleneck of your application, then you should consider moving over

to a regular C array since this offers superior access performance to both of the variable array classes, as demonstrated by the figures in Table 13.4.

Matching Color Depths

If the color depth of a window is not the same as the color depth of the physical screen then a translation of colors has to occur when the window is displayed on screen. Similarly, if the color depth of a bitmap does not match that of the window, the color depth must be adjusted. On mobile phones, we have observed that mismatched color depths are around 10 times slower than matching color depth bitmaps.

13.4 Other Considerations

13.4.1 Foreground and Background Events

The window server generates numerous events, such as key events, pointer events, redraw events and foreground and background events. Most applications do not see the `TWsEvent` directly; the framework contains significant amounts of code to process these low-level events. Applications typically see keys through the `OfferKeyEventL()` method, pointer events through the `HandlePointerEventL()` method, and foreground and background events through the `HandleForegroundEventL()` method.

So what should your application do with foreground and background events? Unfortunately there are no universal rules since applications vary significantly. However when receiving a background event, your application should make every effort to reach a quiescent state, release any unnecessary resources and remain quiescent, performing as little processing as possible. In doing so, you release as many resources as possible back to the system, including the processor resource. The overall system remains responsive and a user is comfortable that your application is not adversely affecting their phone, particularly the battery life. Here are examples of how a variety of applications may handle events:

- A game should stop at the current point when it goes into the background and resume in some fashion when it comes into the foreground. It would be quite reasonable to enter a paused mode on background and require the user to resume on foreground. The one thing you should not do is continue running the game, or any polling loop within the game. By the time the user returns to the application, the level or game is likely to have finished.

- A push email solution might not take any action at all. If the user is interacting with the view architecture, the view may automatically shut down. In such a case, if a new email arrives, part of your application needs to know that there is no UI and therefore not attempt to update it.
- If an application needs to perform a very long calculation, there is no absolute requirement to stop processing; however in such a scenario you should ensure that the processing you are performing does not materially affect the performance of the remainder of the phone. In particular, you should ensure that you are able to respond to any shut-down events that may be automatically generated by the system. In practice, this is no different to ensuring that you can respond to any user-generated events when in the foreground.
- An application that uses a non-sharable resource, such as a camera, should release the resource when it goes into the background to ensure other applications that wish to use that resource can do so. The `SignedAppPhase3` application in Chapter 11 demonstrates how you might go about owning and releasing the camera resource.

Because Symbian OS is a multi-tasking operating system by design, you generally have to do very little. For example, unless you have gone out of your way to stop the window server acting correctly, the process priority is adjusted when your application moves between the foreground and the background. This ensures that the application visible to the user is responsive, even if your application is computationally intensive when it goes into the background. It does not, however, stop poorly implemented applications needlessly running in the background and potentially draining the battery.

13.4.2 Backup and Restore

Applications are responsible for declaring their own backup and restore policy. Your application can have a zero backup policy, in which case neither the application data nor the application is backed up; therefore it will not be restored either.

In most cases, we want our application and its associated data to be backed up, such that when the user restores their phone, both the application and data are restored. This is the policy we have adopted for our Symbian Signed compatible application, `SignedAppPhase1`, presented in Chapter 10. The application-specific policy is defined within the `backup_registration.xml` file included in an application installation package. This file must be present in your installer to pass Symbian Signed.

Some applications need to participate actively in backup and restore, particularly those that wish to perform synchronization of data with alternative data sources as opposed to a simple backup of the entire application data.

A full discussion of active backup is beyond the scope of this book. The document ***developer.symbian.com/main/downloads/papers/pc_connectivity/PlatSec_PC_Connectivity.pdf*** provides comprehensive details on all aspects of passive and active backup, along with numerous examples of `backup_registration.xml` files and the full XML DTD.

13.4.3 Starting up and Exiting Applications

The Symbian Signed Test Criteria discussed in Chapter 14 give some guidelines on this functionality. In particular, C++ applications should present some visual feedback that they are starting within five seconds of the application being launched. Applications should also be prepared to exit, for example, if the user chooses 'Close' from the task manager. In general, enough of the UI needs to be made available within the `ViewActivatedL()` methods to allow the application to present some kind of user interface. Any slow operations required to fully populate a specific view can be run as separate active objects, the completion of which causes the UI to be updated.

14

Symbian Signed

In Chapters 10 and 11, we built an application, `SignedAppPhase3`, that we have Symbian Signed. We indicated some of the places where you need to consider Symbian Signed during your development and design. In this chapter, we provide an overview of capabilities and the Symbian Signed process. We examine the test cases, with particular reference to the `SignedAppPhase2` example. Finally, we provide some tips and advice based on our experience of taking applications through the process.

Symbian Signed is a scheme run by Symbian and backed by the mobile phone industry. Symbian Signed applications follow industry-agreed quality guidelines and support Network Operator requirements for signed applications.

Symbian Signed aims to ensure applications behave correctly in three main ways:

- It confirms the application's origin and ensures that the application does what it says it does.
- It controls access to protected APIs. These are APIs that allow sensitive operations to be performed by an application such as those that initiate billable events or access private and personal user information.
- It specifies key test cases to which all Symbian OS applications should conform and, depending on the signing option chosen, may require independent testing of conformity.

In the context of Symbian Signed, signing is the process of encoding a tamper-proof digital certificate into a SIS file which guarantees the origin of the contents of the SIS file. The certificate also grants access to selected protected APIs within Symbian OS.

14.1 Symbian Signed Options

Symbian Signed was updated in late 2007 to provide three basic options for getting your applications signed:

- Open Signed makes it easy for you to sign applications for limited deployment to known devices, either for testing or for personal use. Deployment is restricted by device IMEI.
- Express Signed offers an effective signing route with minimal restrictions. There is no requirement for independent testing if you own a Publisher ID and are releasing commercial software. Freeware and shareware developers without a Publisher ID can also access this signing option via publisher partners.
- For mainstream commercial software developers, the full testing and accreditation regime of Certified Signed will be the preferred option, with entitlement to use the Symbian OS logo to aid differentiation and brand building.

Changes compared to the previous Symbian Signed scheme include:

- TC Trustcenter introduced as new certificate authority for Symbian Signed
- simplification of the signing options and the signing process available to developers with the introduction of Open Signed, Express Signed and Certified Signed
- simplification of the Test Criteria
- reduction in the reliance upon independent test houses.

This chapter is based upon the updated scheme introduced by Symbian, although some details are still in draft at the time of writing. We also explain some of the differences that have been introduced and cover a few cases from the previous scheme.

The Symbian Signed website, ***www.symbiansigned.com***, provides further information. Sony Ericsson and Motorola both provide specific information relating to Symbian Signed for their particular handsets.

14.2 Further Considerations

Even if your application does not require access to protected APIs, you should still consider obtaining Symbian Signed certification. Your decision whether to Symbian Sign an application or not has several downstream consequences that you should consider prior to proceeding with application development.

Some distribution channels may only take applications that have been Symbian Signed. You should check the requirements of whichever channels you intend to use to distribute and sell your application.

All applications that have passed Symbian Signed are listed within the Symbian Signed applications catalog. This catalog is made available to channel partners, such as operators, giving added exposure to your application.

Some mobile phones default to only allowing signed applications to be installed. Attempting to install an unsigned application on such a mobile phone reports a `Certificate Error`. The user may request technical assistance or may give up and decide not to install the application.

At the time of writing, most UIQ 3 phones are unrestricted for the User capability set. This means that users can choose to install unsigned applications that need User capabilities, however the general trend is to restrict installation to only those that have been signed.

Support problems can arise for users who have installed unsigned applications. For example, a mobile operator may ask if any unsigned applications have been installed and request un-installation to be sure that the issue is not caused by the unsigned application.

14.3 Application Origin

When installing an application, UIQ presents a number of informative and warning dialog boxes to the user.

Firstly, the application (Software) name, version and supplier name are presented. The first dialog box (see Figure 14.1) presented to the user

Figure 14.1 Typical version and vendor details

takes its information from the vendor name and package header fields within the package (PKG) file used to create the SIS.

```
; Localized vendor name
%{"Sony Ericsson Mobile Communications"}

; Standard PKG file header
#{"LayoutManager1"}, (0xEDEAD004), 1, 0, 0, TYPE=SA
```

On the face of it, this seems fine but, with an unsigned SIS file, this information is not guaranteed. As developers we could put any information we desire in here. The user has no way of knowing if the information is valid or not (see Figure 14.2). If your application is unsigned, the installer goes on to display the warning in Figure 14.3.

Install software
Software name:
Any name I want!
Version:
12.34(567)
Supplier:
I can put anything in here that I want to. Unless the application is signed it will not be verified
Install

Figure 14.2 Unsigned applications can have any vendor details

Control panel
Install
Install
Themes
Uninstall
Security information
The identity of the software supplier cannot be guaranteed.
Do you want to proceed?
Yes
No

Figure 14.3 Vendor identity warning dialog

Many users are very conscious of malicious applications in the desktop and laptop computer world. Some users may decide not to install your application because of this warning. If your application is Symbian Signed, however, the user should feel that they can trust the supplier details and no security warning is presented.

14.4 Capabilities

Symbian OS Platform Security uses capabilities to control access to protected functions. Examples of such protected functions are:

- access to contact or calendar information, in order to protect the user's private data
- sending an SMS, because the user must pay
- controlling the power state of the phone, since this could cause the user to miss incoming calls.

Capabilities are simply a set of attributes an application can possess. They grant an application the privilege to access otherwise protected system resources and sensitive data through Symbian OS APIs. Around 40 % of the APIs in Symbian OS v9 are classified as sensitive and require that applications have the appropriate capability to use them. The remaining 60 % have no such requirement.

To access the user's private calendar entries, your application must have the `ReadUserData` capability. Once granted, your application can also read contact information, because `ReadUserData` gives access to a number of APIs.

The capabilities are divided into three groups, User, System and Device Manufacturer (see Table 14.1).

Table 14.1 Capability groups

Capability Group	Capabilities	No Signing	Open Signed		Express Signed	Certified Signed
			Online	Offline		
User	**LocalServices** **Location** **NetworkServices** **ReadUserData** **WriteUserData** **UserEnvironment**	✓	✓	✓	✓	✓
System	**PowerMgmt** **ProtServ** **ReadDeviceData** **SurroundingsDD** **SwEvent** **TrustedUI** **WriteDeviceData**	X	✓	✓	✓	✓
	CommDD **DiskAdmin** **MultimediaDD** **NetworkControl**	X	X	✓	X	✓
Device Manufacturer	**AllFiles** **DRM** **TCB**	X	X	✓	X	✓

Signing is not required for User capabilities, however it is recommended, as discussed in Section 14.1.2. Device Manufacturer capabilities must be approved by the device manufacturer which can only approve for its own mobile phones.

Prior to the changes in Symbian Signed during late 2007, the capability groups were slightly different. User capabilities were known as Basic capabilities and System capabilities were called Extended capabilities. Also, `CommDD`, `DiskAdmin`, `NetworkControl` and `MultimediaDD` were in the Device Manufacturer group.

14.4.1 User Capabilities

User capabilities are designed to be meaningful to mobile phone users. Depending on mobile phone manufacturer security policies, users may be able to grant blanket or single-shot permission to applications which use these capabilities. You can get User capabilities via any of the signing options.

The capabilities in the User group cover the following areas of functionality:

- `LocalServices`: Local USB, infrared or Bluetooth network. In general, these services do not incur any financial cost to the user but they could compromise privacy.
- `Location`: Location-based services offered by a mobile phone, for example determining the current network the phone is on. The user may wish location to be kept private.
- `NetworkServices`: Functionality associated with the network, for example, sending an SMS or using GPRS. Usually such services incur a financial cost to the user.
- `ReadUserData`: Read data that has been marked as private to the user. For example, the contacts engine requires you have this capability before you can read contacts.
- `WriteUserData`: Update user data stored within system components. For example, the contacts engine requires you have this capability if you want to update a contact entry.
- `UserEnvironment`: Access to information about the user and their current environment, such as using the camera.

If you use any of the capabilities from the User group then your application does not have to be signed. You do, however, need to consider what happens when the unsigned application is installed. To show what this means in practice, we have modified the `LayoutManager1` application so that the capability `LocalServices` is declared in the MMP file. Although the application does not require any capabilities protected by `LocalServices`, by declaring it in the MMP file we trigger the software installer to check to see if the application is signed using the correct certificate.

We then built the application for the emulator and packaged it in an unsigned SIS file. During installation, the user is presented with a dialog asking for the capability to be granted (see Figure 14.4).

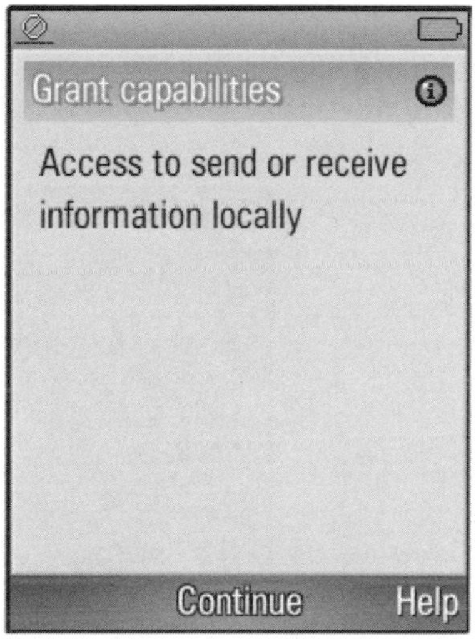

Figure 14.4 Dialog asking to be granted a single capability

Symbian offers a configurable policy that allows handset manufacturers the ability to state which capabilities a user can grant (rather than requiring a digital signature). Currently the manufacturers follow Symbian's recommendation and limit this to the User group. Capabilities from the System or Device Manufacturer group must be signed for. There is no guarantee that this will stay static in the future and some manufacturers, or even operators, may adopt a different policy.

Returning to our example, remember that the MMP was modified so that the application requires the single capability `LocalServices` and that the application does not use any APIs protected by this capability.

Pro tip

When you declare the capabilities that the application requires in the application's MMP file, the software build tools do not check to see if your application actually uses these capabilities. It is up to you to ensure that you declare the correct capabilities for your application.

If multiple capabilities are required, all are listed within the single dialog so the user only has the option to grant all of them or none of them. Figure 14.5 shows the same application, however this time it declares that it requires the capabilities `Location`, `NetworkServices` and `ReadUserData`.

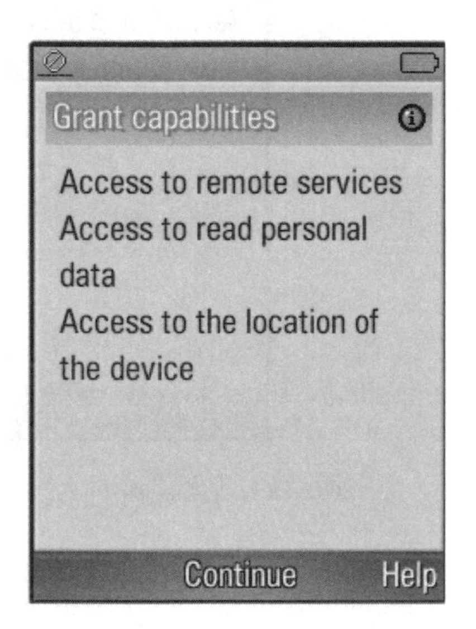

Figure 14.5 Dialog asking to be granted multiple capabilities

You may have noticed that the user only has the option to continue with the installation. You may be asking 'What if the user does not want to grant capabilities?' Within the emulator there is no easy way to do anything else apart from continue. However, if we look at actual hardware and take two examples (see Figure 14.6), the Sony Ericsson P990i (Pen Style) offers the standard Cancel button on the button bar and the M600i (Softkey Style Touch) offers the Cancel button on the title bar. In both cases, selecting Cancel stops the software from being installed.

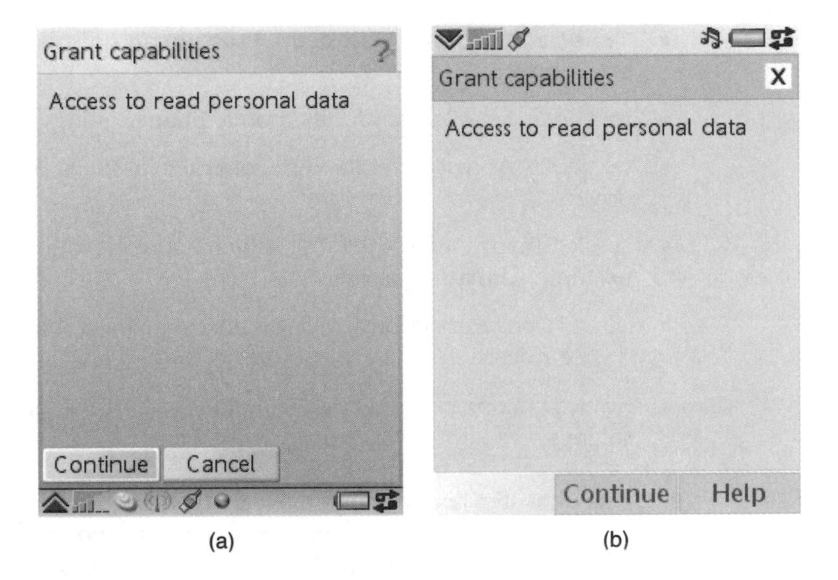

Figure 14.6 Canceling the grant capabilities dialog on a) P990i and b) M600i

Even if it only requires User capabilities, signing your application means that the user will be presented with a trusted application and will not be asked to grant capabilities.

14.4.2 System Capabilities

If your application uses one or more capabilities from the System group, then it must be Symbian Signed (Open, Express or Certified Signed). Unless a mobile phone manufacturer changes the phone configuration, the end user cannot grant the following capabilities:

- `PowerMgmt`: Set power states of the phone, such as standby modes or switch off.
- `ProtServ`: Symbian OS server processes that wish to register themselves with protected names. Is also required by plug-ins to such servers, for example ECOM-based recognizers.
- `ReadDeviceData`: Access the operator, manufacturer and phone settings.
- `SwEvent`: Generate and capture key presses and pen events.
- `SurroundingsDD`: Access the device drivers that provide information about the surroundings within which the phone exists.

- `TrustedUI`: Create a trusted UI session and display dialogs in that environment.
- `WriteDeviceData`: Modify mobile phone settings.
- `CommDD`: Access directly the communication subsystem's device drivers.
- `DiskAdmin`: Access the file system administration functions, for example, formatting a disk.
- `MultimediaDD`: Priority or direct access to the multimedia subsystem device drivers.
- `NetworkControl`: Access or modify any of the network protocol settings.

Note that the last four System capabilities, `CommDD`, `DiskAdmin`, `MultimediaDD`, and `NetworkControl`, are *not* available via the Express Signed option or via Open Signed without a Publisher ID (see Table 14.1). To deploy an application that uses them to a phone for testing, you need to use Open Signed with a Publisher ID. To release your software, you will need to sign using Certified Signed, which requires independent testing by a test house.

14.4.3 Device Manufacturer Capabilities

The final group of capabilities is the Device Manufacturer group, which is also known as the Licensee or Channel group. For the purpose of this chapter, the term Manufacturer has been chosen to imply that ultimately the phone manufacturer (Sony Ericsson, Motorola, etc.) is responsible for approving or rejecting the use of these capabilities. The manufacturer only approves or rejects the use of these capabilities for their particular phones. If one phone manufacturer grants the use of a manufacturer capability then there is no guarantee that another phone manufacturer will also grant its use.

The capabilities in this group are:

- `AllFiles`: Read access to the entire file system and write access to the private folders of other applications.
- `DRM`: Access to content protected by Digital Rights Management (DRM).
- `TCB`: Trusted Computing Base (TCB). Access to install software, create processes and set capabilities.

You must use the Open Signed with Publisher ID option for development and testing and the Certified Signed method for the final product.

Using Open Signed with a Publisher ID enables you to download a developer certificate from the Symbian Signed portal. When you request a certificate with one or more Device Manufacturer capabilities, it is not immediately available. The manufacturer is notified of your request and the individual or group responsible will review your request and reject it, approve it or ask for further information or clarification.

This means that you must work with a manufacturer to gain approval for your Symbian Signed application. You must justify your request and provide supporting information such as:

- the capabilities required by your application
- why these capabilities are needed
- alternatives using less-sensitive functions and why these cannot meet your needs.

When you submit your application for Symbian Signing via Certified Signed, the test house will contact the manufacturer on your behalf and request approval for the use of the capabilities. Again, the individual or group responsible within the manufacturer will handle this request and update the test house with the decision. Granting of capabilities is not guaranteed. We have seen situations where the use of certain capabilities has been refused and, due to poor design, it has required months of work before the application could be re-submitted.

Pro tip

Very few applications need to have Device Manufacturer capabilities. However, if your application does fall into that category, you must have a very strong relationship with a mobile phone manufacturer and an extremely compelling reason to have the capability.

14.5 Routes to Symbian Sign an Application

Symbian Signed gives you three options for getting your application signed. Depending on the capabilities that your application uses and the options that you choose, you can trade off speed and convenience against fewer restrictions in terms of API access and scale of deployment. The three routes are:

- Open Signed replaces the concept of developer certificate (DevCert) signing. It is available to developers with and without access to a Publisher ID, although it is more powerful and flexible if you do have a Publisher ID. Its aim is to provide signing for applications that require limited deployment to known phones during testing, development and evaluation.

- Express Signed is available to any developer who has access to a Publisher ID certificate. Testing may be done in-house rather than at an external test house. As noted earlier, Device Manufacturer capabilities cannot be obtained using this method, and the most sensitive System capabilities are also not available.
- Certified Signed provides the maximum level of application quality assurance and trust to end users. It is also required if the developer requires access to the most sensitive capabilities.

A Publisher ID is a key concept in Symbian Signed. It is a digital certificate that is issued by a third-party trusted authority. It verifies your identity to the test house and to end users installing your application. TC TrustCenter is the certificate authority for Symbian Signed.

14.5.1 Open Signed

Open Signed enables you to sign applications for limited deployment to known mobile phones (based on IMEI) for development and testing. It is similar to the developer certificate in the previous Symbian Signed scheme. Open Signed is available online without a Publisher ID and offline with a Publisher ID.

Without a Publisher ID

Without a Publisher ID, signing is performed wholly by the Symbian Signed website portal. You fill in online forms, upload your application and receive an email when your signed application is ready to download. The main features of this signing option are:

- No Publisher ID is required.
- No Symbian Signed account is required.
- All User and most System capabilities (excluding `CommDD`, `Disk Admin`, `NetworkControl`, `MultimediaDD` and Device Manufacturer set) are available.
- It is restricted to a single mobile phone (one IMEI).
- There is no cost to developers.
- It is for testing and personal use only and is not suitable for commercial distribution.
- A signed application is valid for 36 months.

With a Publisher ID

Open Signed with a Publisher ID provides you with a developer certificate that you can download and use locally to sign applications offline. The main features of this option are:

- A Publisher ID is required (US$200 per year at the time of writing).
- A Symbian Signed account is required.
- All User and System capabilities may be requested.
- Device Manufacturer capabilities are subject to approval.
- The application can be installed on up to 1000 mobile phones, based on a list of their IMEIs.
- The developer certificate is downloaded and used to sign applications locally using `signsSIS.exe`.
- The developer certificate can be used to sign any number of applications during its lifetime.
- It is for testing and personal use only and is not suitable for commercial distribution.
- A signed application is valid for 36 months.

Unlike the online option, the above procedure uses a combination of PC-based tools and the Symbian Signed website, as well as a downloaded developer certificate. Since we use a developer certificate to build the `SignedApp` example for deployment to real mobile phones, we explain developer certificates in more detail.

Note that you can specify up to 1000 IMEI numbers, making deployment to large groups more manageable. You must specify each IMEI so it is recommended that good hardware tracking and management is enforced within your software development projects.

Obtaining a Developer Certificate

This section describes the process required to obtain a developer certificate. It considers the scenario of a certificate for an application that requires a Device Manufacturer capability as this is the most complex. We highlight where this differs from the process of obtaining a certificate for the User set or System set. You need a Symbian Signed account and a Publisher ID.

The process of obtaining a developer certificate is a straightforward one; however, for first time users it can be confusing, especially if you require Device Manufacturer capabilities.

Download the latest version of the `DevCertRequest` tool from the Symbian Signed website. The `DevCertRequest` tool is an application wizard for Microsoft Windows that generates a Symbian development certificate request file (CSR) (see Figure 14.7). The certificate request file represents your requirements which includes the IMEIs and capabilities for which the certificate will sign. This file is subsequently uploaded to your Symbian Signed account and used to generate your certificate.

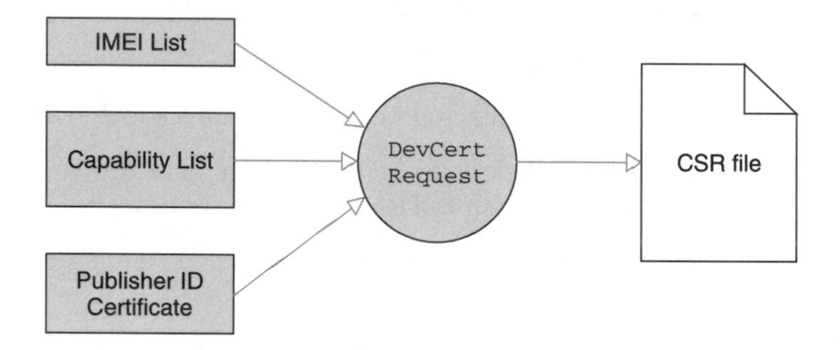

Figure 14.7 Creating a certificate request file

`DevCertRequest` provides a GUI for you to select the required capabilities and identify if Device Manufacturer capabilities are required. Once created, the CSR file cannot be modified; if it contains any errors or omissions then a new file must be generated.

To obtain the developer certificate, you upload your CSR file to your Symbian Signed account. The portal creates a certificate and makes it available for you to check and download.

If you request Device Manufacturer capabilities, you must provide extra information with your CSR file upload. Your request will be routed to the selected manufacturer for approval. Once approved, you can download your certificate.

Once you have your certificate, you can use it to sign your application any number of times during the test and development phase. Your certificate has a lifetime of 36 months so after this period you need to request another development certificate. Note, too, that the applications you sign with it will only install during that 36-month lifetime. If you change the capabilities or IMEIs required then you also need to request a new developer certificate.

Comparison with Previous Scheme

Prior to the Open, Express and Certified Signed schemes introduced in late 2007, the term *Developer Certificate Signing* was used for signing applications during the development, test and evaluation stages.

Old development certificates were valid for only six months and for a maximum of 100 IMEI numbers. The previous scheme used ACS Publisher

IDs from VeriSign. For the time being, these remain valid for Open Signed offline. New Publisher IDs should be obtained from TC TrustCenter.

Requesting a Developer Certificate in Carbide.c++

One recent addition to the process of obtaining a development certificate is the request plug-in tool available for the Carbide.c++ IDE (see Figure 14.8). This allows you to load the request tool directly from your IDE.

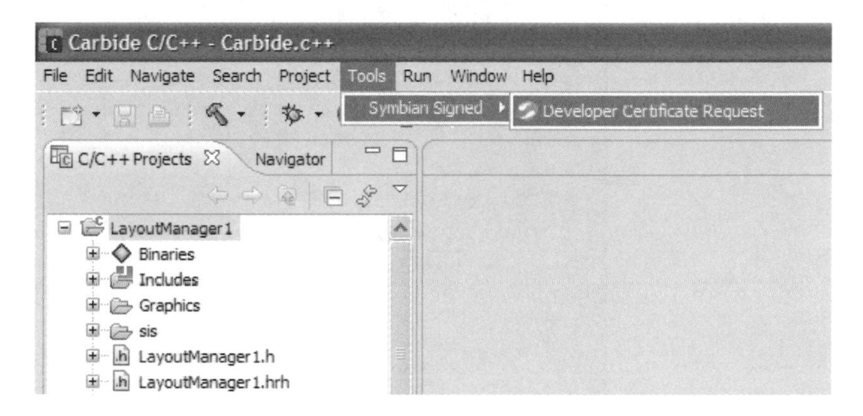

Figure 14.8 Carbide.c++ Tools menu

Carbide.c++ provides a form for you to enter the required information (see Figure 14.9). This example lists the capabilities according to the groups that applied before the introduction of Open, Express and Certified signing.

Once the certificate is created, you must still upload the certificate request file to your Symbian Signed account.

Signing with a Developer Certificate

Once you have obtained your developer certificate, you can use it to sign your application. To do this, you use the `signSIS` tool which is available in the UIQ SDK. You need your developer certificate CER file and password.

For detailed guides to signing and installing application SIS files, refer to the Software Installation Toolkit guide and reference, under Tools and Utilities in the Symbian OS Library documentation in the UIQ SDK or online from ***developer.symbian.com***.

You can sign your application any number of times; there is no restriction on the number of development–build–sign cycles.

We advise you to keep a careful note of the developer certificate expiration date, which is 36 months after the creation date. If your

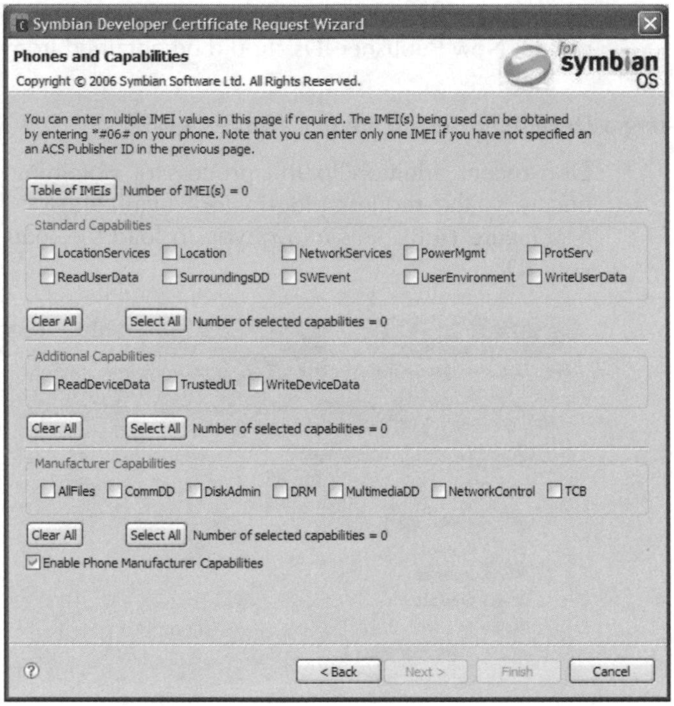

Figure 14.9 Developer certificate request wizard in Carbide.c++

application suddenly fails to install, an expired developer certificate is very likely to be the problem. The error message you see does not say that the certificate has expired; you are simply told that a component has failed a security check.

Once you have your signed SIS file, it can be deployed to your mobile phone in the usual manner. Remember that the certificate restricts this distribution so it will only install on devices whose IMEI is declared in the certificate (this was done when the certificate request file was generated). There is no difference in the way a SIS file signed with a development certificate is deployed compared to unsigned or Symbian Signed SIS files. The SIS file can be installed using PC-based software or sent to the device via email, MMS, infrared or Bluetooth. With all of these methods, it is the software installed on the phone that validates the certificate and ensures all requirements (IMEI, capabilities signed for, certificate expiration, etc.) are satisfied.

Deploying an Application

When testing is complete and your application is ready for final signing, you must move on to use either Express Signed or Certified Signed.

14.5.2 Express Signed

Express Signed is intended for commercial and non-commercial applications that will be released to the public. The main features of this signing option are:

- A Publisher ID is required (US$200 per year at the time of writing).
- A Symbian Signed account is required.
- All User and most System Capabilities (excluding `CommDD`, `DiskAdmin`, `NetworkControl`, `MultimediaDD`) are available.
- Deployment is unrestricted.
- No independent testing is required; you must however pass the test criteria and may be audited for compliance.
- A signed application is valid for 10 years.

Express signing is not free although the cost model is simple and predictable. As well as the annual cost for the Publisher ID, you need to purchase a number of *content IDs* from the certificate authority, TC TrustCenter. Each signing submission uses one content ID. Although the cost for each content ID may change with time, it should be around US$20.

The basic procedure follows these steps:

1. Test and debug your application to meet the Test Criteria.
2. Sign your application using `SignSIS` and your Publisher ID.
3. Make sure you have sufficient content IDs in your Symbian Signed account.
4. Submit the signed SIS, PKG and documentation to the portal.
5. Select options and supply further information to the submission page.
6. Portal validates the application.
7. Download your signed application once it has been accepted.

All applications submitted through Express Signed should be tested against the Symbian Signed Test Criteria and it is the responsibility of the person or organization submitting to ensure that this has been done. Symbian carries out random auditing of submitted applications to verify that they meet the standards set by the test criteria.

Freeware and software publishers without a Publisher ID can access this signing option via publisher partners. In general, these are companies that sign, publish and distribute applications. Different publisher partners have different terms and conditions which dictate how the testing costs are recouped and may have commercial restrictions on what you can do with your signed application.

Prior to the introduction of Express Signed there was a service called *Self Signing*. Express Signing is very similar to Self Signing and is in some ways a replacement. The significant difference is that it is not possible to access Device Manufacturer capabilities through Express Signing.

14.5.3 Certified Signed

For mainstream commercial software developers, the full testing and accreditation regime of Certified Signed is the preferred option. It also entitles you to use the 'for Symbian OS' logo to aid differentiation and brand building for your application.

The main features of this signing option are:

- A Publisher ID is required (US$200 per year at the time of writing).
- A Symbian Signed account is required.
- All User and System Capabilities are available; Device Manufacturer capabilities can be granted with manufacturer approval.
- Deployment is unrestricted.
- Applications are tested by an independent Symbian Signed accredited test house as part of the process. You must pay for this testing.
- A signed application is valid for 10 years.
- Your application may display the Symbian OS logo and branding.

If your application requires any of the Device Manufacturer capabilities (`AllFiles`, `DRM` or `TCB`), or the most sensitive of the System capabilities (`CommDD`, `DiskAdmin`, `MultimediaDD` and `NetworkControl`), Certified Signed is the only option available to you when you need to release for wide-scale deployment.

Because Certified Signing requires independent testing, it is considered to be the most trusted of all signing options. Symbian recommends that this should be considered the natural choice for all commercial and enterprise software intended for wide-scale deployment.

The procedure for Certified Signed is as follows:

1. Sign your application using `SignSIS` and your Publisher ID.
2. Submit the signed SIS, PKG and documentation to the portal.
3. Select options and supply further information to the submission page.
4. Select your chosen test house and accept the legal agreement.

5. Track the progress and test results via the portal.
6. Download your signed application once it has been accepted.

The test house may require access to external servers or to be loaned SIM cards, hardware or other facilities in order to test your application.

14.6 Procedural Impact

Symbian Signed should be part of your software design, implementation and testing procedure. We recommend that you decide as early as possible to include Symbian Signed in your plans. This enables you to do the right thing at various points in the development process and avoid rework later on.

In the design phase, you need to assess whether any functions must make use of protected APIs, for which capabilities are required (see Section 14.4).

Symbian Signed provides a Test Criteria document detailing the tests to be performed and the expected results. The tests are updated periodically, so check that you are working to the latest version, especially if you are working on a long software project. You should check your application design against this document to make sure that the tests can be passed. In Section 14.9.2, we examine our `SignedAppPhase2` example application in respect of these criteria.

If you are using Certified Signed, you need to submit your application to a test house. This adds at least one week to your development schedule, perhaps longer. We recommend that you allow at least two weeks for your first application. For both Express and Certified Signed, you need to allow time and resource for understanding and complying with the test criteria in order to maximize your chances of successfully passing the tests.

If your application requires a waiver for a particular test case, then the test house will also have to contact the appropriate manufacturer for a decision on how to proceed. This will require extra time.

Failing Symbian Signed does not always mean disaster for your software project. On many occasions, a supplier is simply required to make small changes, such as updating the package file. The application is then re-submitted, which only results in minimal delays to your project.

14.7 Getting Started with Symbian Signed

14.7.1 Publisher ID

For all types of signing except Open Signed online, you must have a Publisher ID. Publisher ID digital certificates form part of the public key infrastructure and are issued by certificate authorities. The certificate authority for Symbian Signed is TC TrustCenter. You can find out more and purchase a Publisher ID at ***www.trustcenter.de/en*** and follow the *Products & Services, TC Certificates links*. At the time of writing the cost is US$200 and the certificate lasts 12 months.

The TC TrustCenter certificate is assigned to companies or organizations and not individuals (an individual person, not working with a company, can be considered to be an organization). If your organization does not have one of these certificates and the application you are developing requires a capability from the System or Device Manufacturer group, then you will need to obtain one.

> **Pro tip**
>
> The process of applying for your Publisher ID and obtaining it can take a significant amount of time which can run in to weeks. You may have to provide proof of identity such as a certificate of company incorporation, which may take time for you to locate within your organization.

The main purpose of this trust certificate is to securely identify the application and authenticate the organization providing the application. This gives the end user the confidence that the application can be traced back to a development source and that the application they are installing is what they expected.

14.7.2 Symbian Signed Account

Creating a Symbian Signed account is free. It is recommended that, where possible, you have only one account and that all UID requests, Development Certificate requests and Symbian Signed submissions are managed from this one account. Once the account is created, there are no real restrictions on the number of development certificates that you can request through this account. This is much simpler to manage than having your developer certificates, IMEI lists and other portal transactions spread across a number of accounts.

> **Pro tip**
>
> There are at least two legal agreements within the Symbian Signed process that you need to accept. The first is required when you sign up to join the Symbian Signed website. The second is only accessible when you attempt to submit an application to a test house. If, like

most organizations, your organization has a legal department that needs to authorize all legal agreements, we strongly recommend that immediately you sign up to Symbian Signed, you also go through the submission process at least to the point where you are presented with the second agreement. By getting these agreements resolved early in the development process, you should not encounter delays later on when your application is ready to submit to the test house.

14.7.3 Symbian Signed and UIDs

Each application on a mobile phone needs to be uniquely identified. This is achieved with a 32-bit unique number known as a UID. Prior to UIQ 3, or strictly speaking, prior to Symbian OS v9, a UID had much less significance than it does now. UIDs are allocated from three ranges of numbers according to the intended use (see Figure 14.10).

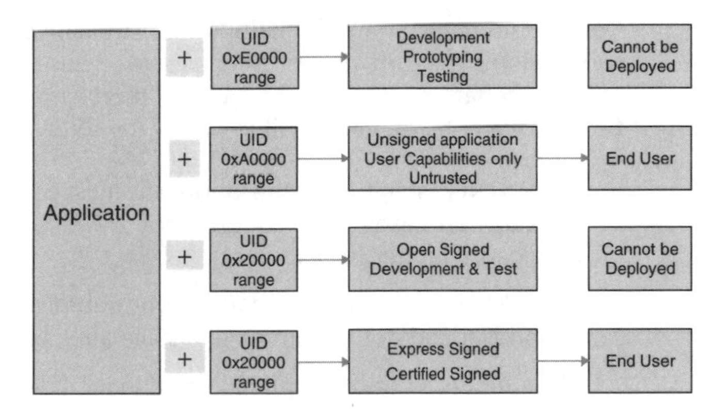

Figure 14.10 UID ranges and Symbian Signed

The example applications in Chapters 6 to 9 used UIDs in the range `0xE0000000–0xEFFFFFFF`. These values are reserved for internal development purposes only. Any applications that are to be distributed should not use these UID values. In particular, there is no formal mechanism to ensure two developers do not use the same values from this range.

If you have chosen not to get your application Symbian Signed, development can proceed more or less as it did with previous versions of Symbian OS, provided that you obtain a UID from the Symbian OS v9 unprotected range `0xA0000000` to `0xAFFFFFFF` for your application UID. You cannot use previously allocated UID values from the `0x10000000` to `0x1FFFFFFF` range. The application installer refuses to install such applications.

You must request the correct UIDs for your application. There is comprehensive information available on the Symbian Signed website covering protected and unprotected UIDs, what range of UIDs to use for test code, etc.

If you have chosen to get your application Symbian Signed, your first requirement is to obtain a UID from the Symbian OS v9 protected range `0x20000000` to `0x2FFFFFFF`. If you later change your mind and decide that you no longer want to go through the Symbian Signed tests, you have to change the UID. Unless they are signed, the application installer refuses to install applications with UID values in the protected range.

UIDs are obtained via the Symbian Signed website, and are registered against the company or organization that requested them. In this way, the test house can verify that the UID of a submitted application belongs to the same developer that has submitted the application.

14.8 Submission and Compliance Criteria

You should review the submission and compliance criteria prior to each submission. The aim of the criteria is to ensure that your submission meets a certain standard and to reduce the chance that it gets rejected. Failure of a criterion can result in one of the following:

- WARNING: signing of the application is permitted, assuming all relevant test cases pass, although the developer is warned that this may cause further issues.
- REVOKED: the application fails during submission or a signed application is found not to comply at a stage after signing.

  The submission and compliance criteria include the following checks:

  - Any embedded SIS file included must already be signed, depending on the contents of the embedded SIS file and the capabilities it requires. Failure of this criterion results in a warning.
  - The SIS file being submitted must be correctly signed with a valid Publisher ID. The Publisher ID you used to sign the SIS file must match the Publisher ID in your Symbian Signed account. Failure of this criterion results in a revocation.
  - The SIS file must include the correct version information which corresponds to the application About box and supporting documentation. If you do not adhere to a consistent naming convention, you may have difficulties in delivering future upgrades and patches. Failure of these criteria results in a warning.
  - At time of submission, the developer must make declarative statements about capability usage with satisfactory rationale. The

capability usage must disclose the APIs used for each declared capability. These statements are retained on file and must be true and accurate. Failure of this criterion results in a revocation.

- The product and platform UID must match the specified target products. Failure of this criterion results in a warning.
- The PKG file's UID must be owned by the developer and be from the correct range. Failure of this criterion results in a revocation.
- All application UIDs must be owned by the developer and be from the correct range. Failure of this criterion results in a revocation.
- If a PKG file VID is specified, it must be from the correct dedicated range. Failure of this criterion results in a revocation.
- There must be adherence to special rules for Central Repository and C32 updates.
- If auto-start functionality is present and the `ProtServ` Capability is required, the application is checked for illegal recognizer usage for auto-start. Failure of this criterion results in a revocation.
- Operator approval is required for use of the `NetworkControl` capability for accessing the SIM. Failure of this criterion results in a revocation.

14.9 Symbian Signed Test Criteria

The full Symbian Signed Test Criteria document is available from the Symbian Signed website. It is updated from time to time and a significant update is in draft at the time of writing this book. We provide an overview of the proposed Test Criteria v3.0.0 and discuss in detail the current Test Criteria v2.11.0.

14.9.1 Test Criteria v3.0.0

With the introduction of Open, Express and Certified Signed, Symbian is undertaking a major review of the test criteria. The new Test Criteria are still in draft at the time of writing.

The tests are broken into three sets:

- Universal Tests
- Specific Symbian OS v9.x Tests and Symbian OS v9.x Capability Access Tests
- Internet Telephony Application Tests.

Universal Tests

All applications must meet these basic criteria:

- They must install correctly, including from memory cards.
- They must uninstall correctly, leaving no installation files behind.
- They must reinstall correctly.
- They must backup and restore successfully, using appropriate file creation locations.
- They must survive low-memory conditions at startup.
- They must handle exceptional events such as out-of-memory conditions.
- They must survive power-down and reboot events.
- They must handle service interruptions and repeated switching.
- They must handle system events correctly and comply with task list behavior guidelines.

Specific Symbian OS v9.x Tests and Symbian OS v9.x Capability Access Tests

All UIQ 3 applications must meet the relevant criteria:

- They must comply with scaleable UI requirements.
- Capability declarations must be correct.
- The policy statement dialog must be displayed telling the user which phone features are used.
- Multimedia applications must not interfere with phone calls.
- Applications which access the network must have network operator approval.

Internet Telephony Application Tests

If your application uses VoIP or replaces the standard phone UI, then the following tests also apply:

- Phone applications must present a UI which gives the user control of the application.
- VoIP applications must present a Device Manufacturer disclaimer and must not interfere with GSM-based telephony functions, including the ability to make emergency calls.

14.9.2 Test Criteria v2.11.0

In this section, we examine the `SignedAppPhase2` application that we developed in Chapter 10 to help explain the nature of the testing, the possible impacts upon our development and how to maximize our chances of passing. Note that the Symbian Signed Test Criteria are updated from time to time, so it is important to keep checking for updates. We have taken Test Criteria v2.11.0 as these were current at the time of writing. As they change over time, the general principle of checking your application against the specification remains the same. The IDs of the revised v3.0.0 Test Criteria are added in brackets where applicable.

The Symbian Signed tests aim to check the general behavior of your application and interoperability with Symbian OS.

The test cases do not aim to:

- comprehensively test all of the functionality of your application
- test that you have implemented correctly against your specification
- ensure that you have followed the UI Style Guide.

PKG-01

The SIS file must be signed with a valid ACS publisher certificate.

Symbian provides a tool called `VerifySymbianSigned`, freely available from the Symbian Signed website, that can analyze a SIS file. For our application, the tool provides a report as below:

```
------------------------------
File format                   : SISX (V9)
SIS file                      : SignedApp_ACSCert_uiq_3_0_v_1_00_01.sis
------------------------------
Symbian Signed                : No
Developer Certificate Signed  : No
ACS Publisher Signed          : Yes
Unknown Certifictaes          : 0
```

This tool confirms that our SIS file is signed with an ACS Publisher Certificate. The application complies with this test.

In Test Criteria v3.0.0, this criterion becomes part of submission and compliance.

PKG-02 (UNI-01)

A Symbian native application is expected to start up within a reasonably short timeframe, presenting a progress indicator if start up takes more than about five seconds.

Our application starts up rapidly, so it does not require any progress information. We should not have a problem with this test.

PKG-03 (UNI-08)

Files should only be created on the drive the application was installed to.

Our application has chosen to store configuration information within an INI file. The application framework creates that file on our behalf, currently always on the internal C drive, irrespective of the disk the application is installed on.

An exception, EX-006, covers the case where the system code creates files on our behalf. To ensure we do not fail this test it must be documented within a `readme.txt` file. This file must be submitted along with the application so that the exception is clear to the tester.

While we recommend that you design commercial applications to meet this criterion, our example application makes use of the exception.

PKG-04 (UNI-08)

Uninstallation removes all files originally installed or created by the application.

On Symbian OS v9 mobile phones, the uninstallation process automatically removes all files within the private application folder. Since our application creates files only in that folder, we should not have any problems with this test. If we later change the application to create files in other folders, we must reconsider this test.

PKG-05

Installation should not be to a specific media drive.

Our PKG file contains statements such as:

```
"\epoc32\release\gcce\urel\SignedApp_0x20000462.exe"-
    "!:\sys\bin\SignedApp_0x20000462.exe"

"\epoc32\data\z\resource\apps\SignedApp_icons_0x20000462.mbm"-
"!:\resource\apps\SignedApp_icons_0x20000462.mbm"

"\epoc32\data\z\resource\apps\SignedApp_0x20000462.mbm"-
    "!:\resource\apps\SignedApp_0x20000462.mbm"

"\epoc32\data\z\private\10003a3f\apps\SignedApp_reg_0x20000462.rsc"-
    "!:\private\10003a3f\import\apps\SignedApp_reg_0x20000462.rsc"
```

These statements inform the installer to place files into the disk called `"!"`. When an application is actually installed and the user chooses a specific media drive, the `!` is replaced by the actual drive letter.

Our PKG file has no statements that require application files be installed to specific media drives, so we should not have a problem with this test.

PKG-06 (UNI-09)

An application can be reinstalled to a device after it has been uninstalled.

We have no specific requirements as to the media drive on which our application is installed. When uninstalled, we remove all the application files so it is highly unlikely that we would have a problem with this test. You should perform this test to ensure compliance.

PKG-07

UID and platform compatibility information are appropriate for the target device.

Our application UID, `0x20000462`, has been allocated correctly by Symbian Signed and is within the correct range of UIDs for applications that can be Symbian Signed. Our package file does not specify a VID, so that defaults to zero, as required.

Our PKG file contains the following statement:

```
; ProductID for UIQ 3.0
; Product/platform version UID, Major, Minor, Build, Product ID
[0x101F6300], 3, 0, 0, {"UIQ30ProductID"}
```

This statement claims that our application supports UIQ from version 3.0 onwards. If your application requires functionality that is only present in UIQ 3.1 or higher, you must update the statement.

Our application is designed to operate on all versions of UIQ 3 from 3.0 onwards, therefore the statements in the PKG file are correct and we should not have a problem with this test.

PKG-08

Version information is consistent.

This test requires that the version information contained within the SIS file is consistent with the version information displayed within your application's About box and with any documentation supplied as part of the submission.

Our PKG file contains the following statement:

```
; What our app name is displayed as
#{"Signed App"},(0x20000462),1,00,01,TYPE=SA
```

This statement declares our application to have a major version number of 1, a minor version of 00 and a build of 01. The About box must present the same values and, in particular, it must present the same number of leading zeros in each of the fields. Otherwise, the application is very likely to fail this test.

You must include an About box within your application to pass this test. Our example application contains an About box so should pass it.

Pro tip

The Symbian Signed Test Criteria state that you must have a major, minor and build number. If your version control system does not actually work like that, we suggest that you map or allocate major and minor version numbers in a simple way which also allows you to identify future versions. The build number would be chosen to match your version control build label. While it is only a 'guideline' within the Test Criteria, our experience is that if you do not have all three fields or do not use exactly two digits for the minor version number, you are highly likely to fail this test.

GEN-01

The application does not affect the use of system features or other applications.

The basis of this test is to ensure that your application behaves responsibly. Symbian OS is a powerful operating system with many features. Abuse of these features can cause a mobile phone to misbehave or affect other applications. For example, you could write an application that monitors incoming email messages and automatically deletes them. Such an application would be considered to be detrimental to the overall user experience of a mobile phone and is unlikely to pass Symbian Signed testing.

Poorly designed or malicious code is likely to fail this test. We also advise careful inter-operability testing, since new or changed features in inter-operability can give rise to failure cases. We do not expect any such problems with our application.

GEN-02

The application can handle exceptional, illegal or erroneous actions.

The test house spends around 15 minutes stress testing your application, for example attempting to enter too many characters into text editors or deleting all entries then selecting all remaining menu options. If the test house manages to cause your application to panic, for whatever reason, it fails this test.

This is probably one of the hardest tests to pass since it implies that there are no bugs in your application. Passing this test requires you to design, implement and test your application as much as you can before submission to a test house. You should design and run your own stress tests; simple functional testing is not enough to pass this criterion.

GEN-03

The application should support various device resolutions and formats appropriately.

The intention of this test is to ensure that your application can be used properly on those mobile phones you intend the application to support. In particular, it is required that the application correctly handles any dynamic screen switching it claims to support.

At the time of writing, only the Sony Ericsson P990i supports dynamic screen switching. You have two choices:

- to support both the flip-open and flip-closed modes
- not to support flip closed.

Flip open and flip closed are usually handled using the view architecture and `QIK_VIEW_CONFIGURATIONS`. In addition, we indicate support for flip-closed mode by the inclusion of a `VIEW_DATA` structure within our application localization file:

```
VIEW_DATA
  {
  uid=EViewIdPrimaryView;
  screen_mode=EQikScreenModeSmallPortrait;
  caption_and_icon =
    {
    CAPTION_AND_ICON_INFO
      {
      }
    };
  }
```

Apart from testing that the example application operates in both flip-open and flip-closed mode, we do not have to perform any special actions to pass this test.

Some applications may choose not to support flip closed on the Sony Ericsson P990i. In this case you should ensure that you do not include a `VIEW_DATA` structure claiming that you do and that you document this in any submission notes.

Pro tip

Sometimes this test is not as exact as stated in the specification. For example, ''All menu controls are fully visible'' does not mean that all menu controls must be fully visible all of the time. In our example, we have a cascade menu and our menu scrolls. By definition, cascade menu items are hidden until the user expands the cascade and scrollable items are off screen and therefore not visible. The primary intention of the test is to ensure that your application is fully usable in all mobile phone configurations. The intention of this test could perhaps be stated as "All menu controls are accessible". Applications that contain cascade or scrollable menus do not fail this test.

As we saw earlier, UIQ allows applications to change the content of menus and distribute commands over other controls, such as command bar buttons, depending on the current mobile phone configuration. Other Command Processing Framework flags allow you to specify various attributes about commands such as:

- `EQikCpfFlagOkToExclude`
- `EQikCpfFlagTouchscreenOnly`
- `EQikCpfFlagNoTouchscreenOnly`
- `EQikCpfFlagPortraitOnly`.

If you use these advanced features, we strongly recommend that you fully document your application's behavior as part of your submission. If you do not, you risk the test house raising questions or comments, or possibly even failing your application.

MEM-01 (UNI-03, UNI-04)

The application handles low-memory situations during application start up.

Currently the tool to perform this test is not available for Symbian OS v9 applications. As a result, neither the test house nor developers are able to perform it. Even if the test tool were available, the change from APP to EXE introduced in OS v9 means that we can now specify the `EPOCHEAPSIZE` within our MMP file. If you specify a minimum heap that is in excess of the amount of memory required to start your application, it is not possible for the application launcher to actually run your application until that amount of memory is available.

MEM-02 (UNI-04)

The application handles low storage during execution.

In a multi-tasking operating system, it is both difficult and inappropriate for one application to control or restrict the actions of another completely independent application. In particular, even if your application checks to see if there is enough space on a disk to store its content, by the time it comes to actually store that content it cannot guarantee that other applications in the system have not used the disk space. It is, however, reasonably easy to simulate low storage situations.

Data on media drives is rarely allocated at a byte level, there is usually a sector size dictated by parameters such as the size of the media and the type of filing system on the drive. The amount of disk space used by your application is usually in excess of the number of bytes saved and is rounded up to the nearest sector size. In general, applications do not and should not need to worry about the detailed internal operation of the file server and its associated media device drivers.

From the practical perspective of passing this test, the important point is to ensure your application can survive errors saving content to disk. In particular, we recommend that your error-handling code for a failure to fully save a file causes the file to be removed from the media drive. Our application uses system-managed INI files. At the application level, we do not have much control over their internal buffer sizes or when they choose to flush buffers to disk. We do, however, know when we have finished writing our application data to a stream and are required to commit the stream then the dictionary to the media drive. The stream store functions (which commit the data) leave on error; our task is to ensure we handle the leave appropriately, typically by deleting the partially formed file.

PHN-01 (UNI-02)

Incoming interruptions pause the application and save the state.

A mobile phone has numerous mechanisms by which the current activity can be interrupted. The classic example is an incoming phone call. Applications need to be written to integrate properly with these types of activity. In general, either a view switch or a background event occurs to handle the interrupting task. On the emulator, you can simulate this by selecting the applications button.

Many applications do not have to perform any special tasks; our example is one such application. If your application contains a significant amount of animation and state changing, as most games do, you need to make your application reach a quiescent state with no further animation running, so the user can resume from the position they left the application. In many applications, this may be as simple as stopping the animation timer; in others, it may be slightly more complicated as the animation is updating the UI to display the new engine state. In such cases, your application needs to be able to cancel (or complete) the animation rapidly.

PHN-02 (UNI-06)

The application handles an unexpected reboot.

This test is to ensure that a mobile phone and your application can survive situations such as the user removing the battery of the phone. There is not a requirement that the application continues from the position it was last at, only that the mobile phone and application can be re-started.

In practice this means that your application needs to be able to handle error conditions when starting up, such as its INI file being incomplete. Our application creates and initializes the engine with a set of default data. It then attempts to load the INI file. Only when that is fully successful will it switch to using those parameters. An error reading the INI file causes the engine to roll back to a stable state, using the default parameters. This should be sufficient to pass this test.

CON-01

The application can be closed via the task list.

On many mobile phones, there is a task manager application. We have already discussed the ability of our application to respond to any close messages sent by the task manager, so we should be able to pass this test. Your application must not panic in this situation.

CON-02

The application contains a privacy statement dialog.

A Symbian OS v9 application must display a dialog informing a user about the usage of various OS functions if such functionality exists in an application. The functionality covers:

- network access, such as using GPRS, SMS or MMS
- local connectivity, such as using Bluetooth or infrared
- multimedia recording
- reading or writing any user data, such as reading Contacts or Agenda entries
- usage of any Location services.

The intention of the test is to inform the user of the type of activity the application may perform. The user can then take an informed decision as to whether they find it acceptable for an application to use these services. Any dialog presented to the user needs to be coherent, however it only needs to summarize the services used rather than attempt to explain in detail exactly which APIs are used, when those APIs are used and why.

Since our application supports Send as, which can send content via MMS, Bluetooth or infrared, our application requires a privacy statement dialog. Without such a dialog our application will fail this test. `SignedAppPhase2` does not have such a dialog, so we add it in `SignedAppPhase3`.

This criterion is expected to be removed in Test Criteria v3.0.0

CON-03

The application handles billable events.

Our application supports Send as. One of the transports that a user can choose on a real mobile phone is MMS. Even though our application passes control to the UIQ Messaging application to send the MMS, a Test House may interpret the specification strictly and require that our application must provide a warning dialog. To be sure of passing this test, we therefore adjust our application to present a billable events dialog prior to using the Send as functionality.

It is acceptable to include a ''Don't tell me about this again'' option in the dialog, so that the user does not have to go through the warning every time.

This criterion is expected to be removed in Test Criteria v3.0.0

CON-04 (UNI-07)

The application complies with backup and restore.

To comply with this Test Criterion, the primary application requirement is to provide a `backup_registration.xml` file as part of the application installation package.

Our `backup_registration.xml` file contains the following content:

```
<?xml version="1.0" standalone="yes"?>
<backup_registration>
<passive_backup>
<include_directory name="\"/>
</passive_backup>
<system_backup/>
<restore requires_reboot="no"/>
</backup_registration>
```

There are two distinct types of backup, one in which an application actively participates (called active backup) and one where the application does not participate (called passive backup). Typically, applications that perform data synchronization perform active backups. Other applications perform passive backups.

As indicated by the XML, our application performs passive backup. All the information of interest is stored within our private folder. The `<include_directory name="\"/>` statement specifies that we want all files within our private folder backed up.

The `<system_backup/>` statement specifies that all executables and resource files defined within the original PKG file should be backed up. We recommend that you include this statement. Otherwise, if a user backs up their mobile phone, performs a master reset and restores the data, your application will be missing and the user will have to reinstall.

The full syntax, or in XML terminology, the backup registration file DTD, exists within the document ***developer.symbian.com/main/downloads/papers/pc_connectivity/PlatSec_PC_Connectivity.pdf***. This document also discusses backup and restore fully.

It is also possible to opt out of backup and restore (that is, the application does not register with the backup–restore server), if it is not necessary for your application.

> **Pro tip**
>
> Unfortunately it is not actually possible to test application backup and restore if an application is signed with a developer certificate. Since it is not possible to install applications signed with a full certificate until they have been Symbian Signed, it is difficult to verify backup and restore compliance. Symbian Signed and the test houses are aware of this issue. Therefore, your application should not be failed if backup and restore compliance is not met, given that it is not testable. The vast majority of applications should simply use the `backup_registration.xml` example file as it is shown above.

Summary of `SignedAppPhase2` Testing

Having gone through each of the Symbian Signed tests, we can be satisfied that our `SignedAppPhase2` application conforms to the vast majority of tests. Although some of the tests appear quite onerous, we do not have to perform any significant modifications to our application in order to conform. The two tests we need to address are CON-02 and CON-03 where we need to add fairly simple dialogs. These are added in `SignedAppPhase3`, described in Chapter 11.

14.10 Lessons Learned

We would like to share with you a number of key lessons about Symbian Signed that we have learned as a result of working with the Symbian and UIQ development communities.

14.10.1 Understand the Capability Rules

Understanding the Symbian capability rules is essential for anyone who is designing and implementing an application for a UIQ phone, so before we go any further we should familiarize ourselves with the rules as set out in the book *Symbian OS Platform Security*, Section 2.4.5:

- Rule 1 – Every process has a set of capabilities (as defined by an EXE) and its capabilities never change during its lifetime.
- Rule 2 – A process can only load a DLL if that DLL has been trusted with at least the capabilities that the process has.
- Rule 2b – The loader will only load a DLL that statically links to a second DLL if that second DLL is trusted with at least the same capabilities as the first DLL.

The Symbian Platform Security model has been designed to control what a process can do which means your application can only perform the tasks (use the APIs) for which capabilities have been approved. When a binary is built (this can be an EXE or DLL), it is with the set of capabilities defined in the application's MMP file. The capabilities defined in the MMP file should match the capabilities that your application uses. When this binary is installed using a SIS file, the software installer checks to see if it is trustworthy and is allowed to use these capabilities. Experience shows that the confusion arises in the different ways that an executable and a DLL are treated.

If you consider your application EXE, when a process is created from this executable, the process runs with the capabilities declared at build time for that executable. Once this process starts, its capabilities cannot change, which means that loading a DLL with higher capabilities does not increase the capabilities of the process (Rule 1).

This differs from a DLL; the capabilities declared within a DLL indicate a level of trust. At run time, a DLL can be loaded into a process with fewer privileges as Rule 2 states that essentially a process can only load a DLL which has a superset of capabilities. The critical point is that many developers believe that their application needs the same capabilities as the DLL it is loading. This is not the case; you only need the capabilities for the APIs that you are using. There are some DLLs that have `all – TCB` (All capabilities except `TCB`). Many vendors believe that to use this DLL they also require `all – TCB` when in fact they only need one, two or three capabilities as they don't use all the functionality provided by the DLL. This problem is often identified when a request is rejected because the vendor or developer is asking for a capability that they do not use.

14.10.2 Integrate the Capability Request

When you submit your application to Symbian Signed and it requires Device Manufacturer capabilities, part of the process requires you to complete a capability request form. This form requires basic information such as organization and contact information. More importantly, it asks you to provide the name and description of the application, the capabilities that it requires, the APIs used for each of these capabilities and the functions implemented using these APIs. It is critical that this information is concise, accurate and up to date because this document is used by handset manufacturers (along with information from the test house) to determine if an application should be granted its required capabilities. It is also used in the case when one or more waivers are required.

If your documents are inaccurate or invalid, the only options the handset manufacturer has are to reject the request or ask for more information before the request can be approved. In the first instance, the test house must mediate between the handset manufacturer and the software vendor. By embedding the capability request form into your project and assigning clear responsibility for managing the submission, you can ensure that the correct people update the documentation (such as Architect, Systems Analysis) as the application is being developed. It also ensures that the documentation is maintained and reflects any changes in design.

14.10.3 Integrate the Symbian Signed Test Criteria

The Symbian Signed test cases are not secret. Despite the fact that they are easily available from the Symbian Signed website, many vendors and developers do not test their application against these test cases in advance of submission. They only find out the outcome when they get the good or bad news from the test house. On many occasions, this is not a big problem, however, sometimes this can cause serious delays. For small developers, the unnecessary additional cost of re-testing may also be significant.

The most successful vendors and developers (successful at getting their application through the signing process as quickly as possible) are those who have the Symbian Signed Test Criteria integrated into their test plan. Usually this means that their application is tested against the criteria at the major project milestones such as an alpha release, beta release and, most significantly, against the application they plan to submit. This means that there are no surprises and it also means that they can have ready all the required documentation such as failure justification and waiver documentation when they actually submit to Symbian Signing.

14.10.4 Ensure Access to Infrastructure Dependencies

This lesson only applies to vendors or developers who are providing a client application that utilizes the services of a back-end infrastructure. For example:

- Instant messaging client
- Push email client

- PBX or VoIP client
- Streaming audio or video client
- Blogging client
- Mobile TV
- VPN client.

The key difference with many of these is the proprietary nature of the service that the client is implementing on the smartphone. When the client is submitted to the test house, it is not just a case of providing an application.

In many cases, user accounts, server access and even hardware, such as SIMs, are needed. Many clients have associated desktop software that puts requirements on the host network (e.g. ports X to Y must be open). It can also be complicated when the network operators available to the test house (limited by geographic location) do not support a specific client because the client only works with specific operators. As the vendor and developer are so committed to getting their application out on time and at the right quality, these small complications are often neglected and it is assumed that the test house has everything in place to support the client.

In almost all but a few extreme cases, the test house is able to accommodate the particular client, however, it does take time for the test house and vendor to get everything in place.

15

Testing, Debugging and Deploying

This chapter covers testing and debugging of your UIQ 3 application and discusses its deployment to end users. Many of the topics discussed are applicable to software testers, project managers or indeed anyone looking to improve procedures and quality within their software projects.

Testing and debugging should be a core part of your software development lifecycle from the point at which you start writing code through to your last maintenance release. These activities are typically most critical between a beta state of your software and final production-quality release.

Deployment is often seen as the small activity at the end of the project where you just get your application on to the user's smartphone.

In this chapter, we summarize and introduce some of the more useful techniques, so that you have a good base to build upon.

15.1 Back to Basics

A good starting point is to take a step back and look at some general behavior characteristics to which all applications should aim to adhere. These have been generated from many lessons learned supporting software suppliers in the mobile phone industry. They should be considered as minimum requirements and you should assess your application against them at every stage of your software development process.

1. Your application is a guest on the phone so it must be well behaved.

UIQ mobile phones are based on an open operating system and are capable of supporting many third-party applications. To understand the variety of applications available for Symbian OS, in March 2007 there were well over 7000 third-party Symbian applications available commercially. In the past there have been instances of third-party applications that take over the mobile phone, suggesting that the developer considered the application to be the only one running. On many occasions, we have seen that the designers, the developers and even the testers were concentrating so hard on getting their application ready to ship that they forgot to consider how their application inter-operated with the rest of the smartphone.

2. Your application is not the only application on the phone.

A UIQ mobile phone offers many functions beyond telephony; it's a business tool, a music player, a camera, a diary, an address book, a games machine and ever more, as technology develops. Users depend upon many of these functions and are very unhappy if they become unreliable. Your application is one part of this and must work in harmony with the mobile phone and its applications.

3. Restarting the phone is not an option.

Mobile phone users frequently leave the phone switched on for days, weeks and even months at a time. There can be no expectation of a daily re-start, as there is for many desktop computers. Therefore, your application must not degrade itself, other applications or the operating environment over time. The need to frequently restart a mobile phone, for example, when a fatal error occurs, is a common cause for users to reject and return a mobile phone to the supplier. This creates a bad impression and loss of money for mobile operators.

4. A smartphone is a limited-resource device.

Despite the significant advances in hardware (memory, processor speed, battery capacity, and so on), the mobile phone should still be considered as a limited-resource device when compared to a desktop computer or laptop. For example, allocating memory because you may need it and not freeing resources when you have finished is bad practice. Symbian provides good documentation including the *C++ Coding Standards* and also the *Coding Idioms* available from ***developer.symbian.com*** which should be followed as closely as possible.

5. Events, Events, Events!

The mobile phone is an event-driven device. Here are just a few of the common events: low battery, new message, incoming call, beamed file, user input, flight mode, no signal, GPRS detachment, memory stick removal, WiFi available, USB connected and key lock. Many of these affect your application, for example it loses the foreground when a message notification is received. You need to ensure that not only do you code for them but you also test for them. Your application may perform well in isolation; however, what happens when you are using it and you receive an incoming call? Does it still function in flight mode? Testing and coding for events, particularly during integration and interoperability testing is crucial to establishing a stable and reliable application.

15.2 An Example of Bad Behavior

Bad behavior does not have to be malicious or damaging; it is sometimes just a small oversight in application design. For example, consider an application that automatically starts up and connects to GPRS in order to access a web service. It may run very well in the development office and test lab and pass all tests. In real life however, it must cope with areas of no service. A Connect dialog may seem like a good idea to inform the user when the connection is initiated, however if that dialog is automatically displayed every time service is lost, it could become very intrusive. The user could be playing a game or reading offline email while in the subway or in flight mode. At the very least, there should be an option to disable the dialog box or switch to manual connection.

15.3 Testing

The quality of your testing process is one factor that can reflect in the quality of the application that you are delivering. Delivering a low-quality application affects the image of the application or service that it provides. More significantly, this can affect your brand image as users always remember the bad things that you do and are happy to tell others about it, particularly if you make a poor first release.

There are so many different methods, types, styles and models of testing software that to cover them all would take too much time and space. We do not intend to cover all test techniques available or, indeed, cover individual techniques in depth. Instead, we highlight some of the key test techniques for UIQ 3 application development.

It is not always a dedicated person or team who carries out testing. Developers complete a certain level of testing during the development process. There will also be users within your organization who are involved in, for example, sales and marketing and management. You may also have external users. The examples below do not refer to one particular group as it should be simple to understand who completes the particular type of testing.

15.3.1 Unit Testing

In theory, you should be testing every unit of your application, with a unit being the smallest part of your application. As we are discussing an object-orientated platform, this unit is a class. If we take a simple C++ application on Windows, this is a relatively simple task to perform. You would most likely create a test harness or test suite and run the application through all the different permutations required to ensure quality and stability. With Symbian OS and UIQ this can be a little more difficult because:

- The UIQ emulator can't accurately emulate performance. If performance is key to your application, then you should model your code performance on target hardware.
- If your class performs certain tasks that require, for example, data connectivity, then again you should run your application on target hardware to get a true feel for how your unit performs.

To do either of these, you should consider developing a test harness UI to allow you to test your different class structures on target hardware. This would bring extra benefits when it comes to regression testing, which we discuss later. If you change or add features to a core or high risk component, then you should ensure that you regression-test it to check that any change you have made does not introduce further errors or even re-introduce a previously fixed problem.

If this is something that you decide to do, you could find that it may not be feasible to write a test harness UI that interrogates all base or super classes. It is up to you to employ good design to test the key features of your engine. It is also important to avoid creating a complex UI that is difficult to navigate. It could be that the complex UI itself leads to mistakes in your testing.

15.3.2 Functional Testing

In short, this is where you test your application against specification and design. At this point you should ensure that the functions or use cases that you planned to implement have been implemented correctly and the application provides the behavior expected of it. It is also important as it confirms that all the components in your application are working together (as opposed to unit testing, which is only concerned with the component level and may be satisfied using a test harness). Functional testing is particularly important with UIQ phones as this is where you are really testing the UI.

One example of functional testing is the Symbian Signed Test Criterion PKG-02:

> *Installation and Startup – The SIS file installs and operates in accordance with user manual and standard application use expectations. The application starts up in a reasonable time period (normally considered to be five seconds) or provides an appropriate indication of its launch progress.*

Although this is not a detailed functional test, it does ensure that the application operates in accordance with the user manual and standard application use expectations. The expected result is that the SIS file installs and operates in accordance with the functional specification.

In this chapter, we reference the Symbian Signed Test Criteria v2.11.0 (see Section 14.9.2).

15.3.3 Interoperability Testing

Interoperability testing is concerned with ensuring that your application can communicate and exchange information as expected with applications with which it interfaces. For example:

- asking the web browser to load a specific URL
- ensuring the image file you create is the correct format and can be viewed in the picture gallery
- ensuring that the sound file you download does not get corrupted and can be played by the music player.

A smartphone is a rich device and you need to ensure that your application is tested against all the applications or services with which it interfaces. The Symbian Signed Test Criterion CON-01 says:

Task List – The application can be closed through the task list.

This test ensures that your application behaves correctly when the task manager asks it to shut down. By behaving correctly it should, at a minimum, save any user data and respond to the shutdown event and close as expected.

15.3.4 Integration Testing

On one level, you can consider the integration of the individual components into your application and how you test this. This may not be necessary as your functional tests should highlight any issues here. What's most important when developing for UIQ is the testing of how your application integrates into the smartphone itself and the other applications' services.

Symbian Signed Test Criterion CON-04 states:

Backup and Restore Compliance – The application should not interfere with a system 'Backup' and should function correctly after a 'Restore'. In addition, on Symbian OS v9 the application should correctly register itself for backup with the system (the developer explicitly opts into this behavior).

To ensure that your application is aware of backup and restore, you must create a valid backup registration file that is used by the backup and restore engine to manage your application and data for you. If this is incorrect then, quite simply, your application will fail on some level.

15.3.5 Regression Testing

The aim of this type of testing is to identify the situation where previously working features and functionality no longer work. This is particularly important when your application is close to a release date and you are making bug fixes. You need to have some way of ensuring that changes you make do not introduce other bugs or break any of the functions in your application. You also need to ensure that these changes have not re-introduced a previously fixed defect. This problem may not occur as a result of coding; it may be to do with a problem in the build process or configuration management process.

15.3.6 Stress Testing

Stress testing checks that your application can handle exceptional, illegal or erroneous actions such as entering incorrectly formatted data or removing the memory card unexpectedly. It is important to ensure that your application survives stress testing and does not cause any damage to the operation of other applications.

Symbian Signed Test Criterion GEN-02 defines a specific stress test that your application must pass. We recommend that you define a more comprehensive stress test for your internal testing.

15.3.7 Acceptance Testing

Acceptance testing is typically run by the customer. Your immediate customer may not be the end user – you may be delivering to an operator, a handset manufacturer, or an intermediate customer (such as a subcontractor). In this case you should aim to agree a level of acceptance testing that ensures what you deliver is correct. On many occasions this is not limited to the application and its functionality; it may also include release documentation, design documentation and user guides.

If you are delivering direct to the end user, then it is important to consider the expectations of your target users. Apart from their expected requirement that all functionality is in place, it is beneficial to consider user-centric requirements. For example, you may ask the following questions:

- Does the UI respond in an acceptable time?
- Does the UI conform to the UIQ Style Guide (where appropriate)?
- Are the buttons in the right place?

15.3.8 Phone-Specific Testing

Some features are specific to particular phones. If you make use of such features, then as well as testing on the target phone, you should test that your application works (or gives a suitable warning) when installed on a different phone that does not have the feature.

UIQ 3 supports a variety of form factors (pen-based, softkey, portrait, landscape). Even though it takes care of re-arranging the layout, you should ensure that your application is tested for each of the form factors that it supports. We advise you to consider phone-specific keys in your testing. For example, what happens when your application is running and the user presses the Camera button?

15.3.9 Testing on the Emulator

The UIQ emulator is a key software tool for the developer. Like many of the topics already discussed, there is a lot of reference information and *how to* information already available. This section will highlight some of the lesser known functions on the emulator to ensure that you can make the most of the tool during testing. For detailed information on these functions you should refer to the UIQ SDK Documentation.

> **Pro tip**
>
> It is advisable to back up the emulator part of your UIQ SDK before you start working with it. As you develop on the emulator, you might get it to a state where settings have changed, directories have been created and files have been altered and you want to remove all of this to test and verify on a clean environment. You can simply rename and then replace (not copy over) the backup using whatever method necessary, for example, a simple copy and paste of the required directories. You then have a clean emulator environment. Note that the UIQ SDK Installer allows you to install just the emulator, which is a good way to isolate the required files.

The following sections discuss some of the key aspects of the emulator and how to modify it for your particular needs.

File Locations

The emulator provides a Symbian file system by mapping drives to your PC file system. In general, the emulator is always configured with at least two drives:

```
Z: = \epoc32\release\<emulator build>\<build>\z
C: = \epoc32\<emulator build>\c\
```

```
; where <emulator build> = WINSCW etc.
;         <build> = UDEB or UREL
```

The Z drive represents the mobile phone's ROM and should be treated as such. When you build your application, it will be deployed to the Z drive and it will be treated as a ROM-based application. The C drive represents the main smartphone writable file system. Any file your phone creates during run time should only appear in the file system from the root of the C drive upwards.

The emulator can be configured to map to other virtual drives on your PC file system. For example, adding the following line in your `\epoc32\data\epoc.ini` file will map the T drive on the emulator to `C:\Temp` on your PC file system:

```
EPOC_DRIVE_T C:\Temp
```

If you look at the file system using `QFileMan` on your emulator (see Figure 15.1), you may see a number of other drives as well as the default Z and C drives plus any drives you have mapped.

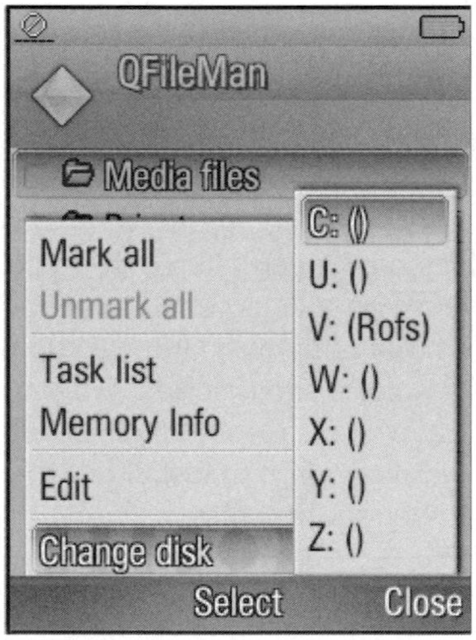

Figure 15.1 `QFileMan`

These drives represent the default drive emulation setup on the emulator. In the case of the UIQ 3 SDK, they represent the following drives:

- U: FAT over emulated NAND flash drive
- V: ROFS over emulated NAND flash drive
- W: LFFS over emulated NOR flash drive

- X: FAT over emulated MMC drive
- Y: FAT over emulated RAM drive.

Rather than mapping to a directory on your PC file system, these drives map to individual BIN files in the `\epoc32\data\media` directory and the configuration of these drives are set in `epoc32\release\<emulator build>\<build>\Z\SYS\DATA\estart.txt`.

Debug Keys

One emulator feature that is not always used but is easily available is the debug key functionality. These are divided into groups which are briefly discussed below.

The resource allocation keys provide information on resources used by your application:

- Ctrl+Alt+Shift+A: The number of cells allocated on the user heap
- Ctrl+Alt+Shift+B: The number of file server resources used by your application
- Ctrl+Alt+Shift+C: The number of window server resources used by your application.

There is also the ability to use the heap failure tool which simulates resource allocation failures and can be used to ensure that your application is handling allocation errors correctly. The tool operates in a number of modes. These are:

- deterministic, which means you can specify the rate at which the failures occur
- random, which means that the failures occur, as expected, in a random manner.

The heap failure tool supports application, window server and file (access) resource failures and these can be switched on independently or all at the same time:

- Ctrl+Alt+Shift+P: Show the heap failure dialog
- Ctrl+Alt+Shift+Q: Turn off the heap failure dialog.

The drawing keys allow you to simulate redraw or window server events:

- Ctrl+Alt+Shift+R: Redraw whole window

- Ctrl+Alt+Shift+F: Enable window server auto flush
- Ctrl+Alt+Shift+G: Disable window server auto flush.

Finally, there are some other debug keys that provide you with useful emulator functionality:

- Ctrl+Alt+Shift+K: Kill the current application with keyboard focus
- Ctrl+Alt+Shift+T: Bring up the task list showing current running applications
- Ctrl+Alt+Shift+V: Turn on or off verbose information messages
- F9: Power on the emulator
- F10: Power off the emulator.

The debug keys can assist you in both development and testing. In particular, tools such as heap failure should be used throughout development to ensure that you have the correct error checking in place so that your application handles any resource problems gracefully. As we can see from the sample application, `Commands1`, with heap failure set to Random and following a number of attempts to add new entries eventually we receive a *Not enough memory* dialog (see Figure 15.2).

Figure 15.2 *Not enough memory* dialog

The application handles this and, as expected (as we have random heap failure), continues to work after the dialog is dismissed.

Debug Output

The emulator outputs certain information to a log file. This log file is located by default in `c:\documents and settings\<your user name>\local settings\temp\epocwind.out`. This file is re-created each time the emulator starts and all entries are logged relative to the emulator starting. The emulator lists configuration information and system messages in this log file. These system messages include:

- application panics
- thread information detailing when threads are created and destroyed
- Platform Security information.

The emulator logs the Platform Security settings at the time it starts up and also when any application attempts to use an API or perform a function, such as writing to a protected directory. For example, we modified `LayoutManager1` so that it tried to create a file in another application's private directory. `LayoutManager1` could only do this if it had the required capability, however, in the case of this example we set the application to have no capabilities.

When this example is run, there are two results to consider. From a code point of view when we actually try to create the file, the `RFile.Create` call returns `KerrPermissionDenied (-46)`. If you then look in the `epocwind.out` file, you will see the following:

```
81.335  *PlatSec* ERROR - Capability check failed - A Message (function
number=0x0000001f) from Thread LayoutManager1[edead004]
0001::LayoutManager1, sent to Server !FileServer, was checked by
Thread EFile.exe[100039e3]0001::Main and was found to be missing the
capabilities: AllFiles .  Additional diagnostic message:
\Private\100051e6\NaughtyFile.txt Used to call: Create File
```

This error message is very detailed; it tells you which application has the error, what function it was trying to call and what capability it needs to complete this call. In this example, it even specifies the file it was trying to create and where it was trying to create it. This sort of extra information is extremely useful as `KerrPermissionDenied` only indicates in the documentation that the request failed because *"the permissions on the file do not allow the requested operation to be performed"*. Naturally you would expect that there are file permission problems and would not automatically realize that your error was caused because your application does not have the required capability.

Emulator Configuration

Epoc.ini

The default configuration file for the emulator is `\epoc32\data\epoc.ini` and this can be used to modify a number of the basic emulator settings. These include screen size and key mappings. Unless you are implementing a bespoke application, it is most likely that you will never need to alter these settings. For example, if you are writing a console application, such as a test harness, then by defining `textshell` in `epoc.ini`, the emulator loads in `textshell` mode; if `textshell` is missing, then the default emulator loads.

Not all possible value–parameter settings are in the emulator. If they are missing, then the emulator picks default values.

- `MegabytesOfFreeMemory` specifies the amount of RAM available to the emulator. The default value is 16 MB. Depending upon what you have running on your emulator, the normal memory usage is around 7.5 MB after startup, so for example, setting this value to around 8 MB will provide you with an emulator that is low in memory. You can then use this setup to test how your application behaves in low-memory situations.
- `JustInTime` allows you to switch just-in-time debugging on and off. With it set to 'on', the just-in-time debugger starts as soon as there is a problem. With it set to 'query', a Windows dialog asks you if you want to start the debugger. As expected, when it is set to none you only receive a panic dialog when the emulator experiences a problem.

`Epoc.ini` also defines the Platform Security settings for the emulator's runtime environment (see Table 15.1). This gives you the ability to turn on and off various Platform Security checks.

We expect that you will use `epoc.ini` to define different emulator runtime settings at different stages of your software development. We recommend, however, that you put all checking on at some point to verify that your application is behaving as expected. This checking should match as closely as possible the phone or phones that your application is targeting.

UIQEnv

`UIQEnv` is a tool used to define common configuration settings for the UIQ SDK and emulator. It is particularly powerful when it comes to

Table 15.1 Platform Security parameters

Parameter	**Definition**
`PlatSecEnforcement`	This setting determines if a Platform Security check or capability check is enforced or not. For example, consider the case where an application is using an API for which it does not have the required capability. With `PlatSecEnforcement` on, the appropriate error is returned or action taken. With `PlatsecEnforcement` off, the application is allowed to continue as if the problem did not exist.
`PlatSecDiagnostics`	With this parameter switched on, any Platform Security error encountered is logged to the emulator log file, `epocwind.out`. Setting it off means that no error is logged.
`PlatSecProcessIsolation`	Some kernel APIs inherited from EKA1 can have insecure uses, such as a thread in one process being allowed to kill a thread in another process. When this flag is set, the kernel provides runtime checks for their correct usage.
`PlatSecEnforceSysBin`	When this parameter is set to 'on', the executable loader only looks for and loads executables from `\sys\bin`.
`PlatSecDisabledCaps`	This parameter gives you the ability to disable one or many capabilities. For example, the following disables `WriteUserData` and `WriteDeviceData`: `PlatSecDisabledCaps WriteUserData+WriteDeviceData`

configuring communications components. Rather than configuring the Symbian `CommsDat` XML database manually, `UIQEnv` can be used to set and modify configuration settings.

`UIQEnv` also provides a quick and easy way to change the screen orientation of the emulator to allow you to, for example, test the behavior in landscape mode. There are shortcuts to set the most common modes:

```
UiqEnv - ui softkey
UiqEnv - ui softkeytouch
UiqEnv - ui pen
```

```
UiqEnv - ui SoftKeySmall
Uiqenv - ui PenLandscape
```

Figure 15.3 shows the result of the final line above.

Figure 15.3 Pen Style Landscape UI configuration

In fact you can, by the use of a bit-wise OR, create the correct configuration to allow the emulator to match the phone you are working on (see Table 15.2).

For example, the following line configures Softkey Style Touch:

```
Uiqenv -ui 17025
```

Table 15.2 Screen control parameters

Parameter	Setting	Hex	Dec.	Constant (`qikon.hrh`)
Screen Mode	Portrait	0x0001	1	`EQikScreenModePortrait`
	Landscape	0x0002	2	`EQikScreenModeLandscape`
	Small	0x0003	4	`EQikScreenModeSmallPortrait`
	Small Landscape	0x0004	8	`EQikScreenModeSmallLandscape`
Touch Screen	No	0x0040	64	`EQikTouchScreenNo`
	Yes	0x0080	128	`EQikTouchScreenYes`
Interaction Style	Menu	0x0100	256	`EQikUiStyleMenubar`
	Softkey	0x0200	512	`EQikUiStyleSoftkey`
Orientation	Normal	0x4000	16384	`EQikOrientationNormal`
	Inverted	0x8000	32768	`EQikOrientationInverted`

SDK configuration When you install the UIQ SDK to your machine, the SDK Configurator tool (see Figure 15.4) is installed by default. This tool allows you to manage a number of different UIQ SDKs through a GUI interface and it also allows you to set emulator styles (see Figure 15.5) and communication settings. This tool is essentially a GUI for `UIQEnv`.

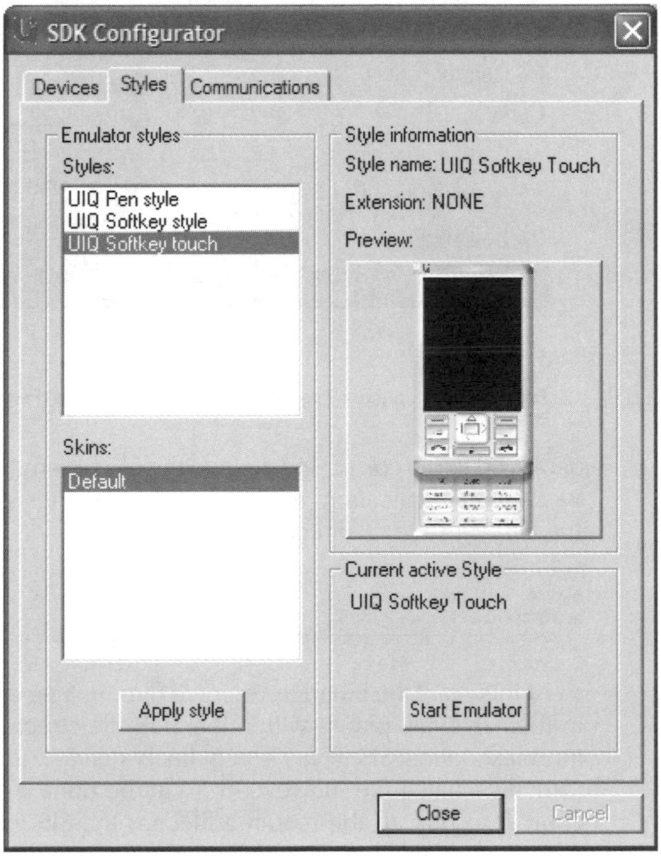

Figure 15.4 SDK Configurator tool

Testing a SIS File in the Emulator (Install)

The actual installation and uninstallation of your SIS file should always be verified to ensure that it enters and leaves the phone correctly. You can use the emulator to verify this and troubleshoot any issues. Using the `LayoutManager1` example application, we walk through the process from building the software to installing it (via SIS) on to the emulator (see Figure 15.6).

The first step is to build debug emulator binaries and check that once the application is installed, we can run it. It is important to ensure that the package file is set up correctly so that the correct files get

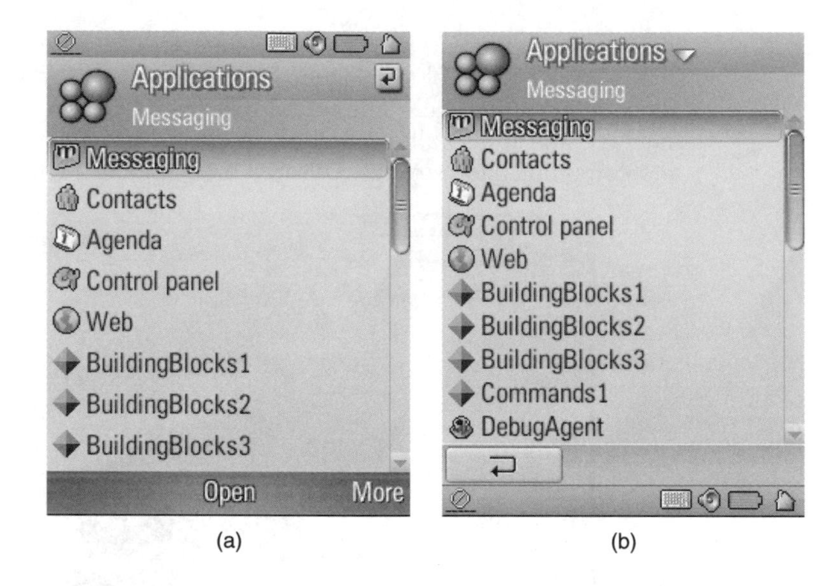

Figure 15.5 Emulator in a) Softkey Style Touch and b) Pen Style UI configurations

deployed to the correct location on the file system. To build the `LayoutManager1` application, the following commands should be run from the `LayoutManager1` directory.

```
bldmake bldfiles
abld build winscw udeb
```

At this point the binaries, resource files, images and so on are deployed in their relevant places within the SDK file structure. When testing and troubleshooting SIS files, it can be beneficial to write a small script (usually a batch file, however this can be done in the language of your choice) to pull all the required files for the SIS into one directory. The advantage of this is to ensure that you always pick up the same files and that you can examine them (date and time stamps and sizes) and validate them (using the Symbian tool `Petran`) to check that the correct files are going into your SIS.

The following copy commands work for `LayoutManager1` and assume that you want to copy them to your current directory. These commands are only concerned with the `WINSCW UDEB` binaries.

```
copy \epoc32\release\winscw\udeb\Z\private\10003a3f\
                              apps\LayoutManager1_reg.rsc .
copy \epoc32\release\winscw\udeb\z\Resource\
                              Apps\LayoutManager1_loc.rsc .
copy \epoc32\release\winscw\udeb\z\Resource\
                            Apps\LayoutManager1_icons.mbm .
```

```
copy \epoc32\release\winscw\udeb\z\Resource\
                                Apps\LayoutManager1.mbm .
copy \epoc32\release\winscw\udeb\z\Resource\
                                Apps\LayoutManager1.rsc .
copy \epoc32\release\winscw\udeb\Layoutmanager1.exe .
```

Note that now you have extracted the required files for the SIS, you will want to clean up your build. Otherwise, when you eventually try to install your SIS file, the software installer correctly tells you that the software is already installed. As we are testing a straight install and not an upgrade to a ROM application, then you need to ensure the output from the build is removed.

```
abld reallyclean
```

The next step is to construct the package file:

```
; Languages supported.
&EN
; Localised vendor name.
%{"Sony Ericsson Mobile Communications"}
; Global vendor name.
:"Sony Ericsson Mobile Communications"
; Standard PKG file header.
#{"LayoutManager1"}, (0xEDEAD004), 1, 0, 0, TYPE=SA
; Platform Dependency.
(0x101F6300), 3, 0, 0, {"UIQ30ProductID"}
; File to install.
"LayoutManager1_reg.rsc"-
    "c:\private\10003a3f\import\apps\LayoutManager1_reg.rsc"
"LayoutManager1_loc.rsc"-"c:\Resource\Apps\LayoutManager1_loc.rsc"
"LayoutManager1_icons.mbm"-"c:\Resource\Apps\LayoutManager1_icons.mbm"
"LayoutManager1.mbm"-"c:\Resource\Apps\LayoutManager1.mbm"
"LayoutManager1.rsc"-"c:\Resource\Apps\LayoutManager1.rsc"
"LayoutManager1.exe"-"c:\Sys\Bin\LayoutManager1.exe"
```

At this point the SIS file can be built:

```
makesis LayoutManager1.pkg LayoutManager1_unsigned.sis
```

Now we have a SIS file to install; however, we can take this one step further and sign the SIS file. The UIQ SDK provides three certificate–key pairs located in `\epoc32\tools\cert`. These are:

- `NoCaps.key` and `NoCaps.cert`
- `AllUserCaps.key` and `AllUserCaps.cert`
- `AllCaps.key` and `AllCaps.cert`.

The names of these certificate and key pairs are quite self-explanatory. `NoCaps` signs your application however it does not allow it access to any capabilities. `AllUserCaps` provides your application with user capabilities which consist of those in the basic set and the extended set. Finally, `AllCaps` provides your application with access to all Platform Security capabilities. By signing with the appropriate certificate, you can test that your final SIS file is accepted as expected by the software installer (see Figure 15.6c).

```
signsis LayoutManager1_unsigned.sis LayoutManager1_signed.sis
\epoc32\tools\cert\AllCaps.cert \epoc32\tools\cert\AllCaps.key
```

Once signed, you are ready to test the full installation of your application on the emulator.

Configuring the Software Installer

The software installer is the gatekeeper of the smartphone. It examines which capabilities an application has declared and then checks to see if the application signature satisfies the requirements.

Configuring the emulator can be done using `\epoc32\release\winscw\udeb\Z\system\data\swipolicy.ini` as this sets the software installer policy. There are a number of settings that can be modified, however, the key line for your testing is:

```
AllowUnsigned = true
```

With this set to `true` unsigned SIS files can be installed. Setting this to `false` means that your SIS file will is checked by the software installer at installation time. In Figure 15.7, `AllowUnsigned` is set to false so that the installer checks for a signature.

In this case, `LayoutManager1` has been built with a capability, however, the SIS file has not been signed. As the software installer is checking the capabilities, it refuses installation.

`AllowGrantUserCapabilities` defines if the user is allowed to grant capabilities. If it is set to true then the user has the ability to accept an application with capabilities defined with the variable `User Capabilities`:

```
AllowGrantUserCapabilities = true
```

In the case of the default `SWIPolicy.ini`, file this is:

```
UserCapabilities = NetworkServices LocalServices
    ReadUserData WriteUserData Location UserEnvironment
```

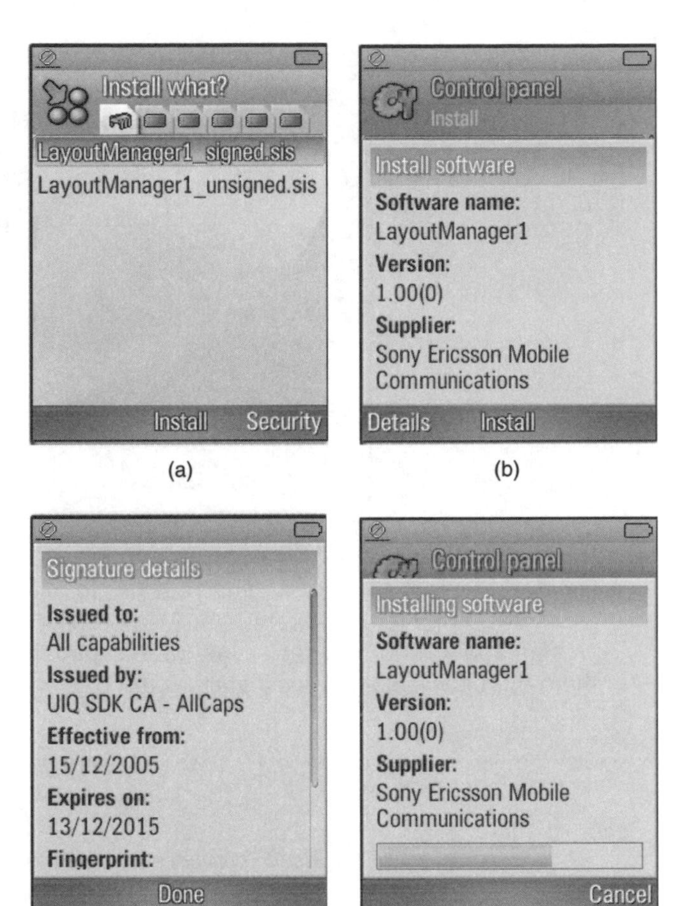

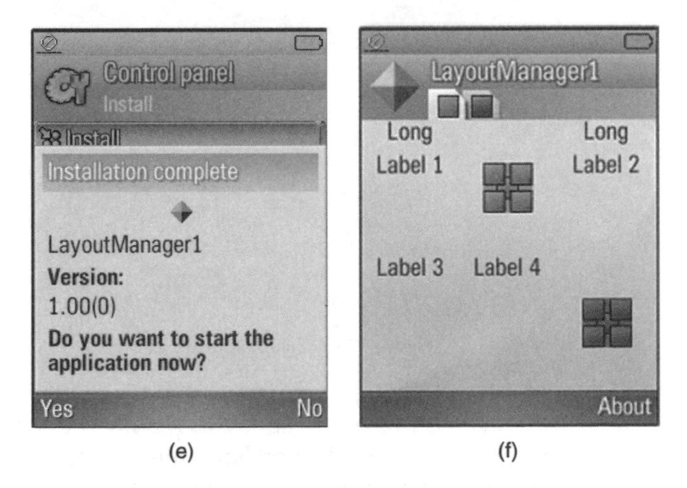

Figure 15.6 SIS install sequence in the emulator

Figure 15.7 Installation with SIS checking enabled

If an application requires any of these capabilities, the user can grant them during installation (see Figure 15.8).

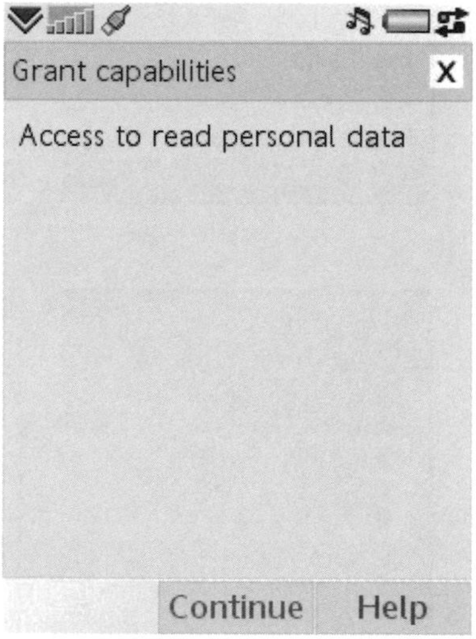

Figure 15.8 Dialog for user to grant User capabilities

Although you have the ability to modify the software installer policy in the emulator, you do not have this ability on mobile phones. The installation policy is set by the individual manufacturer and there is no guarantee that all manufacturers set the same policy.

15.3.10 Testing Tools

LeaveScan

Symbian has developed a lightweight and simple exception-handling mechanism that allows methods to *leave*. For example, a method that has code which may result in an exception can leave and is called a *leaving* method. As a result of this there is a convention within Symbian that states that the name of any leaving method must end in `L`. For example:

```
CqikAppUI::HandleCommandL(CQikCommand &aCommand)
```

This application function takes command objects when a menu item is selected. Depending upon which menu item is selected, for example, *Save,* the appropriate function is called. As a developer, you can see that `HandleCommandL()` is a function that may leave. If it does hit an exception, then the exception will propagate up to a point where it is caught by a trap harness.

If the naming convention of putting the `L` suffix at the end of the function is not followed, then any users of the function will not expect that it may leave. This means that the user of the function will program to handle, for example, a potential memory leak.

`LeaveScan` is a tool provided by Symbian to help you track down problems with your leaving functions and ensures that your code follows the convention. It checks that any function that `Leaves`, or calls a function that `Leaves` (other than in a `TRAP` harness), must have a name ending in `L`. `LayoutManager1` was modified to include a new utility function which unfortunately did not follow the naming convention:

```
void CAppSpecificUi::BadLeave (CLeakClass* aLeakClass)
  {
  CLeakClass * leakExample1 = new(ELeave) CLeakClass;
  // Some code here.
  delete (leakExample1);
  }
```

Everything is all right as long as `BadLeave` can successfully create a new `CLeakClass` and returns it to the calling function. However, if the new `CLeakClass` can not be created, an unexpected leave occurs. Running `LeaveScan` over the CPP file that contains this code highlights the problem:

```
V:\SampleApp2\LayoutManager1>leavescan LayoutManager1.cpp
LayoutManager1.cpp(262) : CAppSpecificUi::BadLeave Calls new(ELeave).
```

The tool simply tells you which line of code has the problem and what the problem is. You should run it over all the code that you produce and any third-party source that you use to ensure that you adhere to this coding standard.

Petran

Petran is a Symbian build tool and it is particularly useful as a debugging tool in that you can check your Symbian OS executable file (E32) to check various aspects of your build binaries. We built the example SignedAppPhase3 using GCCE for target. This resulted in SignedApp_0x20000462.exe being built. We then used Petran to confirm the header of this executable. Note that Petran outputs a large amount of information about the executable. Command-line parameters can be used to reduce this:

```
petran -dump h SignedApp_0x20000462.exe
PETRAN - PE file preprocessor V02.01 (Build 549)
Copyright (c) 1996-2005 Symbian Software Ltd.
E32ImageFile 'SignedApp_0x20000462.exe'
V2.00(505)      Time Stamp: 00e1156f,669cc1c0
EPOC Exe for ARMV5 CPU
Priority Foreground
Secure ID: 20000462
Vendor ID: 00000000
Capabilities: 00000000 00000000
Uids:           1000007a 00000000 20000462 (ecfcaed6)
Header CRC:     6a28e743
File Size:      0000dea4
Code Size:      0000cd74
Data Size:      00000000
Compression:    101f7afc
Min Heap Size:  00001000
Max Heap Size:  00100000
Stack Size:     00005000
```

For the purpose of this example, we have removed some of the information, otherwise the output would span many pages. We can see from the information above that we can obtain the Secure and Vendor IDs, UIDs, capability declaration and heap size. In complex build environments, it can be very useful to ensure that the binaries which you are distributing are as expected. Petran is not limited to returning just these values; it can also analyze the code and data section, provide export information and examine the import table.

15.4 Debugging

15.4.1 Panics

A panic within Symbian OS indicates a programming error of some sort. At the point of a panic, the thread in which the panic occurs is stopped so that the thread can no longer continue to run. A panic is critical to your application. However, depending on where it occurs and taking into account that Symbian OS terminates the thread that causes the problems, this means that your application can not continue to run and cause further problems to the operating system and phone.

A panic does not always have the same result. If it is in the main thread of the application then the process in which that thread runs will close. If it is in a sub-thread (spawned from the main thread) then the main thread survives and, depending on the application structure and design, the main thread may be able to recover. If however the panic occurs in a thread that is identified as a system thread (for example, essential for the operating system) then the panic may cause the phone to restart.

The following piece of code is a purposefully bad example. Firstly, it attempts to create a file in the private directory of another application. Secondly, the developer has been particularly lazy and has not performed complete error checking. It is not a particularly useful piece of code; it simply aims to demonstrate operating system panics.

```
_LIT(KFileName, "C:\\Private\\100051e6\\NaughtyFile.txt");
User::LeaveIfError(iFsSession.Connect());
iFile.Create(iFsSession,KFileName,EFileShareExclusive);
iFile.Size(fileSize);
iFsSession.Close();
```

As the application does not have the correct capability, `AllFiles`, to create the file, the `iFile.Create` call fails. As there is no error checking on the `Create` call, the `iFile.Size` call causes a panic. This is one of the most common panic reasons that you will see: `KERN_EXEC`. However this panic has over 50 associated reason codes.

On most occasions when a panic occurs, a dialog box is displayed with the title Program Closed (see Figure 15.9). Three further pieces of information are displayed:

- the program name (process name)
- the reason code (panic category)
- the reason number.

The panic categories are well defined within the Symbian documentation and should be understood when debugging general application problems. There are too many reason codes (categories) to list here; refer to the SDK for full details. The explanation of the example in Figure 15.9 is as follows:

- Reason code: `KERN-EXEC` – These panics represent program errors which are detected by the Kernel Executive. Typically, they are caused

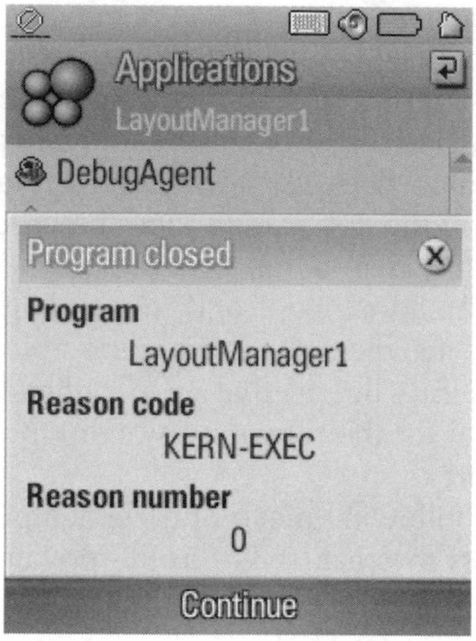

Figure 15.9 KERN_EXEC panic

by passing bad or contradictory parameters to functions. Threads which cause exceptions also raise a KERN-EXEC panic.

- Reason number: 0 – This panic is raised when the kernel cannot find an object in the object index for the current process or current thread using the specified object index number (the raw handle number).

Although this does not state exactly where the error occurs, it does give you an indication of what to look for. The object that it can't find is the handle to the file object. Depending on the complexity of your application, it may not be that obvious from the panic where the problem lies. Further debugging techniques discussed later may help you track this down.

15.4.2 Assertions

Assertions are a standard C++ method to detect errors as soon as they occur. If you are familiar with C++ but not Symbian OS C++, you should be familiar with using assertion macros. Within Symbian OS, there are two key assertions defined in \epoc32\include\e32def.h.

```
#define __ASSERT_ALWAYS(c,p) (void)((c)||(p,0))
#if defined(_DEBUG)
#define __ASSERT_DEBUG(c,p) (void)((c)||(p,0))
#endif
```

As you can see from this, assertions can be used in both debug and non debug builds. From the macro statements you can also see that the OR means that if parameter c is false then p is called. You, as a developer, define what p is; the assert statements do not handle this for you. We recommend that you should always terminate the running code and flag up the failure, rather than return an error or leave.

Taking the example from Section 15.4.1, we can enhance the badly implemented example to help us to track down the root cause of our problem.

```
User::LeaveIfError(iFsSession.Connect());
TInt err =
 iFile.Create(iFsSession,KFileName,EFileShareExclusive);
_LIT(KMyFileSession, "FileCreation");
TInt myFileSessionError = 999;
__ASSERT_DEBUG((KErrNone == err),
 User::Panic(KMyFileSession, myFileSessionError));
iFile.Size(fileSize);
iFsSession.Close();
```

When the new code is implemented, we receive a different error (see Figure 15.10).

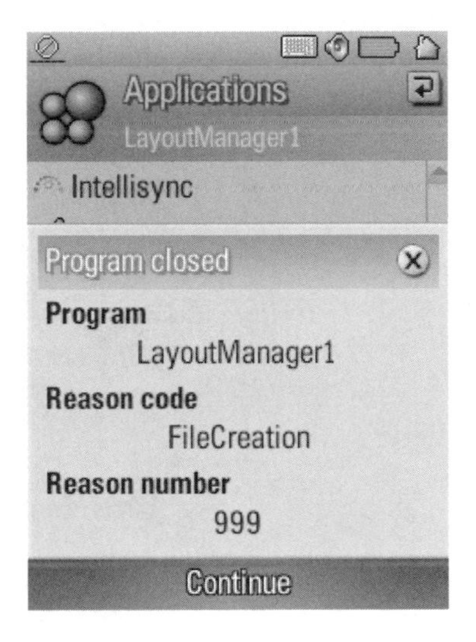

Figure 15.10 Panic error dialog with enhanced information

Of course you will say that this can be handled more efficiently using good error checking and you would be right in saying so. This code snippet is particularly extreme in order to show a specific example. You may not always be in the position where you can easily check all the code logic in, for example, a code review. The file creation process may be abstracted away from you and if so you would rightly assume that the part of the code responsible for creating the file is checking for errors correctly. In this case, it would be right to see this as an exception for the file not to be available and check using an assertion macro.

15.4.3 Debug Macros

When developing for UIQ phones, we keep reminding ourselves that we are developing for phones with limited resources, especially memory. To assist developers, Symbian OS provides a number of tools to help you check for, track down and test for memory management issues.

There are two types of heap-checking macro, the first can stimulate out of memory scenarios and the second can check that your application is not leaking memory. These macros are only built in to your debug builds so, unlike assertion macros, they do not increase the size or affect the performance of your release code.

The following example creates an object of type `CLeakClass` on the heap:

```
CLeakClass : public Cbase {};
__UHEAP_MARK
CLeakClass * leakExample =  new(ELeave) CLeakClass;
// Some more code here.
__UHEAP_MARK_END
```

As delete is not called, the heap check macro results in the panic shown in Figure 15.11.

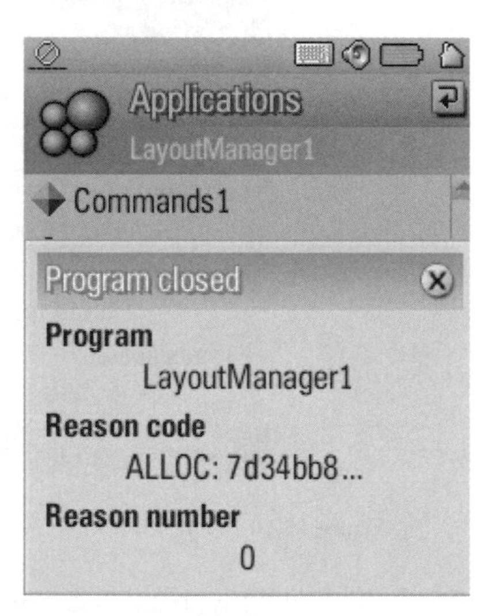

Figure 15.11 Panic on calling heap check macro

The reason code is not that clear; `ALLOC` indicates that this panic has been caused because a memory leak has been detected. The value following `ALLOC` is a hexadecimal pointer to the first orphaned heap cell. You will notice that this is truncated, however, the debug output file for the emulator (`c:\documents and settings\<your user name>\local settings\temp\epocwind.out`) logs the panic in full:

```
Thread LayoutManager1::LayoutManager1 Panic ALLOC: 73e4bb8
```

One method to track down the cause of your memory leak (assuming that you have multiple instances of heap checking) is to switch on just-in-time debugging in the emulator. In this case, the dialog in Figure 15.12 is presented when the error occurs.

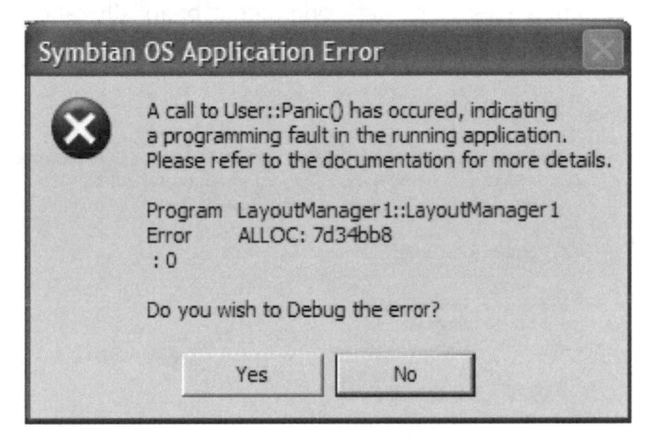

Symbian OS Application Error
A call to User::Panic() has occured, indicating
a programming fault in the running application.
Please refer to the documentation for more details.
Program LayoutManager1::LayoutManager1
Error ALLOC: 7d34bb8
: 0
Do you wish to Debug the error?
Yes
No

Figure 15.12 Application error dialog in the emulator

Choosing 'Yes' loads the just-in-time debugger and the call stack can be used to track down the cause of your problem (see Figure 15.13).

Figure 15.13 Just-in-time debugging in the emulator

This is again a simplified example and it may not always be this easy to track down the root cause. It may be necessary to put a watch on the memory address, run further tests or add extra logging to find the troublesome code.

In addition to checking for memory leaks, the macro `__UHEAP_FAILNEXT` can be used to force memory allocations to fail. It is called specifying a `Tint` parameter that indicates which particular allocation fails. The following code sets a heap allocation failure on the second allocation attempt:

```
__UHEAP_FAILNEXT(2);

CLeakClass * leakExample1 =  new(ELeave) CLeakClass;
CLeakClass * leakExample2 =  new(ELeave) CLeakClass;
CLeakClass * leakExample3 =  new(ELeave) CLeakClass;

// Some more code here.

delete (leakExample1);
delete (leakExample2);
delete (leakExample3);

__UHEAP_RESET;
```

This results in a standard *Not enough memory* dialog when the second allocation is attempted (see Figure 15.14).

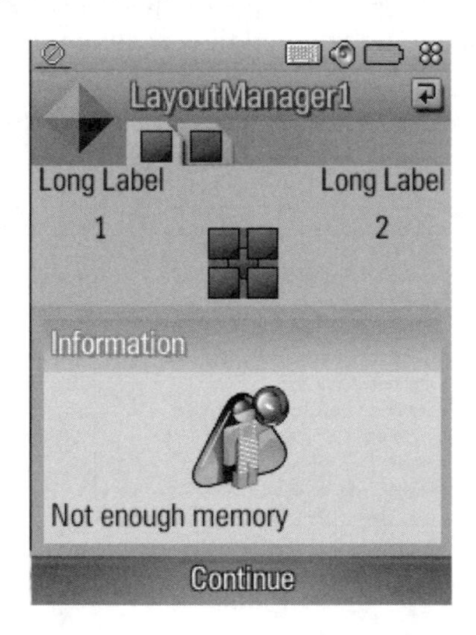

Figure 15.14 *Not enough memory* dialog

We discussed earlier that the emulator supports a similar debug function in the form of the simulated heap failure tool. Unlike the emulator heap failure option, the heap failure macros can be set in specific parts of your code to ensure that specific allocation failures are handled correctly by your application. It is important to note that using `__UHEAP_FAILNEXT` at the same time as the emulator heap failure tool can cause confusing results, so ensure that you use the tools wisely.

15.5 Deploying

15.5.1 Hardware Testing Is Required

Although you have tested your application in the emulator and fixed most of the software defects and usability issues, you still have further testing to do. At this stage, your application should work as expected on the emulator. Despite this, you can not make the assumption that it will also do so on target hardware. The emulator will do as it says and emulate your target hardware, however, there will always be differences that can only be experienced by using real hardware.

For the purpose of the examples below, we assume that we are distributing an application that uses one or more Device Manufacturer capabilities. The tools used in the signing process are documented in detail within your UIQ SDK.

File System Support

The size and performance of emulated drives, for example `ROFS/FAT` over `NAND` flash, are unlikely to be the same as the size and performance on target hardware. For a start, your application may be deployed to numerous different UIQ phones, each with differing hardware platforms. There is configuration support within the emulator to help you achieve a closer emulation matching that of the specific hardware that you are targeting:

- Adjust disk access speeds by modifying the read and write speeds within `epoc.ini`.
- Create a partitioned drive on your Windows machine that matches the storage size of target hardware. You can then map the emulator C drive to the partitioned drive and ensure that low memory situations can be handled.

Further details on each of these methods (and others) are available in your SDK documentation.

Floating-Point Management

Intel x86 processors have floating-point hardware which means that the compiler generates instructions that use this hardware directly. If this facility is not available in hardware then the compiler must implement the calculations in software, which is obviously much slower. The emulator benefits from this hardware support; however, your target hardware may not have a floating-point unit. If your application relies heavily on this type of arithmetic then in this scenario you may see performance differences between the emulator and the target.

Processes and Threads

The emulator runs in a single `Win32` process on your Windows PC. You can see this by looking at task manager on your computer; you will see a single `epoc.exe` process running that represents the entire emulator. Within this process, each Symbian OS executable is run as an individual thread. This differs from target hardware where each Symbian OS executable is launched as a separate process that has a single main thread.

This difference can cause a problem because `Win32` process threads share their writable memory. The emulated processes run as threads; they do not have a process boundary to provide them with memory protection. It could be the case that on the emulator a process accesses another's address space, be it on purpose or by mistake. This is very undesirable, and on target hardware it will result in an access violation.

Stack and Heap

The default heap size on the emulator is 64 MB, which means that when you load your emulator, it emulates a device with a total heap size of 64 MB. The advantage with heap size is that it can be modified by changing the following setting in `\epoc32\data\epoc.ini`.

```
MegabytesofFreeMemory  64
```

This allows you to emulate more precisely a specific target handset.

This `MegabytesofFreeMemory` value sets the total available heap available to all processes (emulated as threads on `Win32`) which mean that they must share this memory. When each emulated Symbian OS executable starts it is given an initial maximum default heap; again this value can be modified by setting the `epocheapsize` within the appropriate MMP file.

The default heap size and the initial maximum default heap are values that you can modify to ensure that your emulator and application match the target environment as closely as possible. This is not possible with the process stack; on the emulator the stack size increases as required to the much larger limits imposed by the Windows operating system. This can result in stack misuse on the emulator being hidden, only to be highlighted when the application is run on target hardware as a stack overflow (KERN-EXEC panics).

The stack size on target hardware is typically small (the default value is 8 KB). Although this can be modified if your application requires it, the epocstacksize keyword in your application's MMP file is not supported for emulator builds.

Further Differences

Further differences in operation between the emulator and target hardware include USB support, timer resolution, serial ports, scheduling and machine word alignment. The key point is that deploying and testing on hardware is another key step in your development process and differences in application behavior must be considered at this stage.

15.5.2 SIS Files

A native UIQ application can be transferred on to the phone using a number of different approaches that include:

- connectivity software
- sending by Bluetooth, infrared, MMS or email
- downloading over the Internet
- copying on to a memory card.

Regardless of the approach that is used, your application can only be installed when built into a SIS file. There is an exception when developing Java-based applications, however this is outside the scope of this book. Some SIS files are referred to as SISX; this name change distinguishes the format change in Symbian OS v9.1.

Unless you are delivering software that will go into a mobile phone manufacturer's ROM then the SIS file is the only mechanism to get your application deployed to target hardware. Figure 15.15 gives a simplified overview of this process.

Any SIS file can contain one or many optional embedded SIS files. There is no difference in the process used to generate embedded SIS files; if you use embedded files you will find that it is a recursive process.

Your package file defines what files (binaries, executables, resource files, text files, etc.) make up your final application. It also defines the application version, vendor, dependency, and language information.

SIS file dependency is used to indicate that your application has a dependency on another component. If the correct version of this component is not available then your application will not install.

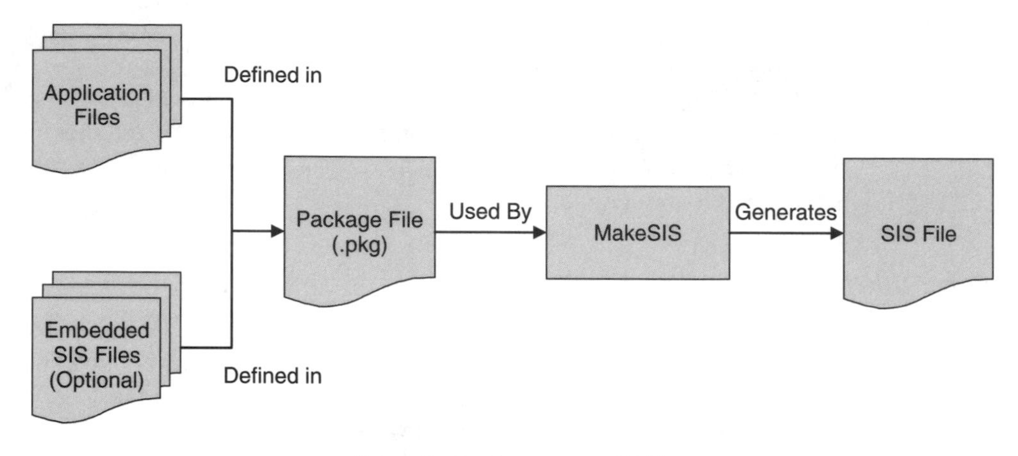

Figure 15.15 Generating a SIS file

Dependency checks can and should also be used to verify the hardware and UI platform of the target phone. Many applications aimed at UIQ 3 are simply deployed with the standard dependency:

```
(0x101f6300), 3, 0, 0,{"UIQ30ProductID"}
```

This informs the software installer that the application is dependent on the UIQ 3 platform which means that the application can be installed to any UIQ 3 phone. If your application has only been developed for a specific UIQ phone then it is most likely that it has only been tested on that phone. As a developer, you should consider the effect on your application of being deployed to phones on which it has not been tested. There are further reasons for restrictions; you may not want to install an application that uses WiFi on a non-WiFi phone. You might want to restrict your application for licensing reasons. Information on how to use dependencies to restrict installation is available from the various handset manufacturers' developer websites.

Once your package file is defined, you can use the `makeSIS` tool to build your SIS file.

Pro tip

If you are submitting to Symbian Signed then any embedded SIS file within your main SIS file must be Symbian Signed before it is embedded. This can lead to complications if you are submitting an embedded SIS file for signing that can not be tested as, for example, it has no UI. The test houses are aware of the use of embedded SIS files and normally sign them on condition that they are only distributed within your main application. Of course, your main application which uses the embedded SIS files must pass Symbian Signed.

At the point where you have your final SIS file you have a number of options depending on the intentions of this SIS file.

- If your application does not use System or Device Manufacturer capabilities and, for whatever reason, you do not wish to have your application signed, your application is ready to be deployed.
- If you want to use the SIS file for testing, debugging, evaluation, or demonstration, you can sign your application with a developer certificate (see Figure 15.16).
- If you are ready for deployment to market, you need to get your application Symbian Signed. Figure 15.17 shows this process using Certified Signed.

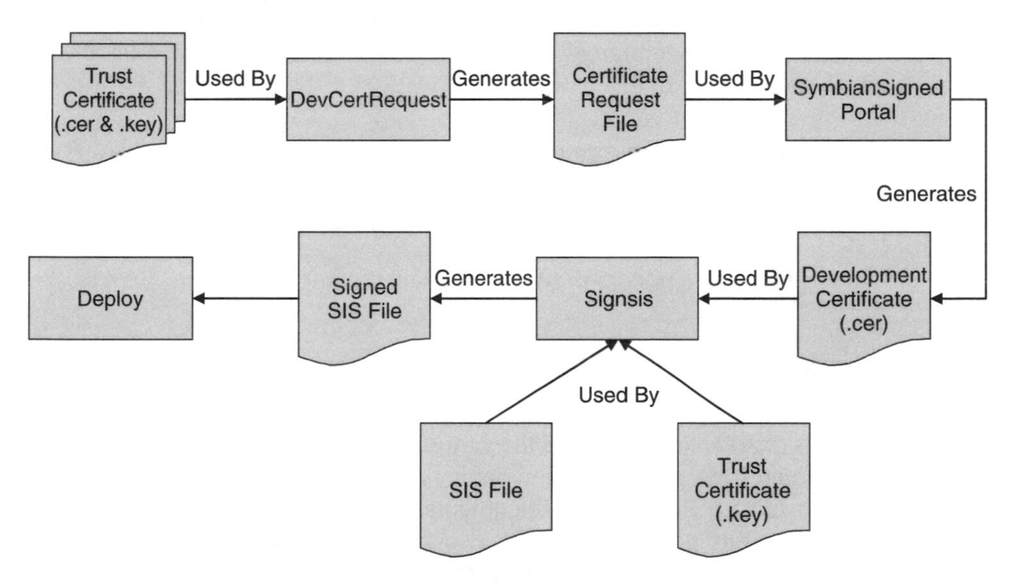

Figure 15.16 Using Open Signed with Publisher ID to deploy for testing

If your SIS file is intended for testing, debugging, evaluation or demonstration then you need to use Open Signed, as described in Chapter 14. If you have a Publisher ID, you can download a developer certificate. This certificate, along with the Trust Key used to generate the certificate request file, is used by `signSIS` to generate a signed SIS file. This SIS file is limited to the target phone IMEIs that were declared when the developer certificate was requested.

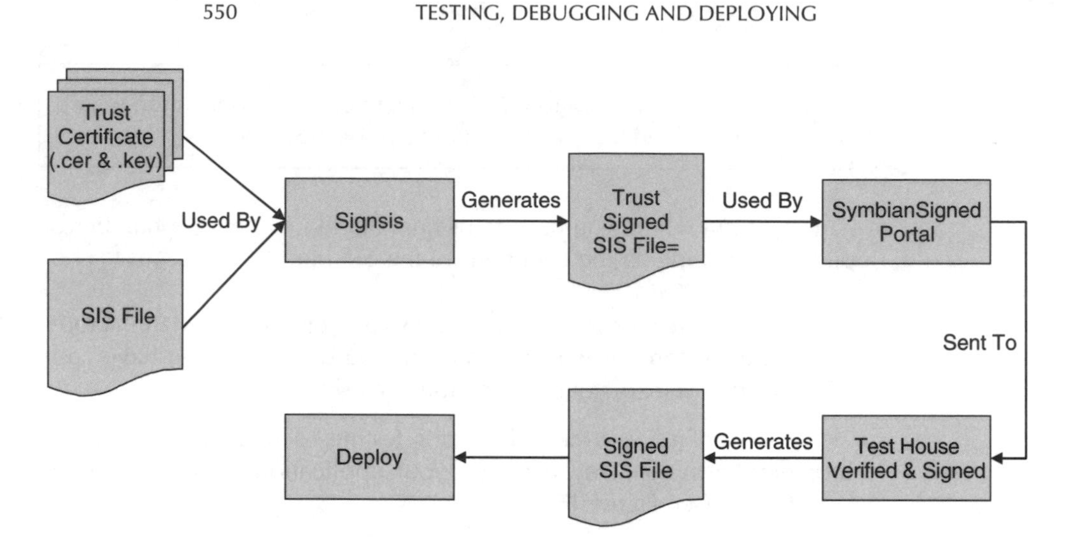

Figure 15.17 Using Certified Signed to deploy to customers

If you use the Certified Signed method, then pre-signing with your Publisher ID (Trust Certificate) is required before your application is submitted to Symbian Signed. Figure 15.17 takes quite a simplistic view. It does not include:

- documentation required by the Test House such as the capability request form and any test case waiver documentation
- the quotation process required to agree on the Test House fee
- handling embedded SIS files that also require signing.

15.6 Summary

This chapter has examined the various methods and approaches available when testing, debugging and deploying your application to a UIQ phone. The UIQ SDK provides documentation and software tools to support you in this process. There is also further material and tools available from the Symbian Developer Network, ***developer.symbian.com***, and UIQ Developer Community, ***developer.uiq.com***. There are many third-party tools available for use; when deciding upon a strategy you should consider all available options and decide upon an approach that suits your application, company, marketplace and end user.

The key points to take from this chapter are:

- Do not forget the basics; your application is a guest on the user's phone, it must be well behaved and respond to the phone-specific features and constraints.

- There are many different testing techniques and approaches available. Due to the nature of the hardware and software platform that your application will run on, test techniques such as interoperability, integration and regression, to name but a few, should feature heavily in your test plans.
- The emulator is a key tool for you to use during development and testing. It is highly configurable and can be adapted to help meet your needs. You should also take into account that, despite the flexibility of the emulator, you still need to test on target hardware.
- All the tools and techniques made available by Symbian and UIQ, for example, assertions and debug macros, should be used where appropriate to help detect and reduce software defects.

Application vendors, along with Symbian, UIQ and handset manufacturers have a vested interest that high-quality applications are developed and delivered to the end user. This will assist in the sustained growth of the UIQ-based handset market. The tools and techniques listed in this chapter go some way to ensure that this goal is achieved.

16

Porting Applications

In almost any software development project, re-use of existing source code is preferable to starting from scratch, particularly if the functionality is well tested. This is true irrespective of whether the source code was originally written for the Symbian platform or an alternative platform. This chapter focuses on porting or re-using C or C++ code originally written for other platforms for the Symbian platform, specifically on UIQ 3.

C++ is the primary development language for Symbian OS but many C++ programmers are initially confused by the differences encountered in attempting to read and understand Symbian OS code. A large part of this is down to style and, in particular, naming conventions. At a high level, Symbian OS supports its own C++ language dialect and a large set of libraries. It is the sheer volume of library calls available that makes the task of porting an application to the Symbian environment look daunting.

Most differences between Symbian OS and other environments, such as Windows, Mac OS and UNIX are a result of the fundamentally different design goals for the operating system. Symbian OS is required to support graphically rich, highly robust, interactive applications on small-screen, battery-powered mobile phones that have comparatively small amounts of memory and low processing power. Mobile phones are expected to be built at low cost, yet still perform tasks such as video telephony or acting as a mobile TV. They must securely store significant quantities of personal data; there can be no expectation of a regular reboot; and incoming calls and messages must be presented reliably and without delay.

Since inception, Symbian OS has been focused on requirements to minimize the size of compiled code, runtime memory usage, power consumption and processor cycles. In addition, the core of the OS was designed before the C++ standard had been formalized, and some features were either not supported by the compiler used to build for target hardware, or were considered to be inefficient on a mobile phone.

As a result, the C++ used to create Symbian OS (and applications to run on it) had some deliberate omissions, including use of standard C++ exceptions, templates and the standard template library. The new kernel architecture used in Symbian OS v9 (found in UIQ 3) and advances in silicon technology (Moore's law) have led to the recent introduction of support for standard C++ exceptions and the potential for an out-of-the-box, standard template library (STL) implementation on Symbian OS.

Symbian OS extends or, perhaps more accurately, complements other standard development patterns. For example, as well as having multi-threaded, multi-tasking capability, Symbian OS uses active objects as a lightweight alternative. Other examples include the use of descriptors instead of standard C strings and leaves as lightweight exceptions.

So, while some compromises and simplifications have been made, compared to an operating system designed for desktop usage, these are generally few and far between. On the other hand, compared to an embedded system, Symbian OS is very sophisticated. It contains file stores, database management, threads, processes, dynamic link libraries, client–server architectures, security models and a large set of other functionality required to make a mobile phone operate.

16.1 Where to Start

If you have little or no Symbian OS experience, you should first read Chapter 4, which provides an introduction to Symbian OS and specifically covers topics that you need to know if you are used to programming for another platform.

If you are completely new to Symbian OS, we recommend these books from Symbian Press:

- *Developing Software for Symbian OS: A Beginner's Guide to Creating Symbian OS v9 Smartphone Applications in C++.*
- *Symbian OS C++ for Mobile Phones Volume 3: Application Development for Symbian OS.*
- *The Accredited Symbian Developer Primer: Fundamentals of Symbian OS.*

In order to follow this chapter, you should have a reasonable grasp of:

- Symbian naming conventions
- panics, TRAPs, leaves and the cleanup stack
- descriptors
- dynamic arrays

- UIDs
- active objects
- C++.

C++ has been listed because Symbian OS tends to make extensive use of advanced C++ constructs, such as implicit type conversion, interface inheritance and thin templates. For example, when first presented with descriptors, some programmers fail to recognize that a method that is declared to take one particular type of descriptor as a parameter can accept any other descriptor whose class hierarchy contains the first descriptor type (implicit type conversion).

If you are familiar with Symbian OS but new to UIQ 3, we suggest that you read Chapters 5 to 10 and work with some of the example applications to familiarize yourself with the UIQ 3 user interface.

16.2 The Aims of Porting

Prior to starting, you should consider the goals for your porting project. In particular, you need to identify which elements of your application can change and which parts must remain the same. For example, if your aim is to recreate an identical copy of your application on a UIQ 3 phone that does not have a touchscreen compared to a Palm or Windows Mobile application that expects a touchscreen, you are forced to make compromises and program for a non-touchscreen GUI.

Several key questions that you should consider during the porting exercise are:

- Should you create an identical Symbian version of the original application?
- Should you recreate the look and feel and functionality of the original application?
- Should you expect the application to support all UIQ 3 phones?

When you examine your project in detail, you may be surprised to find that mobile applications need more change than originally envisaged, in order to work best on a different platform. Unlike desktop applications, it is not quite as simple as recreating the same application on a different operating system.

Exactly recreating your application in UIQ 3 usually requires you to rewrite large quantities of code, particularly for the UI. You should embrace UIQ 3 as best you can for your application, for example, working with building blocks and the Command Processing Framework. You will still need to rewrite a large amount of the UI, but by adapting to UIQ 3 techniques, you do not need to waste time trying to recreate unnecessary characteristics of your original application.

Rewriting is not really porting, even if it is implementing the same algorithm. Rewriting the UI is often quicker and cheaper than attempting to port UI code from another platform. Chapters 5 to 10 of this book tell you all about creating applications in UIQ 3, so this chapter concentrates on porting in the strict sense of moving code from another platform to UIQ 3.

Recreating the functionality also has advantages and disadvantages. For example, you may be porting a desktop application that contains functionality appropriate to the desktop such as printing. Most mobile phones do not have much support for printing; indeed many users do not have an expectation of being able to print from their mobile. Therefore, it may be reasonable not to support the same functionality.

Different UIQ 3 phones have different characteristics and are targeted at different market segments. For example, you may have an application that requires extensive data entry. In this case it is probably best to concentrate on supporting UIQ 3 phones that have a hardware keyboard. While the application may run on other UIQ 3 phones, it may not be particularly usable. Often this type of issue is resolved at the UI design phase, particularly with bespoke systems where the customer has selected the style or model of mobile phone.

A core objective when porting software is to make the minimum number of changes possible to the source code. The process is less elegant when compared to writing code for a particular platform or when compared to designing the code to be highly portable between a known set of target environments. You have to make compromises, especially when limited time is available. All approaches described have advantages and disadvantages and need to be appropriately applied to the conditions and goals within which the porting takes place. There is no single right or wrong way; however, many companies have preferred techniques and procedures that promote portability of code.

Prior to considering some of the ways in which porting can be achieved, we look at the general environments within which we are working.

16.3 General Porting Considerations

This section presents some of the most common issues you will meet, irrespective of the source platform from which you are attempting to port code. Later sections deal with more platform-specific issues and demonstrate some approaches you may choose to port your application.

16.3.1 Standard Data Types

In common with many other platforms, Symbian OS defines its own basic data types. The full set is defined in `e32def.h`. Ultimately, the compiler only knows about the language-defined data types. An `unsigned long int`, a `TUint32` and a `ULONG` are almost identical to a compiler, irrespective of the name your platform uses.

When you port an application that contains platform-specific data types, you will need to select a suitable approach:

- change data type declarations to the target platform data type
- change data type declarations to a neutral data type
- cause the source platform data types to be recognized in the new environment.

Table 16.1 presents some of the common data type declarations for each of the Symbian, Windows and Palm OS platforms. The table is not intended to be comprehensive, it merely demonstrates that there is more or less a one-to-one correlation between the different data type declarations on different platforms.

While we have not listed the full set of data type declarations, particularly from the Windows platform, there are not many data types that need some kind of conversion.

Table 16.1 Common data type declarations

C++ data types	e32def.h	Windef.h	PalmTypes.h
void	void	VOID	void
signed char	TInt8	CHAR	Int8
unsigned char	TUint8	UCHAR	UInt8
short int	TInt16	SHORT	Int16
unsigned short int	TUint16	USHORT	UInt16
long int	TInt32	LONG	Int32
unsigned long int	TUint32	ULONG	UInt32
signed int	TInt	INT	Int32
unsigned int	TUint	UINT	UInt32
float	TReal32	FLOAT	float

Pro tip

Table 16.1 can be used as a basis for changing your source code to have platform-neutral data types or changing the data types to the current platform data type. Relatively few data types require conversion. Applications that incorporate the platform-specific data types will, at some point in their `#include` file chain, reference the platform-specific header files containing the definitions. These header files do not exist on other platforms, however, that does not mean you cannot supply your own version to perform the data type declarations. By doing so you may be able to leave your source code untouched!

16.3.2 Platform Size

The Symbian OS platform is a 32-bit platform; that is, an integer is represented by 32 bits. Most desktop operating systems, such as Mac OS, Linux and Windows, including Windows Mobile are also 32-bit platforms whereas Palm OS is a 16-bit platform. (It is possible to build 32-bit applications on Palm OS, but that is not the default environment.) Migrating from 16- to 32-bit platforms is usually not too difficult. It is rare for any part of an application to rely on integer values being restricted to 16 bits or to use any 16-bit integer wrap around or overflow effects.

Occasionally issues may arise when using constructs such as `sizeof (int)` and `sizeof(long)` since the size will vary depending on the target platform.

While not widespread, 64-bit platforms are now available. Porting applications targeted at 64-bit platforms to 32-bit platforms requires you to understand the code more thoroughly. For example, you need to check the range of values a variable may contain and choose an appropriate data type. Symbian OS supports `TInt64` and `TUint64` should your application require them.

16.3.3 Structure and Data Type Alignment

Any processor has numerous characteristics. One that is important in the Symbian environment is the inability of ARM processors to read integers from anything other than a four-byte boundary and, in the case of some ARM processors, short integers from anything other than a two-byte boundary. If your application attempts to exert specific control over the size of data structures, or makes specific assumptions about the memory layout of a structure, you are likely to have some porting problems.

For example, consider the following data structure:

```
typedef struct
  {
  char oneByte;
  int fourBytes;
  }
```

You should not rely on the size of the structure being a specific value, or rely on the position of elements within the structure being in specific places. For example, on a real Symbian phone this structure is eight bytes in size and the integer starts at offset four.

Pro tip

The Symbian OS emulator uses the Intel processor natively; your code is compiled for a different processor compared to a target build. Due to the different characteristics of the processor, the above structure will be five bytes in size and the integer will be at offset one within the emulator environment on an emulator build.

In most cases, the compiler fills any data structures as required for the target environment. If you simply access data structure members by referencing the field names, you will not have a problem. It is only applications that attempt to map memory onto a data structure, often communications-based applications, that will have difficulties.

Pro tip

Early versions of Symbian OS allowed the ARM processor to throw an exception if your application attempted to access misaligned data. This feature is often disabled in modern Symbian devices; consequently, the problem has become much harder to track down. Code inspection and the use of `ASSERT`s will help.

16.3.4 Stack Size

In contrast to many platforms, the default stack size of an application on Symbian OS is 8 KB. Although this can be controlled to some extent, the stack size is static, that is, once defined within your build it will not dynamically shrink or grow. Some care is required when porting your application so as not to exceed the stack size that you have given your application. If you receive an `unresolved external symbol` __chkstk error message when building your application, this indicates that you are attempting to declare too many stack-based variables.

Stack size is discussed in more detail in Section 10.2.1.

16.3.5 Unicode

Symbian OS is built as a Unicode environment only. This is primarily because Symbian OS supports many languages that cannot be handled by a single-byte character set. For developers, a clear distinction needs to be made between textual content that is to be displayed to a user and any other textual content processed by your application.

For example, assume your application is required to read a text file and display the content to a user. In this scenario, you need to understand what your own application text file contains. In an English-language application it contains ASCII characters with Carriage Return (CR) and Line Feed (LF) line delimiters. The content of the text file can be read, however it will be eight-bit data. Prior to being displayed to the user, it needs to be converted to Unicode. Fortunately, due to the design of Unicode, this is quite trivial.

If, however, your application enables the user to edit your text file you need to be aware that many languages contain characters outside the ASCII character set. You should consider whether this has any consequences for your application.

16.3.6 Standard C Library

In its native form, Symbian OS does not contain the set of functions that make up the standard C library. There are external libraries that you can link against that contain the majority of the standard POSIX C library functions as well as some POSIX style threading.

In the Symbian environment, there are a number of functions within the standard C library that need to be used with care. For example, the standard string functions assume ASCII or UTF-8 (with certain caveats). As previously described, text in Symbian OS is normally processed as Unicode. Attempting to use the standard C string functions on Unicode requires the usage of the wide character set functions. Few applications use the wide character string functions as they assume ASCII text. Therefore they will have to be changed to port successfully.

In general, the standard C library does not contain a large number of functions and most applications use only a small number. For example, many of the standard character input–output functions have limited value in a GUI application; printf and its variants are not useful in a graphically rich display environment. However, if your application contains a lot of ASCII text and string processing the standard C library has considerable value in any porting task.

16.3.7 C++ Standard Template Library

At the time of writing, the standard template library (STL) is not provided as part of Symbian OS or UIQ 3, although some unofficial versions are available. Prior to Symbian OS v9, it was difficult to implement the STL because standard C++ exceptions were not available, and dynamic cast was not supported. Symbian OS v9 now makes it possible, although it must still be carefully coded to avoid the bloat associated with template code. It is probable that an official STL implementation will be available on UIQ 3 at some point in the future.

If you are migrating code that uses the STL heavily, and the size of the binary is not a factor (for example, you are not providing your product as an over-the-air download) then you may wish to investigate the current STL implementations available for use on UIQ 3.

16.3.8 Symbian C++ and Naming Conventions

Symbian OS uses very distinct naming conventions, as Chapter 4 describes. The prime objective is to provide as much information as possible to the reader without having to document the details. For example, if you consider the following code, an experienced Symbian programmer can easily distinguish between the three different types of variable being used: local variables (`k` and `i`), a member variable (`iVal`) and method parameters (`aCount` and `aList`):

```
k=aCount/2;
for (i=0;i<k;i++)
  iVal+=aList[i];
```

If you compare this to the following fragment of code, it demonstrates that it is much harder to distinguish between local variables (although most programmers would be surprised if at least `i` and probably `k` are not local variables), method parameters and object property. A reader will almost certainly have to reference some other part of the code to determine which variables are method parameters or object property.

```
k=count/2;
for (i=0;i<k;i++)
  val+=list[i];
```

Naming conventions are simply that, conventions. Your code will not suddenly stop compiling and running because of what you choose to call things.

Pro tip

In general, naming conventions are of benefit to anyone who reads the code. Over time, and perhaps tens of thousands of lines of code later, any hints within the code as to its operation becomes invaluable. We recommend that you adopt a writing standard that encourages self documentation and easy maintenance.

16.3.9 Symbian C++ and Exception Handling

Prior to Symbian OS v9 there was no support for C++ exception handling. Symbian OS had an equivalent, based on a cleanup stack, TRAP handler and leaves. Due to the evolution of the C++ language, C++ compilers and Symbian OS, C++ exceptions are now supported. There are still some restrictions on their usage and applications should not mix C++ exception handling with Symbian TRAPs and leaves. In particular, applications should not attempt to throw an error to a `catch` statement that jumps out of a `TRAP()`, or attempt to `Leave()` when inside a `try...throw...catch` block. Doing so is likely to break the error-handling mechanisms since both techniques are manipulating the stack resulting in an application crash.

The `Porting1` example shows a simple `try...throw...catch` C++ exception-handling statement in action.

Pro tip

Our experience is that it is unimportant as to whether you use Symbian OS keywords or C++ exception-handling keywords to perform error handling. Neither solves the harder issues of rolling back to a consistent state or cleaning up resources correctly in all conditions. As a programmer, you still have to perform the hard work. Support for C++ exception handling simply provides a few more familiar tools to perform these tasks.

16.3.10 Symbian C++ and Writable Global Variables

Writable static data (WSD) is any per-process modifiable variable that exists for the lifetime of the process. In practice, this means any globally scoped data declared outside a function or any function-scoped static variables.

Global data is only constant (i.e. non-writable) if it is one of the built-in types or an object of a class with no constructor. For example, this data is constant:

```
static const TUid KUidFooDll = { 0xF000C001 };
static const TInt KMinimumPasswordLength = 6;
```

The following definitions have non-trivial class constructors, which require the objects to be constructed at run time, so while they may they appear to be constant, they are not. Although the memory for the object is pre-allocated in code, it isn't initialized until the constructor has run:

```
// At build time, each of these is a
// non-constant global object.
static const TPoint KGlobalStartingPoint(50, 50);
static const TChar KExclamation('!');
static const TPtrC KDefaultInput =_L("");
```

Symbian OS has always supported the use of writable static data (WSD) in EXEs. WSD could not be used in DLLs (applications were previously DLLs) built for target hardware on UIQ 2.0 or UIQ 2.1 but this has now been made possible in UIQ 3 because of the new kernel architecture. In order to enable global writable static data on EKA2, the `EPOCALLOWDLLDATA` keyword must be added to the MMP file of a DLL. See Section 4.2 for more information.

Symbian still recommends that, where possible, you avoid the use of WSD in DLLs because it can lead to inefficient memory use and has limited support on the Symbian OS Windows emulator. It should only be used as a last resort, for example when porting code written for other platforms that uses writable static data heavily. You should consider implementing workarounds such as the following.

Thread-Local Storage

Thread-local storage (TLS) is simply a 32-bit pointer specific to each thread that can be used to refer to an object which simulates global writable static data. All the global data must be grouped within this single object, which is allocated on the heap on creation of the thread.

TLS can be accessed through class `UserSvr` – the pointer to the object is saved using `UserSvr::DllSetTls()` and accessed using `UserSvr::DllTls()`. On destruction of the thread, the data is destroyed too.

Singleton using `CCoeStatic`

The `CCoeStatic` base class can be used to create a singleton class which is stored by the CONE environment (`CCoeEnv`) in TLS. You should derive from this class and implement the singleton. Once it has been created, it may be accessed through `CCoeEnv::Static()`.

For more details, please see the Symbian Developer Library (Application Framework Guide) in the UIQ 3 SDK.

Client–Server Framework

Symbian OS supports writable global static data in EXEs. In UIQ 2, a common porting strategy was to wrap the code in a Symbian server (which is an EXE), and expose its API as a client interface. Since applications on Symbian OS 9 are now EXEs in their own right they can contain writable global static data. This eases the porting burden as the Symbian application model is now much closer to other platform application models.

Embed Global Variables into Classes

It is often possible to move global data into existing classes or create new containers, as long as these themselves are not global.

16.3.11 Symbian C++ and Multiple Inheritance

Symbian OS experts disregard full multiple inheritance, preferring only to use interface inheritance. The origins of this go back to the support for multiple inheritance offered by the compilers available at the time. There are ongoing discussions as to whether full multiple inheritance is necessary or desirable. For example, neither C# or Java programming languages support multiple inheritance.

Since Symbian OS continues to use industry-standard compilers, it does not explicitly ban the use of full multiple inheritance. This is demonstrated by our `Porting1` example application. There are, however, philosophical and comprehension issues associated with multiple inheritance and you may have specific views on the matter. On the whole, Symbian OS only uses interface inheritance, not least of which many objects derive from the same base class `CBase`.

16.3.12 Platform Security

Symbian OS implements a layer of Platform Security that imposes restrictions on the way some APIs can be used. For example, if your application comprises a series of plug-in modules which are dynamically searched for and presented to the user as application functionality, you may need to revise your product specification and/or implementation. In particular, if a plug-in comprises code, it has to be placed in the `\sys\bin` folder for Symbian OS to load it. However, Symbian OS does not easily allow applications to search the `\sys\bin` folder to locate the plug-ins! Potential solutions are to use the Ecom plug-in framework, to register your plug-ins via a public folder or to use the `\private\<SID>\import` folder of the primary application.

Platform Security has other implications, particularly associated with application deployment. We provide an overview of Symbian OS Platform Security in Section 4.14.

16.4 General Porting Techniques

Your initial application design and your porting requirements dictate some of the techniques that can be employed to port your application. No one technique can solve all porting issues: some techniques may not be applicable at all whereas others may not adhere to your own or company standards. This section simply outlines some of the techniques you may consider using to port your application.

16.4.1 Component Isolation

When planning your porting task, one of your first considerations is to determine how easily you can port each component of your application. This, in turn, is often dictated by the application design and testing strategies that were employed during the original application development.

For example, if you have a data-processing engine and you originally abstracted out the data supply and data consumption components, perhaps to ensure your data processing engine could be tested automatically, it should be possible simply to replace the data supply and consumption components with platform-specific components. In that way, you may be able to port the data-processing engine as a whole unit (see Figure 16.1).

As a specific example, let us assume your application has an XML-processing engine. At one level, XML is simply a set of text strings that need to be processed. Further, let us assume the application normally receives the XML content from an HTTP component and delivers its results to a display component. If you originally wrote this application so that you can replace the HTTP and display components, enabling you to write automatic test code for the XML processing engine, then it should be a relatively simple task to replace the HTTP and display components with UIQ 3 specific components.

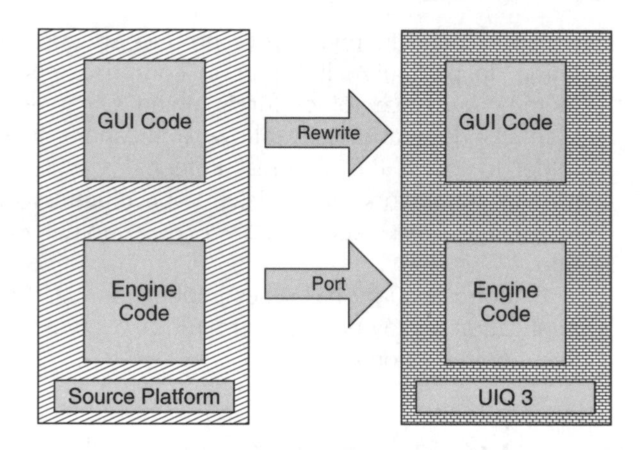

Figure 16.1 Porting and rewriting

In such an example, the components specific to UIQ 3 may have to massage the data between different data formats, introducing a layer of inefficiency; however, in most cases such inefficiencies are not significant. For example, Symbian OS frequently uses descriptors, while your data-processing engine may be processing C-like strings. Translating between the two entities is not difficult, for example to construct an eight-bit descriptor from a zero-terminated, C-style, text string requires the following code:

```
TPtrC8 descriptor(aZTString);
```

Depending on the parameters within which your porting project is operating, it may be possible for you to change a small amount of code in order to generate components that can be isolated.

Using a text file processing example, if your processing engine was written in C++ and included the file-reading code, you may be able to extract that code into a separate object, and then define a virtual interface to the new object. Porting then involves small changes to the original source and the implementation of two versions of a file-reading object, one of which contains the source code you already have.

If the code was originally written in C, you may be able to move all the file-reading code into a few functions and define a 'virtual interface' to those functions. Porting then comprises of small changes to the original source and the implementation of two versions of the file-reading functions, one for each platform.

In both cases, you would use the project definition files, such as MMP or make files, to include only those source modules relevant to each platform.

16.4.2 Using C++ Object Hierarchies

If you have implemented some code in a C++ hierarchy and split its functionality into numerous methods, you may be able to adjust the method declarations and split the source over several files to achieve portability.

In writing an application to process text files, it is likely that the file-reading code will have been isolated using only a few methods, particularly if you have implemented some file buffering. In this case, you may be able to achieve portability by changing the methods to be pure virtual and require that a platform-specific implementation of the object be instantiated.

16.4.3 Breaking up Source Modules

C++ does not require all the methods of a particular class to exist in the same source file. Similarly, the C programming language places little requirement on the programmer to put associated functions in the same source file. It is possible to break up a class or set of functions over multiple source modules and reference different source modules on different platforms. As long as an implementation of a method or function exists somewhere within the source code of your project, the linker will successfully work.

In our text file processing example, if you can easily isolate the file reading code, it is possible to create a new source module containing the platform-specific code. For each supported platform, you have a separate source file. You simply link with the appropriate source for the platform for which you are building the application.

16.4.4 `#define`

The much maligned `#define` is a very powerful tool when it comes to dealing with cross-platform porting issues and can be put to several uses.

Conditional Compilation

In our text file processing example, if you have a small number of functions containing the file-reading code, you may be able to place the source code within `#defines`, causing platform-specific code to be referenced dependent on the compiler `#define` list.

Redefining Entities

In some porting scenarios, you may have a requirement to map one entity to another. For example, you may have some code on Palm OS that calls

StrLen() to determine the length of a C-style string. The equivalent in Symbian OS is User::StringLength(). It is possible to use a #define to map between these two functions. This technique means your source code can remain the same. The only testing required is to ensure that the functions you map between perform the same task.

The Porting1 example application shows how you might use #defines to cause Palm (or Windows) code to build on Symbian OS. The following code fragments can remain identical on Symbian OS without requiring the implementation of any functions or using standard C libraries:

```
CHAR* WindowsLikeFunctions::AppendString(CHAR* str1,CHAR* str2)

  // Create a new string being the concatenation
  // of the two original strings.
  {
  CHAR* result=(CHAR*)malloc(strlen(str1)+strlen(str2)+1);
  strcpy(result,str1);
  strcat(result,str2);
  return(result);
  }

Char* PalmLikeFunctions::AppendString(
                Char* str1,Char* str2)

  // Create a new string being the concatenation
  // of the two original strings.
  {
  Char* result=(Char*)MemPtrNew(StrLen(str1)+StrLen(str2)+1);
  StrCopy(result,str1);
  StrCat(result,str2);
  return(result);
  }
```

16.5 Porting from a Standard C/POSIX Environment

When Symbian OS was originally conceived, POSIX support was not a priority because the hardware it was designed to run on had limited resources, and the emphasis was on support for lightweight mobile applications rather than the potential for porting mainstream desktop services. The evolution of Symbian OS, and the hardware it runs on, now makes it possible to provide a set of libraries to supplement the native Symbian OS C++ APIs with a POSIX-compliant layer. These are intended to make programming for Symbian OS smartphones accessible to more developers, allowing them to bring existing desktop and server code to the platform more easily.

These extension libraries are generically called P.I.P.S. (**P**.I.P.S **I**s **P**OSIX on **S**ymbian OS). They provide a reasonably complete industry-standard POSIX-compliant API. The P.I.P.S. libraries do not provide a UNIX application environment (source code will need to be rebuilt and may need some modification to run) but they do make software development for Symbian OS handsets more accessible to developers used to programming using the C language.

At the time of writing, there are no graphics libraries available, so you will need to consider how to present your application UI, should you need to do so. You can mix or incorporate Symbian native code into an application that links against the P.I.P.S. libraries, but it is a non-trivial task to add the functionality supported by the Symbian native application framework code.

The P.I.P.S. libraries are appropriate if you want to port existing code, or write portable code from scratch. However, since the libraries are a layer over the native APIs, if you require tight, efficient code, a UI or direct integration with the hardware, you should use the native Symbian OS APIs in preference.

16.5.1 P.I.P.S. Libraries

The first version of P.I.P.S. comprises the following libraries:

- `libc`, the standard C and POSIX API:
  - `stdio` and `fileio`, including `printf()`, `scanf()`
  - `stdlib`, including environment variable support, etc.
  - string manipulation and character APIs
  - locale and system services
  - searching, sort and pattern matching
  - IPC mechanisms including shared memory, pipe,
  - FIFO and message queues
  - process creation including `popen()`, `posix_spawn()` and `system()`
  - networking and socket APIs.
- `libm`: mathematical and arithmetic APIs.
- `libpthread`: thread creation and synchronization mechanisms.
- `libdl`: standard C dynamic loading and symbol lookup APIs.

The previous standard C library, `stdlib` (`estlib.dll`), only provided a partially compliant and limited subset of the standard C/POSIX APIs sufficient for the implementation of a Java virtual machine on

Symbian OS. In contrast, P.I.P.S. provides a much more complete and standards-compliant solution. You should note that these two sets of libraries are not compatible, so you should only choose one for your application porting task. The existing `stdlib` library will be deprecated once P.I.P.S. has become more established, so it is recommended that any new porting tasks utilize the P.I.P.S. libraries in preference to the `stdlib` libraries.

16.5.2 POSIX Compliance of P.I.P.S

The P.I.P.S. libraries are compliant with ISO/IEC 9945 where this is possible within the limitations of a mobile operating system.

P.I.P.S. is not fully compliant because of some limitations in what Symbian OS can support (for example, versions of Symbian OS prior to v9.3 only support ordinal lookup in DLLs and cannot offer symbolic lookup, thus `dlsym()` in `libdl` cannot be used with a symbolic name). However, workarounds are suggested in the P.I.P.S. documentation.

Therefore, you should establish whether the current P.I.P.S. functionality is sufficient for your project. Further information can be found in the *PIPS Booklet*, available from the Symbian Developer Network, as described below.

16.5.3 Using P.I.P.S

Assuming you have previously installed the UIQ 3 SDK then, before you can use P.I.P.S., you need to download the P.I.P.S. SDK plug-in. The SDK and documentation are freely available from the Symbian Developer Network: ***developer.symbian.com/wiki/display/oe/P.I.P.S.+Home***. Few phones currently contain the libraries pre-installed, therefore you will need to add the P.I.P.S. environment to your application installation package. The website provides a number of examples of P.I.P.S. in action.

16.6 Porting from Palm OS and Windows Mobile

In this section, we look at some of the characteristics of Palm OS and Windows Mobile applications and show how these are mapped on to Symbian OS applications.

16.6.1 Programming Environment

The native Palm OS environment is the C programming language. C is also the native language for Windows Mobile, though other mainstream development environments are supported.

In contrast, Symbian OS is built using C++. All communication with operating system functionality is through objects of some kind. Individual function calls don't really exist and operations are usually closely associated with the data they manipulate.

Palm OS Environment

By definition, the C programming language is procedural and leads to a particular style of programming. While it is entirely possible to write C++ code and run it on Palm OS, any use of the operating system functionality will result in regular C function calls.

Symbian OS uses most C++ idioms quite heavily. For example, parameters are almost always passed by reference and not by pointer; templates are widely used, as is implicit type conversion. When moving from a Palm environment, you need to understand a different set of library functions, you have to change programming languages and you need to change your programming approach.

As an example, consider the point and rectangle support on both platforms.

On Palm OS you have:

```
typedef struct PointType {
  Coord x;
  Coord y;
} PointType;

typedef struct RectangleType {
  PointType  topLeft;
  PointType  extent;
  } RectangleType;
```

The equivalent in Symbian OS is:

```
class TPoint
  {
public:
  TInt iX;
  TInt iX;
  }

class TRect
  {
public:
  TPoint iTl;
  TPoint iBr;
  }
```

As you can, see these are practically identical. The major difference arises in how they are used in your application. In Palm OS, you have a set of functions that take various `PointType` or `RectangleType` parameters and manipulate the passed data structures. In Symbian OS, the functions are methods that belong to the class definition and tend to manipulate the property belonging to the class itself. The methods have not been shown here for clarity.

For example, to determine the intersection of two rectangles, Palm OS has the following function:

```
void RctGetIntersection (const RectangleType *r1P,
    const  RectangleType *r2P, RectangleType *r3P)
```

This is usually called in code with something like:

```
RectangleType r1;
RectangleType r2;
RectangleType r3;
RctGetIntersection(&r1,&r2, &r3);
```

In contrast, Symbian OS has the following method declared as part of the `TRect` class:

```
void Intersection(const TRect& aRect);
```

This is usually called in code with something like:

```
TRect r1;
TRect r2;
r1.Intersection(r2);
```

So, the method call is using the property belonging to the object as input data and storing the result in its own property. You may also notice that the compiler is passing `r2` by reference and not by value, since the method declaration states that is what should occur.

None of this is particularly special to Symbian OS, it is simply the type of difference you should expect to encounter when moving from a C to a C++ or object-oriented environment.

Windows Mobile Environment

In addition to native C programming, Windows Mobile also supports mainstream development using managed code environments such as Visual Basic or C# within the NET framework and the Model–View–Controller (MVC) environment that uses C++. The MVC environment is really only a layer that interfaces directly with the native `Win32`, C-like operating system functions. As such, MVC is a hybrid between C and C++.

If you are moving from a C-style Windows environment, not only do you need to work with a different set of library functions, you also have to change programming languages, and hence approach, in quite a substantial way. In contrast, if you are moving from an MVC-type C++ environment, then whilst the details will change, the general approach to programming is more or less the same.

However, if you are moving from a managed environment such as C#, the problems you will encounter converting your application to Symbian OS are similar to those you would encounter converting your application to Windows-style MVC C++ code. Indeed, if you are attempting to port an application from C# you may be better off choosing Java as your target language rather than C++.

As an example of some of the similarities between Windows Mobile C/C++ and Symbian C++ consider the point and rectangle example below.

While not the exact class declaration, on Windows Mobile you have something like:

```
class CPoint : public tagPoint
{
public:
  LONG  x;
  LONG  y;
}

class CRect  : public tagRect
{
public:
  LONG    left;
  LONG    top;
  LONG    right;
  LONG    bottom;
} ;
```

In Symbian OS, it looks like this:

```
class TPoint
  {
public:
  TInt iX;
  TInt iX;
  }

class TRect
  {
public:
  TPoint iTl;
  TPoint iBr;
  }
```

These are very similar; a point comprises two integer values and a rectangle four. However, Symbian OS tends to support a more object-oriented approach to manipulation of data structures. For example, to determine the intersection of two rectangles, Windows Mobile has the following method associated with the `CRect` class:

```
void IntersectRect(LPCRECT lpRect1,LPCRECT lpRect2);
```

The method is often called in code something like:

```
CRect r1;
CRect r2;
CRect r3;
r1.IntersectRect(&r2, &r3);
```

Two rectangles are passed by pointer into the class. The result of the intersection of the rectangles is usually stored within a third rectangle object.

Symbian OS has the following method declared as part of the `TRect` class:

```
void Intersection(const TRect& aRect);
```

It is usually called in code like:

```
TRect r1;
TRect r2;
r1.Intersection(r2);
```

The method call is using the property belonging to the object as input data and stores the result in its own property. The compiler is passing `r2` by reference and not by value since the method declaration states that is what should occur.

The task of porting from MVC or Windows-style C++ to Symbian C++ is not very different from porting from MVC or Windows-style C++ to any other C++ variant. If your code does not use many of the MVC classes, you will have a much easier task compared to code that makes extensive use of such classes. In general, you may be able to port some of your data-processing code; however, it will be difficult to port much of the UI code.

From C# to Symbian C++

If you have been using C# for Windows Mobile development, you need to change to using the C++ or Java programming languages. Since this book is concerned with programming in C++ on the Symbian platform, we do not cover using Java as an alternative to C#.

While C# and C++ are similar in many respects, a number of important differences should be noted:

- C# compiles to an intermediate language. This usually requires a virtual machine of some description. Apart from the size of such intermediate virtual machines, an additional layer can affect the performance of applications. It also requires that the virtual machine be present on all phones you are targeting. C++ compiles directly to native code.
- C# performs automatic memory management, C++ does not. You will need to become familiar with the manual techniques employed within Symbian OS such as the use of the cleanup stack. In contrast, automatic garbage collection has inherent problems particularly if pointers are used within applications.
- In C# there are no header files, all code is written inline. In that respect, C# is very similar to Java. C++ has header files declaring classes; Symbian OS has literally thousands of header files within the `epoc32\include` folder. In practice, the task of determining how to use a specific class is not particularly different in either environment as both support a rich and comprehensive set of objects. Whether you look at header files, source modules or documentation there is a large volume of information to digest.
- C# arrays are naturally bounds checked. Pure C++ arrays are not. While you can clearly implement regular C++ arrays in Symbian OS, this is not usually recommended. Symbian OS contains objects called descriptors that add properties such as bounds checking, for the exact reasons that Java and C# have added them within the language itself.

C# and Symbian C++ data types are very similar:

C# data types	Symbian OS data types
`sbyte`	`TInt8`
`short`	`TInt16`
`int`	`TInt32 (TInt)`
`long`	`TInt64`
`byte`	`TUint8`
`ushort`	`TUint16`
`uint`	`TUint32 (TUint)`
`ulong`	`TUint64`
`float`	`TReal32`
`double`	`TReal64`
`char`	`TChar`

In general, the task of porting from C# to Symbian C++ is rather like porting from C# to any other C++ variant. Neither is particularly easy in the sense of being able to re-use any existing code. Most porting projects would really be rewriting the original in a new language.

16.6.2 Application Design

Although Palm OS, Windows Mobile and UIQ 3 are all designed for mobile devices, there are significant differences in the implementation of the UI. You are unlikely to be able to cause much of your Palm OS or Windows Mobile UI code to work on the UIQ 3 platform. However, some of your design work may be re-usable. For example, your application may be organized as a list or summary view with a details view to display data in full. Your application design work can be applied to UIQ 3 applications as they are commonly organized this way.

16.6.3 Application Architecture

Each platform has its own application architecture dictated by the platform vendor.

A typical Palm application contains start-up code such as:

```
static void EventLoop (void)
  {
  Word error;
  EventType event;

  do
    {
    EvtGetEvent (&event, evtWaitForever);
    if (!SysHandleEvent (&event))
      if (!MenuHandleEvent (NULL, &event, &error))
        if (!ApplicationHandleEvent (&event))
          FrmDispatchEvent (&event);
    }
  while (event.eType != appStopEvent);
  }

DWord PilotMain(Word launchCode, Ptr cmdPBP,
                            Word launchFlags)
  {
  if (cmd == sysAppLaunchCmdNormalLaunch)
    {
    FrmGotoForm(MyFirstForm);
    EventLoop();
    StopApplication();
    }
  }
```

A basic Windows application contains start-up code similar to this:

```
int APIENTRY WinMain(HINSTANCE      hInstance,
```

```
                   HINSTANCE  hPrevInstance,
                   LPSTR          lpCmdLine,
                   int             nCmdShow)
{
MSG msg;
HACCEL hAccelTable;

// Initialize global strings
LoadString(hInstance, IDS_APP_TITLE, szTitle,
                          MAX_LOADSTRING);
LoadString(hInstance, IDC_WIN322, szWindowClass,
                             MAX_LOADSTRING);
MyRegisterClass(hInstance);

// Perform application initialization:
if (!InitInstance (hInstance, nCmdShow))
  {
  return FALSE;
  }
// Main message loop:
while (GetMessage(&msg, NULL, 0, 0))
  {
  if (!TranslateAccelerator(msg.hwnd, hAccelTable, &msg))
    {
    TranslateMessage(&msg);
    DispatchMessage(&msg);
    }
  }

return msg.wParam;
}
```

Windows Mobile MVC and C# applications both embed the above type functionality within their application architecture and most applications do not need to be exposed to it. The above functionality is more visible in MVC-type applications since it comprises a number of objects, each with message maps to translate between Windows-type messages and methods.

In contrast, a typical UIQ 3 application has an entry point, `E32Main()`, and three application-specific C++ objects. The vast majority of the functionality of these three objects is provided by the operating system framework code. Your application is really only filling in a few blanks with a bit of syntax. In particular, you should observe that objects tend to do little more than create application-specific versions of the `CQikApplication`, `CQikDocument` and `CQikAppUi` `objects`:

```
void CAppSpecificUi::ConstructL(void)

  // Normal primary entry point to a Symbian App.
  {
  CQikAppUi::ConstructL();
  }

//////////////////////////////////////////////////////////////
```

```
CAppSpecificDocument : public CQikDocument
  {
protected:
  CQikAppUi* CreateAppUiL();
public:
  CAppSpecificDocument(CQikApplication& aApp);
  static CAppSpecificDocument* NewL(CQikApplication& aApp);
  };

CAppSpecificDocument::CAppSpecificDocument(
                   CQikApplication& aApp) :
                         CQikDocument(aApp)
  {}

CAppSpecificDocument* CAppSpecificDocument::NewL(
  CQikApplication& aApp)
  {
  return(new(ELeave)CAppSpecificDocument(aApp));
  }

CQikAppUi* CAppSpecificDocument::CreateAppUiL()
  {
  return(new(ELeave)CAppSpecificUi);
  }

////////////////////////////////////////////////////////////
class CAppSpecificApplication : public CQikApplication
  {
protected:
  TUid AppDllUid() const;
  CApaDocument* CreateDocumentL();
  };

TUid CAppSpecificApplication::AppDllUid() const
  {
  return(KAppSpecificUid);
  }

CApaDocument* CAppSpecificApplication::CreateDocumentL()
  {
  return(CAppSpecificDocument::NewL(*this));
  }

////////////////////////////////////////////////////////////
LOCAL_C CApaApplication* NewApplication()
  {
  return(new CAppSpecificApplication);
  }

GLDEF_C TInt E32Main()
  {
  return(EikStart::RunApplication(NewApplication));
  }
```

All of the example applications provided with this book contain application start-up code similar to the above.

Those familiar with a Model–View–Controller type of application architecture should recognize this model within the UIQ 3 framework.

Similarly those familiar with the MVC `CApplication`, `CDocument` and `CView` objects should be able to see how that maps onto the UIQ 3 framework.

16.6.4 Event Model

Palm OS, Windows Mobile and Symbian OS are event driven. When an event occurs somewhere in the system, it is delivered to the application, the application processes it and returns to wait for the next event.

In Palm OS, each application implements its own copy of the event loop, receives all events, and distributes those events to the different event handlers. Palm OS has a constant set of event handlers coded within your application `EventLoop()` function. The set comprises the system event handlers `SysHandleEvent()` and any application-specific event handlers. Changing the set of event handlers depending on application state is a non-trivial programming task.

As with Palm OS, each Windows Mobile application implements its own copy of the central event loop. Unlike Palm OS the distribution of events is performed by system code. Events are dispatched to one of many potential callback functions registered with the system.

16.6.5 Where Has my Central Event Loop Gone?

In Symbian OS, the application framework code implements the central event loop. It is called the *Active Object Scheduler*. All GUI-based applications require an active scheduler to operate since various parts of the framework code assume such a component exists.

Events in Symbian OS can be considered as having several properties: the event type, an event owner and an associated counting semaphore. With this information, the active scheduler can rapidly determine who the event is for and inform the correct component, without having to obtain event-specific data or pass events to other potential event recipients for inspection.

What Is the Significance of the Term 'Active'?

In Symbian OS, events are represented as objects. In particular, events are represented by objects that derive from the `CActive` class. The term **active object** is widely used in Symbian OS and simply refers to objects that derive from the `CActive` class.

The active scheduler maintains a priority-ordered linked list of active objects. You may think of this as a priority-ordered list of pointers to functions. When an event occurs, the active scheduler searches all the active objects to determine which is associated with that event. Once located, it calls a method belonging to that active object, called `RunL()`, informing the active object that an event for which it was waiting has occurred.

In pseudo code, the active scheduler event loop looks something like this:

```
void CActiveScheduler::Start()
  {
  TDblQueIter<CActive> q(iActiveQueue);
  TInt level=iLevel++;
  while (iLevel>level)
    {
    // Your app should spend all its time in here.
    User::WaitForAnyRequest();

    q.SetToFirst();
    FOREVER
      {
      CActive *ao=q++;
      if (ao->IsActive() && ao->iStatus != KRequestPending)
        { // found object to accept the event
        ao->iActive=EFalse;
        TRAPD(errNo,ao->RunL());
        if (errNo != KErrNone)
          {
          errNo = ao->RunError(errNo );
          if (errNo != KErrNone)
            Error(errNo);
          }
        break;
        }
      }
    }
  }
```

The code `User::WaitForAnyRequest()` simply waits until the counting semaphore associated with the thread is signaled. Since this only occurs when an event is delivered to a thread, the function call only completes when an event is available for processing. If no events are outstanding, the operating system suspends the calling thread. When all threads are suspended, the OS can enter power-saving mode because no events are outstanding.

This is also discussed in Section 4.11.6

Palm OS Event Loop

The call to `User::WaitForAnyRequest()` is broadly equivalent to the Palm OS function `EvtGetEvent(&event, evtWaitForever)`.

The first difference is that the `WaitForAnyRequest()` cannot time out in the way the `EvtGetEvent()` function can. In Symbian OS, timer expiry is handled as an event in its own right. A timer active object would exist to represent the timer event, and would be added to the active scheduler queue. When a timer event occurs, for example, a specified elapsed time passes, the counting semaphore is signaled, any event-specific data is generated and the active scheduler calls the `RunL()` method of the timer active object.

A second difference is that no event-specific data is delivered to or processed by the central event loop. In Palm OS, an `EventType` event structure is defined. This is necessarily a union of all the different event structures that exist within the system. In Symbian OS, each of the active objects is free to design, in co-operation with its associated event source, any data structure they consider fit for their purpose. Typically, an active object making an event request either passes a pointer to some memory it has prepared to accept the event-specific data or fetches the event data when it has been told there is some available.

A third major difference between the Palm approach and the Symbian approach is the flexibility in event handlers. In Symbian OS, event handlers are added and removed from the central event loop as and when required. This typically occurs when active objects are created and deleted. The change in event handlers does not require any change in the application code or central event loop. Indeed, many system components add active objects automatically to perform their tasks. For example, the window server generates keyboard, pointer and draw request events. These are delivered to an active object created by the application framework. This active object performs various tasks before calling much more familiar methods such as `OfferKeyEvent()`, `HandlePointerEvent()` or `Draw()`.

Windows Mobile Event Loop

The Symbian OS call to `User::WaitForAnyRequest()` is broadly equivalent to the Windows function `GetMessage(&msg, NULL, 0, 0)`.

The remaining code is broadly similar to the `TranslateMessage()` and `DispatchMessage()` functions where the framework locates some code that is prepared to handle the event. As can be seen from the `CActiveScheduler::Start()` pseudo code presented earlier, the Symbian code searches through a list of entities looking for a method to call. In particular, it is searching through the priority-ordered linked list of active objects to find one that was active and whose event has been delivered. Once located, the `RunL()` method of that object is called to process the event. Windows has to perform a very similar task to locate the call back function that has been registered.

You may notice that no event-specific data is delivered to or processed by the central event loop. In Windows, an MSG structure is defined which contains a number of fields. These fields tend to have different meanings depending on the context of the message. This design requires events to format their information into the general-purpose fields. Extracting the parameters often requires the use of macros such as `LOWORD(lParam)` or `GEM_WM_COMMAND_ID(wParam,lParam)`.

In Symbian OS, there are no general-purpose fields; each active object is free to design, in co-operation with its associated event source, any data structure they deem fit for their purpose. Typically an active object making an event request either passes a pointer to some memory within which it is prepared to accept the event-specific data or fetches the event data when it has been told there is some available. Few, if any macros, are required to extract the event data as an event-specific object is delivered.

16.6.6 Application Lifecycle

In Palm OS, applications are typically launched when they come into the foreground and are expected to exit when they go into the background.

In both Windows Mobile and Symbian OS, an application is launched the first time it is selected by the user. In both cases, the style guidelines recommend that applications do not contain any close or exit menu options. When a user tasks away to a second application, the first usually goes into the background but continues executing. Applications tend only to be closed when memory within the mobile phone becomes low.

On all the platforms, when applications exit they should preserve as much of their state as is required to give the user the impression that they are continuing from where they left off when they are next started.

16.6.7 Palm OS Launch Codes

Symbian applications do not support launch codes through the equivalent to the Palm OS `PilotMain()` application entry point. This is mainly due to the fact that Palm applications have usually exited when not in the foreground whereas Symbian applications rarely exit.

Launch codes are often used to enable interaction with other system components or to enable an application to provide a set of services to another part of the system. Symbian OS is a multi-tasking operating system. If a set of services is to be provided to multiple clients, this is normally implemented through the client–server architecture. Each client would register with the server and as events occur within the server or the server is informed of state changes that are of interest to a client, then an event would be sent to each registered client.

UIQ 3 applications can provide services to other applications through the view architecture. For example, the Contacts application provides the ability for applications to view and edit contact database entries. These services are supported through the dynamic navigation link (DNL) architecture.

16.6.8 Threads and Processes

At the application level, Palm OS is a single-threaded OS. All application-specific functionality is run within a single thread and there is no support for applications to create other threads or processes. Behind the scenes, Palm OS has other threads and processes running. Applications interact with these system components through standard Palm OS library calls. Whether these calls result in a client–server type of interaction is largely hidden from the applications programmer.

Symbian OS is a multi-threaded, multi-process operating system. Applications can create threads and spawn other processes if required. In practice, very few Symbian applications actually need to create other threads or processes to perform their task. Any work that you have put into your Palm application design to partition computationally intensive tasks to ensure your Palm application remains responsive to any user events is also applicable to Symbian OS.

Windows Mobile is also a multi-threaded, multi-process operating system. Unlike Windows Mobile, Symbian OS discourages the use of threads, preferring applications to use active objects instead. Both threads and active objects have advantages and disadvantages. If the application that you are porting contains threads, you are not forced to convert it to use active objects; you can continue to use threads. You will have to do some work to translate between the Windows Mobile thread management functions and the Symbian OS equivalents.

Similarly, if your application architecture requires the use of more than one process, this model can be accommodated by Symbian OS. Again, you must translate between Windows Mobile process management and Symbian OS process management functions

The client–server pattern is a familiar one in Symbian OS since it offers a good level of security and robustness. Most Symbian OS applications use some form of client–server interface to perform their tasks. For example, access to files is via the file server, access to the screen and keyboard is via the window server and associated processes, and access to the communications subsystems is via the communications server.

16.6.9 Error Handling

Symbian OS distinguishes between two types of errors: ones that are most likely to be programming errors and ones that are caused by abnormal runtime conditions.

For example, attempting to put more data into a descriptor (a memory buffer) than there is memory declared to exist, writing to a `NULL` pointer or attempting to write to a file that has not been opened are considered programming errors. These types of error result in your application being immediately terminated via a panic and reason code; it is said to have *panick*ed. The Symbian OS philosophy is to terminate any application as soon as it detects what can only be a programming error, usually resulting in a more robust system and facilitating debugging.

The Symbian OS SDK defines hundreds of panic reason codes. These are listed within the system panic reference section of the Symbian Developer Library found in the UIQ SDK.

In contrast, errors caused by abnormal runtime conditions, for example, attempting to open a file that does not exist, running out of memory or errors validating user input, result in a recoverable error condition that applications should be able to handle.

Palm OS and native Windows Mobile usually handle this in the traditional way, returning error codes which applications need to test for and propagate back to the caller, performing error handling and rollback en route. Studies have shown that this technique tends to significantly add to the volume of code written, and can make the code much harder to follow. Symbian OS has a strong tendency to leave should an error occur.

If you are familiar with the standard C `setjmp/longjmp` mechanism, then Symbian OS effectively runs a stack of `setjmp` buffers. The `TRAPD()` macro is broadly equivalent to a `setjmp`, while a `User::Leave()` is broadly equivalent to a `longjmp`. For those familiar with MVC, `TRAPD()` is broadly equivalent to the `TRY` and `CATCH` macros and a leave is broadly equivalent to throwing a `CException` that contains a single integer value. See Section 4.5.4 for more information.

Typical Palm OS, native Windows Mobile or standard C code might look like this:

```
int ret=FunctionOne();
if (ret==0)
  {
  ret=FunctionTwo();
  if (ret==0)
    {
    ret=FunctionThree();
    if (ret==0)
      {
      // All ok to continue.
      }
    else
      {
      TidyUpFunctionTwoResources();
      TidyUpFunctionOneResources();
      }
    }
  else
    {
    TidyUpFunctionOneResources();
    }
  }
```

In contrast, Symbian OS code would look something like:

```
{
TRAPD(ret,
  FunctionOneL();
  FunctionTwoL();
  FunctionThreeL();

    // All ok to continue.

  );
}
```

Note that we have not deliberately missed out the `TidyUp` functions to try to make it look a lot simpler. In general, the methods being called would register any resources that need tidying up with a component called the `CleanupStack`. Should an error occur, the cleanup stack automatically cleans up the resources. Some of the real world complexity has not been shown in this short example, for example, registering resources with the cleanup stack, however, the general principle is to move most error handling out of the primary algorithm and let the framework code perform as much tidying up as you can.

Nothing in Symbian OS stops you from writing C-like, error-propagation code, particularly if you do not call any native API functions. However, rather than returning error codes, a significant number of the native API functions will leave. Symbian OS naming conventions usually inform you if this is the case, by adding an upper case `L` to the name of methods that leave.

16.6.10 Text

Strings within the C programming language are arrays of bytes with a trailing null. Unfortunately, there are many languages where the limit of 256 unique values in a single byte is not sufficient to represent all the characters in the language. For example, latest estimates suggest that Chinese contains around 50,000 characters.

Unicode was developed as a standard representation of character sets. Unicode is designed to be compatible with ASCII in the sense that the characters in the ASCII table have the same numerical value as their equivalent in the Unicode table. The Unicode standard defines three encoding schemes, one of which is called UTF-8. In this scheme, characters are represented as one, two, three or four bytes. As such, many platforms can continue to use standard string-processing functions with some restrictions. For example, the number of characters in a UTF-8 string is not necessarily the same as the number of bytes occupied by the string. Similarly indexing characters in a UTF-8-encoded string is not as simple as using the array indexing operator `[]`, you need to decode the string from the start.

Most Palm OS applications are written using single-byte character sets to contain any text presented to the user. Symbian OS chooses to use Unicode to represent text throughout. However, rather than using the UTF-8 encoding scheme, it uses the UTF-16 encoding scheme, as does Windows Mobile. Here, each character is represented by two bytes. This has a number of advantages over UTF-8, for example, the character at position X within a string can be indexed directly using the operator `[]`. A disadvantage is that strings now occupy two bytes for every character, however this is a relatively minor problem given the large amount of RAM available in today's smartphones.

From a programming perspective, most of the standard C string functions as defined in `<string.h>`, or their Palm OS equivalents, cannot be used to process text displayed to and collected from a user. The wide character equivalents would be required. Similarly string constants cannot simply be quoted strings such as "`A standard C string`", they need to be their equivalent wide character versions, `L`"`A wide char string`" or perhaps more frequently written as `TEXT("A wide char string")` in Windows Mobile.

Symbian OS uses a set of classes collectively called descriptors to process textual data. Buffer overrun errors are a common cause of problems in systems. Standard C strings and buffers are prone to this type of problem as there is no natural mechanism of limiting the amount of content a programmer can attempt to place into such a container. Being objects, Symbian descriptors also contain information about the size of buffer they represent. They can therefore determine if the amount of data you expect the object to contain is greater than the size of the buffer you have declared.

Rather than using the wide character string definition directly you will commonly see `_LIT()` within Symbian code. Not only are these declaring the strings to be wide character ones, they are also constructing descriptor objects to wrap the strings.

16.6.11 Descriptors

As Chapter 4 discussed, descriptors come in two types, one for eight-bit data and one for 16-bit data. The 16-bit variants are normally used to process UTF-16-encoded text, as collected from and displayed to a user. The eight-bit variants are normally used to process binary data and perhaps ASCII text.

Some typical binary data processing to count the number of occurrences of a specific value on the Palm OS or Windows Mobile platform might look like this:

```
{
unsigned char buffer[100];
...

int bufSize=50; // Amount of live data within buffer[].
int i;
int count=0;
for (i=0;i<bufSize;i++)
  {
  if (buffer[i]==val)
    count++;
  }
}
```

The equivalent using descriptors in Symbian OS may look like this:

```
{
TBuf<100> buffer;
...

TInt bufSize=buffer.Length();
TInt i;
TInt count=0;
for (i=0;i<bufSize;i++)
  {
  if (buffer[i]==val)
    count++;
  }
}
```

Notice that the data structures being manipulated are quite different but the code is remarkably similar. Some further examples of the similarities you can expect to observe are presented in the `Porting1` example application.

Pro tip

Descriptors are perhaps the most difficult objects for newcomers to Symbian OS to understand. A day or two spent studying the object hierarchy, understanding standard C++ implicit type conversion and writing some example code of your own will save you an enormous amount of time later on. Besides the examples given in Chapter 4, the `SymbianCompatible\Base\BufsAndStrings` folder of the UIQ 3 SDK installation contains some additional sample applications you can review.

You do not have to use descriptors. You can continue to use standard C-style buffers, however, at some point you will probably want to call a native API function that only accepts a descriptor as a parameter. If you are writing new code for the Symbian platform, we recommend that you embrace descriptors. If you have existing code fragments or want to write code compatible with regular C-style buffers, you can create temporary pointer-type descriptors (ones that simply reference content as opposed to containing the content itself), when it comes to interfacing to such a function.

For example, supposing you have a string in a buffer that you wish to write to a file. The `RFile` class requires a descriptor. We can create a temporary pointer-type descriptor as follows:

```
{
unsigned char ztsBuffer[100]
...

// Some pseudo code generates a zero-terminated string.
ztsBuffer[]="String 1";
...

// Create a temporary descriptor from the zts and
// write to file.
file.Write(TPtrC8(&ztsBuffer[0]));
}
```

16.6.12 Templates and Templated Functions

Symbian OS makes extensive use of thin templates. Writing thin template objects can be quite complex but using them is usually straightforward. Few applications in Symbian OS can be written without using descriptors or variable array types of classes, both of which are based on thin templates.

For example, you might declare a descriptor to be capable of containing up to 50 bytes of data as `TBuf8<50> buffer`, rather than as a regular array, `unsigned char buffer[50]`. While there is some difference in syntax, the meaning of both is reasonably clear.

Pro tip

Fully comprehending how the code that you are writing actually works is highly desirable but the use of templated objects is one area where you need to place a significant amount of trust in the system. With familiarity in using templated classes comes a gradual understanding of exactly what is going on. The syntax required to define templated classes is a bit obscure but the above example should demonstrate that using such classes is relatively easy to understand.

16.6.13 Files and Databases

Palm OS uses database structures extensively; even application code is stored in database records! Palm OS databases are based on flat record structures. Symbian OS does not have the direct equivalent to Palm OS databases but perhaps the nearest equivalent is a dictionary stream store.

Symbian OS supports files at numerous levels. At the basic level you are provided with raw file access to open, read, write, seek and close files as required. These functions impose no structure on the file. It is entirely up to your application how it chooses to read and interpret content.

A second layer of file access is provided through the stream store classes. These classes provide functionality to present your file as a series of one or more logical sections or streams which you can read and write. Stream stores are typically used by Symbian applications to serialize their object structure. If you are familiar with using the serialization features of `CArchive` and `CObject` in Windows Mobile, then, along with the >> and << operators, you should be able to translate any serialization code into their equivalent Symbian OS classes. Unfortunately, it is unlikely you will be able to use the original code due to the substantial differences in available functionality on the two platforms.

At a higher level, Symbian OS supports a general-purpose relational database model through the DBMS subsystem. Such a database will have multiple tables, indices and views. Views can be created via an SQL query on the underlying database tables. Since Symbian OS is a multi-threaded OS, the database management system also provides transaction-based services to facilitate multiple clients simultaneously accessing the same database. The functionality available is broadly similar to that provided by the MVC `CDaoDatabase` class, however, you are unlikely to be able to use the original code.

16.6.14 Windows Mobile Registry

Symbian OS does not contain the direct equivalent to a registry. The registry is often used to store application-specific information, such as registration keys, trial mode status information, most recently used file lists and perhaps application settings. In Symbian OS, applications typically store this type of information in an application-specific file, usually one based on a stream store. The Symbian OS framework contains direct support for saving configuration information within INI files. It should be noted that these are NOT the same as a typical Windows INI file.

A further use of the registry is to publish an association between data files and applications. Windows knows which application to launch when a file is selected by looking at these registry entries.

Symbian OS strongly prefers to associate data files with applications based on the content of a file. Symbian OS files normally contain a header comprising a number of UIDs describing the content of the file, including which application should be associated with the content. If this is not possible Symbian OS uses a system of recognizers to associate file types with applications.

A further use of the registry is to share information between co-operating applications. Symbian OS has an equivalent to this, called *Publish and Subscribe*. In the Symbian OS model, one application publishes the value of a variable. A second application can obtain the current value and register an interest in being informed when a particular variable changes value.

16.6.15 Summary

This section has looked at some of the common characteristics of Palm OS and Windows Mobile and their equivalents within Symbian OS. While there are substantial differences, particularly in the detail, there are enough similarities to facilitate the direct porting of some software components between these platforms.

16.7 Porting from S60 3rd Edition

Since both S60 3rd edition and UIQ 3 are based on Symbian OS, the majority of issues encountered when attempting to port applications from platforms not based on Symbian OS to Symbian OS should not apply. For example, programmers should already be familiar with descriptors, `TRAP()` harnesses, leaves, cleanup error handling, active objects and other idioms that are specific to Symbian OS.

This section covers the porting of applications from S60 3rd Edition to UIQ 3. Both platforms are based on Symbian OS v9, so this section does not consider the issues related to porting applications from earlier platforms such as S60 2nd Edition. Further information on this task can be found in the UIQ Developer Community FAQ and the document *Porting to UIQ 3 from S60: A practical approach.*

16.7.1 Target Phone Differences

Prior to starting your porting activity, you should carefully consider the differences in the target phones that are currently supported by the two platforms, and how this might affect a user who needs to use your application. At the time of writing, there are no S60 devices with touch-sensitive screens and, while they are expected soon, the number of non-touchscreen phones will outnumber touchscreen phones for some time to come. For phones that do not have a touch-sensitive screen, user interaction is very similar, if not identical. However, touchscreen mobile phones are typically used in different ways, not least of which the user expects to be able to tap on displayed components to cause actions.

For example, a chess game must provide a way for the user to move the pieces around the board. In a touchscreen implementation, the pen can be used to drag them. Otherwise, the user needs a way of selecting a piece and then moving it around using, say, arrow keys on the keypad and finally confirming the new location.

Apart from the touchscreen, the two platforms have many things in common:

- Screen size. Both S60 and UIQ support a range of screen resolutions. In particular, both platforms support a 240x320 (QVGA) screen resolution.
- Processing capability. Both S60 and UIQ phones are capable of significant data processing. Our experience shows they are broadly equivalent – if you have a processing bottleneck on one phone, the same bottleneck will occur on the other.
- Available memory. Current S60 and UIQ phones contain comparable amounts of RAM, typically many tens of megabytes. Very few applications should need to worry about the differences here.
- Keyboard availability. S60 supports a range of phones, some of which contain physical keyboards. Likewise, some UIQ 3 mobile phones present more of a keyboard than others. This may or may not be a consideration for your specific application.

16.7.2 Application Metafiles

Symbian applications include a number of metafiles which describe various aspects of the application. When porting from S60 to UIQ 3 there are a number of differences in the content of these metafiles as described below.

Application Icons

An S60 application icon comprises a single SVG image. A UIQ 3 application icon comprises a set of three bitmap-based images with associated mask. UIQ version 3.1 introduces the ability to use SVG format icons.

In S60, your application icon is compiled to a MIF file via the `mifconv` utility program outside of the primary MMP file build. Therefore S60 MMP files do not make any reference to an application icon. In contrast, UIQ 3 icons are compiled into an MBM file via the `bmconv` utility. The makefile statement to cause this is contained within the MMP file:

```
// Create application icon.
START BITMAP  Porting1_icons.mbm
SOURCEPATH    Graphics
TARGETPATH    \Resource\Apps
SOURCE        c16 AppIcon18x18.bmp
SOURCE        1 AppIcon18x18mask.bmp
SOURCE        c16 AppIcon40x40.bmp
SOURCE        1 AppIcon40x40mask.bmp
SOURCE        c16 AppIcon64x64.bmp
SOURCE        1 AppIcon64x64mask.bmp
END
```

In both platforms the icon is referenced from the application registration file or, more precisely, from the localization file included by the application registration file.

A typical S60 application `CAPTION_AND_ICON_INFO` statement would be:

```
CAPTION_AND_ICON_INFO
  {
  caption = "Porting1";
  number_of_icons = 1;
  icon_file = "\\Resource\\Apps\\ Porting1_icons.mif";
  }
```

The UIQ 3 equivalent would be:

```
CAPTION_AND_ICON_INFO
  {
  caption = "Porting1";
  number_of_icons = 3;
  icon_file = "\\Resource\\Apps\\Porting1_icons.mbm";
  }
```

Applications Appearing in Softkey Style Small Modes

UIQ 3 applications need to inform the application launcher explicitly that they can be listed within the Softkey Style Small view of any phone that supports such a mode. While there is no direct equivalent S60 phone at this time, numerous phones support multiple display sizes, for example, the internal and external screens found on a Nokia E90. S60 applications are expected to support all display sizes of a particular phone.

In UIQ 3, the `LOCALISABLE_APP_INFO` resource would typically contain the following field to indicate support for both a default size UI configuration and a Softkey Style Small:

```
view_list =
  {
```

```
VIEW_DATA
  {
  uid=EViewIdPrimaryView;
  screen_mode=0;
  caption_and_icon =
    {
    CAPTION_AND_ICON_INFO
      {
      }
    };
  },
VIEW_DATA
  {
  uid=EViewIdPrimaryView;
  screen_mode=EQikScreenModeSmallPortrait;
  caption_and_icon =
    {
    CAPTION_AND_ICON_INFO
      {
      }
    };
  }
};
```

On S60 this field would not be defined within the resource and therefore takes the system default of zero entries in the `view_list` array.

Application UID

The same application UID can be used for S60 and UIQ 3 versions of an application. This includes applications that will be submitted to Symbian Signed for testing. If you include the application UID in any source filename for the purposes of Symbian Signed you clearly need to be able to distinguish between files targeted at S60 and those targeted at UIQ 3. Your version control software should be capable of handling this.

Backup Registration Files

Assuming you are adopting the same backup policy on both platforms, the same `backup_registration.xml` file can be used to declare the backup policy of your application.

Package File

Apart from any changes in the list of files to install (for example, on S60 3rd edition you have a MIF file containing your icon, on UIQ 3 you have an MBM file), the major change to apply to your PKG file is the declaration of platform or phone compatibility.

An S60 PKG file would typically contain the following statement:

```
; ProductID for S60 3rd Edition
[0x101F7961], 0, 0, 0, {"S60ProductID"}
```

The equivalent for UIQ 3 is as follows:

```
; ProductID for UIQ 3.0
[0x101F6300], 3, 0, 0, {"UIQ30ProductID"}
```

The most important part is the UID declaration. This determines what platforms an application can be installed on without users being presented with warnings about possible incompatibilities.

MMP File

Most of your application attributes remain the same, for example your application needs the same stack size and capabilities; however there are some differences.

In general, you need to reference a different set of source modules and library files. One goal of porting is to try to re-use as many source modules as possible. However, the content of numerous files, for example, your application start up module, application UI class and your application resource file are likely to change considerably. An S60 application references some S60-specific libraries such as `avkon.lib`. The equivalent UIQ 3 libraries need to be referenced instead.

Using Folders and Version Control

A traditional Symbian application places the metafiles within a group folder but remember that this is only a naming convention. The build tools still work if you use a different folder name. If you create `GroupS60` and `GroupUIQ` folders (see Figure 16.2), you can partition your application instead of having to use `#define` statements to manage, say, a single MMP file.

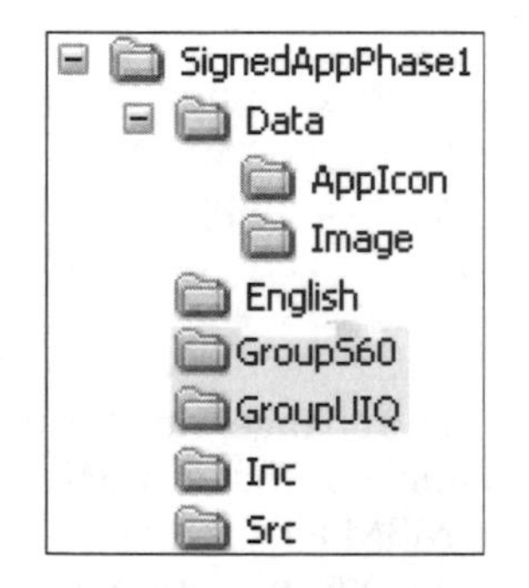

Figure 16.2 Folder structure

Don't forget to consider all the tools available to you when it comes to managing the porting process. Since S60 and UIQ 3 projects exist at different locations on a hard disk, your version control software may be able to help with sharing the source code that remains identical across platforms, while allowing you to use the same folder and file names for different versions of logically equivalent files.

16.7.3 Application Source Files

A prime objective of porting is to re-use as many of the application source files as possible. If you have followed previous Symbian recommendations regarding application architecture, in particular ensuring that engine and UI code are separated into different source modules, you stand a good chance of being able to re-use a significant proportion of your application. If your engine and UI code is intermingled, you have a much harder task. In this case it is often preferable to spend some time splitting the original source code into re-usable and non re-usable components. Be sure to test that the newly split code compiles and works on S60 before you start porting it to UIQ 3.

Assuming you have split up your application into engine and UI classes, porting between S60 and UIQ 3 frequently becomes a task of replacing S60 UI components (`Avkon`) with their equivalent UIQ 3 UI components. There are however, a number of architectural changes between the platforms that also need to be considered. The GUI in UIQ 3 has specific features that enable applications to run on different types of hardware, for example, with or without a touchscreen. We recommend that you spend some time going through the chapters and example applications presented earlier in this book so that you are familiar with the following UIQ 3 features:

- menus and command handling
- view configurations
- layout managers
- building blocks.

Learning how to use these components effectively enables you to port your application so that it works well on different UIQ 3 mobile phones.

Application Framework and Startup

The fundamental application framework is the same on both S60 and UIQ 3. On both platforms you have an application class, a document class, an application user interface class and a set of views. Events are represented by active objects and the framework code runs an active scheduler on behalf of your application.

The application startup code on both platforms is very similar, the code fragments below show the typical S60 and UIQ 3 class hierarchy and application startup code.

```
class CAppSpecificUi : public CAknViewAppUi
  {
protected:

  // From CEikAppUi.
  void ConstructL();

  ... remaining methods + property
  };
void CAppSpecificUi::ConstructL()

  // Normal primary entry point to a Symbian App on S60.
  {
  BaseConstructL(CAknAppUi::EAknEnableSkin);

  ... add views here.
  }

////////////////////////////////////////////////////////////
class CAppSpecificDocument : public CAknDocument
  {
protected:
  CEikAppUi *CreateAppUiL();
public:
  CAppSpecificDocument(CEikApplication& aApp);
  static CAppSpecificDocument* NewL(CEikApplication& aApp);
  };

CAppSpecificDocument::CAppSpecificDocument(
                   CEikApplication& aApp) :
                        CEikDocument (aApp)
  {
  }

CAppSpecificDocument* CAppSpecificDocument::NewL(CEikApplication& aApp)
  {
  return(new(ELeave)CAppSpecificDocument(aApp));
  }

CEikAppUi* CAppSpecificDocument::CreateAppUiL()
  {
  return(new(ELeave)CAppSpecificUi);
  }

////////////////////////////////////////////////////////////
class CAppSpecificApplication : public CAknApplication
  {
protected:
  TUid AppDllUid() const;
  CApaDocument* CreateDocumentL();
  };
```

```
TUid CAppSpecificApplication::AppDllUid() const
  {
  return(KAppSpecificUid);
  }

CApaDocument* CAppSpecificApplication::CreateDocumentL()
  {
  return(CAppSpecificDocument::NewL(*this));
  }

////////////////////////////////////////////////////////////
LOCAL_C CApaApplication* NewApplication()
  {
  return(new CAppSpecificApplication);
  }

GLDEF_C TInt E32Main()
  {
  return(EikStart::RunApplication(NewApplication));
  }
```

In contrast, the typical UIQ 3 class hierarchy and application startup code is as follows:

```
class CAppSpecificUi : public CQikAppUi
  {
protected:
  // From CEikAppUi.
  void ConstructL();
  };

void CAppSpecificUi::ConstructL()

  // Normal primary entry point to a Symbian App.
  {
  CQikAppUi::ConstructL();

  ... add views here.
  }

////////////////////////////////////////////////////////////
class CAppSpecificDocument : public CQikDocument
  {
protected:
  CQikAppUi* CreateAppUiL();

public:
  CAppSpecificDocument(CQikApplication& aApp);
  static CAppSpecificDocument* NewL(CQikApplication& aApp);
  };

CAppSpecificDocument::CAppSpecificDocument(
                   CQikApplication& aApp) :
                         CQikDocument(aApp)

  {
  }
```

```
CAppSpecificDocument* CAppSpecificDocument::NewL(CQikApplication& aApp)
  {
  return(new(ELeave)CAppSpecificDocument(aApp));
  }

CQikAppUi* CAppSpecificDocument::CreateAppUiL()
  {
  return(new(ELeave)CAppSpecificUi);
  }

///////////////////////////////////////////////////////////////
class CAppSpecificApplication : public CQikApplication
  {
protected:
  TUid AppDllUid() const;
  CApaDocument* CreateDocumentL();
  };

TUid CAppSpecificApplication::AppDllUid() const
  {
  return(KAppSpecificUid);
  }

CApaDocument* CAppSpecificApplication::CreateDocumentL()
  {
  return(CAppSpecificDocument::NewL(*this));
  }

///////////////////////////////////////////////////////////////
LOCAL_C CApaApplication* NewApplication()
  {
  return(new CAppSpecificApplication);
  }

GLDEF_C TInt E32Main()
  {
  return(EikStart::RunApplication(NewApplication));
  }
```

As you may see, apart from sub-classing slightly different objects, the structure is identical and the details are almost the same. This suggests that the code may easily be re-used with appropriate `#defines` to cover the differences. However, we recommend that you maintain two separate versions, one for each platform, so that each is clear and easy to read.

Resource File Content

Both S60 and UIQ 3 applications use resource files. Apart from language localization, the two platforms use resource files in completely different ways and have little in common. If you have adopted the practice of splitting your language text strings into a separate file, often called a LOC file in S60 and an RLS file on UIQ 3, these files can remain the same, although you are likely to have a slightly different set of strings between the platforms. For example, an S60 LOC file will typically contain the

`Options`, `Exit` and `Back` strings. UIQ 3 applications tend not to display such text. The equivalent to the `Options` text, `More`, is automatically displayed by the UIQ 3 framework, `Exit` is not a typical application option and `Back` is usually handled by a dedicated phone key.

Statements such as the following are effectively the same:

```
rls_string STR_R_CMD_NEW "New"

#define STR_R_CMD_NEW "New"
```

When a resource file is compiled, a textual substitution between the token and value pairs occurs. Therefore, it is not particularly relevant if you have more statements than are actually used within a particular application version. The token is never located in the resource file so no substitution is required. Combining both S60 and UIQ 3 token–value pairs in a single file, with no pre-processor directives, has little effect on the output for either platform.

Menus and Commands

The S60 platform supports a very traditional approach to the definition of menus within resource files. On S60, a menu is usually defined through a resource structure such as the following:

```
RESOURCE MENU_BAR r_application_menubar
  {
  titles =
    {
    MENU_TITLE
      {
      menu_pane=r_ application _menu;
      }
    };
  }

RESOURCE MENU_PANE r_ application _menu
  {
  items=
    {
    MENU_ITEM
      {
      command=EAppCmdHelp;
      txt=STR_R_CMD_HELP;
      },
    MENU_ITEM
      {
      command=EAppCmdAbout;
      txt=STR_R_CMD_ABOUT;
      }
    };
  }
```

The content, placement and ordering of commands is firmly defined and fixed by the resource definition. In contrast, UIQ 3 does not have strict menu definitions; you simply define a set of commands which the application framework distributes around the user input controls available on a specific phone:

```
RESOURCE QIK_COMMAND_LIST r_application_commands
  {
  items=
    {
    QIK_COMMAND
      {
      id = EAppCmdHelp;
      type = EQikCommandTypeHelp;
      groupId = EAppCmdMiscGroup;
      priority = EAppCmdHelp Priority;
      text = STR_R_CMD_HELP;
      },
    QIK_COMMAND
      {
      id = EAppCmdAbout;
      type = EQikCommandTypeScreen;
      groupId = EAppCmdMiscGroup;
      priority = EAppCmdAboutPriority;
      text = STR_R_CMD_ABOUT;
      }
    };
  }
```

Your application can have a reasonable level of influence upon the placement of controls by using attributes such as `groupID` and `priority`. We recommend that you allow the UIQ 3 framework to take care of the commands, rather than trying to dictate too much in your application.

A primary task of a menu is to allow the user to choose which command they wish the application to perform. On both platforms, a command ID is defined in the resource and is delivered to the `HandleCommandL()` method of the current view. However these methods are different on each platform.

The S60 version is delivered a single integer value:

```
void CAppSpecificAknView::HandleCommandL(TInt aCommand)

  // Menu commands come here on S60.
  {
  switch (aCommand)
    {
    case EAppCmdHelp:
    case EAppCmdAbout:
      break;
    }
  }
```

In contrast a `CQikCommand` object is delivered on UIQ 3:

```
void CAppSpecificQikView::HandleCommandL(
  CQikCommand& aCommand)

  // Handle the commands coming in from the controls
  // that can deliver cmds, menus etc.
  {
  switch (aCommand.Id())
    {
    case EAppCmdHelp:
    case EAppCmdAbout:
      break;

    default: // Exit button
      CQikViewBase::HandleCommandL(aCommand);
      break;

    }
  }
```

Since each platform has a different view class hierarchy it is not particularly easy to re-use the code from one platform on the other. In general, the more functionality you can push into an engine-like component, even if it is processing UI-centric data elements, the more portability you can achieve. There is however, a limit to this process and at some point it becomes quicker, easier and more maintainable simply to write new code.

Views

Both S60 and UIQ 3 contain a view architecture. They are similar in many respects: views have IDs, they can be activated and deactivated, are informed of dynamic screen changes and are the primary entry point for processing user commands.

They do however differ quite significantly, particularly when you look at the detailed implementation. For example, views in S60 subclass `CAknView`, while views in UIQ 3 subclass `CQikViewBase` or `CQikMultiPageViewBase`. Different data is delivered to the `HandleCommandL()` method.

At a more fundamental level, views on S60 have minimal resource definition via the `AVKON_VIEW` resource. This only describes the menu and CBA. In contrast, a UIQ 3 view is defined via a `QIK_VIEW_CONFIGURATIONS` resource. Not only does this resource define the menu, via the command list, it also defines how to construct the view depending on the current phone type and details which controls should be placed within the view.

The following section gives an example of the differences between views on the two platforms in the context of an application that displays a simple list box.

List Boxes and Views

List boxes are commonly used in S60 and UIQ 3 applications. Both platforms provide a substantial list box framework within which applications can work; however, the details differ significantly.

The following code fragments demonstrate a list box view being created for the S60 platform. The code is not intended to be complete and we have extracted out the relevant sections to demonstrate the differences between the platforms.

Firstly we have the S60 view resource definition:

```
RESOURCE AVKON_VIEW r_application_list_view
  {
  cba=r_ application _list_cba;
  menubar=r_ application _list_menubar;
  }

RESOURCE CBA r_ application _list_cba
  {
  buttons =
    {
    CBA_BUTTON
      {
      id=EAknSoftkeyOptions;
      txt="Options";
      },
    CBA_BUTTON
      {
      id=EAknSoftkeyBack;
      txt="Exit";
      }
    };
  }

RESOURCE MENU_BAR r_ application _list_menubar
  {
  titles =
    {
    MENU_TITLE
      {
      menu_pane=r_application_list_menu;
      }
    };
  }

RESOURCE MENU_PANE r_application_list_menu
  {
  items=
```

```
{
  MENU_ITEM
    {
    command=EAppCmdAbout;
    txt="About";
     }
  };
}
```

In S60, views typically comprise two objects, the view and a container. This code fragment demonstrates a typical class definition and implementation of the container:

```
class CListContainer : public CCoeControl,
               public MEikListBoxObserver,
                       public MDesCArray
  {
protected:

  // From CCoeControl.
  void SizeChanged();
  TInt CountComponentControls() const;
  CCoeControl* ComponentControl(TInt aIndex) const;

  // From MEikListBoxObserver.
  void HandleListBoxEventL(CEikListBox* aListBox,
           MEikListBoxObserver::TListBoxEvent aEventType);

  // From MDesCArray.
  TInt MdcaCount() const;
  TPtrC MdcaPoint(TInt aIndex) const;

public:
  ~CListContainer();
  CListContainer(CAppEngine* aEngine);
  void ConstructL(const TRect& aRect);

protected:
  CAppEngine* iEngine;
  CAknSingleGraphicStyleListBox* iListBox;
  };

CListContainer::CListContainer(CAppEngine* aEngine) :iEngine(aEngine)
  {
  }

CListContainer::~CListContainer()
  {
  delete(iListBox);
  }

TInt CListContainer::MdcaCount() const
  // Report number of items in our list - our engine knows this.
  {
  return(iEngine->ListCount());
  }
```

```
TPtrC CListContainer::MdcaPoint(TInt aIndex) const

  // Report text to display in the list -
  // our engine generates correct format str.
  {
  return(iEngine->ListDataAt(aIndex));
  }

void CListContainer::ConstructL(const TRect& aRect)

  // We have our own window in which to display everything.
  {
  CreateWindowL();
  SetRect(aRect);

  iListBox=new(ELeave)CAknSingleGraphicStyleListBox();
  iListBox->ConstructL(this,0);
  iListBox->SetListBoxObserver(this);

  iListBox->Model()->SetItemTextArray(this);
  iListBox->Model()->SetOwnershipType(ELbmDoesNotOwnItemArray);

  iListBox->CreateScrollBarFrameL(ETrue);
  iListBox->ScrollBarFrame()->SetScrollBarVisibilityL(
       CEikScrollBarFrame::EOff,CEikScrollBarFrame::EAuto);

  iListBox->SetRect(Rect());
  iListBox->ActivateL();

  ActivateL();
  }

TInt CListContainer::CountComponentControls() const

  // Report the number of sub-component controls we have.
  {
  return(1);
  }

CCoeControl* CListContainer::ComponentControl(
  TInt aIndex) const

  // Report handle of 'aIndex' control.
  {
  return(iListBox);
  }
```

Finally the S60 view itself:

```
class CListAknView : public CAknView
  {
protected:
  // From CAknView
```

```
  void HandleCommandL(TInt aCommand);
  void DoActivateL(const TVwsViewId& aPrevViewId,
                              TUid aCustomMessageId,
                   const TDesC8& aCustomMessage);
  void DoDeactivate();

public:
  // From CAknView
  TUid Id() const;

  // New methods
  CListAknView(CAppEngine* aEngine);
  ~CListAknView();

protected:
  CAppEngine* iEngine;
  CSessionListContainer* iContainer;
  };

CListAknView::CListAknView(CAppEngine* aEngine) :
  iEngine(aEngine)
  {
  }

CListAknView::~CListAknView()
  {
  delete(iContainer);
  }

void CListAknView::HandleCommandL(TInt aCommand)
  {
  switch (aCommand)
    {
    ... process commands here
    }
  }

TUid CListAknView::Id() const
  {
  return(KUidListView);
  }

void CListAknView::DoActivateL(
  const TVwsViewId& aPrevViewId,
  TUid aCustomMessageId,
  const TDesC8& aCustomMessage)
  {
  CListContainer* q=new(ELeave)CListContainer(iEngine);
  CleanupStack::PushL(q);
  q->SetMopParent(this);
  q->ConstructL(ClientRect());
  AppUi()->AddToStackL(*this,q);
  CleanupStack::Pop();  // q
  iContainer=q;
  }
```

```
void CListAknView::DoDeactivate()

  // The view is being deactivated; lose any components.
  {
  if (iContainer)
    {
    AppUi()->RemoveFromStack(iContainer);
    delete(iContainer);
    iContainer=NULL;
    }
  }
```

In contrast, an equivalent UI on UIQ 3 might be described by the following resource and classes. The resource and code has been extracted from the `ListView1` example application supplied as part of this book:

```
RESOURCE QIK_VIEW_CONFIGURATIONS r_list_view_configurations
  {
  configurations =
    {
    QIK_VIEW_CONFIGURATION
      {
      ui_config_mode = KQikPenStyleTouchPortrait;
      command_list = r_list_view_pen_style_commands;
      view = r_list_view_pen_style_view;
      },

    QIK_VIEW_CONFIGURATION
      {
      ui_config_mode = KQikSoftkeyStyleTouchPortrait;
      command_list = r_list_view_key_style_commands;
      view = r_list_view_key_style_view;
      }
    };
  }

RESOURCE QIK_COMMAND_LIST r_list_view_pen_style_commands
  {
  items=
    {
   QIK_COMMAND
      {
      id = EAppCmdZoom;
      type = EQikCommandTypeScreen;
      groupId = EAppCmdMiscGroup;
      priority = EAppCmdZoomPriority;
      text = "Zoom";
      }
    };
  }

// Define the view to contain a set of pages.
RESOURCE QIK_VIEW r_list_view_pen_style_view
  {
  pages = r_list_view_pen_style_pages;
  }
```

```
// Defines the pages of a view - we have 1.
RESOURCE QIK_VIEW_PAGES r_list_view_pen_style_pages
  {
  pages =
    {
    QIK_VIEW_PAGE
      {
      page_id = EAppSpecificListViewPageId;
      page_content = r_list_view_page_control;
      }
    };
  }

RESOURCE QIK_CONTAINER_SETTINGS r_list_view_page_control
  {
  layout_manager_type = EQikRowLayoutManager;
  layout_manager = r_row_layout_manager_default;
  controls =
    {
    QIK_CONTAINER_ITEM_CI_LI
      {
      unique_handle = EAppSpecificListViewListId;
      type = EQikCtListBox;
      control = r_app_listview_listbox;
      layout_data = r_row_layout_data_fill;
      }
    };
  }

RESOURCE QIK_ROW_LAYOUT_DATA r_row_layout_data_fill
  {
  vertical_alignment = EQikLayoutVAlignFill;
  vertical_excess_grab_weight = 1;
  }

RESOURCE QIK_LISTBOX r_app_listview_listbox
  {
  view = r_app_listview_listbox_view_default;
  layouts = { r_app_listview_normal_layout_pair };
  }

RESOURCE QIK_LISTBOX_ROW_VIEW r_app_listview_listbox_view_default
  {
  }

RESOURCE QIK_LISTBOX_LAYOUT_PAIR r_app_listview_normal_layout_pair
  {
  standard_normal_layout = EQikListBoxLine;
  }
```

In UIQ 3 views are typically made up of a single object, the view. This code fragment demonstrates a typical class definition and implementation of the view. The full source is available in the `ListView1` example application:

```
class CAppSpecificListView : public CQikViewBase,
                        public MQikListBoxObserver
  {

protected:
  // From CQikViewBase.
  TVwsViewId ViewId() const;
  void HandleCommandL(CQikCommand& aCommand);
  void ViewConstructL();
  void ViewDeactivated();
  void ViewActivatedL(const TVwsViewId& aPrevViewId,
                        const TUid aCustomMessageId,
                      const TDesC8& aCustomMessage);

  // From MQikListBoxObserver.
  void HandleListBoxEventL(CQikListBox* aListBox,
                     TQikListBoxEvent aEventType,
                   TInt aItemIndex,TInt aSlotId);

  void AddItemsL(void);

public:
  CAppSpecificListView(CAppSpecificUi& aAppUi,
                         CAppEngine* aEngine);

protected:
  CAppEngine* iEngine;
  };

CAppSpecificListView::CAppSpecificListView(
              CAppSpecificUi&aAppUi,CAppEngine* aEngine) :
          CQikViewBase(aAppUi,KNullViewId),iEngine(aEngine)
  {
  }

TVwsViewId CAppSpecificListView::ViewId() const
  {
  return(KViewIdListView);
  }

void CAppSpecificListView::HandleCommandL(CQikCommand& aCommand)
  {
  switch (aCommand.Id())
    {
    ... process commands here.
    }
  }

void CAppSpecificListView::ViewConstructL()
  {
  ViewConstructFromResourceL(R_LIST_VIEW_CONFIGURATIONS);

  // We want to HandleListBoxEventL() - so observe the listbox.
  LocateControlByUniqueHandle<CQikListBox>(
            EAppSpecificListViewListId)->
                SetListBoxObserver(this);
  }
```

```
void CAppSpecificListView::ViewDeactivated()
  {
  CQikListBox*
  listbox=LocateControlByUniqueHandle<CQikListBox>(
                      EAppSpecificListViewListId);
  TRAPD(junk,listbox->RemoveAllItemsL());
  }

void CAppSpecificListView::AddItemsL(void)
  {
  CQikListBox*
  listbox=LocateControlByUniqueHandle<CQikListBox>(
                      EAppSpecificListViewListId);

  // Get the listbox model.
  MQikListBoxModel& model(listbox->Model());
  model.ModelBeginUpdateLC();

  const TInt count=iEngine->ListItemCount();
  for (TInt i=0;i<count;i++)
    {
    MQikListBoxData* lbData=model.NewDataL(
            MQikListBoxModel::EDataNormal);
    CleanupClosePushL(*lbData);

    lbData->AddTextL(iEngine->ListItem(i),
                EQikListBoxSlotText1);
    lbData->SetItemId(i);

    // Removes the listboxData from the
    // stack and calls close on lbData.
    CleanupStack::PopAndDestroy(lbData);
    }

  model.ModelEndUpdateL();
  }

void CAppSpecificListView::ViewActivatedL(
           const TVwsViewId& aPrevViewId,
             const TUid aCustomMessageId,
            const TDesC8& aCustomMessage)
  {
  AddItemsL();
  }
```

Comparing the two, you observe that considerably more of the UIQ 3 list box view is described by the resource. Indeed the C++ code has more or less been reduced to setting the list box to contain application-specific content. All the management of the main UI, such as creating a window and creating and configuring the list box, is handled by the framework code.

One subtle difference you should note is that on S60 the view container is destroyed, and therefore the list box is destroyed with it, when the view is deactivated. Views are considered very lightweight wrappers on S60. In contrast the UI components such as the list box are not destroyed when a view is deactivated on UIQ 3. If the list is displaying content from a data source, such as an engine, and that list can be changed within another view, it is important to realize that you may have to re-build the content of your list box when the UIQ 3 view is re-activated to display the correct content. An S60 list box is always re-built since it does not exist until the view is re-activated.

In our example, we have chosen to reset the UIQ 3 list box to reduce the overall amount of memory used when the list is deactivated. A side effect is that this forces us to re-build the list when the UIQ 3 view is re-activated, removing this potential issue.

Dialogs

Both platforms support the display of dialogs to a user. S60 3rd edition provides many fully defined dialogs, most of which comprise of a single input control. For example, S60 contains a text query dialog and a number query dialog where applications only need to provide the application-specific data plus a small resource. The following resource and code fragment shows a typical S60 text editor dialog:

```
RESOURCE DIALOG r_ name_edit_dialog
  {
  flags = EGeneralQueryFlags;
  buttons = R_AVKON_SOFTKEYS_OK_CANCEL;
  items =
    {
    DLG_LINE
      {
      type = EAknCtQuery;
      id = EGeneralQuery;
      control = AVKON_DATA_QUERY
        {
        layout = EDataLayout;
        label = "Name";
        control = EDWIN
          {
          ... config omitted for clarity.
          };
        };
      }
    };
  }

  {
  _LIT(KUIQ,"UIQ");
  TBuf<32> name(KUIQ);
  CEikDialog* dlg=((new(ELeave) CAknTextQueryDialog(
                        name,CAknQueryDialog::ENoTone);
  dlg->ExecuteLD(R_ NAME_EDIT_DIALOG))
  }
```

The S60 class `CAknTextQueryDialog` ultimately derives from the `CAknDialog` class which in turn derives from `CEikDialog`.

The S60-specific classes do not exist on the UIQ 3 platform. Furthermore, the UIQ 3 platform has deprecated the CEikDialog class. While CEikDialog still exists, works on currently available phones and has even been updated to support UIQ 3, it will be phased out in the future. Any new code for UIQ 3 should not use CEikDialog, but should use the CQikSimpleView or CQikViewBase type classes.

A UIQ dialog resource uses a QIK_DIALOG structure as opposed to a DIALOG structure, for example:

```
RESOURCE QIK_DIALOG r_about_dialog
  {
  title = "About";
  configurations =
    {
    QIK_DIALOG_CONFIGURATION
      {
      ui_config_mode = KQikPenStyleTouchPortrait;
      container = r_about_container;
      command_list = r_about_commands;
      },
    QIK_DIALOG_CONFIGURATION
      {
      ui_config_mode = KQikPenStyleTouchLandscape;
      container = r_about_container;
      command_list = r_about_commands;
      },
    QIK_DIALOG_CONFIGURATION
      {
      ui_config_mode = KQikSoftkeyStyleTouchPortrait;
      container = r_about_container;
      command_list = r_about_commands;
      },
    QIK_DIALOG_CONFIGURATION
      {
      ui_config_mode = KQikSoftkeyStylePortrait;
      container = r_about_container;
      command_list = r_about_commands;
      },
    QIK_DIALOG_CONFIGURATION
      {
      ui_config_mode = KQikSoftkeyStyleSmallPortrait;
      container = r_about_container;
      command_list = r_about_commands;
      }
    };
  }

RESOURCE QIK_COMMAND_LIST r_about_commands
  {
  items=
    {
    QIK_COMMAND
      {
      id = EAppCmdContinue;
      type = EQikCommandTypeDone;
```

```
      text = "Continue";
      }
    };
  }

RESOURCE QIK_SCROLLABLE_CONTAINER_SETTINGS r_about_container
  {
  controls =
    {
    };
  }
```

As can be seen, this effectively describes one dialog for each view configuration available. In this example, all view configurations reference the same container (or data display and input controls) and command list. While this would be true in general, there is the option to split the commands and data input controls across multiple dialogs should that be appropriate for a specific view configuration.

Our experience is that very little dialog code can easily be ported between these two platforms. About the best that can be achieved is providing an engine class that ultimately owns the data being manipulated by dialogs.

16.8 Summary

This chapter has investigated some of the approaches you may wish to adopt when porting applications to the UIQ 3 platform. We then looked at several competing mobile platforms and discussed how you might approach mapping the features and functions supported by each of those platforms to the UIQ 3 platform.

While any porting task is difficult, our experience is that porting to the UIQ 3 platform is no more difficult than any other porting task. Much depends on the original structure of the application and the skill set of the people entrusted with the task. If you are not a C++ programmer, you will have more problems porting to a rich C++ environment such as UIQ 3 than a fluent C++ programmer would have. Similarly if the original application has most of the engine within the UI, the porting task is that much harder.

References and Resources

Wiki Site

A wiki based on this book is available at ***books.uiq.com***. You can also download the example code from there.

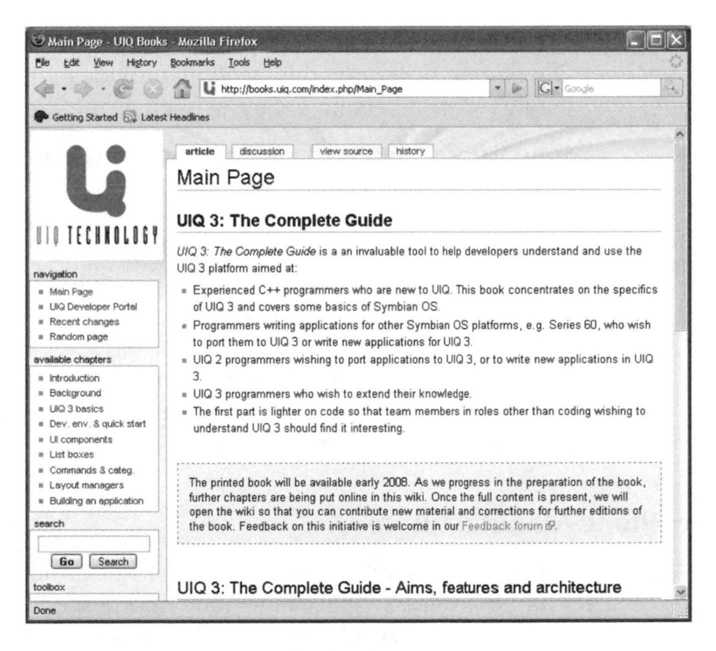

books.uiq.com

Developer Resources

UIQ Developer Community

The UIQ Developer Community has many resources invaluable to developers creating applications, content and services for all UIQ 3 mobile phones. Free membership registration is necessary for some content and to post messages in the developer forums.

You can join the UIQ Developer Community at ***developer.uiq.com***.

- **Documentation and code:** in this section, you will find all kinds of documentation, ranging from reference documentation to more tutorial- and guide-style documentation to tips and code snippets. Learn with the whitepapers and find tips in the knowledgebase.
- **Discussion forum:** this section is where the community hangs out and where you can get help or help others to benefit from your experience. Always give as much detail as possible so that others have all the information needed to give you a helpful reply.
- **News and events:** here you can find the latest news regarding the UIQ developer community and the list of upcoming and past events.
- **SDK and tools:** the place to download the UIQ Software Development Kits and other tools necessary to develop for UIQ 3.
- **Developer program:** this section provides further information on fee-based membership programs such as the Premium membership.
- **Support:** for registered members to purchase professional support tickets.

Symbian Resources

The **Symbian Developer Network** (***developer.symbian.com***) provides support for developers working with all aspects of Symbian OS.

The **Symbian Signed portal** (***www.symbiansigned.com***) provides information, Test criteria, UID allocation and Symbian Signed procedures. Registration is free of charge.

Mobile Phone Manufacturers

Mobile phone manufacturers also provide developer support for their UIQ-based products. This can include phone-specific APIs.

Sony Ericsson Developer World (***developer.sonyericsson.com***) supports phone-specific features of UIQ 3 phones from Sony Ericsson.

Motorola MOTODEV (***developer.motorola.com/technologies/UIQ/***) supports phone-specific features of UIQ 3 phones from Motorola.

IDE

The recommended Integrated Development Environment for UIQ 3 is Carbide.c++. It is available from ***www.forum.nokia.com/carbide***.

References

Symbian Press Books

The following Symbian Press books are referenced in this book:

Developing Software for Symbian OS, Second Edition, Babin, John Wiley & Sons.

Symbian OS C++ for Mobile Phones, Vol 3, Harrison, Shackman *et al.*, John Wiley & Sons.

Symbian OS Communications Programming, Second Edition, Campbell *et al.*, John Wiley & Sons.

Symbian OS Explained, Stichbury. John Wiley & Sons.

Symbian OS Platform Security, Heath *et al.* John Wiley & Sons.

The Accredited Symbian Developer Primer, Stichbury and Jacobs, John Wiley & Sons.

Online Documents

Coding idioms for Symbian OS: ***developer.symbian.com/main/downloads/papers/coding_idioms/2002_10_09_codingSymbianOS.pdf***

Eliminating Memory Leaks in Symbian OS C++ Projects: ***developer.symbian.com/main/downloads/papers/TracingMemoryLeaks/MemLeakTracking_Rev2.pdf***

PC Connectivity: How To Write Backup Aware Software for Symbian OS v9: ***developer.symbian.com/main/downloads/papers/pc_connectivity/PlatSec_PC_Connectivity.pdf***

Porting to UIQ 3 from S60 - A practical approach: ***developer.uiq.com/downloads/public/Taskalavista_S60_to_UIQ3.zip***

Programmer's Guide to New Features in UIQ 3: ***developer.uiq.com/downloads/public/ProgrammersGuide.pdf***

Software Installation Toolkit – Reference: ***www.symbian.com/developer/techlib/v9.2docs/doc_source/ToolsAndUtilities/Installing-ref/index.html***

Special Interest Paper – Porting and Porting DreamConnect to UIQ 3: ***developer.uiq.com/downloads/public/Interview_DreamSpring.pdf***

Symbian OS C++ coding standards: ***developer.symbian.com/main/learning/press/books/pdf/coding_standards.pdf***

Symbian Signed Test Criteria: ***developer.symbian.com/wiki/display/sign/Symbian+Signed+Test+Criteria***
UIQ Migration Quick Guide: ***developer.uiq.com/downloads/public/MigrationQuickGuide.pdf***
UIQ Product Datasheet and UIQ Product Description: ***www.uiq.com/platform_documents.html***
UIQ Style Guide: ***developer.uiq.com/downloads/public/StyleGuidePreview.pdf***
Using Symbian OS: P.I.P.S.: ***developer.symbian.com/wiki/download/attachments/1411/PIPS_Essential_Booklet.pdf***

Glossary

Here is an glossary of the abbreviations and acronyms that are used in this book.

A useful *Symbian OS Glossary* can be found at ***developer.uiq.com/devlib/uiq_31/sdkdocumentation/doc_source/doc_source/Global Glossary/index.html***.

2G	Refers to the second generation of mobile phone systems, when fully digital operation was achieved. GSM is the most widely deployed 2G system.
3G	Refers to the third generation of mobile phone systems. UMTS is a 3G system.
3GPP	The 3rd Generation Partnership Project. Collaboration of telecommunications standards bodies which oversees GSM, GPRS, EDGE and Universal Terrestrial Radio Access.
3GSM	Major congress event for the mobile phone industry, formerly called 3GSM, now known as the Mobile World Congress, ***www.mobileworldcongress.com***.
A2DP	Advanced Audio Distribution Profile A Bluetooth profile which describes how stereo audio can be streamed from a media source to a sink.

AAC	Advanced Audio Coding. A format for compressing audio media.
ACS	Authenticated Content Signing. A Publisher ID certificate provided by a certificate authority.
ADSL	Asymmetric Digital Subscriber Line. A technology for delivering broadband IP connectivity over the fixed network.
AIF	Application Information File.
AMR	Adaptive Multi-Rate. A voice codec used in GSM.
AMR-NB	Adaptive Multi-Rate Narrow Band. A voice codec used in GSM.
API	Application Program Interface.
APN	Access Point Name. Used in GPRS, EDGE and UMTS networks to specify the service to which a connection is made.
App	Application.
ARM	Term used to refer to a processor with 16/32 bit embedded RISC from ARM, ***www.arm.com***.
ASCII	American Standard Code for Information Interchange. A character-set coding standard.
AU	Audio. A format for audio data.
BAFTA	British Academy of Film and Television Arts.
BAT	File extension for batch files.
BMP	Bitmap. An image format.
bps	Bits per second – data throughput speed.
BSc	Bachelor of Science. A type of university degree.
BT	British Telecommunications plc.

CeBIT	Very large trade fair held in Hannover, Germany, covering computer and telecoms industries.
CER	CERtificate file. The file containing a digital certificate.
CONE	The Symbian OS CONtrol Environment. The framework responsible for graphical interaction.
CPF	Command Processing Framework. The component of UIQ that manages application commands in different UI Configurations.
CPP	C Plus Plus. The file extension for C++ files.
CPU	Central Processing Unit.
CR	Carriage Return. Non-printing character to return to the beginning of the line.
CSD	Circuit-switched data.
CSR	File extension for the certificate request file produced by the `DevCertRequest` utility.
DBCS	Double-byte character set.
DBMS	DataBase Management System. The component that controls the organization, storage and retrieval of data in a database.
DHCP	Dynamic Host Configuration Protocol. Supplies configuration information such as IP number to devices on a network.
DLL	Dynamic Link Library.
DNL	Direct Navigation Link. A mechanism in UIQ to move from one view to another, for example from Contacts detail view to Phone.
DNS	Domain Name Server. Resolves a domain name, such as ***www.uiq.com*** to its IP address.
DRM	Digital Rights Management. Generic term for techniques to control distribution of multimedia content.

DTD	Document Type Definition. Defines the content of an XML file.
DTMF	Dual-Tone Multi Frequency. Also known as *touch tone*.
ECam	Symbian OS Camera API.
ECOM	A generic Symbian OS framework for use with plug-in modules.
EDGE	Enhanced Data rates for GSM Evolution. An enhancement to GPRS networks which increases the data throughput speed.
EDR	Enhanced Data Rate. Bluetooth feature which provides faster data transfer.
EKA1	EPOC Kernel Architecture 1. Systems prior to Symbian OS v9 are based on EKA1.
EKA2	EPOC Kernel Architecture 2. From Symbian OS v9 onwards, all systems are based on EKA2.
EMS	Enhanced Messaging Service. EMS is an extension to SMS which allows simple images, sounds and text formatting to be sent using SMS as the bearer.
EPOC	The name for Symbian OS prior to Symbian OS v6.0. Its use is deprecated.
ESOCK	Symbian OS network socket server process.
ETEL	The Symbian OS telephony framework.
EXE	Executable (file type). A type of binary which, when loaded, is used as the basis for a new process.
EXIF	EXchangeable Image File Format.
FAQ	Frequently Asked Questions.
FAT	File Allocation Table. A type of file system.
FEP	Front End Processor. Allows the input of characters by a user, for example, by handwriting recognition or voice, and converts the input transparently into text.

FIFO	First In First Out. A storage queue where the first item stored in is also the first returned out.
FM	Frequency Modulation. Commonly refers to radio services in the 88–108 MHz frequency band.
FSY	File extension for a filesystem plug-in.
FTP	File Transfer Protocol.
GCCE	GNU Compiler Collection for Embedded. GCCE builds application code for ARM processors.
GIF	Graphical Interchange Format. A file format for images.
GPRS	General Packet Radio Service. A packet-based data service available in most GSM networks.
GSM	Global System for Mobile communications, the most widely used digital mobile phone system and the de facto wireless telephone standard in Europe.
GUI	Graphical User Interface.
HRH	File extension for header files.
HSCSD	High Speed Circuit-Switched Data. Feature available in some GSM networks providing up to 57.6 kbps downlink speed.
HSDPA	High Speed Downlink Packet Access. UMTS feature providing download throughput of up to 14.4 Mbps.
HSPA	High Speed Packet Access. Generic term for HSDPA and HSUPA.
HSUPA	High Speed Uplink Packet Access. UMTS feature providing upload throughput of up to 5.76 Mbps.
HTML	HyperText Mark-up Language.
HTTP	Hypertext Transfer Protocol. A method used to transfer information on the World Wide Web.
IAP	Internet Access Point.

ICL	Image Conversion Library.
ICO	ICOn image file format.
IDE	Integrated Development Environment.
IEEE	Institute of Electrical and Electronics Engineers, Inc, ***www.ieee.org***.
IM	Instant Messaging. Generic term for applications and services for textual chat and real time data transfer.
IMAP4	Internet Message Access Protocol. A protocol for working with a remote email server.
IMEI	International Mobile Equipment Identification. A unique number which identifies a mobile phone.
iMelody	A format for mobile phone ringtones.
IMPS	Instant Messaging and Presence Service. An IM specification managed by the OMA.
INI	Type of file used as a stream store, typically to initialize an application.
IP	Internet Protocol.
IPC	Inter-Process Communication.
IPSec	Internet Protocol SECurity.
IPv4	Internet Protocol Version 4.
IR	Infrared.
IrCOMM	Provides emulation of Serial and Parallel ports over infrared.
IrDA	Infra Red Data Association. This body sets the standards used in infra-red communications.
IrOBEX	Infrared OBject EXchange protocol. Used to exchange data objects such as vCards and vCals.
IrTranP	An image transfer protocol using infrared.

ISM	Industrial Scientific and Medical. A frequency band around 2.4 GHz.
ISO	International Organization for Standardization, ***www.iso.org***.
ISV	Independent Software Vendor.
Java ME	Java platform, Micro Edition.
JPEG	Joint Photographic Experts Group. Defines a form of image file compression.
JPG	File extension commonly used for JPEG compressed image files.
JSR	Java Specification Request.
kbps	Kilobits per second – data transfer throughput speed.
LAN	Local Area Network.
LF	Line Feed. Non-printing character to advance to the next line.
LFFS	Logging Flash File System. Provides robust persistent file storage using naked FLASH memory.
LM-IAS	Link Management Information Access Service.
LOC	Type of file containing language strings.
LSK	Left SoftKey.
MB	Megabyte – 1024 bytes. Unit of memory or storage.
MBCS	Multi-Byte Character Set.
MBG	Generated MBM header file. This file defines which bitmaps exist within the corresponding MBM file.
MBM	Multi-BitMap. A Symbian OS file format for bitmaps.
Mbps	Megabits per second – data throughput speed.
MEng	Master of Engineering. A type of university degree.

MFC	Microsoft Foundation Classes.
MIDI	Musical Instrument Digital Interface.
MIME	Multipurpose Internet Mail Extensions. A protocol whereby an Internet mail message can be composed of several independent items, including binary and application-specific data.
MMC	MultiMedia Card. A type of flash memory card.
MMF	MultiMedia Framework.
MMP	File extension for the application master project file.
MMS	Multimedia Messaging Service. Provides transmission of images, video clips, sound files and text messages over a mobile network.
MNG	Multiple-image Network Graphics. An image format supporting animation.
MP3	MPEG-1 Audio Layer 3. A common format for compressed audio content.
MPEG4	Moving Picture Experts Group. MPEG4 is a standard for compressed video content.
MTM	Messaging Type Module. A component of the Symbian OS messaging architecture. MTMs enable additional messaging types to be supported within the messaging system.
NAND	A type of flash memory.
NAT	Network Address Translation.
NOR	A type of flash memory.
OBEX	Object Exchange. A set of high-level protocols allowing objects such as vCard contact information to be exchanged using either infrared or Bluetooth wireless technology.
OEM	Original Equipment Manufacturer.
OMA	Open Mobile Alliance, ***www.openmobilealliance.org***.

OPP	Object Push Profile. A Bluetooth profile for exchange of data.
OS	Operating system.
OTA	Over the Air. Refers to sending settings and content to a mobile phone over the mobile network.
P.I.P.S.	PIPS Is POSIX on Symbian OS.
PAN	Personal Area Network.
PBX	Private Branch eXchange. For example, an office telephone system.
PC	Personal Computer.
PDA	Personal Digital Assistant. Generic term for a portable device running contacts, diary and other personal applications.
PKG	File extension for package files.
PNG	Portable Network Graphics. An image file format.
POP3	Post Office Protocol 3. A protocol for fetching email from a server.
POSIX	Portable Operating System Interface. See ***www.pasc.org***.
PRT	File extension for a protocol module plug-in.
Publisher ID	Digital certificate which verifies the identity of a software publisher. Used in Symbian Signed.
Qikon	The name of the component building the upper layer of the application framework. This part is specific to UIQ.
QVGA	Quarter-VGA. Refers to a display that is quarter the size of VGA, that is, 240 pixels wide by 320 pixels high.
R&D	Research and Development.
RAD	Rapid Application Development.

RAM	Random Access Memory.
RDS	Radio Data System. A method for sending small amounts of data with FM radio signals, common in Europe and Latin America.
reg	_reg file. The application registration resource file.
RFC	Request For Comments. Documents defining Internet standards are known as RFC, for example RFC 2616 defines the HTTP protocol.
RFCOMM	An interface that allows an application to treat a Bluetooth link in a similar way as if it were communicating over a serial port.
RISC	Reduced Instruction Set Computer.
RLS	Resource file used to hold translated language strings.
ROAP	Rights Object Acquisition Protocol.
ROFS	Read Only File System.
ROM	Read Only Memory.
RSG	A resource header file.
RSK	Right SoftKey.
RSS	File extension for resource files.
RTT	Round Trip Time.
RVCT	RealView Compilation Tools.
SCBS	Single-Byte Character Set.
SDK	Software Development Kit.
SIBO	Generic name for the operating system and architecture of the Psion Series 3 Pocket Computers.
SID	Secure IDentifier.
SIR	Slow InfraRed.

SIS	Software Install Script. A package format for delivering applications to the phone in installable form.
SISX	Software Install Script. New file format introduced in Symbian OS v9.1
SMS	Short Message Service.
SMSC	Short Message Service Center.
SMTP	Simple Mail Transfer Protocol. A protocol for interacting with a server to send email.
SQL	Structured Query Language. A standard language for querying and modifying relational databases.
SSID	Service Set ID. Identifies a Wireless LAN Access Point.
SSL	Secure Socket Layer.
STL	Standard Template Library.
SVG	Scalable Vector Graphics.
TC	TrustCenter, ***www.trustcenter.de***. Provides Publisher ID certificates for Symbian Signed.
TCB	Trusted Computing Base.
TCE	Trusted Computing Environment.
TCP	Transmission Control Protocol.
TIFF	Tagged Image File Format. A file format for images.
TLS	Transport Layer Security.
TSY	An ETEL extension module that handles the interaction between the ETEL server, and a particular telephony device or family of devices.
UDP	User Datagram Protocol.
UGC	User Generated Content.
UI	User Interface.

UID	Unique Identifier. A globally unique 32-bit number used in a compound identifier to uniquely identify an object, file type, etc. When users refer to *UID* they often mean UID3, the identifier for a particular program.
Uikon	The name of the component building the middle layer of the application framework.
UMIST	University of Manchester Institute of Science and Technology.
UMTS	Universal Mobile Telecommunications System.
Unicode	ISO 10646-1 defines a *universal character code* which uses either 2 or 4 bytes to represent characters from a large character set. In Symbian OS, 2-byte UNICODE support is built deep into the system.
URI	Uniform Resource Identifier.
URL	Uniform Resource Locator.
URN	Uniform Resource Name.
USB	Universal Serial Bus.
USIM	UMTS SIM. A SIM card configured for UMTS services.
UTF	Unicode Transformation Format. Format for representing characters. Encodings include 7-, 8-, 16- and 32-bit versions.
VGA	Video Graphics Array. Term used to refer to a display that is 640 pixels wide by 480 pixels high.
VID	Vendor ID. Specifies the vendor of the application executable.
VoIP	Voice over Internet Protocol. A protocol that enables the transmission of voice traffic over packet-based networks.
VPN	Virtual Private Network.
WAP	Wireless Application Protocol.
WAV	Waveform Audio Format.

WBMP	Wireless Application Protocol Bitmap Format.
WEP	Wired Equivalent Privacy. An encryption method used in wireless LANs.
WiFi	Industry association promoting wireless LANs, ***www.wi-fi.org***.
WINSCW	WINdows Single process with CodeWarrior. Compiler used to build code to run in the emulator.
WLAN	Wireless Local Area Network.
WMF	Windows MetaFile.
WPA	WiFi Protected Access. An encryption method used in Wireless LANs.
WSD	Writeable Static Data.
WSERV	Window Server.
XML	Extensible Markup Language.

Index